Preface

The year 2006 is the centenary of Alois Alzheimer's presentation to a meeting of German psychiatrists held in Tübingen, Germany. In 1906, Alzheimer described the results of his studies on a female patient known as Auguste D., who had suffered from a progressive presenile dementia. In 1907, Alzheimer published this study in a paper entitled "Über eine eigenartige Erkrankung der Hirnrinde" in *Allgemeine Zeitschrift für Psychiatrie und psychisch-gerichtliche Medizin*. This paper was a landmark in our understanding of the disease that now bears his name. The paper described the major lesions that are now known to be common to all forms of Alzheimer's disease.

After 100 years it is time to reflect upon the enormous progress that has been made since Alois Alzheimer's first observations were reported. The chapters within this book describe some of the major conceptual advances of the last few years, particularly in understanding Alzheimer's disease pathogenesis, and the research that may lead to successful therapies. Central to the story of Alzheimer's disease is the β-amyloid protein or Aβ, a 4-kDa polypeptide that is intimately involved in the pathogenic cascade. Increasingly it is recognized that Aβ is a causative agent that plays a key role in disease pathogenesis.

The chapters in this book are written by experts in their respective fields, and each author provides individual insight into the role of Aβ in the pathogenesis of Alzheimer's disease. The chapters contain innovative ideas on the biochemical, cellular, and behavioral pathogenesis of Alzheimer's disease that should propel research over the next few years.

Colin J. Barrow, PhD
Ocean Nutrition Canada
Dartmouth, Nova Scotia
Canada

David H. Small, PhD
Monash University
Clayton, Victoria
Australia

Contents

Contributors

Marie-Isabel Aguilar, BSc, PhD
Laboratory of Molecular Neurobiology
Department of Biochemistry and Molecular Biology
Monash University
Clayton, VIC, Australia

Maria Ankarcrona, PhD
Karolinska Institutet
Neurotec, Section for Experimental Geriatrics
Huddinge, Sweden

Kevin J. Barnham, PhD
Department of Pathology
The University of Melbourne
Parkville, VIC, Australia;
The Mental Health Research Institute of Victoria
Parkville, VIC, Australia

Colin J. Barrow, BSc(Hons), PhD, MBA
Ocean Nutrition Canada
Dartmouth, NS, Canada

D. Allan Butterfield, PhD
Department of Chemistry
Center for Membrane Sciences and Sanders-Brown Center on Aging
University of Kentucky
Lexington, KY, USA

Roberto Cappai, BSc(Hons), PhD
Department of Pathology, and Centre for Neuroscience
The University of Melbourne
Melbourne, VIC, Australia;
Mental Health Research Institute
Parkville, VIC, Australia

Robert A. Cherny, PhD
Department of Pathology
The University of Melbourne
Parkville, VIC, Australia;
The Mental Health Research Institute of Victoria
Parkville, VIC, Australia

Giuseppe D. Ciccotosto, BSc, PhD
Department of Pathology
The University of Melbourne
Parkville, VIC, Australia;
Mental Health Research Institute
Parkville, VIC, Australia

Peter Clifford, MS
New Jersey Institute for Successful Aging
University of Medicine and Dentistry of New Jersey – SOM
Stratford, NJ, USA

Joanna M. Cordy, PhD
Department of Pharmacology
Boston University School of Medicine
Boston, MA, USA

Cyril C. Curtain, PhD, DSc
School of Physics
Monash University
Clayton, VIC, Australia;
Department of Pathology
The University of Melbourne
Parkville, VIC, Australia

Della C. David, PhD
Brain and Mind Research Institute (BMRI)
University of Sydney
Camperdown, NSW, Australia

Piet Eikelenboom, PhD, MD
Department of Neurology
Academic Medical Center
University of Amsterdam
Amsterdam, The Netherlands;
Department of Psychiatry
Vrije Universiteit Medical Center
Amsterdam, The Netherlands

Dwight C. German, PhD
The University of Texas Southwestern Medical Center at Dallas
Dallas, TX, USA

Jürgen Götz, PhD
Brain and Mind Research Institute (BMRI)
University of Sydney
Camperdown, NSW, Australia

Gillian C. Gregory, PhD
Prince of Wales Medical Research Institute and the University of New South Wales
Sydney, NSW, Australia

Glenda M. Halliday, BSc, PhD
Prince of Wales Medical Research Institute and the University of New South Wales
Ranwick, Sydney, NSW, Australia

Jeroen J.M. Hoozemans, PhD
Department of Neuropathology
Academic Medical Center
University of Amsterdam
Amsterdam, The Netherlands;
Department of Psychiatry
Vrije Universiteit Medical Center
Amsterdam, The Netherlands

Lars M. Ittner, MD
Brain and Mind Research Institute (BMRI)
University of Sydney
Camperdown, NSW, Australia

Jack H. Jhamandas, MD, PhD
Department of Neurology
University of Alberta
Edmonton, AB, Canada

Doreen Kabogo, BSc
Department of Psychiatry
University of Alberta
Edmonton, AB, Canada

Satyabrata Kar, PhD
Departments of Medicine (Neurology) and Psychiatry
University of Alberta
Edmonton, AB, Canada

Josef Karkos, MD
Clinical Studies CCN
Institut "Methodenforum"
Berlin, Germany

Mary Kosciuk, PhD
New Jersey Institute for Successful Aging
University of Medicine and Dentistry of New Jersey – SOM
Stratford, NJ, USA

David MacTavish, Dls
Department of Neurology
University of Alberta
Edmonton, AB, Canada

Colin L. Masters, MD, FRCPA
Department of Pathology
The University of Melbourne
Parkville, VIC, Australia;
The Mental Health Research Institute of Victoria
Parkville, VIC, Australia

Robert G. Nagele, PhD
New Jersey Institute for Successful Aging
University of Medicine and Dentistry of New Jersey – SOM
Stratford, NJ, USA

B. Elise Needham, PhD
Department of Pathology
The University of Melbourne
Parkville, VIC, Australia;
Mental Health Research Institute
Parkville, VIC, Australia

Judy Ng, MSc
Laboratory of Molecular Neurobiology
Department of Biochemistry and Molecular Biology
Monash University
Clayton, VIC, Australia

Kathy E. Novakovic, BSc
Department of Nuclear Medicine
Centre for PET
Austin Hospital
Melbourne, VIC, Australia;
Department of Pathology
The University of Melbourne
Parkville, VIC, Australia

Carlos Opazo, PhD
Department of Pathology
The University of Melbourne
Parkville, VIC, Australia;
The Mental Health Research Institute of Victoria
Parkville, VIC, Australia

Laszlo Otvos, Jr, PhD, DSc, CBA
The Wistar Institute
Philadelphia, PA, USA

Christopher C. Rowe, MD, FRACP
Department of Nuclear Medicine
Centre for PET
Austin Hospital
Melbourne, VIC, Australia;
Department of Pathology
The University of Melbourne
Parkville, VIC, Australia

Annemieke J.M. Rozemuller, MD, PhD
Department of Neuropathology
Academic Medical Center
University of Amsterdam
Amsterdam, The Netherlands

Wiep Scheper, PhD
Neurogenetics Laboratory
Academic Medical Center
University of Amsterdam
Amsterdam, The Netherlands

Claire E. Shepherd, BSc, PhD
Prince of Wales Medical Research Institute and the University of New South Wales
Sydney, NSW, Australia

Gilbert Siu, BSc, PhD
New Jersey Institute for Successful Aging
University of Medicine and Dentistry of New Jersey – SOM
Stratford, NJ, USA

David H. Small, BSc, PhD
Department of Biochemistry and Molecular Biology
Monash University
Clayton, VIC, Australia

Mee-Sook Song, PhD
Department of Psychiatry
University of Alberta
Edmonton, AB, Canada

Willem A. van Gool, MD, PhD
Department of Neurology
Academic Medical Center
University of Amsterdam
Amsterdam, The Netherlands

Rob Veerhuis, PhD
Department of Psychiatry
Vrije Universiteit Medical Center
Amsterdam, The Netherlands

Venkat Venkataraman, PhD
Department of Cell Biology
University of Medicine and Dentistry of New Jersey – SOM
Stratford, NJ, USA

Victor L. Villemagne, MD
Department of Nuclear Medicine
Centre for PET, Austin Hospital
Melbourne, VIC, Australia;
Department of Pathology
The University of Melbourne
Parkville, VIC, Australia;
The Mental Health Research Institute of Victoria
Parkville, VIC, Australia

Z. Wei, PhD
Department of Psychiatry
University of Alberta
Edmonton, AB, Canada

Benjamin Wolozin, MD, PhD
Department of Pharmacology
Boston University School of Medicine
Boston, MA, USA

1
A Brief Introduction to the History of the β-Amyloid Protein (Aβ) of Alzheimer's Disease

David H. Small and Colin J. Barrow

Alzheimer's disease (AD) is the most common cause of dementia in the elderly. Typically, the disease progresses in a prolonged, inexorable manner [1]. Patients initially show symptoms of mild cognitive impairment, which may include some memory loss. As the disease progresses, more severe memory loss occurs (e.g., retrograde amnesia) leading to confusion and lack of orientation. The patient is often institutionalized in this period, as it becomes increasingly difficult for family members to cope with the constant requirements of care. In later stages of the disease, apathy and stupor can occur, and the patient becomes bedridden.

The histopathology of AD is characterized by gliosis and tissue atrophy caused by both synaptic and neuronal loss, which are most pronounced in the frontal and temporal cortices [2]. Proteinaceous deposits are seen in both the intracellular and extracellular compartments of the brain, typically in the hippocampus and neocortex. The intracellular deposits consist of neurofibrillary tangles that are made up of paired helical filaments of a hyperphosphorylated form of the cytoskeletal protein tau [3]. Extracellular amyloid plaques are found most commonly in the hippocampus and neocortex and may be diffuse or compact in nature [4]. Amyloid is also deposited as cerebral amyloid angiopathy within small- to medium-sized arterioles [5]. Although neurofibrillary tangles are associated with a number of different types of neurodegenerative disease, the presence of numerous compact or neuritic amyloid plaques is a hallmark feature of Alzheimer's disease. For this reason, it may be argued that accumulation of the β-amyloid protein (Aβ) is a key step in the pathogenic mechanism of Alzheimer's disease. In contrast, although the density of neurofibrillary tangles correlates more closely with the cognitive symptoms, it is now commonly thought that tangles are a secondary feature or the underlying disease process [6].

1.1 The Role of Aβ in AD

Glenner and Wong [7] first identified the major protein component of vascular amyloid, which was a low-molecular-weight, 4-kDa polypeptide, now referred to as the β-amyloid protein (Aβ). Subsequent studies established that the same protein was the major component of amyloid plaques [8]. The complete amino acid sequence of Aβ led to the identification of its precursor, the β-amyloid precursor protein (APP) [9].

APP has features of an integral type I transmembrane glycoprotein, with a large ectodomain containing the N-terminus and a small cytoplasmic domain containing the C-terminus (Fig. 1.1). Multiple mRNA splicing of exons can generate several different isoforms of APP that lack domains homologous to Kunitz-type protease inhibitors (KPI domain) and the OX-2 antigen as well as a domain encoded by an exon that regulates O-linked glycosylation by chondroitin sulfate. The Aβ sequence itself comprises part of the ectodomain of the protein and extends into, but not all the way through, the transmembrane domain [9, 10].

Soon after its identification, APP was shown to undergo ectodomain shedding by an enzyme

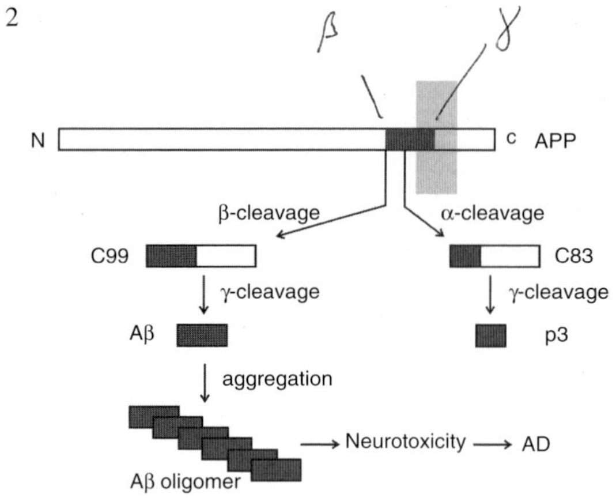

FIGURE 1.1. Proteolytic processing of APP and the role of Aβ in AD. APP can be proteolytically cleaved via two different processing pathways. Cleavage by α-secretase and γ-secretase generates C-terminal fragments known as C83 and p3, whereas cleavage by β-secretase (BACE1) and γ-secretase generates C99 and Aβ. According to the amyloid hypothesis, Aβ aggregates into amyloid fibrils or low-molecular-weight oligomers that are neurotoxic. The resulting neurotoxicity causes neurodegeneration and leads to dementia.

dubbed the α-secretase. The α-secretase cleaves APP within the Aβ sequence, adjacent to lysine-16, thereby destroying the sequence [11, 12]. Recently studies suggest that enzymes of the ADAM family of metalloproteases are responsible for this activity [13, 14]. Other studies have demonstrated that APP can also be cleaved at the N- and C-terminal ends of the Aβ sequence by enzymes dubbed β- and γ-secretase, respectively, to generate the full-length Aβ sequence [15]. Amyloidogenic processing by β- and γ-secretase is a normal, albeit minor, pathway of APP processing. The β-secretase has been unequivocally identified as an aspartyl protease termed BACE1 (an acronym for β-site APP cleaving enzyme-1) [16–19]. The γ-secretase comprises a complex of several proteins including presenilin-1, presenilin-2, Aph1, Pen2, and nicastrin. However, other protein components of this complex may also exist [15].

There is considerable evidence that the accumulation of Aβ in the brain is toxic to neurons and that this toxicity underlies the neurodegeneration that occurs in AD (Fig. 1.1) [20]. Aβ peptides are toxic to cells in culture [21], and this toxicity is associated with aggregation of the peptide [22]. Recent studies support the view that the most toxic species are the low-molecular-weight, soluble oligomers of Aβ [23].

Despite many studies that have shown that Aβ can disrupt biochemical events within neurons, direct proof that the accumulation of Aβ is the cause of AD has been lacking. Nevertheless, evidence that this is the case has slowly been accumulating. Some of the strongest evidence that Aβ accumulation is the cause, rather than an epiphenomenon, of AD has come from the finding of familial AD mutations present in the APP gene [24]. All of these mutations have been found to cluster around the Aβ sequence, and all of them have so far been shown to directly or indirectly cause an increase in forms of Aβ that aggregate [25]. For example, although the most commonly produced form of Aβ contains 40-amino-acid residues (Aβ40), a minor form containing 42 residues is also formed. This minor form aggregates into amyloid fibrils much more readily than Aβ40 [26]. The first mutation to be identified in the APP gene, the London mutation, involves a single base change at codon 717, which encodes a form of APP that is more readily cleaved to produce Aβ42. To date, at least 10 familial AD mutations are known to occur in APP [27].

The direct involvement of APP and Aβ in the pathogenesis of AD is also strongly supported by studies on transgenic mice. A number of transgenic lines have been developed in which human APP is expressed [28]. Many of these mice develop amyloid plaques. In addition, other features of AD pathology such as neuritic dystrophy, abnormal tau phosphorylation, gliosis, synaptic loss, and behavioral abnormalities have been observed. Although human APP mice do not develop neurofibrillary tangles, this is probably due to differences between mouse tau and human tau isoforms. Indeed, in double transgenic mice expressing both mutant human tau and APP, Aβ is seen to increase tau deposition [29].

Mutations in the APP gene account for only a very small percentage of all familial Alzheimer's disease (FAD) cases. Shortly after the identification of the first familial AD mutation in the APP gene, mutations were identified in two other genes, *PS1* encoding presenilin-1 and *PS2* encoding presenilin-2, located on chromosomes 14 and 1, respectively [30, 31]. Both presenilin proteins are components of the γ-secretase complex, and familial AD mutations within the PS1 and PS2 genes alter γ-secretase processing in a way that leads to the production of more Aβ42 [32].

In general, mutations in the APP, PS1, and PS2 genes lead to early-onset forms of AD. In contrast, the apolipoprotein E (apoE) gene located on chromosome 19 is a risk factor for late-onset AD [33]. There are three forms of apoE, termed E2, E3, and E4, encoded by three allelic variants ε2, ε3, and ε4. The ε4 variant is a risk factor for late-onset AD, whereas the ε2 may be protective. Although the reason for this is still unknown, it is undoubtedly related to Aβ production, aggregation, or clearance from the brain. Individuals with the ε4 allele have more Aβ deposition within the brain [34]. In addition, APP x apoE knockout transgenic mice develop little amyloid deposition in their brains, unlike normal APP mice [35]. Thus, studies on the role of apoE in AD provide strong support for the Aβ hypothesis.

1.2 Anti-Aβ Therapies for AD

The idea that Aβ is a primary causative agent in AD leads inevitably to the view that an effective therapy based on inhibiting the production, aggregation, clearance, or toxicity of Aβ may be achievable. One of the most promising but controversial approaches in recent years has been Aβ immunization. Studies show that in transgenic mice, immunization with Aβ42 leads to the generation of an immune response [36]. Anti-amyloid antibodies bind to amyloid plaques and appear to facilitate their removal from the brain, leading to an improvement in cognitive performance compared with nonimmunized control animals. Unfortunately, clinical trials of this approach in humans have been halted because a small percentage of individuals immunized with Aβ have developed a severe meningoencephalitis [37]. Nevertheless, there is some evidence that patients who develop a strong immune response to Aβ without the associated brain inflammation may benefit from this approach [38].

1.3 Current Status of the Aβ Hypothesis of AD

There is now very strong evidence that accumulation of oligomeric or fibrillar Aβ in the brain is a key event in the pathogenesis of AD. Perhaps the most important unresolved question is the mechanism by which Aβ causes its neurotoxic effect. It is also unclear what form of aggregated Aβ is the most neurotoxic. Another major question is how many unidentified genetic risk factors there are and how these risk factors affect Aβ production, aggregation, or clearance. If anti-Aβ therapies can be used successfully for the treatment of AD, then the remaining concerns about the role of Aβ in the pathogenesis of AD will have been answered.

References

1. Storey E, Kinsella GJ, Slavin MJ. The neuropsychological diagnosis of Alzheimer's disease. J Alzheimers Dis 2001; 3:261-285.
2. Probst A, Langui D, Ulrich J. Alzheimer's disease: a description of the structural lesions. Brain Pathol 1991; 1:229-239.
3. Iqbal K, Alonso Adel C, Chen S, et al. Tau pathology in Alzheimer's disease and other tauopathies. Biochim Biophys Acta 2005; 1739:198-210.
4. Wisniewski HM, Wegiel J, Kotula L. Review. David Oppenheimer Memorial Lecture 1995: Some neuropathological aspects of Alzheimer's disease and its relevance to other disciplines. Neuropathol Appl Neurobiol 1996; 22:3-11.
5. Castellani RJ, Smith MA, Perry G, Friedland RP. Cerebral amyloid angiopathy: major contributor or decorative response to Alzheimer's disease pathogenesis. Neurobiol Aging 2004; 25:599-602.
6. Small DH, McLean CA. Alzheimer's disease and the amyloid beta protein: What is the role of amyloid? J Neurochem 1999; 73:443-449.
7. Glenner GG, Wong CW. Alzheimer's disease: initial report of the purification and characterization of a novel cerebrovascular amyloid protein. Biochem Biophys Res Commun 1984; 120:885-890.
8. Masters CL, Simms G, Weinman NA, et al. Amyloid plaque core protein in Alzheimer's disease and Down syndrome. Proc Natl Acad Sci U S A 1985; 82:4245-4249.
9. Kang J, Lemaire HG, Unterbeck A, et al. The precursor of Alzheimer's disease amyloid A4 protein resembles a cell-surface receptor. Nature 1987; 325:733-736.

10. Wilquet V, De Strooper B. Amyloid-beta precursor protein processing in neurodegeneration. Curr Opin Neurobiol 2004; 14:582-588.
11. Weidemann A, Konig G, Bunke D, et al. Identification, biogenesis, and localization of precursors of Alzheimer's disease A4 amyloid protein. Cell 1989; 57:115-126.
12. Esch FS, Keim PS, Beattie EC, et al. Cleavage of amyloid beta peptide during constitutive processing of its precursor. Science 1990; 248:1122-1124.
13. Buxbaum JD, Liu KN, Luo Y, et al. Evidence that tumor necrosis factor alpha converting enzyme is involved in regulated alpha-secretase cleavage of the Alzheimer amyloid protein precursor. J Biol Chem 1998; 273:27765-27767.
14. Lammich S, Kojro E, Postina R, et al. Constitutive and regulated alpha-secretase cleavage of Alzheimer's amyloid precursor protein by a disintegrin metalloprotease. Proc Natl Acad Sci U S A 1999; 96: 3922-3927.
15. Nunan J, Small DH. Regulation of APP cleavage by alpha-, beta- and gamma-secretases. FEBS Lett 2000; 483:6-10.
16. Vassar R, Bennett BD, Babu-Khan S, et al. Beta-secretase cleavage of Alzheimer's amyloid precursor protein by the transmembrane aspartic protease BACE. Science 1999; 286:735-741.
17. Lin X, Koelsch G, Wu S, et al. Human aspartic protease memapsin 2 cleaves the beta-secretase site of beta-amyloid precursor protein. Proc Natl Acad Sci U S A 2000; 97:1456-1460.
18. Sinha S, Anderson JP, Barbour R, et al. Purification and cloning of amyloid precursor protein beta-secretase from human brain. Nature 1999; 402:537-540.
19. Yan R, Bienkowski MJ, Shuck ME, et al. Membrane-anchored aspartyl protease with Alzheimer's disease beta-secretase activity. Nature 1999; 402:533-537.
20. Small DH, Mok SS, Bornstein JC. Alzheimer's disease and Abeta toxicity: from top to bottom. Nat Rev Neurosci 2001; 2:595-598.
21. Yankner BA, Dawes LR, Fisher S, et al. Neurotoxicity of a fragment of the amyloid precursor associated with Alzheimer's disease. Science 1989; 245:417-420.
22. Pike CJ, Walencewicz-Wasserman AJ, Kosmoski J, et al. Structure-activity analyses of beta-amyloid peptides: contributions of the beta 25-35 region to aggregation and neurotoxicity. J Neurochem 1995; 64:253-265.
23. Walsh DM, Selkoe DJ. Deciphering the molecular basis of memory failure in Alzheimer's disease. Neuron 2004; 44:181-193.
24. Hardy JA, Higgins GA. Alzheimer's disease: the amyloid cascade hypothesis. Science 1992; 256:184-185.
25. Scheuner D, Eckman C, Jensen M, et al. Secreted amyloid beta-protein similar to that in the senile plaques of Alzheimer's disease is increased in vivo by the presenilin 1 and 2 and APP mutations linked to familial Alzheimer's disease. Nat Med 1996; 2:864-870.
26. Jarrett JT, Lansbury PT, Jr. Seeding "one-dimensional crystallization" of amyloid: a pathogenic mechanism in Alzheimer's disease and scrapie? Cell 1993; 73:1055-1058.
27. Bertram L, Tanzi RE. The current status of Alzheimer's disease genetics: what do we tell the patients? Pharmacol Res 2004; 50:385-396.
28. Hock BJ Jr., Lamb BT. Transgenic mouse models of Alzheimer's disease. Trends Genet 2001; 17:S7-12.
29. Lewis J, Dickson DW, Lin WL, et al. Enhanced neurofibrillary degeneration in transgenic mice expressing mutant tau and APP. Science 2001; 293:1487-1491.
30. Sherrington R, Rogaev EI, Liang Y, et al. Cloning of a gene bearing missense mutations in early-onset familial Alzheimer's disease. Nature 1995; 375:754-760.
31. Levy-Lahad E, Wasco W, Poorkaj P, et al. Candidate gene for the chromosome 1 familial Alzheimer's disease locus. Science 1995; 269:973-977.
32. Saunders AM, Strittmatter WJ, Schmechel D, et al. Association of apolipoprotein E allele epsilon 4 with late-onset familial and sporadic Alzheimer's disease. Neurology 1993; 43:1467-1472.
33. Schmechel DE, Saunders AM, Strittmatter WJ, et al. Increased amyloid beta-peptide deposition in cerebral cortex as a consequence of apolipoprotein E genotype in late-onset Alzheimer's disease. Proc Natl Acad Sci U S A 1993; 90:9649-9653.
34. Bales KR, Verina T, Dodel RC, et al. Lack of apolipoprotein E dramatically reduces amyloid beta-peptide deposition. Nat Genet 1997; 17:263-264.
35. Schenk D, Barbour R, Dunn W, et al. Immunization with amyloid-beta attenuates Alzheimer's-disease-like pathology in the PDAPP mouse. Nature 1999; 400:173-177.
36. Nicoll JA, Wilkinson D, Holmes C, et al. Neuropathology of human Alzheimer's disease after immunization with amyloid-beta peptide: a case report. Nat Med 2003; 9:448-452.
37. Hock C, Konietzko U, Streffer JR, et al. Antibodies against beta-amyloid slow cognitive decline in Alzheimer's disease. Neuron 2003; 38:547-554.

2 The Aβcentric Pathway of Alzheimer's Disease

Victor L. Villemagne, Roberto Cappai, Kevin J. Barnham, Robert A. Cherny, Carlos Opazo, Kathy E. Novakovic, Christopher C. Rowe, and Colin L. Masters

(Melbourne, Australia)

2.1 Introduction

Alzheimer's disease (AD), the leading cause of dementia in the elderly, is an irreversible, progressive neurodegenerative disorder clinically characterized by memory loss and cognitive decline [1], leading invariably to death, usually within 7–10 years after diagnosis. The dominant risk factor for sporadic AD is increasing age.

In the absence of biologic markers, direct pathologic examination of brain tissue derived from either biopsy or autopsy remains the only definitive method for establishing a diagnosis of AD [2]. The typical macroscopic picture is gross cortical atrophy. Microscopically, there is widespread cellular degeneration and neuronal loss that affects primarily the outer three layers of the cerebral cortex, initially affecting more the temporal and frontal cortical regions subserving cognition than the parietal and occipital cortices. These changes are accompanied by reactive gliosis, diffuse synaptic and neuronal loss, and by the presence of the pathological hallmarks of the disease, intracellular neurofibrillary tangles (NFT) and extracellular amyloid plaques [3, 4].

Neurofibrillary tangles are intraneuronal bundles of paired helical filaments. The main structural component of NFT is a normal constituent of cellular microtubules, but present in AD is an abnormally phosphorylated form, known as tau protein [5, 6]. They are most easily identified in the hippocampus. NFT are not specific to AD and are found in a variety of other neurodegenerative conditions such as frontotemporal dementia, subacute sclerosing panencephalitis, Hallervorden-Spatz disease, Parkinson dementia complex, and dementia pugilistica [2, 7]. Tau is a widely expressed phosphoprotein from the microtubule associated family, the main function of which is to maintain microtubule stability [8]. In AD, hyperphosphorylated tau aggregates reduce its ability to bind microtubules [9], leading to cytoskeletal degeneration and neuronal death [10–12]. A number of in vitro and in vivo studies have shown Aβ protein to be directly toxic to neurons, leading to the aggregation and secondary phosphorylation of the tau protein [13].

Amyloid plaques are extracellular aggregates of β-amyloid peptide (Aβ) of about 50–100 μm in diameter intimately surrounded by dystrophic axons and dendrites, reactive astrocytes, and activated microglia. Though mainly located in the amygdala and hippocampus, they are present throughout the cortex [6].

The progressive nature of neurodegeneration suggests an age-dependent process that ultimately leads to synaptic failure and neuronal damage [14] in cortical areas of the brain essential for memory and higher mental functions.

Currently, the clinical diagnosis of AD is based on progressive impairment of memory and decline in at least one other cognitive domain and on excluding other diseases that might also present with dementia such as frontotemporal dementia, dementia with Lewy bodies, stroke, brain tumor, normal pressure hydrocephalus, or depression [15, 16]. A variable period of up to 5 years of prodromal decline in cognition characterized by a relatively isolated impairment in long-term memory that may also be accompanied by impairments of working memory, known as mild cognitive impairment (MCI), usually precedes the formal diagnosis of AD. These deficits presumably relate to damage to the medial temporal lobe and/or specific prefrontal–temporal lobe circuits. About 40–60% of carefully characterized subjects with MCI will subsequently progress to meet criteria for AD over a 3- to 4-year period [17–19].

Briefly, the Aβ hypothesis postulates that the progressive rise, either by increased production or decreased clearance, in Aβ cerebral levels is the central event in the pathogenesis of AD [20]. Genetic evidence not only indicates that the metabolism of Aβ is clearly linked to the disease [21] but also points to specific metabolic pathways with the potential for developing diagnostic and therapeutic agents, and though there is a poor correlation between the density of deposits and disease severity, there is a correlation between the levels of soluble Aβ and cognitive impairment [22]. Even though synthetic Aβ is toxic to neuronal cells [10, 23], the precise mechanism(s) of action and the nature of the toxic Aβ species remain to be identified [24].

2.1.1 In Illo Tempore

November 4, 1906

On entering [he] looked at me blear eyed and vacuous, [. . .] now and again pulled his tangled wits together, and hints and sparkles of intelligence came and went in his eyes. There they crouched by the fire, [. . .] at the end of their days, old and withered and helpless. [He] rocked back and forth in a slow and hopeless way, and regularly once every five minutes he emitted a low groan. It was not so much a groan of pain as of weariness. [He] seemed singing back into his senility.

The preceding extract does not belong to Dr. Alois Alzheimer's presentation to his colleagues in Tübingen that very same day but is rather an excerpt of *The White Man's Way*, a short story by Jack London, first published on the crepuscular shore of the Atlantic that same November 4, 1906, in the Sunday Magazine of the *New York Tribune*. The audience present at the Conference in Tübingen witnessed Alzheimer's very first description of the neuropathology of AD, with the silver stained "miliary foci" and the "tangled bundle of fibrils." Alzheimer's presentation of Auguste D.'s case was published the following year on the *Allegmeine Zeitschrift für Psychiatrie und Phychisch-Gerichtliche Medizin* [25].

In 1910, Gaetano Perusini [26], in a depiction that has now become the everyday ritual for millions of AD caretakers around the world, published extracts of the clinical history of Auguste D., admitted by Alzheimer to the Hospital of Mentally Ill and Epileptic in Frankfurt in 1901. Perusini transcribes: "she becomes excited again and screams terribly" (Nov. 30, 1901). "She is in a state of fright, anxious and completely disoriented, violent towards everything. She lies in bed in a strange way" (Feb. 1902). "Completely rebellious, screams and stamps her feet when someone goes near her. She refuses to be examined, screams spontaneously and often for hours" (June 1902). "Her legs are drawn up to her chest. She does not speak but continues to mutter. She must be helped to eat" (Oct. 1905). On April 8, 1906, she died.

Four years later, Emil Kraepelin, the leading German psychiatrist, wrote: "That the involutional processes, known in man as old age, can also influence mental health seriously is most clearly demonstrated by the well-known fact of senile dementia which in certain circumstances can lead to a progressive transformation and, finally, to the destruction of the personality in the last decades of life" [27]. In the same book, Kraepelin graciously bestowed on the disease the eponymy of his colleague [27]. A new disease was born.

2.1.2 The Weight of Time

Age is the dominant risk factor in AD. The increase in the number of new cases of AD is the consequence of an improvement in life expectancy. AD is just another tragic adverse side effect of progress.

The research that dawned with the 20th century gathered momentum with the passing of the decades. New pathological approaches were developed, histochemical and cytochemical techniques were tested, and though a magnificent increase in AD research was seen on the 1980s, it seems not only that there is no slowdown but also that renewed efforts are dedicated to further characterize the pathogenic mechanisms of this devastating disease, one discovery after another leading to more elegant, refined, and sophisticated studies.

Epidemiological approaches, assisted by pioneering genetic evaluation, contributed to establishing the prevalence of the disease [28–32]. In 1968, Tomlinson, Blessed, and Roth [33] published a seminal work showing that 62% of the brains of deceased demented elderly patients presented the same pathological hallmarks described by Alzheimer 50 years earlier. AD was no more considered an unusual disease—a special case of *Senium Praecox*—but became the leading cause of dementia in the elderly [34]. By 1976, on the heels of the discovery of dopamine deficits in Parkinson disease, decreases in cholinergic neurons in the basal forebrain areas of AD patients were described in AD [35].

From the early days, controversy centered around the identification of the lesion(s) or substance(s) responsible for the neuronal death. The "drusige Entartung," or plaque-like degeneration, proposed by Scholz as the origin of the plaques [36] has been known since 1954 as congophilic angiopathy [37]. Now, as it was then, the controversies do not lie in the description of the neuropathological lesions but in the discrepant views on their role in the pathogenesis of AD. The introduction of the electron microscope in the 1960s allowed new insight into the disease, leading to the description of the structure of the senile plaques [38] and to the realization that NFT were composed of pairs of abnormal intertwined filaments [39].

Alzheimer's original description "these fibrils can be stained with dyes different from the normal neurofibrils, a chemical transformation of the fibril substance must have taken place" [25] proved to be accurate when NFT were shown to contain a hyperphosphorylated form [40] of a normal constituent of cellular microtubules: the tau protein [41–45]. Due to the stubborn insolubility of NFT [46], research mainly focused on plaques, specially on its main component: the amyloid protein. One hundred fifty years earlier, Virchow, at the zenith of the 19th century [47], called the waxy substance he likened to starch "amyloid" (from *amylum* or amylose). The term stuck.

2.1.3 Aggregated Time

By the mid-1980s a cascade of discoveries was triggered by the isolation and characterization of the amyloid protein. Glenner, who specialized in studying amyloidosis, first isolated an enriched sample of amyloid out of vessels from an AD brain [48, 49]. The following year, Masters and Beyreuther characterized amyloid from plaques in the brains of AD and Down syndrome patients [50]. The realization that whole families developed, generation after generation, the same symptoms of *Senium Praecox*, that patients with Down syndrome developed the same pathognomonic neuropathological features of AD, and that a protein played a key role in the composition of the plaques triggered the quest to identify the gene or genes involved in AD. The first candidate was chromosome 21, though it proved not to be as straightforward as initially thought [51–53]. By 1987, almost the whole sequence of the gene encoding the amyloid precursor protein (APP) was published [54–56]. Gene mutations were subsequently identified [57–59] and linked to increased production of Aβ [60]. Other chromosomes, such as chromosome 14, were also found to be associated with familial early-onset AD [61–64]. A fragment of APP, Aβ, was shown to be toxic to neurons [10, 65]. Aβ toxicity was shown to be linked to Aβ aggregation into fibrils [66, 67] and, furthermore, that transition metals were involved in Aβ aggregation [68, 69]. Aβ was found to bind to apolipoprotein E (ApoE) and that ApoE was genetically associated with late-onset AD [70–72]. The presenilins were eventually identified and cloned on chromosomes 14 and 1, respectively [73, 74]. Individuals with presenilin mutations were shown to have increased production of Aβ [75].

But despite all the tremendous corpus of knowledge of genetics, epidemiology, risk factors, and neuropathological mechanisms, there is no cure for AD.

2.2 Aβ: The Theory Behind the Hypothesis

Through the years, several hypotheses have been postulated to explain the molecular mechanisms leading to AD [76–82], but the Aβ theory is the dominant etiologic paradigm at this time [83] because it is the only one that can best or most comprehensively articulate the current available knowledge regarding the cellular, molecular, and functional alterations observed in AD. Not only is there a wealth of histopathological, biochemical, genetic, animal model, and functional neuroimaging data that support the key role of Aβ in the pathogenesis of AD, but no alternative hypothesis has emerged in the past two decades of intensive AD research. Genetic mutations within the APP gene cause rare cases of early-onset familial AD, and other causative mutations within genes associated with the secretase complex (presenilin 1, 2) are the most compelling evidence that Aβ production is the key factor at the center of AD pathogenesis.

In short, the hypothesis states that an imbalance between the production and removal of Aβ leads to its progressive accumulation triggering a series of reactions leading to synaptic dysfunction, microgliosis, and neuronal loss, clinically manifested with memory loss and impaired cognitive functions. The loss of synaptic function seems to be the critical factor in cognitive decline.

Aβ, the primary component of the characteristic plaques in the brain of AD patients, is a self-aggregating, 39- to 43-amino-acid metalloprotein (4 kDa) product of the proteolytic cleavage of APP by β- and γ-secretases [50, 84–86]. Aβ is not only found within senile plaques but is also present around cortical arterioles as a congophilic angiopathy. Aβ can also be assessed in cerebrospinal fluid, plasma, and even in neuronal cultures [87–89].

Aβ was first identified and sequenced from meningeal blood vessels of AD and Down syndrome patients 20 years ago [48–50]. The aggregation process that converts soluble Aβ into amyloid fibrils is thought to be a nucleation-dependent process [90] requiring structural transitions of Aβ [91]. The peptide is referred to as "beta" amyloid due to its secondary structure of β-pleated sheets. On electron microscopy, amyloid fibrils are composed of multiple protofibrils wrapped around each other forming a crossed β-pleated sheet [92, 93].

Much of the controversy derives from the use of the term *amyloid*. The broad term can be applied not only to Aβ but also to several unrelated extracellular deposits of fibrillar protein, such as β2-microglobulin, amylin, or serum amyloid A, each one of them associated with a specific disease [94–96].

The earliest structural, microscopically visible pathological changes in AD are diffuse Aβ deposits. These deposits are also observed in normal ageing individuals, but the density is lower than in AD patients [97, 98], indicative of an immature or not yet toxic form of Aβ [99]. The presence of extracellular Aβ in highly specialized cortical brain regions implicated in memory and cognition precede the other pathognomonic pathological features of AD, indicating that increases in Aβ are involved in the early presymptomatic stages of the disease. Being the earliest phenotypical marker of disease has crucial implications for neuroimaging and treatment. The increase in Aβ-deposition is accompanied by decreases in Aβ in CSF. Presymptomatic carriers of missense mutations of APP or PS present elevated Aβ42 in plasma and skin fibroblasts indicating again that increases in Aβ are the earliest signs of the disease.

Recent studies have detected stable intraneuronal pools of insoluble Aβ deposits, generated in the endoplasmic reticulum [100], indicating that Aβ is also produced intracellularly, suggesting that Aβ might not be the end result of the abnormal cleavage of APP but a protein with a specific physiological role [88, 89, 101, 102] and that only the alteration of its metabolism, leading to its increase, precipitates the neurotoxic effects, leading to synaptic loss and cell death.

Though extracellular amyloid plaques are the hallmark brain lesions of sporadic AD, the distribution and density of both diffuse and Aβ plaques at the light microscopic level [22] have not been consistently shown to correlate with the degree of cognitive impairment [103, 104]. The best correlation occurs with soluble levels of Aβ, measured biochemically [22, 105–108]. Soluble Aβ is in equilibrium with insoluble Aβ in the plaques. The significance of these aggregates can be interpreted as they either are a reservoir for the soluble oligomers or represent the sequestered pool of soluble and now precipitated Aβ, therefore fulfilling a "protective" function, or just the end stage or final product of the Aβ cascade.

One of the criticisms of the amyloid hypothesis has come from some of the interpretations of the work of Braak and Braak [109], who stated that neurofibrillary degeneration of cell bodies and their neurites not only predate morphologically detectable amyloid plaques but they also increase gradually with age. However, as Hardy and Selkoe point out [110], the postmortem cases used to establish the Braak Stage I neuropathology criteria were nondemented older individuals, in whom it is impossible to determine whether their neurofibrillary changes represents early stages of AD or a different process altogether [111], because it has been well established in patients with Down syndrome that Aβ deposition predates the formation of neurofibrillary tangles [112, 113].

2.3 Insights into the Genetics of Aβ

The Aβ hypothesis is further supported by genetic data [114–118]. Though it is highly probable than additional genes are associated with AD, to date only four different genes, associated with either Aβ production or removal, are implicated in the pathophysiology of AD and have been described in patients with the rare early-onset familial AD [119–121]: mutations of the APP gene [59, 60, 122–125] on chromosome 21, mutations in the presenilin 1 (PS1) [73, 126] and presenilin 2 (PS2) [127] genes on chromosome 14 and 1, respectively [73–75, 128–130], and polymorphism of the apolipoprotein E (ApoE) on chromosome 19 [70, 71, 131]. Three of them—PS1, PS2, and APP—have a clear-cut autosomal dominant pattern with a penetrance above 85%; whereas the other, APOE, despite being the most prevalent of these risk factors for AD, has a weaker susceptibility factor. The main feature of mutations on APP, PS1, and PS2 involved in different steps of APP processing pathway is the increased production and elevated plasma levels of Aβ specially Aβ42 [75, 129, 130, 132]. These various genetic mutations, all manifesting as a similar clinical entity, all leading to increased levels of Aβ and to Aβ buildup in the brain before AD symptoms arise, further support the Aβ theory of AD [72, 104, 129, 133–136].

2.3.1 APP

The Aβ hypothesis was further supported by the cloning and sequencing of the APP gene [54, 137–139] and its localization to chromosome 21 [54, 55, 140, 141], the chromosome involved in Down syndrome, a condition that invariably develops the typical AD neuropathology by age 50 [142] though they start getting amyloid plaques as early as age 12, long before they get NFT and other AD lesions [112, 143].

APP is a 751–770 residue ubiquitously expressed glycosylated transmembrane protein with a large hydrophilic aminoterminal extracellular domain, a single hydrophobic transmembrane domain consisting of 23 residues, and a small carboxy-terminal cytoplasmic domain [144].

~85kDa

The majority of APP is degraded in the endoplasmic reticulum and only a small fraction enters the secretase cleavage pathway [145, 146]. While APP is usually proteolytically cleaved by α-secretase, mutations on the APP gene were shown to be associated with increased Aβ self-aggregation [57–59, 147–150] and Aβ production by the sequential cleavage by β- and γ-secretases [60, 123, 124].

The free N-terminus of Aβ, considered the first critical step in amyloid formation [151], is derived from the APP by proteolytic cleavage by β-secretase. Several lines of evidence demonstrate that β-secretase cleavage of APP is required for Aβ generation [152, 153]. Generation of the N-terminus is followed by C-terminal cleavage by γ-secretase to release the final Aβ-product from the β-secretase cleavage fragment C99. Cleavage by γ-secretase occurs within the transmembrane region of APP yielding mainly 40- and 42-amino-acid Aβ C-terminal variants, Aβ40 and Aβ42 (Fig. 2.1).

APP can also undergo nonamyloidogenic processing by α-secretase, which cleaves APP within the Aβ domain to generate α-APPs (the ectodomain of APP ending at the α-secretase cleavage site) [119] and C83 (the C-terminal tail of APP), which can then undergo γ-secretase cleavage leading to the release of p3 (Fig. 2.1), a shortened, probably non-pathogenic, form of Aβ [75].

Although the function of APP is unknown, recent evidence suggests it functions as a kinesin-1 cargo receptor mediating the targeting of several synaptic proteins to the nerve terminals [154] and as part of a complex metal-transport systems essential in maintaining cellular Cu and Fe homeostasis [155, 156] by delivering Cu and Fe to metalloenzymes and proteins, such as superoxide dismutase 1 (SOD1) [157] and the Cu ATPase [158]. Overexpression of the Aβ containing carboxyl-terminal fragment of APP in transgenic mouse models results in significantly reduced brain Cu, but not Fe levels [159], whereas APP knockout mice have increased Cu levels in both brain and liver [160]. Cu modulates APP processing [161, 162] with higher Cu levels resulting in a reduction in Aβ production and a consequential increase in the nonamyloidogenic p3 form of the peptide [163]. Independent Cu-binding sites have been identified on both Aβ and APP. The Cu-binding domain of

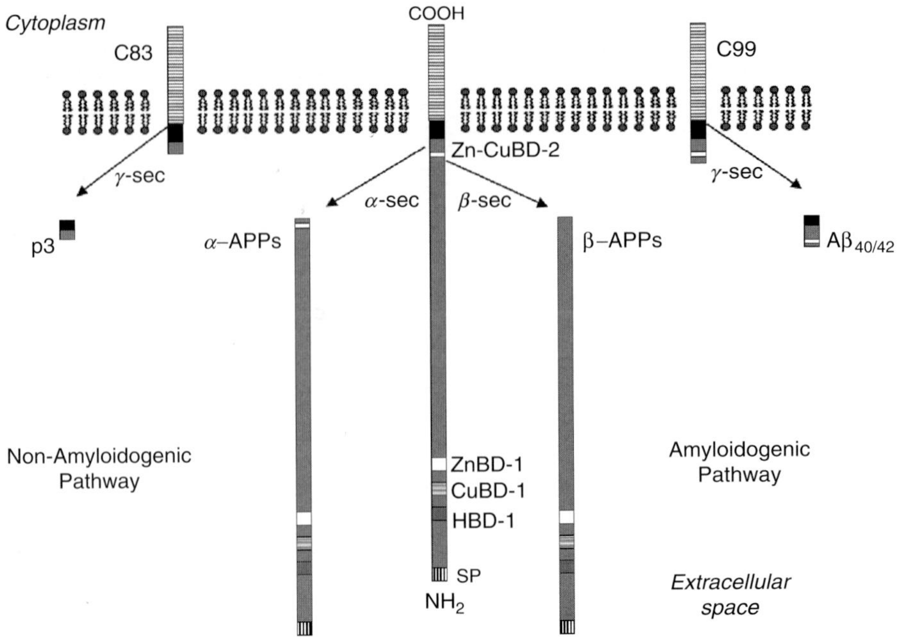

FIGURE 2.1. Schematic diagram of amyloidogenic and non-amyloidogenic proteolytic pathways of APP and production of Aβ. APP is cleaved by either α-secretase (α-sec) or β-secretase (β-sec) yielding α-APPs or β-APPs, respectively. The C-terminal C83 fragment produced by α-secretase, and the C-terminal C99 fragment produced by β-secretase, are then further cleaved by γ-secretase (γ-sec) into P3 or Aβ40/42, respectively.

APP, homologous to copper chaperones, contains a tetrahedral binding site consisting of two histidine residues (147, 151), a tyrosine (168) and methionine (170) that favors Cu(I) coordination [164].

2.3.2 Presenilins

There is also more genetic evidence coming from mutations of PS1 and PS2 [75] and from the cloning of presenilin proteins [73–75, 129, 130, 132] that affect secretases [165, 166]. The majority of early onset familial AD cases are linked to mutations within the PS genes. More than 40 mutations have been described in the gene for PS1 that can subsequently result in AD. Mutations in both genes selectively increase the production of Aβ42 in cultured cells and in the brains of transgenic mice and are associated with early onset familial AD [73, 120, 151, 166]. Some PS mutations associated with increases in Aβ metabolism instead of presenting AD symptoms show large plaques and special symptoms such as spastic paraparesis [167–171].

Presenilins are two proteins, presenilin 1 (PS1) and presenilin 2 (PS2), encoded by two closely related genes PS1 and PS2, and located in intracellular membranes [172]. They are ubiquitously expressed within the brain, primarily in neurons. PS1 and PS2 contain multiple transmembrane domains, with both amino and carboxy terminus as well as a large hydrophilic loop. Both proteins, the 46 kDa PS1 and 55 kDa PS2, share 67% amino acid identity [132]. The exact functions associated with PS protein have not been fully elucidated yet. PS1 is involved in normal neurogenesis and formation of the axial skeleton, as well as in γ-secretase activity and binding of PS to APP. Gene deletion of PS1 shows that it is indispensable for the generation of Aβ [166]. Two transmembrane aspartate residues in PS1 are essential for Aβ production, indicating that PS1 is either an essential cofactor for γ-secretase or maybe γ-secretase itself [173]. PS2 also contains two transmembrane residues critical for γ-secretase activity.

A growing list of proteins, including tau, have been identified as interacting directly or indirectly with PSs [174–176]. PS proteins have also been proposed to function in the control of apoptosis. While PS2 appears to play a direct role in fas-mediated apoptosis [177], mutations in PS1, through the activity of related kinases and phosphatases [178] and destabilized calcium homeostasis [175], may present a higher predisposition to neuronal apoptosis [177]. Par-4, a protein implicated in apoptosis, is overexpressed in AD brain and mutated PS-1 transfected cells [179].

2.3.3 ApoE

Genetic variability in Aβ catabolism and clearance increase the risk for late-onset AD [180–184]. In contrast to the rare, early-onset autosomal dominant forms, the only consistent marker for both the early-onset familial and late-onset nonfamilial form of dementia is the polymorphism of ApoE allele on chromosome 19 [185, 186]. Encoded on the long arm of chromosome 19, ApoE is a 34-kDa lipid transport protein considered the major genetic risk factor in the pathogenesis of AD [187, 188]. ApoE is normally present in oligodendroglia, astrocytes, and microglia. ApoE is a lipid carrier protein involved in the transport of cholesterol and phospholipids, believed to play an important role in synaptic plasticity and neuronal repair mechanisms. ApoE protects neuronal-glial cells cultures against H_2O_2 oxidative injury from by reducing secondary glutamate excitotoxicity in vitro [189]. ApoE is both directly and indirectly involved in oxidative mechanisms in the brain [190]. ApoE interacts directly with Aβ and with APP through the carboxy-terminal domain of ApoE. The association of ApoE and Aβ inhibits fibril formation [191] and also attenuates glial activation by Aβ [192]. ApoE exists in three allelic variants: ε2 (8%), ε3 (77%), and ε4 (15%). The presence of the ApoE4 allele increases fourfold the risk of AD and much more if the allelic variant is inherited from both parents. The ε4 allele is absent in approximately 30–40% of patients with AD and present in about 30% of healthy subjects [193], as well as in patients with Down syndrome [194, 195]. In carriers of ApoE4 allele, Aβ deposition responsible for the congophillic angiopathy [196, 197] could play an important role in contributing to the chronic cortical hypoperfusion typically observed in neuroimaging studies of patients with AD [198]. While the ε4 allele is associated increased risk for AD, the ε2 allele is believed to represent no increased or decreased risk, while the ε3 allele may confer some protection against Aβ-induced toxicity [71] through its antioxidant and membrane stabilizing properties and via complexation and internalization of Aβ through ApoE receptors [199].

Furthermore, ApoE is also a metal chelator, and the ε4 allele variant binds more rapidly to Aβ while at the same time displaying the weakest chelator affinity [200].

2.3.4 Transgenic Mice Models

Further insight was gained through the development of transgenic mice models of AD. Transgenic mice models with mutations in APP and PS genes lead to increase production and progressive aggregation of Aβ, reproducing the major features of AD: Aβ plaques, associated with neuronal and microglial damage [201–203]. The absence of human tau molecules in transgenic mice might explain why despite the progressive Aβ deposition [201, 203], there are no NFT and very little neuronal loss [204, 205]. Other reasons to be considered are species differences in neuronal vulnerability, the relatively short duration of exposure to Aβ, and the lack of certain cytokines necessary for a full complement inflammatory response.

Mutations in tau protein leading to large deposits of tau in intracellular NFT is not associated with amyloid deposits and is clinically manifested as frontotemporal dementia with parkinsonism [206–209], indicating that the NFT in AD are secondary to Aβ production [210] and probably triggered by Aβ [13, 211].

While the density of NFT correlates better than Aβ aggregates with the degree of dementia [212], and the hyperphosphorylation of tau leading to the formation of NFT has neurotoxic consequences in and of itself, mutations in tau are associated not with familial AD but with frontotemporal dementia [206]. Furthermore, in patients with the rare PS1 mutations or in individuals with Down syndrome who died prematurely from other diseases, Aβ either as diffuse deposits or typical plaques precede the appearance of NFT [213, 214].

Transgenic mice overexpressing both mutant human tau and mutant human APP while showing the same number and structure in their amyloid plaques present a significant higher number of tau-positive NFT than transgenic mice overexpressing only mutant human tau [215] indicating that the mutant APP and the consequent Aβ production precede and promote the formation of NFT [211].

The offspring of ApoE-deficient mice crossed with APP transgenic mice showed a significant reduction in Aβ deposition [216] supporting the role played by ApoE in the metabolism of Aβ [71].

2.4 Aβ Is Toxic

A common factor in the postulated mechanisms of Aβ toxicity is the oligomerization of Aβ, whether as dimers or trimers [217, 218], protofibrils [219], or fully formed fibrils [220, 221]. Despite several attempts, the main obstacle to the full validation of the Aβ hypothesis lies in the identification in vivo of the specific neurotoxic Aβ soluble oligomer. There is an inverse relationship between amyloid burden and oxidative damage in vivo as assessed by 8-OH guanosine levels in AD-affected tissue [222–224]. Several lines of evidence demonstrate that diffusible soluble Aβ oligomers, but not monomers or insoluble amyloid fibrils, are toxic to cultured neurons and responsible for the neurotoxicity and synaptic dysfunction present in AD [225, 226]. Microinjection into rats of culture medium containing soluble oligomers of human Aβ (in the absence of monomers and amyloid fibrils) inhibits long-term potentiation in the hippocampus [218]. Aβ fibrils injected into the brain of aged primates induces local gliosis and neuronal loss [8]. Similar changes are observed in young APP transgenic mice before plaque formation [227, 228], though the diversity and unstable nature of Aβ intermediates, from monomers to mature fibrils, makes it difficult to identify the specific species responsible for the neurotoxic effects.

2.5 Mechanisms of Aβ Toxicity

As a result of its high lipid content and high oxygen consumption, the brain is particularly susceptible to oxidative stress [229]. Several mechanisms have been proposed to explain Aβ neurotoxicity: production of reactive oxygen species (ROS) such as hydrogen peroxide, nitric oxide, superoxide, highly reactive hydroxyl radicals and nitric oxide (NO), excitotoxicity with intracellular calcium accumulation, decreased membrane fluidity, energy depletion, alteration of the cytoskeleton, and inflammatory processes [110, 156, 177, 230–234]. All of these events converge into similar pathways of necrosis or apoptosis, leading to progressive dysfunction and loss of specific neuronal cell populations [156] (Fig. 2.2).

2.5.1 Generation of ROS

Extra- and intracellular production of ROS initiate and promote neurodegeneration in AD [235–239]. Evidence of oxidative stress in AD is manifested through higher levels of oxidized proteins [238, 240], advanced glycation [241], lipid peroxidation products [188, 242], formation of toxic species, such as peroxides, alcohols, aldehydes, ketones, cholesterol oxide (toxic to microglial cells) [243], cholestenone [244], altered gene expression [245], damaged DNA [246], and induced apoptosis [247]. Aβ induces lipoperoxidation of membranes and lipid peroxidation products [248]. Lipids are modified by ROS and there is a high correlation between lipid peroxides, antioxidant enzymes, amyloid plaques, and NFT in AD brain [249]. Markers of oxidative DNA damage have been localized in plaques and NFT [241, 250–253].

Several breakdown products of oxidative stress including 4-hydroxy-2,3-nonenal (HNE) [254, 255], acrolein, malondialdehyde, and F2-isoprostanes have been observed in AD brains when compared with age-matched controls [256]. HNE modifies proteins resulting in a multitude of effects, including inhibition of neuronal glucose and glutamate transporters [257], Na-K ATPases [258], plus activation of kinases and dysregulation of intracellular calcium signaling that ultimately induce an apoptotic cascade [259–266].

Catalase, superoxide dismutase (SOD), glutathione peroxidase, and glutathione reductase, indicators of cellular defense mechanisms against oxidative stress, are increased in the hippocampus and amygdala of AD patients [267, 268].

DNA bases are vulnerable to oxidative stress damage involving hydroxylation [269], protein carbonylation, and nitration. ROS-induced calcium influx, via activation of glutamate receptors, triggering an excitotoxic response leading to cell death have also been observed in AD brains [266, 270].

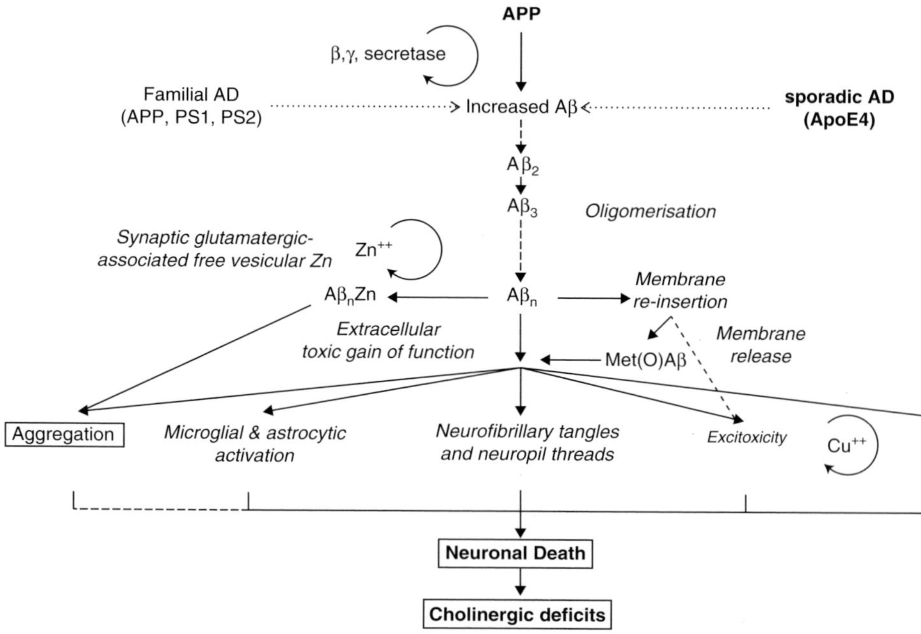

FIGURE 2.2. Schematics showing the role of Aβ in Alzheimer's disease (AD) pathogenesis. Increased production or reduced clearance of Aβ leads to aggregation, deposition, and neuronal injury through a variety of neurotoxic mechanisms, such as generation of oxygen and nitrogen radicals (H_2O_2, OH·, NO), transition metal ion interactions, excitotoxicity, tau hyperphosphorylation into neurofibrillary tangles, inflammatory response via microglia, and astrocytic activation leading to synaptic deficits and cell death.

ROS are also generated when oxygen reacts with unregulated redox-active metals. Metalloproteins such as Aβ in AD might abnormally present Cu or Fe for inappropriate reaction with O_2 are implicated in several age-dependent neurodegenerative disorders [156].

2.5.2 Generation of RNS

NO induced neurotoxicity has been extensively studied. NO is synthesized by NO synthases (NOS), and the three isoforms of NOS, endothelial (eNOS), neuronal (nNOS), and inducible (iNOS), are present in the brain [271]. NO synthesis is activated by glutamate release accompanied by excess calcium ion influx through activation of the NMDA [272] and AMPA receptor [273]. Aβ induces NO production by interacting with glial cells or by disrupting Ca^{++} homeostasis through NMDA receptor [272, 274].

NO combines with superoxide anion forming peroxynitrite, and the resultant RNS can induce significant oxidative stress leading to lipid peroxidation, damaged DNA, and neuronal death [275]. NO also promotes the over expression of metalloproteinases, particularly MMP9 enzymes, that disrupt the extracellular matrix [276, 277].

2.5.3 Activation of Inflammatory Processes

Aβ fibrils are toxic for cultured neurons and activate microglia. Blocking Aβ fibril formation prevents this toxicity [220, 221, 278, 279]. Astrocytes and microglial cells are involved in the chronic inflammatory responses in AD through the upregulated expression of phospholipase A2, leading to increased arachidonic acid/prostaglandin inflammatory pathway activity by secreting interleukin-1 [280], activation of complement pathways [281], and by producing a variety of potentially neurotoxic compounds, including superoxides, glutamate, and NO [282, 283].

2.5.4 Altered Energy Metabolism

Intermediate metabolism is essential to maintain signaling activities and depends on mitochondrial function. Disturbed energy metabolism and the appearance of degenerating mitochondria in axonal terminals are an early feature of AD [284, 285].

ROS production, calcium ion uptake, and mitochondrial membrane depolarization have been linked to neuronal apoptosis [286, 287] by disrupting the normal mitochondrial functioning, through the uncoupling of oxidative phosphorylation and impairment of cellular respiration, compromising energy production [288]. The mitochondrial electron transport chain specifically, cytochrome C oxidase or complex IV, is altered in AD [289, 290] maybe secondary to mutated and oxidatively damaged mitochondrial DNA [253, 291, 292]. This is supported by results with cytoplasmic hybrid or cybrid cells [290] that resemble electron transport chain defects observed in AD [289].

2.5.5 Altered Metal Homeostasis

The evidence not only supports Aβ as the key factor in the pathogenesis of AD [21, 50, 54, 293], but it also points to the fact that brain metal homeostasis, specially Zn and Cu, is significantly altered in AD [101, 294–297]. The progressive synaptic disruption and ultimately neuronal loss observed in AD might be secondary to toxic oxidative stress from excessive free-radical generation favored by transition metals bound to Aβ [101, 156, 294, 297–300]. The generation of ROS usually requires the reaction of O_2 with a redox metal ion such as Cu or Fe. Aβ is a metalloprotein with high in vitro affinity for Cu (highest), and Fe and Zn (lowest) [101, 301–303]. Aβ coordinates transition metal ions through bridging histidine residues at positions 6, 13, and 14, similar to the ones found in the active site of superoxide dismutase [156]. When Aβ binds Cu and Fe, extensive redox chemical reactions take place [156, 224, 294, 304–307]. Isolated senile plaques generate ROS in a manner dependent upon Cu and Fe [300, 306].

Several lines of evidence point to the participation of transition metals in Aβ neurotoxicity. Brain copper and iron concentrations increase with age [159, 308, 309]. Very high concentrations of Cu (400 μM), Zn (1 mM), and Fe (1 mM) have been found in plaques of AD-affected brains [298, 310]. Genetic ablation of the zinc transporter 3 protein, required for zinc transport into synaptic vesicles, reduced plaque formation in Tg2576 transgenic mice [311]. There are two methods of inducing aggregation of Aβ, metal induced cross-linking leading to amorphous aggregates and fibril formation, or lowering the pH [312]. Zn, Cu, and Fe induce Aβ aggregation in vitro [302, 313]. Soluble oxidized Aβ accumulates within the synaptic cleft, at which high concentrations of Zn (300 μM) and Cu (30 μM) are released during neurotransmission, which could coordinate with soluble Aβ, promoting its toxicity, explaining the synaptic loss observed in AD [311, 314]. The high Zn concentrations also promotes the aggregation of the Cu/Fe-metallated Aβ, creating a reservoir of potentially toxic Aβ that is in equilibrium with the soluble pool. The large polymeric deposits of misfolded proteins do not only represent the end result of the aggregation process but they may mainly act as inactive reservoirs in equilibrium with the small diffusible oligomeric toxic species responsible for the neurodegenerative pathology. Paradoxically, some emerging data suggest that Aβ might have a role as an antioxidant, a function that may wane with aging [101, 315].

Addition of Cu or Zn to Aβ causes a conformational change from β-sheet to α-helix, generating an allosterically ordered membrane-penetrating oligomer [222]. The extensive oxidative damage associated with Aβ [299, 307, 316, 317] may involve calcium dysregulation, caused by either the formation of membrane calcium channels [318] or modulation of an existing channel [319]. In the normal brain, most Aβ will form a hexamer that is embedded in the cell membrane [222, 320–322], but reactions of Cu with Aβ lead to the oxidative modification of the methionine 35 (Met35) [323] producing covalent cross-linking of Aβ yielding soluble oligomers [22, 303, 323, 324] that are released from the membrane with a toxic gain of function and that resist clearance [156]. Met(O)Aβ, which has been isolated from AD amyloid brain deposits [325, 326], is toxic to neuronal cells, toxicity attenuated by clioquinol and completely rescued by catalase. Unlike the unoxidized peptide, Met(O)Aβ is unable to penetrate lipid membranes to form ion channel-like structures and alters the aggregation profile of the peptide such that the formation of Aβ trimers and tetramers is attenuated [327] and fibril formation inhibited [328]. Met(O)Aβ production contributes to the elevation of soluble Aβ seen in the brain in AD [323]. These abnormally soluble toxic forms are correlated with cognitive and memory decline [22]. Spectroscopy studies have shown that Zn and Cu are coordinated to the histidine residues of the deposited Aβ in the senile plaque and that the Met35 of Aβ is oxidized [329]. Aβ toxicity is enhanced in the presence of Cu [306] and inhibited by extracellular catalase [306, 307]. Association of soluble Aβ with both Fe and Cu produces H_2O_2, which is neurotoxic in vitro [224, 304, 305], while complexing of Aβ with redox-inert Zn causes precipitation of the soluble metalloprotein complex [69]. The Zn associated to the aggregated amyloid partly reduces H_2O_2 production [224], which might explain the poor correlation between plaque amyloid burden and cognitive decline, while soluble Aβ levels correlate well with clinical severity [22].

2.6 Prospects for Treatment and Neuroimaging

The insight into the molecular mechanism of AD pathogenesis has not only opened new opportunities for the successful development of neuroprotective treatment strategies aimed at the prevention of Aβ generation but also for new neuroimaging approaches [330].

2.6.1 Therapeutic Strategies

2.6.1.1 Traditional Therapeutic Approaches

To date, no current therapy has been shown to halt or reverse the underlying disease process, and these remain confined to symptomatic palliative interventions [331]. Given the neuronal degeneration with impairment in cholinergic transmission in hippocampal and basal forebrain, areas associated with memory and cognition [332], as well as decreased levels of the cholinergic markers choline acetyltransferase and acetyl cholinesterase [333], most treatment strategies are based in increasing intrasynaptic ACh levels. Though now approved for AD, the cholinesterase inhibitors tacrine, donepezil, rivastigmine, and galantamine only provide patients with modest relief to their symptoms [334]. Recently, the noncompetitive NMDA antagonist memantine has been proposed as a safe and effective symptomatic treatment of AD patients [335–338]. Other approaches to alter the progression of AD involve the use of estrogen, antioxidants (alone or in combination with selegiline), or nonsteroidal anti-inflammatory drugs (NSAIDs) (Fig. 2.3).

Compounds with the ability to inactivate ROS might have therapeutic potential in the treatment of AD, and some cell culture toxicity studies have shown beneficial effects [339], though there has been limited clinical evaluation of antioxidants The classical lipophilic free-radical scavenger, α-tocopherol (vitamin E), has been evaluated in both AD and Parkinson disease (PD), and though it showed some encouraging results in AD patients [340], especially when combined with ascorbic acid [229, 341], it was found to have no beneficial effects in PD [342]. Upregulation of ROS-scavenging enzyme capacities through neurotrophins [343] may provide a mechanism for the prevention of neurotoxicity. Cholinergic drugs are routinely used in the treatment of AD to improve cognitive functions and in association with antioxidants have been proposed to be more effective in the treatment of AD than the individual agents alone [237]. There is a growing interest in the use of polyphenolic antioxidants to reverse age-related decline in neuronal signal transduction and cognitive and motor behavior deficits [344, 345].

ROS generation triggers glutamate-mediated excitotoxicity. Memantine, which targets the NMDA receptor, slows the development of the disease and is of modest benefit to patients in the moderately severe to severe range of the disease [335, 336, 338]. Use of coenzyme Q10, L-carnitine, and creatine might prevent mitochondrial oxidative damage and mitochondrial mutations [285, 346, 347]. Another potential therapeutic strategy proposes the use of brain-derived neurotrophic factor or nerve growth factor [348]. Estrogens have been shown not only to modulate neurotransmission but also to act as free-radical scavengers, activating nuclear estrogen receptor in intracellular signaling [349] and preventing Aβ formation by promoting the α-secretase APP non-amyloidogenic pathway [350].

2.6.1.2 Novel Therapeutic Approaches

If, as postulated, AD pathology is the consequence of a chronic imbalance between Aβ production and clearance, the most rational strategy to treat the disease would involve either retarding, halting, or

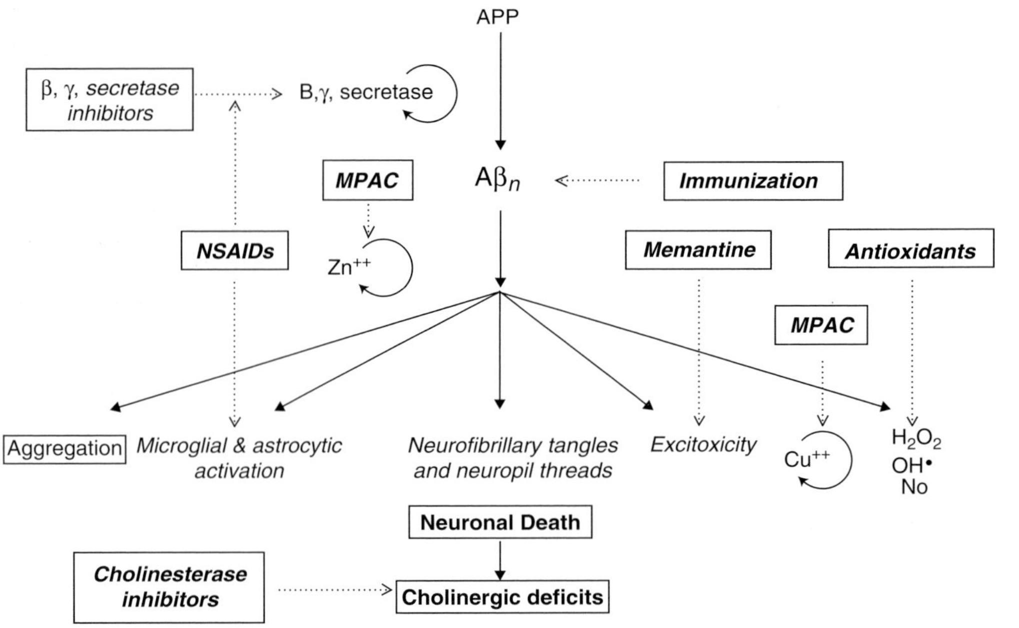

FIGURE 2.3. Schematic representation of therapeutic strategies for Alzheimer's disease. The therapeutic interventions are boldface and set in boxes, and the dotted arrows indicate the target(s). Abbreviations: $A\beta_n$, Aβ oligomers; MPAC, metal–protein attenuating compound; NSAIDs, nonsteroidal anti-inflammatory drugs.

even reversing the process that leads to increase production of Aβ [331, 334, 351, 352].

The most promising strategy for neuroprotection might be reducing formation of Aβ by partially inhibiting either β- or γ-secretase (Fig. 2.3), which generate Aβ from APP, and/or stimulation of α-secretase activity [151, 353–358]. Total inhibition of either β- or γ-secretase should block Aβ production completely. There are vigorous attempts to identify small lipophillic inhibitors of β-secretase. There are already potent γ-secretase inhibitors available [359–365] and undergoing human trials.

Given the evidence that levels of soluble Aβ correlate with disease severity [22, 108] and that the Aβ amyloid is probably the main neurotoxic factor in the development of AD, the development of agents inhibiting Aβ oligomerization should be more effective than those that merely block Aβ deposition [366]. Two basic strategies have been proposed in order to reduce or remove Aβ from the brain: immunization [367–371] breaking the pathway that leads to Aβ deposition [372] by precipitating an active immune response against the Aβ [370, 373, 374], or the passive administration of specific anti-Aβ antibodies [375–377] promoting microglial clearance [370, 375] and/or by redistribution of Aβ into the systemic circulation [376] (Fig. 2.3). Active immunization with synthetic Aβ was effective in APP transgenic mice without detectable toxicity, [375] though recent human trials resulted in the development of encephalic inflammatory reactions that precluded further human evaluation [378–381].

The use of anti-inflammatory medications is not only aimed at reducing the Aβ-elicited cellular inflammatory response [382], but it has also been shown to have direct effects on the cleavage of APP by γ-secretase, an effect that is independent of their inhibition of cyclooxygenase and other inflammatory mediators [383, 384] (Fig. 2.3). Some such drugs reduce cytopathology in APP transgenic mice [385, 386].

Another approach postulates modulating cholesterol homeostasis. The use of cholesterol-lowering drugs has been shown to reduce pathology in APP transgenic mice [387] and has been associated with lower incidence of AD [388, 389] while high-cholesterol diets increase Aβ pathology in experimental animals [390, 391] through a yet not elucidated effect of cholesterol on APP processing [392, 393].

Based on the role that metal ions such as Cu, Fe, and Zn play in the biochemical processes associated with Aβ deposition and neurotoxicity [69, 156, 224, 295, 302–305, 310, 321], a further therapeutic strategy using the metal binding sites of Aβ lead to the design and development of molecules, known as metal–protein attenuating compound (MPAC) [156] (Fig. 2.3), which inhibit the deleterious effects of aberrant metal interactions through competition with the target protein for the metal ions, leading to a normalization of metal homeostasis. MPAC not only inhibit the in vitro generation of hydrogen peroxide but also have been shown to reverse the precipitation of Aβ in vitro and in postmortem human brain specimens [394], reducing Aβ burden by a direct solubilization and by reducing toxic oxidative stress [372]. Clioquinol (CQ), 5′-chloro-7-iodo-8-hydroxyquinoline, is a hydrophobic quinoline Zn and Cu chelator that freely crosses the blood–brain barrier [395]. After initial studies showed that CQ increased soluble phase Aβ by more than 200% in a concentration-dependent fashion in homogenized postmortem human brain samples, its efficacy was tested in transgenic Tg2576 mice expressing mutant APP protein and which develop Aβ amyloid deposits and showed a dramatic 49% decrease in brain Aβ deposition after 9 weeks of oral treatment [372]. CQ was chosen to be tested as an Aβ amyloid solubilizing and antitoxic agent in a randomized, double-blind, placebo-controlled pilot Phase II clinical trial [396]. The effects of oral CQ treatment was statistically significant in preventing cognitive deterioration in the moderately severe AD patient group, with no evidence of toxicity [396].

2.6.2 Funtional Neuroimaging Approaches

When in his 1907 [25] report Alzheimer wrote, "there exist many more mental diseases than our textbooks indicate. In many such cases, a further histological examination must be effected to determine the characteristics of each single case," he stated what for the past century remained the gold standard for the diagnosis of AD. We are now at the threshold of a new era of noninvasive, in vivo diagnosis through molecular imaging. The same way neuropathology was boosted by the techniques and dyes introduced by visionary pioneers like Cajal and Nissl, we are now seeing some derivatives of those histological dyes finding their way into emission tomography [198, 397] and magnetic resonance imaging [398, 399].

Modern functional neuroimaging techniques such as positron emission tomography (PET), single photon emission tomography (SPECT), magnetic resonance spectroscopy (MRS), functional magnetic resonance imaging (fMRI), and magnetoencephalography (MEG) have been developing new approaches not only to determine if an individual suffers from a particular form of dementia but also to delve into the molecular mechanisms of AD [400–402].

PET allows in vivo quantification of radiotracer concentrations, where either the radiotracer bears the same biochemical structure or is an analogue, or is a substrate of the chemical process being evaluated, allowing the in vivo assessment of the molecular process at their sites of action [403] permitting detection of disease processes at asymptomatic stages when there is no evidence of anatomic changes on CT and MRI.

Several studies have evaluated regional cerebral glucose metabolism with fluorodeoxyglucose (FDG) and PET. A typical pattern of reduced temporoparietal FDG uptake with sparing of the basal ganglia, thalamus, cerebellum, and primary sensorimotor cortex is typical of AD [404, 405]. FDG-PET might improve diagnostic and prognostic accuracy, thereby reducing both disease and treatment-related morbidity of patients with AD [406] due to its high sensitivity (94%) for detecting temporoparietal hypometabolism in patients with probable AD [405, 407, 408]. In a multicenter study, the prognostic value of FDG-PET showed a high degree of sensitivity (93%) and moderate specificity (73%) for prediction of progressive dementia [409].

Though clinical criteria together with current structural neuroimaging techniques (CT or MRI) are sensitive and specific enough for the diagnosis of AD at the mid or late stages of the disease, the development of a reliable method of assessing Aβ amyloid burden in vivo may permit early diagnosis at presymptomatic stages, more accurate differential diagnosis, while also allowing treatment follow-up.

Extracellular amyloid plaques are the hallmark brain lesions of sporadic AD. These microscopic Aβ aggregates [22] are well beyond the resolution of the usual neuroimaging techniques used for the evaluation of patients with AD. Furthermore, current techniques focus on nonspecific features derived mainly from neuronal loss and atrophy, which are late features in the progression of the disease, and are secondary to the basic functional alteration. Because Aβ is at the center of pathogenesis of AD, and because we are now approaching a point at which several pharmacological agents aimed at reducing levels of Aβ in the brain are being developed and tested, many efforts are focused on developing radiotracers that allow Aβ in vivo imaging [198, 397].

Several compounds have been evaluated as potential Aβ probes: derivatives of histopathological dyes such as Congo red, Chrysamine-G, Thioflavin S and T, and acridine orange [410–438] (Fig. 2.4), NSAID derivatives [439–445], as well as self-associating Aβ fragments [446–452] and anti-Aβ monoclonal antibodies [453, 454], serum amyloid P, and basic fibroblast growth factor [455].

The criteria for the diagnosis, management, and early detection of dementia [456–458] published by the American Academy of Neurology Quality Standards Subcommittee supports the use of CT and MRI in the work-up of the patient with dementia while recommending further research to determine the utility of other neuroimaging modalities such as PET and to a lesser degree SPECT [456]. Though FDG PET is mainly used in the differential diagnosis of AD, it is the neuroimaging technique that has been shown to yield the highest prognostic value for providing a diagnosis of presymptomatic AD 2 or more years before the full dementia picture is manifested [409, 459–461]. Given the growing evidence, PET will likely come to be at the forefront of the AD neuroimaging tools both as a diagnostic as well as a prognostic tool, providing new insights into the spatial and temporal pattern of disease progression.

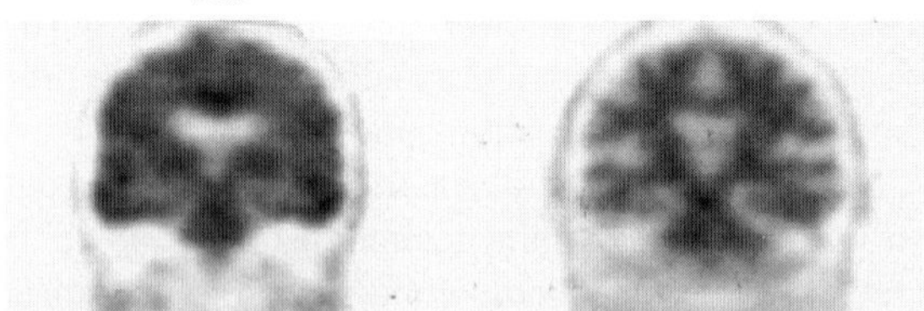

FIGURE 2.4. Coronal PET images showing the regional uptake of a thioflavin derivative, ^{11}C-PIB, reflecting Aβ burden in the brain. The images demonstrate a marked difference in ^{11}C-PIB regional distribution between an Alzheimer's disease patient (AD) and an age-matched healthy control (AC) subject, with high uptake of ^{11}C-PIB in gray matter areas in AD but only nonspecific uptake in white matter in AC. Images were obtained at Centre for PET, Austin Hospital, Melbourne, Australia.

Because new treatment strategies to prevent or slow disease progression through early intervention are being developed and implemented, there is an urgent need for early disease recognition, which is reflected in the necessity of developing sensitive and specific biomarkers, specific for a particular trait underlying the pathological process, as adjuncts to clinical and neuropsychological tests.

But the emphasis should not be limited to the ability of early diagnosis. With new therapeutic approaches being developed that either prevent the deposition of Aβ or increase its solubilization—agents that could delay the onset of dementia—the role of imaging and quantifying Aβ amyloid in vivo is becoming crucial. The ability to detect preclinical or early-stage disease through clinical, laboratory, and neuroimaging tests, combined with anti-Aβ amyloid in the at-risk patient, or the patient with MCI, may prevent or delay functional and irreversible cognitive losses, allowing one at the same time to customize and monitor treatment.

2.7 Conclusions

Alzheimer's disease is a neurodegenerative disorder characterized by a slow but relentless progressive cognitive decline and memory loss. It has a devastating effect not only on the sufferer but also on their caregivers, with a tremendous socioeconomic impact not only on families but also on the health care system that will only increase in the upcoming years.

The neuropathologic hallmarks of the disease are extracellular deposits of Aβ in senile plaques, NFT, with selective neuronal and synaptic loss in cortical areas of the brain associated with cognitive and memory functions.

Aβ is the main component of the amyloid plaques. All the available evidence points at the breakdown of the economy of Aβ as playing the key role in AD pathogenesis. Genetic studies have shed light on the pathogenesis and progression of AD. To date, four genes have been linked to autosomal dominant, early-onset familial AD: APP, PS1, PS2, and ApoE. All mutations linked to APP and PS proteins lead to an increase in Aβ production. Aβ not only aggregates into amyloid plaques but is toxic *per se* while having an effect on intracellular tangle formation and other factors (e.g., cytokines, neurotoxins, etc.) that also play an important role in the neurotoxic progression of AD.

Aβ is neurotoxic through a number of possible mechanisms including oxidative stress, excitotoxicity, energy depletion, inflammatory response, and apoptosis, and while the exact mechanism by which Aβ might produce synaptic loss and neuronal death is controversial, it is believed that a toxic oxidative interaction between various metal species and Aβ triggers an oxidative response with free-radical production leading to progressive disruption of neuronal function and ultimately to cell death.

At this point, there is no cure for AD. A deeper understanding of the molecular mechanism of Aβ formation, degradation, and neurotoxicity is being translated into new neuroimaging and therapeutic approaches. Most of the approved palliative treatment regimens involve the use of acetylcholinesterase inhibitors, glutamatergic agents, nonsteroidal anti-inflammatory drugs, as well as antioxidants. The most promising approaches focus on either reducing Aβ formation through secretase inhibitors or increasing its removal either by immunotherapy or MPAC aiming at blocking the formation of Aβ oligomers and fibrils therefore inhibiting neurotoxicity.

Like the attendees at Alois Alzheimer's presentation 100 years ago, we might be at the threshold of groundbreaking developments.

Acknowledgments. Supported in part by grants from the National Health and Medical Research Council of Australia, Prana Biotechnology, and Schering AG.

References

1. Khachaturian, Z.S., Diagnosis of Alzheimer's disease. Arch Neurol, 1985. 42(11):1097-1105.
2. O'Brien, J., Ames, D., and Burns, A., Dementia (2nd Ed). 2000, Arnold: London.
3. Jellinger, K., Morphology of Alzheimer's disease and related disorders, in Alzheimer's disease: epidemiology, neuropathology, neurochemistry, and clinics. K. Maurer, P. Riederer, and H. Beckmann, Editors. 1990, Springer-Verlag: Berlin. 61-77.
4. Selkoe, D.J., Alzheimer's disease: genotypes, phenotypes, and treatments. Science, 1997. 275(5300): 630-631.
5. Michaelis, M.L., Dobrowsky, R.T., and Li, G., Tau neurofibrillary pathology and microtubule stability. J Mol Neurosci, 2002. 19(3):289-293.
6. Jellinger, K.A. and Bancher, C., Neuropathology of Alzheimer's disease: a critical update. J Neural Transm Suppl, 1998. 54:77-95.
7. Perl, D.P., Neuropathology of Alzheimer's disease and related disorders. Neurol Clin, 2000. 18(4):847-864.
8. Geula, C., Wu, C.K., Saroff, D., Lorenzo, A., Yuan, M., and Yankner, B.A., Aging renders the brain vulnerable to amyloid beta-protein neurotoxicity. Nat Med, 1998. 4(7):827-831.
9. Lu, M. and Kosik, K.S., Competition for microtubule-binding with dual expression of tau missense and splice isoforms. Mol Biol Cell, 2001. 12(1):171-184.
10. Yankner, B.A., Duffy, L.K., and Kirschner, D.A., Neurotrophic and neurotoxic effects of amyloid beta protein: reversal by tachykinin neuropeptides. Science, 1990. 250(4978):279-282.
11. Lovestone, S. and Reynolds, C.H., The phosphorylation of tau: a critical stage in neurodevelopment and neurodegenerative processes. Neuroscience, 1997. 78(2):309-324.
12. Frank, R.A., Galasko, D., Hampel, H., et al., Biological markers for therapeutic trials in Alzheimer's disease. Proceedings of the biological markers working group; NIA initiative on neuroimaging in Alzheimer's disease. Neurobiol Aging, 2003. 24(4):521-536.
13. Geula, C., The early diagnosis of Alzheimer's disease, in Pathological diagnosis of Alzheimer's disease. L.F.M. Scinto and K.R. Daffner, Editors. 2000, Humana: Totowa, NJ. 65-82.
14. Isacson, O., Seo, H., Lin, L., et al., Alzheimer's disease and Down's syndrome: roles of APP, trophic factors and ACh. Trends Neurosci, 2002. 25(2):79-84.
15. Cummings, J.L., Vinters, H.V., Cole, G.M., and Khachaturian, Z.S., Alzheimer's disease: etiologies, pathophysiology, cognitive reserve, and treatment opportunities. Neurology, 1998. 51(1 Suppl 1): S2-17; discussion S65-17.
16. Larson, E.B., Edwards, J.K., O'Meara, E., et al., Neuropathologic diagnostic outcomes from a cohort of outpatients with suspected dementia. J Gerontol A Biol Sci Med Sci, 1996. 51(suppl 6): M313-M318.
17. Petersen, R.C., Mild cognitive impairment: transition between aging and Alzheimer's disease. Neurologia, 2000. 15(3):93-101.
18. Petersen, R.C., Smith, G.E., Ivnik, R.J., et al., Apolipoprotein E status as a predictor of the

development of Alzheimer's disease in memory-impaired individuals. JAMA, 1995. 273:1274-1278.
19. Petersen, R.C., Smith, G.E., Waring, S.C., et al., Mild cognitive impairment: clinical characterization and outcome. Arch Neurol, 1999. 56:303-308.
20. Selkoe, D.J., Alzheimer's disease: genes, proteins, and therapy. Physiol Rev, 2001. 81(2):741-766.
21. Hardy, J., Amyloid, the presenilins and Alzheimer's disease. Trends Neurosci, 1997. 20(4):154-159.
22. McLean, C.A., Cherny, R.A., Fraser, F.W., et al., Soluble pool of Aß amyloid as a determinant of severity of neurodegeneration in Alzheimer's disease. Ann. Neurol, 1999. 46(6):860-866.
23. Harkany, T., Hortobagyi, T., Sasvari, et al., Neuroprotective approaches in experimental models of beta-amyloid neurotoxicity: relevance to Alzheimer's disease. Prog Neuropsychopharmacol Biol Psychiatry, 1999. 23(6):963-1008.
24. Harkany, T., Abraham, I., Konya, C., et al., Mechanisms of beta-amyloid neurotoxicity: perspectives of pharmacotherapy. Rev Neurosci, 2000. 11(4):329-382.
25. Alzheimer, A., Uber eine eijenartige Erkrankung der Hirnride. Allg Z Psychiatr, 1907. 64:146–148.
26. Perusini, G., Uber klinisch und histologisch eigenartige psychische Erkankungen des spateren Lebensalters, in Histologische und Histolopathologische Arbeiten, F. Nissl, and A. Alzheimer, Editors. 1910, Gustav Fischer: Jena. 297-351.
27. Kraepelin, E., Das senile und prasenile Irresein, in Psychiatrie: Ein Lehrbuch fur Studierende und Arzte. E. Kraepelin, Editor. 1910, Verlag von Johann Ambrosius Barth: Leipzig. 533-554; 593-632.
28. Neumann, M.A. and Cohn, R., Incidence of Alzheimer's disease in a large mental hospital: relation to senile psychosis and psychosis with cerebral arteriosclerosis. Arch Neurol Psychiatr, 1953. 69:615-636.
29. Rorsman, B., Hagnell, O., and Lanke, J., Prevalence and incidence of senile and multi-infarct dementia in the Lundby study: a comparison between the time periods 1947-1957 and 1957-1972. Neuropsychobiology, 1986. 15:122-129.
30. Kay, D.W.K., Beamish, P., and Roth, M., Old age mental disorders in Newcastle upon Tyne. Part I: a study of prevalence. Br J Psychiatry, 1964. 110:146-158.
31. Sjogren, T., Sjogren, H., and Lindgren, G.H., Morbus Alzheimer and morbus Pick: a genetic, clinical and patho-anatomical study. Acta Psychiatr Neurol Scand, 1952. 82(Suppl):1-152.
32. Larsson, T., Sjogren, T., and Jacobsen, G., Senile dementia: a clinical, sociomedical and genetic study. Acta Psychiatr Scand, 1963. 167(Suppl): 1-259.
33. Tomlinson, B.E., Blessed, G., and Roth, M., Observations on the brains of non-demented old people. J Neurol Sci, 1968. 7(2):331-356.
34. Evans, D.A., Funkenstein, H.H., Albert, M.S., et al., Prevalence of Alzheimer's disease in a community population of older persons. Higher than previously reported. JAMA, 1989. 262(18):2551-2556.
35. Davies, P. and Maloney, A.J.F., Selective loss of central cholinergic neurons in Alzheimer's disease. Lancet, 1976. 2:1400-1403.
36. Scholz, W., Studien zur Pathologie der Hirngefasse. II Die drusige Entartung der Hirnarterien und Capillaren. Z Gesamte Neurol Psychiatr, 1938. 162:694-715.
37. Pantelakis, S., Un type particulier d'angiopathie senile du systeme nerveux central: l'angiopathie congophile. Topographie et frequence. Monat Psychiatr Neurol, 1954. 128:219-256.
38. Terry, R.D., Gonatas, N.K., and Weiss, M., Ultrastructural studies in Alzheimer's presenile dementia. Am J Pathol, 1964. 44:269-287.
39. Kidd, M., Paired helical filaments in elctron microscopy of Alzheimer's disease. Nature, 1963. 197:192-193.
40. Kosik, K.S., Tau protein and neurodegeneration. Mol Neurobiol, 1990. 4(3-4):171-179.
41. Nukina, N. and Ihara, Y., One of the antigenic determinants of paired helical filaments is related to tau protein. J Biochem (Tokyo), 1986. 99(5):1541-1544.
42. Kosik, K.S., Joachim, C.L., and Selkoe, D.J., Microtubule-associated protein tau (tau) is a major antigenic component of paired helical filaments in Alzheimer's disease. Proc Natl Acad Sci U S A, 1986. 83(11):4044-4048.
43. Goedert, M., Wischik, C.M., Crowther, R.A., et al., Cloning and sequencing of the cDNA encoding a core protein of the paired helical filament of Alzheimer's disease: identification as the microtubule-associated protein tau. Proc Natl Acad Sci U S A, 1988. 85(11):4051-4055.
44. Wischik, C.M., Novak, M., Thogersen, H.C., et al., Isolation of a fragment of tau derived from the core of the paired helical filament of Alzheimer's disease. Proc Natl Acad Sci U S A, 1988. 85(12):4506-4510.
45. Kosik, K.S., Duffy, L.K., Dowling, M.M., et al., Microtubule-associated protein 2: monoclonal antibodies demonstrate the selective incorporation of certain epitopes into Alzheimer neurofibrillary tangles. Proc Natl Acad Sci U S A, 1984. 81(24): 7941-7945.
46. Selkoe, D.J., Ihara, Y., and Salazar, F.J., Alzheimer's disease: insolubility of partially purified paired helical filaments in sodium dodecyl sulfate and urea. Science, 1982. 215(4537):1243-1245.
47. Virchow, R., Zur cellulosefrage, in Virchows Arch Pathol Anat Physiol, 1854. 416–426.
48. Glenner, G.G. and Wong, C.W., Alzheimer's disease: initial report of the purification and characterization

of a novel cerebrovascular amyloid protein. Biochem Biophys Res Commun, 1984. 120(3):885-890.
49. Glenner, G.G., Wong, C.W., Quaranta, V., and Eanes, E.D., The amyloid deposits in Alzheimer's disease: their nature and pathogenesis. Appl Pathol, 1984. 2(6):357-369.
50. Masters, C.L., Simms, G., Weinman, N.A., et al., Amyloid plaque core protein in Alzheimer's disease and Down syndrome. Proc Natl Acad Sci U S A, 1985. 82(12):4245-4249.
51. St George-Hyslop, P.H., Tanzi, R.E., Polinsky, R.J., et al.., The genetic defect causing familial Alzheimer's disease maps on chromosome 21. Science, 1987. 235(4791):885-890.
52. Tanzi, R.E., St George-Hyslop, P.H., Haines, J.L., et al., The genetic defect in familial Alzheimer's disease is not tightly linked to the amyloid beta-protein gene. Nature, 1987. 329(6135):156-157.
53. Van Broeckhoven, C., Genthe, A.M., Vandenberghe, A., et al., Failure of familial Alzheimer's disease to segregate with the A4-amyloid gene in several European families. Nature, 1987. 329(6135):153-155.
54. Kang, J., Lemaire, H.G., Unterbeck, A., Salbaum, J.M., et al., The precursor of Alzheimer's disease amyloid A4 protein resembles a cell-surface receptor. Nature, 1987. 325(6106):733-736.
55. Tanzi, R.E., Gusella, J.F., Watkins, P.C., et al., Amyloid beta protein gene:cDNA, mRNA distribution, and genetic linkage near the Alzheimer locus. Science, 1987. 235(4791):880-884.
56. Robakis, N.K., Wisniewski, H.M., Jenkins, E.C., et al., Chromosome 21q21 sublocalisation of gene encoding beta-amyloid peptide in cerebral vessels and neuritic (senile) plaques of people with Alzheimer's disease and Down syndrome. Lancet, 1987. 1(8529):384-385.
57. Van Broeckhoven, C., Haan, J., Bakker, E., et al., Amyloid beta protein precursor gene and hereditary cerebral hemorrhage with amyloidosis (Dutch). Science, 1990. 248(4959):1120-1122.
58. Levy, E., Carman, M.D., Fernandez-Madrid, I.J., et al., Mutation of the Alzheimer's disease amyloid gene in hereditary cerebral hemorrhage, Dutch type. Science, 1990. 248(4959):1124-1126.
59. Goate, A., Chartier-Harlin, M.C., Mullan, M., et al., Segregation of a missense mutation in the amyloid precursor protein gene with familial Alzheimer's disease. Nature, 1991. 349(6311):704-706.
60. Citron, M., Oltersdorf, T., Haass, C., et al., Mutation of the beta-amyloid precursor protein in familial Alzheimer's disease increases beta-protein production. Nature, 1992. 360(6405):672-674.
61. Schellenberg, G.D., Bird, T.D., Wijsman, E.M., et al., Genetic linkage evidence for a familial Alzheimer's disease locus on chromosome 14. Science, 1992. 258(5082):668-671.
62. Mullan, M., Houlden, H., Windelspecht, M., et al., A locus for familial early-onset Alzheimer's disease on the long arm of chromosome 14, proximal to the alpha 1-antichymotrypsin gene. Nat Genet, 1992. 2(4):340-342.
63. St George-Hyslop, P., Haines, J., Rogaev, E., et al., Genetic evidence for a novel familial Alzheimer's disease locus on chromosome 14. Nat Genet, 1992. 2(4):330-334.
64. Van Broeckhoven, C., Backhovens, H., Cruts, M., et al., Mapping of a gene predisposing to early-onset Alzheimer's disease to chromosome 14q24.3. Nat Genet, 1992. 2(4):335-339.
65. Yankner, B.A., Dawes, L.R., Fisher, S., et al., Neurotoxicity of a fragment of the amyloid precursor associated with Alzheimer's disease. Science, 1989. 245(4916):417-420.
66. Pike, C.J., Walencewicz, A.J., Glabe, C.G., and Cotman, C.W., In vitro aging of beta-amyloid protein causes peptide aggregation and neurotoxicity. Brain Res, 1991. 563(1-2):311-314.
67. Pike, C.J., Walencewicz, A.J., Glabe, C.G., and Cotman, C.W., Aggregation-related toxicity of synthetic beta-amyloid protein in hippocampal cultures. Eur J Pharmacol, 1991. 207(4):367-368.
68. Bush, A.I., Multhaup, G., Moir, R.D., et al., A novel zinc(II) binding site modulates the function of the beta A4 amyloid protein precursor of Alzheimer's disease. J Biol Chem, 1993. 268(22):16109-16112.
69. Bush, A.I., Pettingell, W.H., Multhaup, G., Paradis, M., et al., Rapid induction of Alzheimer Aß amyloid formation by zinc. Science, 1994. 265(5177):1464-1467.
70. Strittmatter, W.J., Saunders, A.M., Schmechel, D., et al., Apolipoprotein E: high-avidity binding to beta-amyloid and increased frequency of type 4 allele in late-onset familial Alzheimer's disease. Proc Natl Acad Sci U S A, 1993. 90(5):1977-1981.
71. Corder, E.H., Saunders, A.M., Strittmatter, W.J., et al., Gene dose of apolipoprotein E type 4 allele and the risk of Alzheimer's disease in late onset families. Science, 1993. 261(5123):921-923.
72. Schmechel, D.E., Saunders, A.M., Strittmatter, W.J., et al., Increased amyloid beta-peptide deposition in cerebral cortex as a consequence of apolipoprotein E genotype in late-onset Alzheimer's disease. Proc Natl Acad Sci U S A, 1993. 90(20):9649-9653.
73. Sherrington, R., Rogaev, E.I., Liang, Y., et al., Cloning of a gene bearing missense mutations in early-onset familial Alzheimer's disease. Nature, 1995. 375(6534): 754-760.
74. Levy-Lahad, E., Wasco, W., Poorkaj, P., et al., Candidate gene for the chromosome 1 familial

Alzheimer's disease locus. Science, 1995. 269(5226): 973-977.
75. Scheuner, D., Eckman, C., Jensen, M., et al., Secreted amyloid beta-protein similar to that in the senile plaques of Alzheimer's disease is increased in vivo by the presenilin 1 and 2 and APP mutations linked to familial Alzheimer's disease. Nat Med, 1996. 2(8):864-870.
76. Selkoe, D.J., Alzheimer's disease is a synaptic failure. Science, 2002. 298(5594):789-791.
77. Selkoe, D.J., Toward a comprehensive theory for Alzheimer's disease. Hypothesis: Alzheimer's disease is caused by the cerebral accumulation and cytotoxicity of amyloid beta-protein. Ann N Y Acad Sci, 2000. 924:17-25.
78. Selkoe, D.J., The genetics and molecular pathology of Alzheimer's disease: roles of amyloid and the presenilins. Neurol Clin, 2000. 18(4):903-922.
79. Masters, C.L. and Beyreuther, K., Alzheimer's disease. BMJ, 1998. 316(7129):446-448.
80. Masters, C.L. and Beyreuther, K., Molecular neuropathology of Alzheimer's disease. Arzneimittelforschung, 1995. 45(3A):410-412.
81. Bartus, R.T., Dean, R.L., 3rd, Beer, B., and Lippa, A.S., The cholinergic hypothesis of geriatric memory dysfunction. Science, 1982. 217(4558):408-414.
82. Bartus, R.T. and Emerich, D.F., Cholinergic markers in Alzheimer's disease. JAMA, 1999. 282(23):2208-2209.
83. Masters, C.L. and Beyreuther, K., Henryk M. Wisniewski and the amyloid theory of Alzheimer's disease. J Alzheimers Dis, 2001. 3(1):83-86.
84. Martins, R.N., Robinson, P.J., Chleboun, J.O., et al., The molecular pathology of amyloid deposition in Alzheimer's disease. Mol Neurobiol, 1991. 5(2-4): 389-398.
85. Beyreuther, K. and Masters, C.L., Amyloid precursor protein (APP) and beta A4 amyloid in the etiology of Alzheimer's disease: precursor-product relationships in the derangement of neuronal function. Brain Pathol., 1991. 1(4):241-251.
86. Cappai, R. and White, A.R., Amyloid beta. Int J Biochem Cell Biol, 1999. 31(9):885-889.
87. Haass, C., Koo, E.H., Mellon, A., et al., Targeting of cell-surface beta-amyloid precursor protein to lysosomes: alternative processing into amyloid-bearing fragments. Nature, 1992. 357(6378):500-503.
88. Seubert, P., Vigo-Pelfrey, C., Esch, F., et al., Isolation and quantification of soluble Alzheimer's beta-peptide from biological fluids. Nature, 1992. 359(6393):325-327.
89. Shoji, M., Golde, T.E., Ghiso, J., et al.., Production of the Alzheimer amyloid beta protein by normal proteolytic processing. Science, 1992. 258(5079): 126-129.
90. Harper, J.D. and Lansbury, P.T., Jr., Models of amyloid seeding in Alzheimer's disease and scrapie: mechanistic truths and physiological consequences of the time-dependent solubility of amyloid proteins. Annu Rev Biochem, 1997. 66:385-407.
91. Soto, C., Castano, E.M., Frangione, B., and Inestrosa, N.C., The alpha-helical to beta-strand transition in the amino-terminal fragment of the amyloid beta-peptide modulates amyloid formation. J Biol Chem, 1995. 270(7):3063-3067.
92. Inoue, S., Kuroiwa, M., Tan, R., and Kisilevsky, R., A high resolution ultrastructural comparison of isolated and in situ murine AA amyloid fibrils. Amyloid, 1998. 5(2):99-110.
93. Cohen, A.S., Shirahama, T., and Skinner, M., Electron microscopy of amyloid, in Electron Microscopy of Proteins. J.R. Harris, Editor. 1982, Academic Press: London. 165–205.
94. Westermark, P., Benson, M.D., Buxbaum, J.N., et al., Amyloid fibril protein nomenclature – 2002. Amyloid, 2002. 9(3):197-200.
95. Westermark, P., Araki, S., Benson, M.D., et al., Nomenclature of amyloid fibril proteins. Report from the meeting of the International Nomenclature Committee on Amyloidosis, August 8-9, 1998. Part 1. Amyloid, 1999. 6(1):63-66.
96. Stevens, F.J. and Kisilevsky, R., Immunoglobulin light chains, glycosaminoglycans, and amyloid. Cell Mol Life Sci, 2000. 57(3):441-449.
97. Perry, E.K., Tomlinson, B.E., Blessed, G., et al.., Correlation of cholinergic abnormalities with senile plaques and mental test scores in senile dementia. Br Med J, 1978. 2(6150):1457-1459.
98. Blessed, G., Tomlinson, B.E., and Roth, M., The association between quantitative measures of dementia and of senile change in the cerebral grey matter of elderly subjects. Br J Psychiatry, 1968. 114(512):797-811.
99. Yamaguchi, H., Hirai, S., Morimatsu, M., et al., Diffuse type of senile plaques in the brains of Alzheimer-type dementia. Acta Neuropathol (Berl), 1988. 77(2):113-119.
100. Skovronsky, D.M., Doms, R.W., and Lee, V.M., Detection of a novel intraneuronal pool of insoluble amyloid beta protein that accumulates with time in culture. J Cell Biol, 1998. 141(4):1031-1039.
101. Bush, A.I., The metallobiology of Alzheimer's disease. Trends Neurosci, 2003. 26(4):207-214.
102. Haass, C., Schlossmacher, M.G., Hung, A.Y., et al., Amyloid beta-peptide is produced by cultured cells during normal metabolism. Nature, 1992. 359(6393):322-325.
103. Mega, M.S., Chu, T., Mazziotta, J.C., et al., Mapping biochemistry to metabolism: FDG-PET

and amyloid burden in Alzheimer's disease. Neuroreport, 1999. 10(14):2911-2917.
104. Greenberg, S.M., Rebeck, G.W., et al., Apolipoprotein E epsilon 4 and cerebral hemorrhage associated with amyloid angiopathy. Ann Neurol, 1995. 38(2): 254-259.
105. McLean, C.A., Beyreuther, K., and Masters, C.L., Amyloid Abeta levels in Alzheimer's disease -A diagnostic tool and the key to understanding the natural history of Abeta? J Alzheimers Dis, 2001. 3(3):305-312.
106. Naslund, J., Haroutunian, V., Mohs, R., et al., Correlation between elevated levels of amyloid beta-peptide in the brain and cognitive decline. JAMA, 2000. 283(12):1571-1577.
107. Lue, L.F., Kuo, Y.M., Roher, A.E., et al., Soluble amyloid beta peptide concentration as a predictor of synaptic change in Alzheimer's disease. Am J Pathol, 1999. 155(3):853-862.
108. Wang, J., Dickson, D.W., Trojanowski, J.Q., and Lee, V.M., The levels of soluble versus insoluble brain Aß distinguish Alzheimer's disease from normal and pathologic aging. ExNeurol, 1999. 158(2):328-337.
109. Braak, H. and Braak, E., Neuropathological stageing of Alzheimer-related changes. Acta Neuropathol (Berl), 1991. 82(4):239-259.
110. Hardy, J. and Selkoe, D.J., The amyloid hypothesis of Alzheimer's disease: progress and problems on the road to therapeutics. Science, 2002. 297(5580): 353-356.
111. Price, J.L. and Morris, J.C., Tangles and plaques in nondemented aging and "preclinical" Alzheimer's disease. Ann Neurol, 1999. 45(3):358-368.
112. Lemere, C.A., Blusztajn, J.K., Yamaguchi, H., et al., Sequence of deposition of heterogeneous amyloid beta-peptides and APO E in Down syndrome: implications for initial events in amyloid plaque formation. Neurobiol Dis, 1996. 3(1):16-32.
113. Mann, D.M., Yates, P.O., Marcyniuk, B., and Ravindra, C.R., The topography of plaques and tangles in Down's syndrome patients of different ages. Neuropathol Appl Neurobiol, 1986. 12(5):447-457.
114. Selkoe, D.J., The molecular pathology of Alzheimer's disease. Neuron, 1991. 6(4):487-498.
115. Hardy, J.A. and Higgins, G.A., Alzheimer's disease: the amyloid cascade hypothesis. Science, 1992. 256(5054):184-185.
116. Checler, F. and Vincent, B., Alzheimer's and prion diseases: distinct pathologies, common proteolytic denominators. Trends Neurosci, 2002. 25(12):616-620.
117. Robinson, S.R. and Bishop, G.M., The search for an amyloid solution. Science, 2002. 298(5595):962-964; author reply 962-964.
118. Robinson, S.R. and Bishop, G.M., Abeta as a bioflocculant: implications for the amyloid hypothesis of Alzheimer's disease. Neurobiol Aging, 2002. 23(6):1051-1072.
119. Mudher, A. and Lovestone, S., Alzheimer's disease—do tauists and baptists finally shake hands? Trends Neurosci, 2002. 25(1):22-26.
120. Selkoe, D.J., Translating cell biology into therapeutic advances in Alzheimer's disease. Nature, 1999. 399(6738 Suppl):A23-31.
121. St George-Hyslop, P.H., Genetic factors in the genesis of Alzheimer's disease. Ann N Y Acad Sci, 2000. 924:1-7.
122. Haass, C., Hung, A.Y., Selkoe, D.J., and Teplow, D.B., Mutations associated with a locus for familial Alzheimer's disease result in alternative processing of amyloid beta-protein precursor. J Biol Chem, 1994. 269(26):17741-17748.
123. Cai, X.D., Golde, T.E., and Younkin, S.G., Release of excess amyloid beta protein from a mutant amyloid beta protein precursor. Science, 1993. 259(5094):514-516.
124. Suzuki, N., Cheung, T.T., Cai, X.D., et al., An increased percentage of long amyloid beta protein secreted by familial amyloid beta protein precursor (beta APP717) mutants. Science, 1994. 264(5163): 1336-1340.
125. Citron, M., Vigo-Pelfrey, C., Teplow, D.B., et al., Excessive production of amyloid beta-protein by peripheral cells of symptomatic and presymptomatic patients carrying the Swedish familial Alzheimer's disease mutation. Proc Natl Acad Sci U S A, 1994. 91(25):11993-11997.
126. Miklossy, J., Taddei, K., Suva, D., et al.., Two novel presenilin-1 mutations (Y256S and Q222H) are associated with early-onset Alzheimer's disease. Neurobiol Aging, 2003. 24(5):655-662.
127. Rogaev, E.I., Sherrington, R., Rogaeva, E.A., et al.., Familial Alzheimer's disease in kindreds with missense mutations in a gene on chromosome 1 related to the Alzheimer's disease type 3 gene. Nature, 1995. 376(6543):775-778.
128. Thinakaran, G., Teplow, D.B., Siman, R., et al., Metabolism of the "Swedish" amyloid precursor protein variant in neuro2a (N2a) cells. Evidence that cleavage at the "beta-secretase" site occurs in the golgi apparatus. J Biol Chem, 1996. 271(16):9390-9397.
129. Citron, M., Westaway, D., Xia, W., Carlson, G., et al.., Mutant presenilins of Alzheimer's disease increase production of 42-residue amyloid beta-protein in both transfected cells and transgenic mice. Nat Med, 1997. 3(1):67-72.
130. Duff, K., Eckman, C., Zehr, C., et al.., Increased amyloid-beta42(43) in brains of mice expressing

mutant presenilin 1. Nature, 1996. 383(6602):710-713.
131. Poirier, J., Davignon, J., Bouthillier, D., et al., Apolipoprotein E polymorphism and Alzheimer's disease. Lancet, 1993. 342(8873):697-699.
132. Borchelt, D.R., Thinakaran, G., Eckman, C.B., et al., Familial Alzheimer's disease-linked presenilin 1 variants elevate Abeta1-42/1-40 ratio in vitro and in vivo. Neuron, 1996. 17(5):1005-1013.
133. Polvikoski, T., Sulkava, R., Haltia, M., et al., Apolipoprotein E, dementia, and cortical deposition of beta-amyloid protein. N Engl J Med, 1995. 333(19):1242-1247.
134. Rebeck, G.W., Reiter, J.S., Strickland, D.K., and Hyman, B.T., Apolipoprotein E in sporadic Alzheimer's disease: allelic variation and receptor interactions. Neuron, 1993. 11(4):575-580.
135. Hyman, B.T., West, H.L., Rebeck, G.W., et al., Quantitative analysis of senile plaques in Alzheimer's disease: observation of log-normal size distribution and molecular epidemiology of differences associated with apolipoprotein E genotype and trisomy 21 (Down syndrome). Proc Natl Acad Sci U S A, 1995. 92(8):3586-3590.
136. Mehta, N.D., Refolo, L.M., Eckman, C., et al., Increased Abeta42(43) from cell lines expressing presenilin 1 mutations. Ann Neurol, 1998. 43(2): 256-258.
137. Beyreuther, K., Dyrks, T., Hilbich, C., et al., Amyloid precursor protein (APP) and beta A4 amyloid in Alzheimer's disease and Down syndrome. Prog Clin Biol Res, 1992. 379:159-182.
138. Masters, C.L. and Beyreuther, K.T., The pathology of the amyloid A4 precursor of Alzheimer's disease. Ann Med, 1989. 21(2):89-90.
139. Dyrks, T., Weidemann, A., Multhaup, G., et al., Identification, transmembrane orientation and biogenesis of the amyloid A4 precursor of Alzheimer's disease. EMBO J, 1988. 7(4):949-957.
140. Goldgaber, D., Lerman, M.I., McBride, O.W., et al., Characterization and chromosomal localization of a cDNA encoding brain amyloid of Alzheimer's disease. Science, 1987. 235(4791):877-880.
141. Robakis, N.K., Ramakrishna, N., Wolfe, G., and Wisniewski, H.M., Molecular cloning and characterization of a cDNA encoding the cerebrovascular and the neuritic plaque amyloid peptides. Proc Natl Acad Sci U S A, 1987. 84(12):4190-4194.
142. Olson, M.I. and Shaw, C.M., Presenile dementia and Alzheimer's disease in mongolism. Brain, 1969. 92(1):147-156.
143. Querfurth, H.W., Wijsman, E.M., St George-Hyslop, P.H., and Selkoe, D.J., Beta APP mRNA transcription is increased in cultured fibroblasts from the familial Alzheimer's disease-1 family. Brain Res Mol Brain Res, 1995. 28(2):319-337.
144. Turner, P.R., O'Connor, K., Tate, W.P., and Abraham, W.C., Roles of amyloid precursor protein and its fragments in regulating neural activity, plasticity and memory. Prog Neurobiol, 2003. 70(1):1-32.
145. Kuentzel, S.L., Ali, S.M., Altman, R.A., et al., The Alzheimer beta-amyloid protein precursor/protease nexin-II is cleaved by secretase in a trans-Golgi secretory compartment in human neuroglioma cells. Biochem J, 1993. 295(Pt 2):367-378.
146. Citron, M., Identifying proteases that cleave AP. Ann N Y Acad Sci, 2000. 920:192-196.
147. Mullan, M., Crawford, F., Axelman, K., et al., A pathogenic mutation for probable Alzheimer's disease in the APP gene at the N-terminus of beta-amyloid. Nat Genet, 1992. 1(5):345-347.
148. Hardy, J., Framing beta-amyloid. Nat Genet, 1992. 1(4):233-234.
149. Hendriks, L., van Duijn, C.M., Cras, P., et al., Presenile dementia and cerebral haemorrhage linked to a mutation at codon 692 of the beta-amyloid precursor protein gene. Nat Genet, 1992. 1(3):218-221.
150. Wisniewski, T., Ghiso, J., and Frangione, B., Peptides homologous to the amyloid protein of Alzheimer's disease containing a glutamine for glutamic acid substitution have accelerated amyloid fibril formation. Biochem Biophys Res Commun, 1991. 179(3):1247-1254.
151. Citron, M., Secretases as targets for the treatment of Alzheimer's disease. Mol Med Today, 2000. 6(10): 392-397.
152. Seubert, P., Oltersdorf, T., Lee, M.G., et al.., Secretion of beta-amyloid precursor protein cleaved at the amino terminus of the beta-amyloid peptide. Nature, 1993. 361(6409):260-263.
153. Citron, M., Haass, C., and Selkoe, D.J., Production of amyloid-beta-peptide by cultured cells: no evidence for internal initiation of translation at Met596. Neurobiol Aging, 1993. 14(6):571-573.
154. Muller, U. and Kins, S., APP on the move. Trends Mol Med, 2002. 8(4):152-155.
155. Andrews, N.C., Mining copper transport genes. Proc Natl Acad Sci U S A, 2001. 98(12):6543-6545.
156. Barnham, K.J., Masters, C.L., and Bush, A.I., Neurodegenerative diseases and oxidative stress. Nat Rev Drug Discov, 2004. 3(3):205-214.
157. Culotta, V.C., Klomp, L.W., Strain, J., et al.., The copper chaperone for superoxide dismutase. J Biol Chem, 1997. 272(38):23469-23472.
158. Waggoner, D.J., Bartnikas, T.B., and Gitlin, J.D., The role of copper in neurodegenerative disease. Neurobiol Dis, 1999. 6(4):221-230.

159. Maynard, C.J., Cappai, R., Volitakis, I., et al., Overexpression of Alzheimer's disease amyloid-beta opposes the age-dependent elevations of brain copper and iron. J Biol Chem, 2002. 277(47): 44670-44676.
160. White, A.R., Reyes, R., Mercer, J.F., et al., Copper levels are increased in the cerebral cortex and liver of APP and APLP2 knockout mice. Brain Res, 1999. 842(2):439-444.
161. Bayer, T.A., Schafer, S., Simons, A., et al., Dietary Cu stabilizes brain superoxide dismutase 1 activity and reduces amyloid Abeta production in APP23 transgenic mice. Proc Natl Acad Sci U S A, 2003. 100(24):14187-14192.
162. Phinney, A.L., Drisaldi, B., Schmidt, S.D., et al., In vivo reduction of amyloid-beta by a mutant copper transporter. Proc Natl Acad Sci U S A, 2003. 100(24):14193-14198.
163. Borchardt, T., Camakaris, J., Cappai, R., Masters, C.L., Beyreuther, K., and Multhaup, G., Copper inhibits beta-amyloid production and stimulates the non-amyloidogenic pathway of amyloid-precursor-protein secretion. Biochem J, 1999. 344(Pt 2):461-467.
164. Barnham, K.J., McKinstry, W.J., Multhaup, G., et al., Structure of the Alzheimer's disease amyloid precursor protein copper binding domain. A regulator of neuronal copper homeostasis. J Biol Chem, 2003. 278(19):17401-17407.
165. De Strooper, B., Saftig, P., Craessaerts, K., et al., Deficiency of presenilin-1 inhibits the normal cleavage of amyloid precursor protein. Nature, 1998. 391(6665):387-390.
166. Wolfe, M.S., Xia, W., Ostaszewski, B.L., et al., Two transmembrane aspartates in presenilin-1 required for presenilin endoproteolysis and gamma-secretase activity. Nature, 1999. 398(6727):513-517.
167. Kwok, J.B., Taddei, K., Hallupp, M., et al., Two novel (M233T and R278T) presenilin-1 mutations in early-onset Alzheimer's disease pedigrees and preliminary evidence for association of presenilin-1 mutations with a novel phenotype. Neuroreport, 1997. 8(6):1537-1542.
168. Crook, R., Verkkoniemi, A., Perez-Tur, J., et al., A variant of Alzheimer's disease with spastic paraparesis and unusual plaques due to deletion of exon 9 of presenilin 1. Nat Med, 1998. 4(4):452-455.
169. Houlden, H., Baker, M., McGowan, E., et al., Variant Alzheimer's disease with spastic paraparesis and cotton wool plaques is caused by PS-1 mutations that lead to exceptionally high amyloid-beta concentrations. Ann Neurol, 2000. 48(5):806-808.
170. Verkkoniemi, A., Kalimo, H., Paetau, A., et al., Variant Alzheimer's disease with spastic paraparesis: neuropathological phenotype. J Neuropathol Exp Neurol, 2001. 60(5):483-492.
171. Smith, M.J., Kwok, J.B., McLean, C.A., et al., Variable phenotype of Alzheimer's disease with spastic paraparesis. Ann Neurol, 2001. 49(1):125-129.
172. Kovacs, D.M., Fausett, H.J., Page, K.J., et al., Alzheimer-associated presenilins 1 and 2: neuronal expression in brain and localization to intracellular membranes in mammalian cells. Nat Med, 1996. 2(2):224-229.
173. Kimberly, W.T., Xia, W., Rahmati, T., et al., The transmembrane aspartates in presenilin 1 and 2 are obligatory for gamma-secretase activity and amyloid beta-protein generation. J Biol Chem, 2000. 275(5):3173-3178.
174. Takashima, A., Murayama, M., Murayama, O., et al., Presenilin 1 associates with glycogen synthase kinase-3beta and its substrate tau. Proc Natl Acad Sci U S A, 1998. 95(16):9637-9641.
175. Buxbaum, J.D., Choi, E.K., Luo, Y., et al., Calsenilin: a calcium-binding protein that interacts with the presenilins and regulates the levels of a presenilin fragment. Nat Med, 1998. 4(10):1177-1181.
176. Shinozaki, K., Maruyama, K., Kume, H., et al., The presenilin 2 loop domain interacts with the mu-calpain C-terminal region. Int J Mol Med, 1998. 1(5): 797-799.
177. Drouet, B., Pincon-Raymond, M., Chambaz, J., and Pillot, T., Molecular basis of Alzheimer's disease. Cell Mol Life Sci, 2000. 57(5):705-715.
178. Wolozin, B., Alexander, P., and Palacino, J., Regulation of apoptosis by presenilin 1. Neurobiol Aging, 1998. 19(1 Suppl): S23-27.
179. Guo, Q., Fu, W., Xie, J., et al., Par-4 is a mediator of neuronal degeneration associated with the pathogenesis of Alzheimer's disease. Nat Med, 1998. 4(8):957-962.
180. Bertram, L., Blacker, D., Mullin, K., et al., Evidence for genetic linkage of Alzheimer's disease to chromosome 10q. Science, 2000. 290(5500):2302-2303.
181. Myers, A., Holmans, P., Marshall, H., et al. Susceptibility locus for Alzheimer's disease on chromosome 10. Science, 2000. 290(5500):2304-2305.
182. Wavrant-DeVrieze, F., Lambert, J.C., Stas, L., et al., Association between coding variability in the LRP gene and the risk of late-onset Alzheimer's disease. Hum Genet, 1999. 104(5):432-434.
183. Ertekin-Taner, N., Graff-Radford, N., Younkin, L.H., et al., Linkage of plasma Abeta42 to a quantitative locus on chromosome 10 in late-onset Alzheimer's disease pedigrees. Science, 2000. 290(5500):2303-2304.
184. Olson, J.M., Goddard, K.A., and Dudek, D.M., The amyloid precursor protein locus and very-late-onset

Alzheimer's disease. Am J Hum Genet, 2001. 69(4):895-899.

185. Saunders, A.M., Strittmatter, W.J., Schmechel, D., et al., Association of apolipoprotein E allele epsilon 4 with late-onset familial and sporadic Alzheimer's disease. Neurology, 1993. 43(8):1467-1472.
186. Rocchi, A., Pellegrini, S., Siciliano, G., and Murri, L., Causative and susceptibility genes for Alzheimer's disease: a review. Brain Res Bull, 2003. 61(1):1-24.
187. Marques, M.A. and Crutcher, K.A., Apolipoprotein E-related neurotoxicity as a therapeutic target for Alzheimer's disease. J Mol Neurosci, 2003. 20(3): 327-337.
188. Ramassamy, C., Krzywkowski, P., Averill, D., et al., Impact of apoE deficiency on oxidative insults and antioxidant levels in the brain. Brain Res Mol Brain Res, 2001. 86(1-2):76-83.
189. Lee, Y., Aono, M., Laskowitz, D., et al., Apolipoprotein E protects against oxidative stress in mixed neuronal-glial cell cultures by reducing glutamate toxicity. Neurochem Int, 2004. 44(2):107-118.
190. Ramassamy, C., Averill, D., Beffert, U., et al.., Oxidative insults are associated with apolipoprotein E genotype in Alzheimer's disease brain. Neurobiol Dis, 2000. 7(1):23-37.
191. Beffert, U. and Poirier, J., ApoE associated with lipid has a reduced capacity to inhibit beta-amyloid fibril formation. Neuroreport, 1998. 9(14):3321-3323.
192. Hu, J., LaDu, M.J., and Van Eldik, L.J., Apolipoprotein E attenuates beta-amyloid-induced astrocyte activation. J Neurochem, 1998. 71(4): 1626-1634.
193. Mayeux, R., Saunders, A.M., Shea, S., et al., Utility of the apolipoprotein E genotype in the diagnosis of Alzheimer's disease: Alzheimer's Disease Centers Consortium on Apolipoprotein E and Alzheimer's Disease. N Engl J Med, 1998. 338(8):506-511.
194. St George-Hyslop, P., McLachlan, D.C., Tsuda, T., et al., Alzheimer's disease and possible gene interaction. Science, 1994. 263(5146):537.
195. Schupf, N., Kapell, D., Lee, J.H., et al., Onset of dementia is associated with apolipoprotein E epsilon4 in Down's syndrome. Ann Neurol, 1996. 40(5):799-801.
196. Kalaria, R.N., Small vessel disease and Alzheimer's dementia: pathological considerations. Cerebrovasc Dis, 2002. 13(Suppl 2):48-52.
197. de Figueiredo, R.J., Oten, R., Su, J., and Cotman, C.W., Amyloid deposition in cerebrovascular angiopathy. Ann N Y Acad Sci, 1997. 826:463-471.
198. Villemagne, V.L., Rowe, C.C., Macfarlane, S., et al. Imaginem Oblivionis: The prospects of neuroimaging for early detection of Alzheimer's disease. J Clin Neurosci, 2005. 12:221-230.
199. Jordan, J., Galindo, M.F., Miller, R.J., et al., Isoform-specific effect of apolipoprotein E on cell survival and beta-amyloid-induced toxicity in rat hippocampal pyramidal neuronal cultures. J Neurosci, 1998. 18(1):195-204.
200. Moir, R.D., Atwood, C.S., Romano, D.M., et al., Differential effects of apolipoprotein E isoforms on metal-induced aggregation of A beta using physiological concentrations. Biochemistry, 1999. 38(14): 4595-4603.
201. Games, D., Adams, D., Alessandrini, R., et al.., Alzheimer-type neuropathology in transgenic mice overexpressing V717F beta-amyloid precursor protein. Nature, 1995. 373(6514):523-527.
202. Masliah, E., Sisk, A., Mallory, M., Mucke, L., et al., Comparison of neurodegenerative pathology in transgenic mice overexpressing V717F beta-amyloid precursor protein and Alzheimer's disease. J Neurosci, 1996. 16(18):5795-5811.
203. Hsiao, K., Chapman, P., Nilsen, S., et al., Correlative memory deficits, Abeta elevation, and amyloid plaques in transgenic mice. Science, 1996. 274(5284):99-102.
204. Irizarry, M.C., McNamara, M., Fedorchak, K., et al., APPSw transgenic mice develop age-related A beta deposits and neuropil abnormalities, but no neuronal loss in CA1. J Neuropathol Exp Neurol, 1997. 56(9):965-973.
205. Irizarry, M.C., Soriano, F., McNamara, M., et al., Abeta deposition is associated with neuropil changes, but not with overt neuronal loss in the human amyloid precursor protein V717F (PDAPP) transgenic mouse. J Neurosci, 1997. 17(18):7053-7059.
206. Poorkaj, P., Bird, T.D., Wijsman, E., et al., Tau is a candidate gene for chromosome 17 frontotemporal dementia. Ann Neurol, 1998. 43(6):815-825.
207. Hutton, M., Lendon, C.L., Rizzu, P., et al., Association of missense and 5′-splice-site mutations in tau with the inherited dementia FTDP-17. Nature, 1998. 393(6686):702-705.
208. Spillantini, M.G., Bird, T.D., and Ghetti, B., Frontotemporal dementia and Parkinsonism linked to chromosome 17: a new group of tauopathies. Brain Pathol, 1998. 8(2):387-402.
209. Spillantini, M.G. and Goedert, M., Tau protein pathology in neurodegenerative diseases. Trends Neurosci, 1998. 21(10):428-433.
210. Hardy, J., Duff, K., Hardy, K.G., et al., Genetic dissection of Alzheimer's disease and related dementias: amyloid and its relationship to tau. Nat Neurosci, 1998. 1(5):355-358.
211. Rapoport, M., Dawson, H.N., Binder, L.I., et al., Tau is essential to beta -amyloid-induced neurotox-

icity. Proc Natl Acad Sci U S A, 2002. 99(9):6364-6369.
212. Terry, R.D., Masliah, E., and Hansen, L.A., Structural basis of the cognitive alterations in Alzheimer's disease, in Alzheimer's disease. R.D. Terry, R. Katzman, and K.L. Bick, Editors. 1994, Raven Press: New York.
213. Wisniewski, K.E., Dalton, A.J., McLachlan, C., et al.., Alzheimer's disease in Down's syndrome: clinicopathologic studies. Neurology, 1985. 35(7): 957-961.
214. Lippa, C.F., Nee, L.E., Mori, H., and St George-Hyslop, P., Abeta-42 deposition precedes other changes in PS-1 Alzheimer's disease. Lancet, 1998. 352(9134):1117-1118.
215. Lewis, J., Dickson, D.W., Lin, W.L., et al., Enhanced neurofibrillary degeneration in transgenic mice expressing mutant tau and AP. Science, 2001. 293(5534):1487-1491.
216. Bales, K.R., Verina, T., Dodel, R.C., et al., Lack of apolipoprotein E dramatically reduces amyloid beta-peptide deposition. Nat Genet, 1997. 17(3): 263-264.
217. Roher, A.E., Chaney, M.O., Kuo, Y.M., et al., Morphology and toxicity of Abeta-(1-42) dimer derived from neuritic and vascular amyloid deposits of Alzheimer's disease. J Biol Chem, 1996. 271(34):20631-20635.
218. Walsh, D.M., Klyubin, I., Fadeeva, J.V., et al., Naturally secreted oligomers of amyloid beta protein potently inhibit hippocampal long-term potentiation in vivo. Nature, 2002. 416(6880):535-539.
219. Harper, J.D., Wong, S.S., Lieber, C.M., and Lansbury, P.T., Jr., Assembly of A beta amyloid protofibrils: an in vitro model for a possible early event in Alzheimer's disease. Biochemistry, 1999. 38(28):8972-8980.
220. Pike, C.J., Burdick, D., Walencewicz, A.J., et al., Neurodegeneration induced by beta-amyloid peptides in vitro: the role of peptide assembly state. J Neurosci, 1993. 13(4):1676-1687.
221. Lorenzo, A. and Yankner, B.A., Beta-amyloid neurotoxicity requires fibril formation and is inhibited by congo red. Proc Natl Acad Sci U S A, 1994. 91(25):12243-12247.
222. Curtain, C.C., Ali, F., Volitakis, I., et al.., Alzheimer's disease amyloid-ß binds copper and zinc to generate an allosterically ordered membrane-penetrating structure containing superoxide dismutase-like subunits. J. Biol. Chem, 2001. 276(23):20466-20473.
223. Bush, A.I. and Goldstein, L.E., Specific metal-catalysed protein oxidation reactions in chronic degenerative disorders of ageing: focus on Alzheimer's disease and age-related cataracts. Novartis Found Symp, 2001. 235:26-38; discussion 38-43.
224. Cuajungco, M.P., Goldstein, L.E., Nunomura, A., et al., Evidence that the ß-amyloid plaques of Alzheimer's disease represent the redox-silencing and entombment of Aß by zinc. J Biol Chem, 2000. 275(26):19439-19442.
225. Lambert, M.P., Barlow, A.K., Chromy, B.A., et al., Diffusible, nonfibrillar ligands derived from Abeta1-42 are potent central nervous system neurotoxins. Proc Natl Acad Sci U S A, 1998. 95(11):6448-6453.
226. Hartley, D.M., Walsh, D.M., Ye, C.P., et al., Protofibrillar intermediates of amyloid beta-protein induce acute electrophysiological changes and progressive neurotoxicity in cortical neurons. J Neurosci, 1999. 19(20):8876-8884.
227. Hsia, A.Y., Masliah, E., McConlogue, L., et al., Plaque-independent disruption of neural circuits in Alzheimer's disease mouse models. Proc Natl Acad Sci U S A, 1999. 96(6):3228-3233.
228. Mucke, L., Masliah, E., Yu, G.Q., et al., High-level neuronal expression of abeta 1-42 in wild-type human amyloid protein precursor transgenic mice: synaptotoxicity without plaque formation. J Neurosci, 2000. 20(11):4050-4058.
229. Reiter, R.J., Oxidative processes and antioxidative defense mechanisms in the aging brain. FASEB J, 1995. 9(7):526-533.
230. Tomidokoro, Y., Ishiguro, K., Harigaya, Y., et al., Abeta amyloidosis induces the initial stage of tau accumulation in APP(Sw) mice. Neurosci Lett, 2001. 299(3):169-172.
231. Zheng, W.H., Bastianetto, S., Mennicken, F., et al., Amyloid beta peptide induces tau phosphorylation and loss of cholinergic neurons in rat primary septal cultures. Neuroscience, 2002. 115(1):201-211.
232. Parihar, M.S. and Hemnani, T., Alzheimer's disease pathogenesis and therapeutic interventions. J Clin Neurosci, 2004. 11(5):456-467.
233. Vajda, F.J., Neuroprotection and neurodegenerative disease. J Clin Neurosci, 2002. 9(1):4-8.
234. Saez, T.E., Pehar, M., Vargas, M., et al., Astrocytic nitric oxide triggers tau hyperphosphorylation in hippocampal neurons. In Vivo, 2004. 18(3):275-280.
235. Multhaup, G., Ruppert, T., Schlicksupp, A., et al., Reactive oxygen species and Alzheimer's disease. Biochem Pharmacol, 1997. 54(5):533-539.
236. Perry, G., Taddeo, M.A., Nunomura, A., et al., Comparative biology and pathology of oxidative stress in Alzheimer and other neurodegenerative diseases: beyond damage and response. Comp

Biochem Physiol C Toxicol Pharmacol, 2002. 133(4):507-513.
237. Prasad, K.N., Hovland, A.R., Cole, W.C., et al., Multiple antioxidants in the prevention and treatment of Alzheimer's disease: analysis of biologic rationale. Clin Neuropharmacol, 2000. 23(1):2-13.
238. Schippling, S., Kontush, A., Arlt, S., et al., Increased lipoprotein oxidation in Alzheimer's disease. Free Radic Biol Med, 2000. 28(3):351-360.
239. Lyras, L., Cairns, N.J., Jenner, A., et al., An assessment of oxidative damage to proteins, lipids, and DNA in brain from patients with Alzheimer's disease. J Neurochem, 1997. 68(5):2061-2069.
240. Smith, C.D., Carney, J.M., Starke-Reed, P.E., et al., Excess brain protein oxidation and enzyme dysfunction in normal aging and in Alzheimer's disease. Proc Natl Acad Sci U S A, 1991. 88(23):10540-10543.
241. Smith, M.A., Sayre, L.M., Vitek, M.P., et al., Early AGEing and Alzheimer's. Nature, 1995. 374(6520):316.
242. Montine, K.S., Olson, S.J., Amarnath, V., et al., Immunohistochemical detection of 4-hydroxy-2-nonenal adducts in Alzheimer's disease is associated with inheritance of APOE4. Am J Pathol, 1997. 150(2):437-443.
243. Chang, J.Y., Chavis, J.A., Liu, L.Z., and Drew, P.D., Cholesterol oxides induce programmed cell death in microglial cells. Biochem Biophys Res Commun, 1998. 249(3):817-821.
244. Bernheimer, A.W., Robinson, W.G., Linder, R., et al., Toxicity of enzymically-oxidized low-density lipoprotein. Biochem Biophys Res Commun, 1987. 148(1):260-266.
245. Allen, R.G. and Tresini, M., Oxidative stress and gene regulation. Free Radic Biol Med, 2000. 28(3): 463-499.
246. Dizdaroglu, M., Oxidative damage to DNA in mammalian chromatin. Mutat Res, 1992. 275(3-6): 331-342.
247. Mattson, M.P., Apoptosis in neurodegenerative disorders. Nat Rev Mol Cell Biol, 2000. 1(2):120-129.
248. Mark, R.J., Fuson, K.S., and May, P.C., Characterization of 8-epiprostaglandin F2alpha as a marker of amyloid beta-peptide-induced oxidative damage. J Neurochem, 1999. 72(3):1146-1153.
249. Lovell, M.A., Ehmann, W.D., Butler, S.M., and Markesbery, W.R., Elevated thiobarbituric acid-reactive substances and antioxidant enzyme activity in the brain in Alzheimer's disease. Neurology, 1995. 45(8):1594-1601.
250. Good, P.F., Werner, P., Hsu, A., et al., Evidence of neuronal oxidative damage in Alzheimer's disease. Am J Pathol, 1996. 149(1):21-28.
251. Smith, M.A., Perry, G., Richey, P.L., et al., Oxidative damage in Alzheimer's. Nature, 1996. 382(6587):120-121.
252. Love, S., Barber, R., and Wilcock, G.K., Apoptosis and expression of DNA repair proteins in ischaemic brain injury in man. Neuroreport, 1998. 9(6):955-959.
253. Mecocci, P., MacGarvey, U., and Beal, M.F., Oxidative damage to mitochondrial DNA is increased in Alzheimer's disease. Ann Neurol, 1994. 36(5):747-751.
254. Selley, M.L., Close, D.R., and Stern, S.E., The effect of increased concentrations of homocysteine on the concentration of (E)-4-hydroxy-2-nonenal in the plasma and cerebrospinal fluid of patients with Alzheimer's disease. Neurobiol Aging, 2002. 23(3): 383-388.
255. Butterfield, D.A., Castegna, A., Lauderback, C.M., and Drake, J., Evidence that amyloid beta-peptide-induced lipid peroxidation and its sequelae in Alzheimer's disease brain contribute to neuronal death. Neurobiol Aging, 2002. 23(5):655-664.
256. Arlt, S., Beisiegel, U., and Kontush, A., Lipid peroxidation in neurodegeneration: new insights into Alzheimer's disease. Curr Opin Lipidol, 2002. 13(3):289-294.
257. Keller, J.N., Pang, Z., Geddes, J.W., et al., Impairment of glucose and glutamate transport and induction of mitochondrial oxidative stress and dysfunction in synaptosomes by amyloid beta-peptide: role of the lipid peroxidation product 4-hydroxynonenal. J Neurochem, 1997. 69(1):273-284.
258. Mark, R.J., Hensley, K., Butterfield, D.A., and Mattson, M.P., Amyloid beta-peptide impairs ion-motive ATPase activities: evidence for a role in loss of neuronal Ca2+ homeostasis and cell death. J Neurosci, 1995. 15(9):6239-6249.
259. Tamagno, E., Robino, G., Obbili, A., et al., H2O2 and 4-hydroxynonenal mediate amyloid beta-induced neuronal apoptosis by activating JNKs and p38MAPK. Exp Neurol, 2003. 180(2):144-155.
260. Lewen, A., Matz, P., and Chan, P.H., Free radical pathways in CNS injury. J Neurotrauma, 2000. 17(10):871-890.
261. Suzuki, Y.J., Forman, H.J., and Sevanian, A., Oxidants as stimulators of signal transduction. Free Radic Biol Med, 1997. 22(1-2):269-285.
262. Neill, S., Desikan, R., and Hancock, J., Hydrogen peroxide signalling. Curr Opin Plant Biol, 2002. 5(5):388-395.
263. Ermak, G. and Davies, K.J., Calcium and oxidative stress: from cell signaling to cell death. Mol Immunol, 2002. 38(10):713-721.
264. LaFerla, F.M., Calcium dyshomeostasis and intracellular signalling in Alzheimer's disease. Nat Rev Neurosci, 2002. 3(11):862-872.
265. Gibson, G.E., Interactions of oxidative stress with cellular calcium dynamics and glucose metabolism

in Alzheimer's disease. Free Radic Biol Med, 2002. 32(11):1061-1070.
266. Mattson, M.and Chan, S.L., Neuronal and glial calcium signaling in Alzheimer's disease. Cell Calcium, 2003. 34(4-5):385-397.
267. Zemlan, F.P., Thienhaus, O.J., and Bosmann, H.B., Superoxide dismutase activity in Alzheimer's disease: possible mechanism for paired helical filament formation. Brain Res, 1989. 476(1):160-162.
268. Pappolla, M.A., Omar, R.A., Kim, K.S., and Robakis, N.K., Immunohistochemical evidence of oxidative [corrected] stress in Alzheimer's disease. Am J Pathol, 1992. 140(3):621-628.
269. Gabbita, S.P., Lovell, M.A., and Markesbery, W.R., Increased nuclear DNA oxidation in the brain in Alzheimer's disease. J Neurochem, 1998. 71(5): 2034-2040.
270. Yamamoto, K., Ishikawa, T., Sakabe, T., et al., The hydroxyl radical scavenger Nicaraven inhibits glutamate release after spinal injury in rats. Neuroreport, 1998. 9(7):1655-1659.
271. Law, A., Gauthier, S., and Quirion, R., Say NO to Alzheimer's disease: the putative links between nitric oxide and dementia of the Alzheimer's type. Brain Res Brain Res Rev, 2001. 35(1):73-96.
272. Parks, J.K., Smith, T.S., Trimmer, P.A., et al., Neurotoxic Abeta peptides increase oxidative stress in vivo through NMDA-receptor and nitric-oxide-synthase mechanisms, and inhibit complex IV activity and induce a mitochondrial permeability transition in vitro. J Neurochem, 2001. 76(4):1050-1056.
273. Blanchard, B.J., Chen, A., Rozeboom, L.M., et al., Efficient reversal of Alzheimer's disease fibril formation and elimination of neurotoxicity by a small molecule. Proc Natl Acad Sci U S A, 2004.
274. Luth, H.J., Holzer, M., Gartner, U., et al., Expression of endothelial and inducible NOS-isoforms is increased in Alzheimer's disease, in APP23 transgenic mice and after experimental brain lesion in rat: evidence for an induction by amyloid pathology. Brain Res, 2001. 913(1):57-67.
275. Su, J.H., Deng, G., and Cotman, C.W., Neuronal DNA damage precedes tangle formation and is associated with up-regulation of nitrotyrosine in Alzheimer's disease brain. Brain Res, 1997. 774(1-2):193-199.
276. Gu, Z., Kaul, M., Yan, B., et al., S-nitrosylation of matrix metalloproteinases: signaling pathway to neuronal cell death. Science, 2002. 297(5584): 1186-1190.
277. Yong, V.W., Power, C., Forsyth, P., and Edwards, D.R., Metalloproteinases in biology and pathology of the nervous system. Nat Rev Neurosci, 2001. 2(7):502-511.
278. Meda, L., Cassatella, M.A., Szendrei, G.I., et al., Activation of microglial cells by beta-amyloid protein and interferon-gamma. Nature, 1995. 374 (6523):647-650.
279. El Khoury, J., Hickman, S.E., Thomas, C.A., et al., Scavenger receptor-mediated adhesion of microglia to beta-amyloid fibrils. Nature, 1996. 382(6593): 716-719.
280. Griffin, W.S., Stanley, L.C., Ling, C., et al., Brain interleukin 1 and S-100 immunoreactivity are elevated in Down syndrome and Alzheimer's disease. Proc Natl Acad Sci U S A, 1989. 86(19):7611-7615.
281. Rogers, J., Schultz, J., Brachova, L., et al., Complement activation and beta-amyloid-mediated neurotoxicity in Alzheimer's disease. Res Immunol, 1992. 143(6):624-630.
282. Brown, G.C. and Bal-Price, A., Inflammatory neurodegeneration mediated by nitric oxide, glutamate, and mitochondria. Mol Neurobiol, 2003. 27(3): 325-355.
283. Colton, C.A., Snell, J., Chernyshev, O., and Gilbert, D.L., Induction of superoxide anion and nitric oxide production in cultured microglia. Ann N Y Acad Sci, 1994. 738:54-63.
284. Harman, D., A hypothesis on the pathogenesis of Alzheimer's disease. Ann N Y Acad Sci, 1996. 786: 152-168.
285. Byrne, E., Does mitochondrial respiratory chain dysfunction have a role in common neurodegenerative disorders? J Clin Neurosci, 2002. 9(5): 497-501.
286. Keller, J.N., Guo, Q., Holtsberg, F.W., et al., Increased sensitivity to mitochondrial toxin-induced apoptosis in neural cells expressing mutant presenilin-1 is linked to perturbed calcium homeostasis and enhanced oxyradical production. J Neurosci, 1998. 18(12):4439-4450.
287. Kruman, II and Mattson, M.P., Pivotal role of mitochondrial calcium uptake in neural cell apoptosis and necrosis. J Neurochem, 1999. 72(2):529-540.
288. Cadenas, E. and Davies, K.J., Mitochondrial free radical generation, oxidative stress, and aging. Free Radic Biol Med, 2000. 29(3-4):222-230.
289. Parker, W.D., Jr., Parks, J., Filley, C.M., and Kleinschmidt-DeMasters, B.K., Electron transport chain defects in Alzheimer's disease brain. Neurology, 1994. 44(6):1090-1096.
290. Swerdlow, R.H., Parks, J.K., Cassarino, D.S., et al., Cybrids in Alzheimer's disease: a cellular model of the disease? Neurology, 1997. 49(4):918-925.
291. Corral-Debrinski, M., Horton, T., Lott, M.T., et al., Mitochondrial DNA deletions in human brain: regional variability and increase with advanced age. Nat Genet, 1992. 2(4):324-329.
292. Wallace, D.C., Lott, M.T., and Brown, M.D., Mitochondrial defects in neurodegenerative

diseases and aging, in Mitochondria and Free radicals in Neurodegenerative Diseases, M.F. Beal, N. Howell and I. Bodis-Walker, Editors. 1997, Wiley-Liss: New York. 283–307.
293. Price, D.L., Tanzi, R.E., Borchelt, D.R., and Sisodia, S.S., Alzheimer's disease: genetic studies and transgenic models. Annu Rev Genet, 1998. 32: 461-493.
294. Bush, A.I., Metals and neuroscience. Curr Opin Chem Biol, 2000. 4(2):184-191.
295. Huang, X., Moir, R.D., Tanzi, R.E., et al., Redox-active metals, oxidative stress, and Alzheimer's disease pathology. Ann N Y Acad Sci, 2004. 1012: 153-163.
296. Huang, X., Cuajungco, M.P., Atwood, C.S., et al., Alzheimer's disease, beta-amyloid protein and zinc. J Nutr, 2000. 130(5S Suppl): 1488S-1492S.
297. Atwood, C.S., Huang, X., Moir, R.D., et al.., Role of free radicals and metal ions in the pathogenesis of Alzheimer's disease. Met Ions Biol Syst, 1999. 36:309-364.
298. Smith, M.A., Harris, P.L., Sayre, L.M., and Perry, G., Iron accumulation in Alzheimer's disease is a source of redox-generated free radicals. Proc Natl Acad Sci U S A, 1997. 94(18):9866-9868.
299. Martins, R.N., Harper, C.G., Stokes, G.B., and Masters, C.L., Increased cerebral glucose-6-phosphate dehydrogenase activity in Alzheimer's disease may reflect oxidative stress. J Neurochem, 1986. 46(4):1042-1045.
300. Sayre, L.M., Perry, G., Harris, P.L., et al., In situ oxidative catalysis by neurofibrillary tangles and senile plaques in Alzheimer's disease: a central role for bound transition metals. J Neurochem, 2000. 74(1):270-279.
301. Bush, A.I., Masters, C.L., and Tanzi, R., E., Copper, beta-amyloid, and Alzheimer's disease: tapping a sensitive connection. Proc Natl Acad Sci U S A, 2003. 100(20):11193-11194.
302. Atwood, C.S., Moir, R.D., Huang, X., et al., Dramatic aggregation of Alzheimer Aß by Cu(II) is induced by conditions representing physiological acidosis. J Biol Chem, 1998. 273(21):12817-12826.
303. Atwood, C.S., Huang, X., Khatri, A., et al., Copper catalyzed oxidation of Alzheimer Aß. Cell Mol Biol, 2000. 46(4):777-783.
304. Huang, X., Atwood, C.S., Hartshorn, M.A., et al., The Aß peptide of Alzheimer's disease directly produces hydrogen peroxide through metal ion reduction. Biochemistry, 1999. 38(24):7609-7616.
305. Huang, X., Cuajungco, M.P., Atwood, C.S., et al., Cu(II) potentiation of Alzheimer Aß neurotoxicity. Correlation with cell-free hydrogen peroxide production and metal reduction. J Biol Chem, 1999. 274(52):37111-37116.
306. Opazo, C., Huang, X., Cherny, R.A., et al., Metalloenzyme-like activity of Alzheimer's disease beta-amyloid. Cu-dependent catalytic conversion of dopamine, cholesterol, and biological reducing agents to neurotoxic H(2)O(2). J Biol Chem, 2002. 277(43):40302-40308.
307. Behl, C., Davis, J.B., Lesley, R., and Schubert, D., Hydrogen peroxide mediates amyloid beta protein toxicity. Cell, 1994. 77(6):817-827.
308. Morita, A., Kimura, M., and Itokawa, Y., The effect of aging on the mineral status of female mice. Biol Trace Elem Res, 1994. 42(2):165-177.
309. Takahashi, S., Takahashi, I., Sato, H., et al., Age-related changes in the concentrations of major and trace elements in the brain of rats and mice. Biol Trace Elem Res, 2001. 80(2):145-158.
310. Lovell, M.A., Robertson, J.D., Teesdale, W.J., et al., Copper, iron and zinc in Alzheimer's disease senile plaques. J Neurol Sci, 1998. 158(1):47-52.
311. Lee, J.Y., Cole, T.B., Palmiter, R.D., et al., Contribution by synaptic zinc to the gender-disparate plaque formation in human Swedish mutant APP transgenic mice. Proc Natl Acad Sci U S A, 2002. 99(11):7705-7710.
312. Yoshiike, Y., Tanemura, K., Murayama, O., et al., New insights on how metals disrupt amyloid beta-aggregation and their effects on amyloid-beta cytotoxicity. J Biol Chem, 2001. 276(34):32293-32299.
313. Huang, X., Atwood, C.S., Moir, R.D., et al., Zinc-induced Alzheimer's Abeta1-40 aggregation is mediated by conformational factors. J Biol Chem, 1997. 272(42):26464-26470.
314. Terry, R.D., The pathogenesis of Alzheimer's disease: an alternative to the amyloid hypothesis. J Neuropathol Exp Neurol, 1996. 55(10):1023-1025.
315. Atwood, C.S., Obrenovich, M.E., Liu, T., et al., Amyloid-beta: a chameleon walking in two worlds: a review of the trophic and toxic properties of amyloid-beta. Brain Res Brain Res Rev, 2003. 43(1):1-16.
316. Multhaup, G., Hesse, L., Borchardt, T., et al., Autoxidation of amyloid precursor protein and formation of reactive oxygen species. Adv Exp Med Biol, 1999. 448:183-192.
317. Butterfield, D.A., Drake, J., Pocernich, C., and Castegna, A., Evidence of oxidative damage in Alzheimer's disease brain: central role for amyloid beta-peptide. Trends Mol Med, 2001. 7(12):548-554.
318. Arispe, N., Rojas, E., and Pollard, H.B., Alzheimer's disease amyloid beta protein forms calcium channels in bilayer membranes: blockade by tromethamine and aluminum. Proc Natl Acad Sci U S A, 1993. 90(2):567-571.

319. Mattson, M.P., Tomaselli, K.J., and Rydel, R.E., Calcium-destabilizing and neurodegenerative effects of aggregated beta-amyloid peptide are attenuated by basic FGF. Brain Res, 1993. 621(1): 35-49.
320. Curtain, C.C., Ali, F.E., Smith, D.G., et al., Metal ions, pH, and cholesterol regulate the interactions of Alzheimer's disease amyloid-beta peptide with membrane lipid. J Biol Chem, 2003. 278(5):2977-2982.
321. Cherny, R.A., Legg, J.T., McLean, C.A., et al., Aqueous dissolution of Alzheimer's disease Aß amyloid deposits by biometal depletion. J Biol Chem, 1999. 274(33):23223-23228.
322. Lau, T.L., Barnham, K.J., Curtain, C.C., et al., Magnetic resonance studies of β-amyloid peptides. Aust J Chem, 2003. 56:349–356.
323. Barnham, K.J., Ciccotosto, G.D., Tickler, A.K., et al., Neurotoxic, redox-competent Alzheimer's beta-amyloid is released from lipid membrane by methionine oxidation. J Biol Chem, 2003. 278(44): 42959-42965.
324. Atwood, C.S., Scarpa, R.C., Huang, X., et al., Characterization of copper interactions with alzheimer amyloid beta peptides: identification of an attomolar-affinity copper binding site on amyloid beta1-42. J Neurochem, 2000. 75(3):1219-1233.
325. Naslund, J., Schierhorn, A., Hellman, U., et al., Relative abundance of Alzheimer A beta amyloid peptide variants in Alzheimer's disease and normal aging. Proc Natl Acad Sci U S A, 1994. 91(18): 8378-8382.
326. Kuo, Y.M., Kokjohn, T.A., Beach, T.G., et al., Comparative analysis of amyloid-beta chemical structure and amyloid plaque morphology of transgenic mouse and Alzheimer's disease brains. J Biol Chem, 2001. 276(16):12991-12998.
327. Palmblad, M., Westlind-Danielsson, A., and Bergquist, J., Oxidation of methionine 35 attenuates formation of amyloid beta -peptide 1-40 oligomers. J Biol Chem, 2002. 277(22):19506-19510.
328. Hou, L., Kang, I., Marchant, R.E., and Zagorski, M.G., Methionine 35 oxidation reduces fibril assembly of the amyloid abeta-(1-42) peptide of Alzheimer's disease. J Biol Chem, 2002. 277(43): 40173-40176.
329. Dong, J., Atwood, C.S., Anderson, V.E., et al., Metal binding and oxidation of amyloid-beta within isolated senile plaque cores: Raman microscopic evidence. Biochemistry, 2003. 42(10):2768-2773.
330. Selkoe, D.J., The early diagnosis of Alzheimer's disease., in The Pathophysiology of Alzheimer's Disease. L.F.M. Scinto and K.R. Daffner, Editors. 2000, Humana: Totowa, NJ. 83-104.
331. Barrow, C.J., Advances in the development of Abeta-related therapeutic strategies for Alzheimer's disease. Drug News Perspect, 2002. 15(2):102-109.
332. Auld, D.S., Kornecook, T.J., Bastianetto, S., and Quirion, R., Alzheimer's disease and the basal forebrain cholinergic system: relations to beta-amyloid peptides, cognition, and treatment strategies. Prog Neurobiol, 2002. 68(3):209-245.
333. Bowen, D.M., Palmer, A.M., Frances, P.T., et al., Classical neurotransmitters in Alzheimer's disease, in Aging and the Brain. R.D. Terry, Editor. 1988, Raven Press: New York. 115–128.
334. Emilien, G., Beyreuther, K., Masters, C.L., and Maloteaux, J.M., Prospects for pharmacological intervention in Alzheimer's disease. Arch Neurol, 2000. 57(4):454-459.
335. Mobius, H.J., Memantine: update on the current evidence. Int J Geriatr Psychiatry, 2003. 18(Suppl 1):S47-54.
336. Winblad, B. and Jelic, V., Treating the full spectrum of dementia with memantine. Int J Geriatr Psychiatry, 2003. 18(Suppl 1):S41-46.
337. Rogawski, M.A. and Wenk, G.L., The neuropharmacological basis for the use of memantine in the treatment of Alzheimer's disease. CNS Drug Rev, 2003. 9:275-308.
338. Reisberg, B., Doody, R., Stoffler, A., et al., Memantine in moderate-to-severe Alzheimer's disease. N Engl J Med, 2003. 348(14):1333-1341.
339. Moosmann, B. and Behl, C., Antioxidants as treatment for neurodegenerative disorders. Expert Opin Investig Drugs, 2002. 11(10):1407-1435.
340. Sano, M., Ernesto, C., Thomas, R.G., et al., A controlled trial of selegiline, alpha-tocopherol, or both as treatment for Alzheimer's disease. The Alzheimer's Disease Cooperative Study. N Engl J Med, 1997. 336(17):1216-1222.
341. Bano, S. and Parihar, M.S., Reduction of lipid peroxidation in different brain regions by a combination of alpha-tocopherol and ascorbic acid. J Neural Transm, 1997. 104(11-12):1277-1286.
342. Effects of tocopherol and deprenyl on the progression of disability in early Parkinson's disease. The Parkinson Study Group. N Engl J Med, 1993. 328(3):176-183.
343. Spina, M.B., Squinto, S.P., Miller, J., et al., Brain-derived neurotrophic factor protects dopamine neurons against 6-hydroxydopamine and N-methyl-4- phenylpyridinium ion toxicity: involvement of the glutathione system. J Neurochem, 1992. 59(1):99-106.
344. Parihar, M.S. and Hemnani, T., Experimental excitotoxicity provokes oxidative damage in mice brain and attenuation by extract of Asparagus racemosus. J Neural Transm, 2004. 111(1):1-12.

345. Parihar, M.S. and Hemnani, T., Phenolic antioxidants attenuate hippocampal neuronal cell damage against kainic acid induced excitotoxicity. J Biosci, 2003. 28(1):121-128.
346. Beyer, R.E., An analysis of the role of coenzyme Q in free radical generation and as an antioxidant. Biochem Cell Biol, 1992. 70(6):390-403.
347. Hemmer, W. and Wallimann, T., Functional aspects of creatine kinase in brain. Dev Neurosci, 1993. 15(3-5):249-260.
348. Mendoza-Ramirez, J.L., Beltran-Parrazal, L., Verdugo-Diaz, L., et al., Delay in manifestations of aging by grafting NGF cultured chromaffin cells in adulthood. Neurobiol Aging, 1995. 16(6):907-916.
349. Behl, C. and Holsboer, F., The female sex hormone oestrogen as a neuroprotectant. Trends Pharmacol Sci, 1999. 20(11):441-444.
350. Xu, H., Gouras, G.K., Greenfield,* J.P., et al., Estrogen reduces neuronal generation of Alzheimer beta-amyloid peptides. Nat Med, 1998. 4(4): 447-451.
351. LeVine III, H., Challenges of targeting Aß fibrillogenesis and other protein folding disorders. Amyloid, 2003. 10:133-135.
352. Conway, K.A., Baxter, E.W., Felsenstein, K.M., and Reitz, A.B., Emerging beta-amyloid therapies for the treatment of Alzheimer's disease. Curr Pharm Des, 2003. 9(6):427-447.
353. Xia, W., Amyloid inhibitors and Alzheimer's disease. Curr Opin Investig Drugs, 2003. 4(1):55-59.
354. Lahiri, D.K., Farlow, M.R., Sambamurti, K., et al., A critical analysis of new molecular targets and strategies for drug developments in Alzheimer's disease. Curr Drug Targets, 2003. 4(2):97-112.
355. Doraiswamy, P.M., Non-cholinergic strategies for treating and preventing Alzheimer's disease. CNS Drugs, 2002. 16(12):811-824.
356. Maiorini, A.F., Gaunt, M.J., Jacobsen, T.M., et al., Potential novel targets for Alzheimer pharmacotherapy: I. Secretases. J Clin Pharm Ther, 2002. 27(3):169-183.
357. Jhee, S., Shiovitz, T., Crawford, A.W., and Cutler, N.R., Beta-amyloid therapies in Alzheimer's disease. Expert Opin Investig Drugs, 2001. 10(4): 593-605.
358. Cutler, N.R. and Sramek, J.J., Review of the next generation of Alzheimer's disease therapeutics: challenges for drug development. Prog Neuropsychopharmacol Biol Psychiatry, 2001. 25(1): 27-57.
359. Haass, C. and De Strooper, B., The presenilins in Alzheimer's disease—proteolysis holds the key. Science, 1999. 286(5441):916-919.
360. Wong, G.T., Manfra, D., Poulet, F.M., et al., Chronic treatment with the gamma-secretase inhibitor LY-411,575 inhibits beta-amyloid peptide production and alters lymphopoiesis and intestinal cell differentiation. J Biol Chem, 2004. 279(13): 12876-12882.
361. King, G.D., Cherian, K., and Turner, R.S., X11alpha impairs gamma-but not beta-cleavage of amyloid precursor protein. J Neurochem, 2004. 88(4):971-982.
362. Lanz, T.A., Hosley, J.D., Adams, W.J., and Merchant, K.M., Studies of Abeta pharmacodynamics in the brain, cerebrospinal fluid, and plasma in young (plaque-free) Tg2576 mice using the gamma-secretase inhibitor N2-[(2S)-2-(3,5-difluorophenyl)-2-hydroxyethanoyl]-N1-[(7S)-5-methyl-6-oxo -6,7-dihydro-5H-dibenzo[b,d]azepin-7-yl]-L-alaninamide (LY-411575). J Pharmacol Exp Ther, 2004. 309(1):49-55.
363. Kornilova, A.Y., Das, C., and Wolfe, M.S., Differential effects of inhibitors on the gamma-secretase complex. Mechanistic implications. J Biol Chem, 2003. 278(19):16470-16473.
364. Lanz, T.A., Himes, C.S., Pallante, G., et al., The gamma-secretase inhibitor N-[N-(3,5-difluorophenacetyl)-L-alanyl]-S-phenylglycine t-butyl ester reduces A beta levels in vivo in plasma and cerebrospinal fluid in young (plaque-free) and aged (plaque-bearing) Tg2576 mice. J Pharmacol Exp Ther, 2003. 305(3):864-871.
365. Takahashi, Y., Hayashi, I., Tominari, Y., et al.., Sulindac sulfide is a noncompetitive gamma-secretase inhibitor that preferentially reduces Abeta 42 generation. J Biol Chem, 2003. 278(20):18664-18670.
366. Wolfe, M.S., Therapeutic strategies for Alzheimer's disease. Nat Rev Drug Discov, 2002. 1(11):859-866.
367. Schenk, D., Amyloid-beta immunotherapy for Alzheimer's disease: the end of the beginning. Nat Rev Neurosci, 2002. 3(10):824-828.
368. McLaurin, J., Cecal, R., Kierstead, M.E., et al., Therapeutically effective antibodies against amyloid-beta peptide target amyloid-beta residues 4-10 and inhibit cytotoxicity and fibrillogenesis. Nat Med, 2002. 8(11):1263-1269.
369. Hock, C., Konietzko, U., Papassotiropoulos, A., et al., Generation of antibodies specific for beta-amyloid by vaccination of patients with Alzheimer's disease. Nat Med, 2002. 8(11):1270-1275.
370. Schenk, D., Barbour, R., Dunn, W., et al., Immunization with amyloid-ß attenuates Alzheimer-disease-like pathology in the PDAPP mouse. Nature, 1999. 400(6740):173-177.

371. Janus, C., Vaccines for Alzheimer's disease: how close are we? CNS Drugs, 2003. 17(7):457-474.
372. Cherny, R.A., Atwood, C.S., Xilinas, M.E., et al., Treatment with a copper-zinc chelator markedly and rapidly inhibits ß-amyloid accumulation in Alzheimer's disease transgenic mice. Neuron, 2001. 30(3):665-676.
373. Weiner, H.L., Lemere, C.A., Maron, R., et al., Nasal administration of amyloid-ß peptide decreases cerebral amyloid burden in a mouse model of Alzheimer's disease. Ann Neurol, 2000. 48(4):567-579.
374. Janus, C., Pearson, J., McLaurin, J., et al., Aß-peptide immunization reduces behavioural impairment and plaques in a model of Alzheimer's disease. Nature, 2000. 408(6815):979-982.
375. Bard, F., Cannon, C., Barbour, R., et al., Peripherally administered antibodies against amyloid ß peptide enter the central nervous system and reduce pathology in a mouse model of Alzheimer's disease. Nat Med, 2000. 6(8):916-919.
376. DeMattos, R.B., Bales, K.R., Cummins, D.J., et al., Peripheral anti-Aß antibody alters CNS and plasma Aß clearance and decreases brain Aß burden in a mouse model of Alzheimer's disease. Proc Natl Acad Sci U S A, 2001. 98(15):8850-8855.
377. Wilcock, D.M., DiCarlo, G., Henderson, D., et al., Intracranially administered anti-Abeta antibodies reduce beta-amyloid deposition by mechanisms both independent of and associated with microglial activation. J Neurosci, 2003. 23(9):3745-3751.
378. Robinson, S.R., Bishop, G.M., Lee, H.G., and Munch, G., Lessons from the AN 1792 Alzheimer vaccine: lest we forget. Neurobiol Aging, 2004. 25(5):609-615.
379. Broytman, O. and Malter, J.S., Anti-Abeta: The good, the bad, and the unforeseen. J Neurosci Res, 2004. 75(3):301-306.
380. Robinson, S.R., Bishop, G.M., and Munch, G., Alzheimer vaccine: amyloid-beta on trial. Bioessays, 2003. 25(3):283-288.
381. Munch, G. and Robinson, S.R., Potential neurotoxic inflammatory responses to Abeta vaccination in humans. J Neural Transm, 2002. 109(7-8):1081-1087.
382. Rogers, J., Webster, S., Lue, L.F., et al., Inflammation and Alzheimer's disease pathogenesis. Neurobiol Aging, 1996. 17(5):681-686.
383. Weggen, S., Eriksen, J.L., Das, P., et al., A subset of NSAIDs lower amyloidogenic Abeta42 independently of cyclooxygenase activity. Nature, 2001. 414(6860):212-216.
384. Beher, D., Clarke, E.E., Wrigley, J.D., et al., Selected non-steroidal anti-inflammatory drugs and their derivatives target gamma -secretase at a novel site-evidence for an allosteric mechanism. J Biol Chem, 2004.
385. Lim, G.P., Yang, F., Chu, T., et al., Ibuprofen effects on Alzheimer pathology and open field activity in APPsw transgenic mice. Neurobiol Aging, 2001. 22(6):983-991.
386. Jantzen, P.T., Connor, K.E., DiCarlo, G., et al., Microglial activation and beta -amyloid deposit reduction caused by a nitric oxide-releasing nonsteroidal anti-inflammatory drug in amyloid precursor protein plus presenilin-1 transgenic mice. J Neurosci, 2002. 22(6):2246-2254.
387. Refolo, L.M., Pappolla, M.A., LaFrancois, J., et al., A cholesterol-lowering drug reduces beta-amyloid pathology in a transgenic mouse model of Alzheimer's disease. Neurobiol Dis, 2001. 8(5):890-899.
388. Wolozin, B., Kellman, W., Ruosseau, P., et al., Decreased prevalence of Alzheimer's disease associated with 3-hydroxy-3-methyglutaryl coenzyme A reductase inhibitors. Arch Neurol, 2000. 57(10): 1439-1443.
389. Jick, H., Zornberg, G.L., Jick, S.S., et al., Statins and the risk of dementia. Lancet, 2000. 356(9242): 1627-1631.
390. Sparks, D.L., Kuo, Y.M., Roher, A., et al., Alterations of Alzheimer's disease in the cholesterol-fed rabbit, including vascular inflammation. Preliminary observations. Ann N Y Acad Sci, 2000. 903:335-344.
391. Refolo, L.M., Malester, B., LaFrancois, J., et al., Hypercholesterolemia accelerates the Alzheimer's amyloid pathology in a transgenic mouse model. Neurobiol Dis, 2000. 7(4):321-331.
392. Fassbender, K., Simons, M., Bergmann, C., et al., Simvastatin strongly reduces levels of Alzheimer's disease beta -amyloid peptides Abeta 42 and Abeta 40 in vitro and in vivo. Proc Natl Acad Sci U S A, 2001. 98(10):5856-5861.
393. Wahrle, S., Das, P., Nyborg, A.C., et al., Cholesterol-dependent gamma-secretase activity in buoyant cholesterol-rich membrane microdomains. Neurobiol Dis, 2002. 9(1):11-23.
394. Bush, A.I., Metal complexing agents as therapies for Alzheimer's disease. Neurobiol Aging, 2002. 23(6):1031-1038.
395. Padmanabhan, G., Klauss, E., and Florey, E.E., Clioquinol, in Analytical Profiles of Drug Substances. K. Florey, Editor. 1989, Academic Press: Orlando, FL. 57-90.
396. Ritchie, C.W., Bush, A.I., Mackinnon, A., et al., Metal-protein attenuation with iodochlorhydroxyquin (clioquinol) targeting Abeta amyloid deposition and toxicity in Alzheimer's disease: a pilot phase 2 clinical trial. Arch Neurol, 2003. 60(12):1685-1691.

397. Sair, H.I., Doraiswamy, P.M., and Petrella, J.R., In vivo amyloid imaging in Alzheimer's disease. Neuroradiology, 2004. 46(2):93-104.
398. Zhang, J., Yarowsky, P., Gordon, M.N., et al., Detection of amyloid plaques in mouse models of Alzheimer's disease by magnetic resonance imaging. Magn Reson Med, 2004. 51(3):452-457.
399. Benveniste, H., Einstein, G., Kim, K.R., et al., Detection of neuritic plaques in Alzheimer's disease by magnetic resonance microscopy. Proc Natl Acad Sci U S A, 1999. 96(24):14079-14084.
400. Rapoport, S.I., Hydrogen magnetic resonance spectroscopy in Alzheimer's disease. Lancet Neurol, 2002. 1(2):82.
401. Schuff, N., Capizzano, A.A., Du, A.T., et al., Selective reduction of N-acetylaspartate in medial temporal and parietal lobes in AD. Neurology, 2002. 58(6):928-935.
402. Petrella, J.R., Coleman, R.E., and Doraiswamy, P.M., Neuroimaging and early diagnosis of Alzheimer's disease: a look to the future. Radiology, 2003. 226(2):315-336.
403. Phelps, M.E., PET: the merging of biology and imaging into molecular imaging. J Nucl Med, 2000. 41(4):661-681.
404. Devanand, D.P., Jacobs, D.M., Tang, M.X., et al., The course of psychopathologic features in mild to moderate Alzheimer's disease. Arch Gen Psychiatry, 1997. 54(3):257-263.
405. Salmon, E., Sadzot, B., Maquet, P., et al. Differential diagnosis of Alzheimer's disease with PET. J Nucl Med, 1994. 35(3):391-398.
406. Silverman, D.H., Cummings, J.L., Small, G., et al., Added clinical benefit of incorporating 2-deoxy-2-[18F]fluoro-D-glucose with positron emission tomography into the clinical evaluation of patients with cognitive impairment. Mol Imaging Biol, 2002. 4(4):283-2893.
407. Kennedy, A.M., Frackowiak, R.S., Newman, S.K., et al., Deficits in cerebral glucose metabolism demonstrated by positron emission tomography in individuals at risk of familial Alzheimer's disease. Neurosci Lett, 1995. 186(1):17-20.
408. Small, G.W., Mazziotta, J.C., Collins, M.T., et al., Apolipoprotein E type 4 allele and cerebral glucose metabolism in relatives at risk for familial Alzheimer's disease. JAMA, 1995. 273(12):942-947.
409. Silverman, D.H., Small, G.W., Chang, C.Y., et al., Positron emission tomography in evaluation of dementia: regional brain metabolism and long-term outcome. JAMA, 2001. 286(17):2120-2127.
410. Zhuang, Z.P., Kung, M.P., Wilson, A., et al., Structure-activity relationship of imidazo[1,2-a]pyridines as ligands for detecting beta-amyloid plaques in the brain. J Med Chem, 2003. 46(2):237-243.
411. Kung, M.P., Hou, C., Zhuang, Z.P., et al., IMPY: an improved thioflavin-T derivative for in vivo labeling of beta-amyloid plaques. Brain Res Bull, 2002. 956(2):202-210.
412. Ono, M., Kung, M.P., Hou, C., and Kung, H.F., Benzofuran derivatives as Abeta-aggregate-specific imaging agents for Alzheimer's disease. Nucl Med Biol, 2002. 29(6):633-642.
413. Ono, M., Wilson, A., Nobrega, J., et al., 11C-labeled stilbene derivatives as Abeta-aggregate-specific PET imaging agents for Alzheimer's disease. Nucl Med Biol, 2003. 30(6):565-571.
414. Kung, M.P., Skovronsky, D.M., Hou, C., et al., Detection of amyloid plaques by radioligands for Abeta40 and Abeta42: potential imaging agents in Alzheimer's patients. J Mol Neurosci, 2003. 20(1): 15-24.
415. Kung, M.P., Zhuang, Z.P., Hou, C., et al., Characterization of radioiodinated ligand binding to amyloid beta plaques. J Mol Neurosci, 2003. 20(3): 249-254.
416. Lee, C.W., Kung, M.P., Hou, C., and Kung, H.F., Dimethylamino-fluorenes: ligands for detecting beta-amyloid plaques in the brain. Nucl Med Biol, 2003. 30(6):573-580.
417. Link, C.D., Johnson, C.J., Fonte, V., et al., Visualization of fibrillar amyloid deposits in living, transgenic Caenorhabditis elegans animals using the sensitive amyloid dye, X-34. Neurobiol Aging, 2001. 22(2):217-226.
418. Klunk, W.E., Debnath, M.L., and Pettegrew, J.W., Development of small molecule probes for the beta-amyloid protein of Alzheimer's disease. Neurobiol Aging, 1994. 15(6):691-698.
419. Bacskai, B.J., Klunk, W.E., Mathis, C.A., and Hyman, B.T., Imaging amyloid-beta deposits in vivo. J Cereb Blood Flow Metab, 2002. 22(9): 1035-1041.
420. Klunk, W.E., Bacskai, B.J., Mathis, C.A., et al., Imaging Aß plaques in living transgenic mice with multiphoton microscopy and methoxy-X04, a systemically administered Congo red derivative. J Neuropath Exp Neurol, 2002. 61(9):797-805.
421. Kung, M.P., Hou, C., Zhuang, Z.P., et al., Radioiodinated styrylbenzene derivatives as potential SPECT imaging agents for amyloid plaque detection in Alzheimer's disease. J Mol Neurosci, 2002. 19(1-2):7-10.
422. Zhuang, Z.P., Kung, M.P., Hou, C., et al., IBOX(2-(4′-dimethylaminophenyl)-6-iodobenzoxazole): a ligand for imaging amyloid plaques in the brain. Nucl Med Biol, 2001. 28(8):887-894.

423. Lee, C.W., Zhuang, Z.P., Kung, M.P., et al., Isomerization of (Z,Z) to (E,E)1-bromo-2,5-bis-(3-hydroxycarbonyl-4-hydroxy)styrylbenzene in strong base: probes for amyloid plaques in the brain. J Med Chem., 2001. 44(14):2270-2275.
424. Klunk, W.E., Wang, Y., Huang, G.F., et al., The binding of 2-(4′-methylaminophenyl)benzothiazole to postmortem brain homogenates is dominated by the amyloid component. J Neurosci, 2003. 23(6): 2086-2092.
425. Klunk, W.E., Wang, Y., Huang, G.F., et al., Uncharged thioflavin-T derivatives bind to amyloid-beta protein with high affinity and readily enter the brain. Life Sci, 2001. 69(13):1471-1484.
426. Bacskai, B.J., Hickey, G.A., Skoch, J., et al., Four-dimensional multiphoton imaging of brain entry, amyloid binding, and clearance of an amyloid-beta ligand in transgenic mice. Proc Natl Acad Sci U S A, 2003. 100(21):12462-12467.
427. Mathis, C.A., Bacskai, B.J., Kajdasz, S.T., et al., A lipophilic thioflavin-T derivative for positron emission tomography (PET) imaging of amyloid in brain. Bioorg Med Chem Lett, 2002. 12(3):295-298.
428. Mathis, C.A., Holt, D.P., Wang, Y., et al., Lipophilic 11C-labelled thioflavin-T analogues for imaging amyloid plaques in Alzheimer's disease. J Label Compd Radiopharm, 2001. 44(Suppl 1):S26-S28.
429. Mathis, C.A., Wang, Y., Holt, D.P., et al., Synthesis and evaluation of 11C-labeled 6-substituted 2-arylbenzothiazoles as amyloid imaging agents. J Med Chem, 2003. 46(13):2740-2754.
430. Wang, Y., Klunk, W.E., Huang, G.F., et al., Synthesis and evaluation of 2-(3′-iodo-4′-aminophenyl)-6-hydroxybenzothiazole for in vivo quantitation of amyloid deposits in Alzheimer's disease. J Mol Neurosci, 2002. 19(1-2):11-16.
431. Wang, Y., Mathis, C.A., Huang, G.F., et al.., Effects of lipophilicity on the affinity and nonspecific binding of iodinated benzothiazole derivatives. J Mol Neurosci, 2003. 20(3):255-260.
432. Helmuth, L., Long-awaited technique spots Alzheimer's toxin. Science, 2002. 297:752-753.
433. Klunk, W.E., Engler, H., Nordberg, A., et al., Imaging brain amyloid in Alzheimer's disease with Pittsburgh Compound-B. Ann Neurol, 2004. 55: 306-319.
434. Zhuang, Z.P., Kung, M.P., Hou, C., et al., Radioiodinated styrylbenzenes and thioflavins as probes for amyloid aggregates. J Med Chem, 2001. 44(12):1905-1914.
435. Skovronsky, D.M., Zhang, B., Kung, M.P., et al., In vivo detection of amyloid plaques in a mouse model of Alzheimer's disease. Proc Natl Acad Sci U S A, 2000. 97(13):7609-7614.
436. Kung, H.F., Lee, C.W., Zhuang, Z.P., et al. Novel stilbenes as probes for amyloid plaques. J Am Chem Soc, 2001. 123(50):12740-12741.
437. Schmidt, M.L., Schuck, T., Sheridan, S., et al., The fluorescent Congo red derivative, (trans, trans)-1-bromo-2,5-bis-(3-hydroxycarbonyl-4-hydroxy)styrylbenzene (BSB), labels diverse ß-pleated sheet structures in postmortem human neurodegenerative disease brains. Am J Pathol, 2001. 159(3):937-943.
438. Shimadzu, H., Suemoto, T., Suzuki, M., et al., A novel probe for imaging amyloid-b: Synthesis of F-18 labelled BF-108, an Acridine Orange analog. J Label Compd Radiopharm, 2003. 46:765-772.
439. Agdeppa, E.D., Kepe, V., Petri, A., et al., In vitro detection of (S)-naproxen and ibuprofen binding to plaques in the Alzheimer's brain using the positron emission tomography molecular imaging probe 2-(1-[6-[(2-[(18)F]fluoroethyl)(methyl)amino]-2-naphthyl]ethylidene)malononitrile. Neuroscience, 2003. 117(3):723-730.
440. Agdeppa, E.D., Kepe, V., Liu, J., et al., Binding characteristics of radiofluorinated 6-dialkylamino-2-naphthylethylidene derivatives as positron emission tomography imaging probes for ß-amyloid plaques in Alzheimer's disease. J Neurosci, 2001. 21(24):RC189.
441. Barrio, J.R., Huang, S.C., Cole, G., et al., PET imaging of tangles and plaques in Alzheimer's disease with a highly lipophilic probe. J Label Compd Radiopharm, 1999. 42:S194-S195.
442. Shoghi-Jadid, K., Small, G.W., Agdeppa, E.D., et al., Localisation of neurofibrillary tangles and ß-amyloid plaques in the brains of living patients with Alzheimer's disease. Am J Ger Psychiatry, 2002. 10(1):24-35.
443. Bresjanac, M., Smid, L.M., Vovko, T.D., et al., Molecular-imaging probe 2-(1-[6-[(2-fluoroethyl) (methyl) amino]-2-naphthyl]ethylidene) malononitrile labels prion plaques in vitro. J Neurosci, 2003. 23(22):8029-8033.
444. Agdeppa, E.D., Kepe, V., Shoghi-Jadid, K., et al., In vivo and in vitro labeling of plaques and tangles in the brain of an Alzheimer's disease patient: a case study. J Nucl Med, 2001. 42(Suppl 1): 65P.
445. Small, G.W., Agdeppa, E.D., Kepe, V., et al., In vivo brain imaging of tangle burden in humans. J Mol Neurosci, 2002. 19(3):323-327.
446. Lee, V.M., Related Amyloid binding ligands as Alzheimer's disease therapies. Neurobiol Aging, 2002. 23(6):1039-1042.
447. Marshall, J.R., Stimson, E.R., Ghilardi, J.R., et al., Noninvasive imaging of peripherally injected Alzheimer's disease type synthetic A beta amyloid in vivo. Bioconjug Chem, 2002. 13(2):276-284.

448. Maggio, J.E., Stimson, E.R., Ghilardi, J.R., et al., Reversible in vitro growth of Alzheimer's disease beta-amyloid plaques by deposition of labeled amyloid protein. Proc Natl Acad Sci U S A, 1992. 89(12):5462-5466.
449. Friedland, R.P., Shi. J, Lamanna, J.C., et al., Prospects for noninvasive imaging of brain amyloid beta in Alzheimer's disease. Ann N Y Acad Sci, 2000. 903:123-128.
450. Ghilardi, J.R., Catton, M., Stimson, E.R., et al., Intra-arterial infusion of [125I]A beta 1-40 labels amyloid deposits in the aged primate brain in vivo. Neuroreport, 1996. 7(15-17):2607-2611.
451. Kurihara, A. and Pardridge, W.M., Abeta(1-40) peptide radiopharmaceuticals for brain amyloid imaging: (111)In chelation, conjugation to poly(ethylene glycol)-biotin linkers, and autoradiography with Alzheimer's disease brain sections. Bioconjug Chem, 2000. 11(3):380-386.
452. Saito, Y., Buciak, J., Yang, J., and Pardridge, W.M., Vector-mediated delivery of 125I-labeled beta-amyloid peptide A beta 1-40 through the blood-brain barrier and binding to Alzheimer's disease amyloid of the A beta 1-40/vector complex. Proc Natl Acad Sci U S A, 1995. 92(22):10227-10231.
453. Majocha, R.E., Reno, J.M., Friedland, R.P., et al., Development of a monoclonal antibody specific for ß/A4 amyloid in Alzheimer's disease brain for application to in vivo imaging of amyloid angiopathy. J Nucl Med, 1992. 33(12):2184-2189.
454. Walker, L.C., Price, D.L., Voytko, M.L., and Schenk, D.B., Labelling of cerebral amyloid in vivo with a monoclonal antibody. J Neuropathol Exp Neurol, 1994. 53(4):377-383.
455. Shi, J., Perry, G., Berridge, M.S., et al., Labeling of cerebral amyloid beta deposits in vivo using intranasal basic fibroblast growth factor and serum amyloid P component in mice. J Nucl Med, 2002. 43(8):1044-1051.
456. Knopman, D.S., DeKosky, S.T., Cummings, J.L., et al., Practice parameter: Diagnosis of dementia (an evidence based review). Report of the Quality Standards Subcommittee of the American Academy of Neurology. Neurology, 2001. 56(9):1143-1153.
457. Doody, R.S., Stevens, J.C., Beck, C., et al., Practice parameter: management of dementia (an evidence-based review)-report of the Quality Standards Subcommittee of the American Academy of Neurology. Neurology, 2001. 56(9):1154-1166.
458. Petersen, R.C., Stevens, J.C., Ganguli, M., et al., Practice parameter: early detection of dementia: mild cognitive impairment (an evidence-based review)-report of the Quality Standards Subcommittee of the American Academy of Neurology. Neurology, 2001. 56(9):1133-1142.
459. Silverman, D.H., Chang, C.Y., Cummings, J.L., et al., Prognostic value of regional brain metabolism in evaluation of dementia. J Nucl Med, 1999. 40(Suppl 1): 71P.
460. Chang, C.Y. and Silverman, D.H., Accuracy of early diagnosis and its impact on the management and course of Alzheimer's disease. Expert Rev Mol Diagn, 2004. 4:63-69.
461. Silverman, D.H., Gambhir, S.S., Huang, H.W., et al., Evaluating early dementia with and without assessment of regional cerebral metabolism by PET: a comparison of predicted costs and benefits. J Nucl Med, 2002. 43(2):253-266.

3
The Function of the Amyloid Precursor Protein Family

Roberto Cappai, B. Elise Needham, and Giuseppe D. Ciccotosto

3.1. The Amyloid Precursor Protein Is a Multidomain Molecule

The purification and sequencing of the β-amyloid peptide (Aβ) [1–3] led to the cloning of the Alzheimer's disease (AD) amyloid precursor protein (APP) gene in the late 1980s [4]. Despite an extensive research effort toward understanding the function of APP, its physiological role remains poorly defined. This review will summarize the key activities associated with APP and its paralogues the amyloid precursor like proteins 1 and 2 (APLP1 and APLP2, respectively).

The human APP gene is encoded by 19 exons located on the long arm of chromosome 21 [4–7] and is ubiquitously expressed in vertebrates [8]. It is highly expressed in the brain, with APP constituting 0.2% of the total mRNA of neurons [9]. APP undergoes extensive alternative splicing of exons 7, 8, and 15 to yield at least 8 isoforms that have cell-specific expression patterns [10, 11].

The primary sequence identified APP as a type I transmembrane glycoprotein with a single transmembrane region, a large extracellular domain, and a short cytoplasmic tail that was suggestive of a cell-surface receptor [4]. A combination of sequence and structural analysis has indicated that APP is organized into distinct domains (Fig. 3.1) [12–15]. The N-terminal signal peptide is followed by a cysteine-rich domain that is composed of two separate smaller domains joined by a short linker [12, 14]. The first N-terminal domain contains a heparin-binding site (HBD) with structural homology to growth factors [12], which is consistent with its neurite outgrowth promoting activity [16]. The second portion of the cysteine-rich region is the metal-binding domain (MBD) with binding sites for copper [17, 18] and zinc [19]. The copper-binding site acts as a modulator of copper homeostasis [20–23], copper-mediated toxicity [24–26], and modulation of APP processing into Aβ [22, 23, 27, 28]. The cysteine-rich domain is followed by an acidic domain that is rich in glutamate and aspartate residues, and these residues constitute nearly 50% of the acidic domain. In some APP isoforms, the acidic domain is then followed by an alternatively spliced exon that is homologous to the Kunitz-type serine protease inhibitor (KPI) family. This is followed by an alternatively spliced 19-residue sequence encoded by exon 8 that lies adjacent to the KPI domain. This sequence is homologous to the immunoregulatory OX2 antigen [29, 30].

The different isoforms of APP are designated by the number of amino acids they contain. There are three major species: APP_{695}, APP_{751}, and APP_{770}. The APP_{695} isoform lacks the KPI and OX-2 exons, while APP_{751} contains the KPI domain and APP_{770} contains both the KPI and OX-2 sequences [31–34]. The C-terminal portion of the ectodomain is glycosylated [30] and is composed of two domains. The first domain of the glycosylated region has been called the central APP domain (CAPPD) [13, 15] or the E2 domain [15]. The NMR and crystal structures of CAPPD/E2 indicate it is composed of six α-helices folded tightly together as a coiled-coil substructure [13, 15]. An E2 dimer was identified in the crystalline state with E2 binding to itself in an antiparallel orientation.

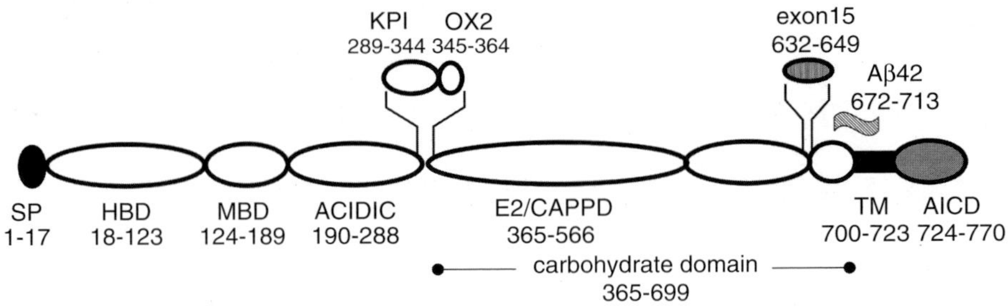

FIGURE 3.1. Schematic representation of the domain structure of APP. SP, signal peptide; HBD, heparin-binding domain; MBD, metal-binding domain, ACIDIC, acidic rich domain; KPI, Kunitz protease inhibitor domain; OX2, immunoregulatory OX-2 antigen domain; E2/CAPPD, extracellular domain 2/central APP domain; exon 15, alternatively spliced exon 15; TM, transmembrane domain; AICD, APP intracellular domain; Aβ42, Aβ42 peptide. The amino acid numbering is based on the APP770 isoform.

The N- and C-terminal ends of the individual subunits were located at a concave surface formed by the dimer. The dimerization of APP via the E2/CAPPD domain may relate to the role of APP in cell-cell adhesion. The CAPPD/E2 region is followed by an unstructured region that contains the α- and β-secretase cleavage sites. Alternative splicing of exon 15 occurs within the sequence, and the omission of exon 15 creates a chondroitin sulfate acceptor site. The large extracellular domain is followed by the transmembrane domain of APP [4, 35]. The γ-secretase cleavage site is located approximately in the middle of the transmembrane domain.

The Aβ peptide is derived from the last 29 C-terminal amino acids of and the first 11–13 amino acids of the transmembrane domain. The final domain of APP is the APP intracellular cytoplasmic domain (AICD) which is released into the cytoplasm following either ε- or γ-secretase cleavage of APP [36]. The AICD has multiple binding partners including Fe65, Jip1b, X11alpha (MINT1), and Tip60 and is transported to the nucleus after its release into the cytoplasm [37–39].

3.2 Expression Patterns of APP Isoforms

The most abundantly expressed isoforms are APP_{695}, APP_{751}, and APP_{770}, and they all contain the Aβ sequence. APP is widely expressed throughout the body in both fetal and adult tissues [40]. Expression of total APP is highest in the brain and kidneys, with lower levels in the spleen, adrenal glands, lungs, and liver [6, 41]. APP is present in the CSF [30, 42], and its expression is increased after traumatic brain injury [43, 44]. The tissue distribution of the various isoforms varies considerably. The APP_{695} species is the most abundant isoform in neurons [4, 6], while the APP-KPI isoforms (APP_{751} and APP_{770}) are predominantly expressed by glial cells [45–47], platelets [48, 49], and peripheral tissue [33, 34]. In AD brains, KPI-containing isoforms are increased approximately twofold as compared with non-AD control brains [50]. The KPI-containing isoforms are the most amyloidogenic [51, 52]. Alternative splicing of exon 15, located in the ectodomain close to the transmembrane domain, occurs resulting in APP isoforms lacking exon 15, termed L-APP, and were initially found in lymphocytes and microglia [53]. Later, cell-associated chrondroitin sulfate proteoglycan (termed appicans) were found to contain L-APP as their core protein [54]. These L-APP isoforms are not detectable in neurons but comprise the majority of APP transcripts in aorta and pancreatic tissue and are also abundant in skeletal muscle [10, 11]. Activation of the Wnt-1 signaling pathway promotes the deletion of exon 15 leading to increased expression of L-APP [55]. This provides a functional link to APP for the association between presenilin and the Wnt-1 pathway [56].

There is considerable evolutionary conservation of the APP-family [57]. APP and the APLPs are found in mammals, whereas homologues to APP have been identified in non-mammalian species including *Drosophila melanogaster*, *Xenopus levis,*

Caenorhabditis elegans, *Narke japonica* (electric ray), *Fugu rubripes*, and *Tetraodon fluviatilis* (both are puffer fish) [58–61]. The evolutionary preservation of proteins homologous to APP strengthens the physiological importance of these proteins.

3.3 Cellular Processing of APP

The metabolism of APP occurs via a complex process involving the activity of three proteases termed secretases. Only a small portion of the total pool of APP is cleaved by the secretases leaving the majority intact as full-length APP [62]. Secretase cleavage can occur via two major pathways, termed the amyloidogenic and non-amyloidogenic pathways [62, 63]. Which pathway is utilized depends on the cell type with neurons and astrocytes producing more amyloidogenic APP compared with glial cells [64]. Importantly, the processing of APP will clearly regulate the actions and ultimately the function of APP.

APP processing via the non-amyloidogenic pathway occurs in the late Golgi compartment or in caveolae (plasma membrane invaginations) [65] and destroys the Aβ sequence and thus prevents amyloid formation. Three members of the ADAM (a disintegrin and metalloprotease) family ADAM 9, 10, and 17 have been proposed to be α-secretases [66–69]. The α-secretase cleavage releases the majority of the ectodomain as a soluble fragment, termed sAPPα, while leaving a C-terminal fragment (CTF-α) in the cell membrane with a truncated Aβ sequence [70, 71]. The CTF-α is then cleaved by γ-secretase, resulting in a truncated 3-kDa Aβ fragment, termed p3 [65, 72].

The amyloidogenic pathway occurs via β-secretase or BACE-1 (beta-site APP-cleaving enzyme-1) [73–76] cleavage on the N-terminal side of the Aβ sequence. This releases a large soluble fragment of the ectodomain, termed sAPPβ, and leaves the membrane-associated CTF-β fragment, which contains an intact Aβ sequence [72, 77, 78]. The CTF-β is then cleaved by the multiprotein complex γ-secretase, which releases Aβ [79, 80]. BACE-1 is a transmembrane aspartyl protease. There are two BACE genes, BACE-1, which is highly expressed in the brain, and BACE-2, which is predominantly expressed in peripheral tissues including the pancreas, stomach, and placenta [81]. BACE-1 expression is upregulated in the brain after an ischemia whereas BACE-2 expression is unchanged [82]. This response in BACE-1 expression coincides with an upregulation in APP expression after ischemia [45]. The γ-secretase is composed of the presenilins, PS1 and PS2, nicastrin, Aph1, and PEN2 (reviewed in [83]).

3.4 The Function of APP

The APP promoter has the sequence elements that are indicative of a housekeeping gene [84, 85]. Such genes are functionally important in all cell types, irrespective of the specialized role of the cells. The actions of APP may depend on the cell type it is expressed in. Given the seemingly ubiquitous nature of its expression throughout the body in both neuronal and non-neuronal tissue, it is not surprising that numerous activities have been attributed to APP, but no single definitive function has been established. APP can affect neuronal survival [86], neurite outgrowth and synaptogenesis [16], cell adhesion [87], inhibition of coagulation factors [49, 88, 89], inhibition of platelet activation [90], and modulation of copper homeostasis [20].

The neuroprotective activity is associated with soluble APP (sAPP), which can protect cell cultures against death from glutamate or Aβ excitotoxicity, and glucose deficiency [86, 91]. This protective effect appears to occur by the lowering of intracellular calcium levels [86, 92]. The neuroprotective activity is mediated by sAPPα and not by sAPPβ as the sAPPα was approximately 100-fold more neuroprotective than sAPPβ [92]. Therefore, the active site is localized to the first 15 amino acids of the Aβ sequence as part of the carboxyl-terminus of sAPPα [86]. The neuroprotective activity of APP also occurred after intraventricular administration of either APP_{695} or APP_{751} in a transient ischemia animal model [93]. The in vivo relevance of this activity is supported by endogenous APP being upregulated after brain injury with strong immunoreactivity being present in both human and experimental models of head injury [43, 94, 95, 96].

APP could be a modulator of synaptogenesis as in both developing and mature neurons, APP is localized primarily to the neurites [97, 98]. In neuronal cultures, APP is predominantly found on cell surface adhesion patches of axons and dendrites

[99]. In the rat brain, the expression levels of APP are highest during the second postnatal week when extensive synaptogenesis occurs [100]. High levels of APP are expressed in the olfactory bulb, the only area of the brain where synaptogenesis continuously occurs in adults [100, 101]. The addition of APP to cell cultures enhances neurite outgrowth [102, 103] presumably via the N-terminal, heparin-binding domain [16, 104]. The interaction with heparin would allow a link between APP and the extracellular matrix. The neurite outgrowth promoting activity of APP varies in an isoform-specific manner with cell-surface expressed APP_{751} and APP_{770} being more active promoters of neurite outgrowth than APP_{695} [105]. This suggests regulation of APP alternative splicing would influence the cell adhesion activity of APP [105]. Moreover, the presenilins and APP are coexpressed and colocalize in the synaptic compartments. Therefore, the synaptogenic activity of APP could be regulated by presenilin-mediated processing [106].

APP is present in non-neuronal peripheral tissues and cells. The KPI-containing isoforms are abundantly expressed in platelets and are released upon platelet activation [48, 49, 107, 108]. The release of APP from platelets is modulated by protein kinase C, rather than by cyclooxygenase [109]. In contrast, Aβ release is independent of either cyclooxygenase or protein kinase C [109]. The KPI containing isoforms can inhibit a range of coagulation factors including IXa, X, and XIa (reviewed in [110]). However regions other than the KPI are necessary for maximal activity as non-KPI species are active. The sAPP can in turn inhibit platelet aggregation and secretion induced by ADP or adrenaline in vitro [90]. In addition, sAPP potently inhibited the activation of washed platelets by low-dose thrombin indicating that the activity does not require plasma cofactors. This occurs via a non-KPI-dependent mechanism as the active site was localized to the N-terminal cysteine rich region [90].

The cytoplasmic and transmembrane domains of APP are capable of complexing with and activating the trimeric G_o protein, a major GTP-binding protein in the brain [111, 112]. This interaction may be a contributing factor to the neurodegeneration of AD as G-protein–associated signaling pathways in AD brains are altered [113–116]. The AICD domain interacts with a number of adaptor proteins including Fe65, Jip1b, X11alpha (MINT1), Dab1, Dab2, Numb, and Tip60 [117]. The binding of these proteins to the AICD generates a transcriptionally active complex that is released from the membrane after ε- or γ-secretase cleavage of APP. The AICD complex translocates to the nucleus and regulates the transcription of APP, BACE, Tip60, GSK3beta, and KAI1 [37, 39, 118–120]. The physiological relevance of this activity is not clear but may reflect the adaptor proteins acting as linkers between APP and its target proteins. This pathway is analagous to NOTCH signaling, which involves the binding of NOTCH to its ligand. This induces a cleavage in the extracellular domain of NOTCH followed by a ε- or γ-secretase cleavage of the cytoplasmic domain (NICD), which then acts as a transcriptional activator. This suggests a functional relationship between APP and NOTCH processing.

3.5 Activities Associated with APLP2

The APLP2 gene is localized to human chromosome 11 [121] and has 71% similarity to APP [122]. APLP2 has two alternatively spliced exons, a KPI-domain exon and a exon equivalent to exon 15 [123], which in APLP2 is exon 14 that provides an chondroitin sulfate attachment site when it is spliced out. The promoter for APLP2, like that for APP, has the features of a housekeeping gene promoter [85].

The expression levels of APLP2, like those of APP, are high in brain, heart, and kidney and lower in the liver and thymus. The APLP2 expression pattern within various brain regions is also similar to that for APP, except APLP2 levels are greater in the thalamus [122]. Interestingly, APLP2 is increased in AD cerebellum samples compared with normal brain, whereas APP levels are decreased. It was proposed that this APLP2 expression is a compensatory response to the decreased APP levels [122, 124]. In contrast with APP, APLP2 is found in the small intestine and lung [122] and the APLP2 isoforms containing the KPI-domain are abundant in both neuronal and non-neuronal tissues [10]. An APLP2 orthologue has been identified in *Xenopus*, which is highly homologous to human APLP2 and contains an a KPI exon and an exon 14, which are alternatively spliced [125]. Similar to its mammalian orthologue, the *Xenopus* APLP2 is ubiquitously expressed.

APLP2 appears to be processed through the same secretory and proteolytic pathways as APP [126]. A number of functions have been attributed to APLP2 including a modulator of synaptogenesis [127], neurite outgrowth [128], and neuronal differentiation [129], which are comparable with the functions associated with APP. APLP2 is also localized to the sensory axons and glomeruli in the olfactory bulb. As olfactory sensory neurons are the only regenerating neuronal population in the adult CNS, the presence of APLP2 within them suggests it has a function in axonal growth or the establishment of synaptic connections [130]. This proposed function is supported by recombinant APLP2 stimulating neurite outgrowth on chick sympathetic neurons [128].

The APLP2 molecule may function within the extracellular matrix and assist in corneal epithelial wound healing as there is a marked increase in APLP2 mRNA and the KPI-contain chondroitin sulfate positive species in the basal epithelial cells that were actively migrating after injury [131]. In contrast, in a skin wound model, APLP2 expression was decreased whereas APP expression increased [132] indicating APLP2 has tissue-specific responses and effects. Retinoic acid can induce APLP2 expression in neuroblastoma cells, indicating it may be involved in neuronal differentiation [133]. Increased expression of APLP2 was also detected in Aβ-treated neuronal cultures, implying that APLP2 expression may be induced by Aβ [134, 135]. Interestingly, the APLP2 gene is the same as the *Cdebp* gene, which encodes a DNA-binding protein thought to be necessary for DNA replication or segregation [136]. This indicates APLP2 may bind DNA, which is an activity not reported for APP [137, 138].

3.6 Activities Associated with APLP1

The human APLP1 gene has a 64% similarity to the APP gene, is located on the long arm of chromosome 19, and consists of 17 exons [139]. APLP1 is not known to have any alternatively spliced transcripts.

Cell culture–based studies showed that APLP1, like APP and APLP2, can undergo both N- and O-linked glycosylation [140]. APLP1 can also be phosphorylated by PKC [141]. APLP1 has been identified in the perinuclear and Golgi regions, which resembles the subcellular distribution of APP [124]. Although limited, proteolysis of APLP1 occurs, resulting in the carboxy-terminal truncated peptide being secreted into the culture medium [140, 142]. The identification of APLP1 in human CSF suggests that its secretion from brain cells also takes place in vivo [143].

APLP1 has a more restricted expression pattern compared with that of APP and APLP2. It is primarily expressed in the CNS, with expression peaking during early embryo development, supporting a role for APLP1 in neurogenesis [124, 144]. The discovery of APLP1 expression in the cerebral cortex postsynaptic density of rats and humans suggests that like APP and APLP2, it has a role in synaptogenesis or synaptic maturation [145]. APLP1 may also be involved in neuronal differentiation [133].

3.7 APP-Family Knockout Mice

3.7.1 APP Knockout Mice

Mice homozygous for a deletion of the entire APP gene ($APP^{-/-}$) are viable and fertile but have reduced body weight, decreased locomotor activity, reduced forelimb grip strength, and reactive gliosis, particularly in the cortex and hippocampus [146]. Aged $APP^{-/-}$ mice display impaired learning abilities [147], cognitive deficits, and impaired long-term potentiation [148]. The mice also have decreased level of synaptic marker proteins at various ages, along with abnormal neuronal morphology and synaptic function [148, 149]. This supports the evidence that APP has an important role in maintaining synaptic function during aging [150]. The generation of APP knockout mice carrying a hypomorphic deletion of APP resulted in impaired spatial learning and increased agenesis of the corpus callosum [151]. They also exhibited reduced postnatal body weight and alterations in sensorimotor development [152]. Furthermore, when these mice with decreased APP expression were compared with the $APP^{-/-}$ mice, both strains were hypersensitive to seizure activity [153] and had reduced brain weight and reduced size of forebrain commissures [154].

Cultured neuronal cells derived from $APP^{-/-}$ mice indicated hippocampal neurons had a decrease in cell viability and neurite development [155]. Studies using cortical or cingulate gyral neurons from APP$^{-/-}$ mice found no differences in their survival or neurite length compared with wild-type mice, even in the presence of various neurotoxic insults including Aβ, glutamate, hydrogen peroxide, or glucose deficiency [134, 156, 157]. This indicates there are cell type–specific responses to APP expression. In contrast, APP$^{-/-}$ cortical cultures grown at low density exhibited less susceptibility to Aβ toxicity, suggesting that APP expression is required for Aβ toxicity [158]. These apparently conflicting results probably reflect experimental differences such as cell culture densities and suggest the response of neurons to APP can be influenced by their growth conditions. The APP$^{-/-}$ cortical neurons did exhibit a clear difference in viability when exposed to neurotoxic levels of copper. The APP$^{-/-}$ cortical neurons were significantly less susceptible to copper-mediated toxicity as compared with wild-type neurons [24]. This correlated with differences in lipid peroxidation between the different genotypes consistent with APP promoting copper reduction and toxicity via a redox-dependent mechanism.

The finding that APP$^{-/-}$ mice exhibit only mild deficits implies that the loss of functional APP is compensated by its paralogues APLP1 and APLP2. The APLPs share many structural similarities with APP and are distributed in a similar manner to APP within brain tissues, providing further evidence for a similarity of function [159].

3.7.2 APLP2 Knockout Mice

Two different lines of APLP2 knockout mice have been described with distinct phenotypes. One line of APLP2 knockout mouse contained an 11.35-kb deletion that removed exons 7 to 14 [138]. The heterozygous APLP2$^{-/+}$ mice developed normally, but in APLP2$^{-/-}$ homozygous mice, there was a clear effect on embryo development, which was arrested before the blastocyst stage [138]. This suggests that APLP2 is involved in the mitotic segregation of the genome and DNA replication. This APLP2$^{-/-}$ mutant mouse varies greatly from another APLP2$^{-/-}$ line that had no obvious abnormalities [160]. This phenotypic variation is most likely due to the size of the genomic deletion as the viable APLP2$^{-/-}$ mice lacked only the APLP2 promoter and exon 1 [160]. Therefore, the smaller deletion may leave sufficient DNA for embryological proteins and cofactors to interact with the DNA-binding regions of the APLP2 gene.

3.7.3 APLP1 Knockout Mice

Mice lacking APLP1 are viable and show a postnatal growth deficit as their only obvious abnormality [157].

3.7.4 APP, APLP1, and APLP2 Combined Knockouts

The lack of an obvious phenotype in the single knockout mice indicates that there may be a redundancy in gene function and that the APP and APLPs supplement for their functions. The generation of double and triple knockout mice has clarified the relationship between the APP-family members. More than 80% of APP and APLP2 double knockout mice die within days of birth [157, 160]. The surviving double knockouts have 20–30% reduced weight and show ataxia, spinning behavior, difficulty in righting, and a head tilt. Similarly, APLP1 and APLP2 double knockouts display postnatal lethality, but APP and APLP1 double knockouts are viable [160]. Together, these results suggest redundancy between APLP2 and both other family members. One allele of APLP2 is not sufficient for survival in APP$^{-/-}$ APLP1$^{-/-}$ APLP2$^{+/-}$ because these mice also die postnatally. Therefore, APLP2 is the most essential member of the family for viability with APP being least necessary.

The lethal double mutants appeared to have no discernible histopathological abnormalities in the brain or any other organ examined [157, 160]. However, there is a defect in the development of neuromuscular synapses [161]. The APP$^{-/-}$ APLP2$^{-/-}$ double knockouts displayed aberrant expression of the presynaptic vesicle protein Syn and a reduction in synaptic vesicle density and excessive nerve terminal sprouting. This resulted in defective neurotransmitter release and a high incidence of synaptic failure. Therefore, the expression of APP and APLP2 is necessary for functional neuromuscular junction formation. An effect by APP on the neuromuscular junction is observed with the *Drosophila* APP orthologue APPL that is expressed in all neuronal cell bodies [162] and modulates the neuromucular junction if overexpressed [163].

Cortical neuronal cultures from the various combined mutant mice showed unaltered survival rates under basal culture conditions or in the presence of glutamate and hydrogen peroxide excitotoxicity [157]. It is interesting that none of the single or combined knockout mice showed basal upregulation of the remaining family members [134, 157, 160].

3.8 Physiological Function of APP as a Cuproprotein

A substantial body of data supports the function of APP as a cuproprotein. APP has a copper-binding domain composed of histidine residues located in the N-terminal cysteine-rich region downstream of the growth factor–like domain [17]. A secondary copper-binding domain is generated in the Aβ peptide after it is proteolytically released from APP [164, 165]. However, it is unclear if the Aβ sequence binds Cu when the Aβ sequence is part of the APP molecule. Both APP and Aβ can strongly bind copper and reduce Cu(II) to Cu(I) in vitro [18, 165]. The first, and most definitive, demonstration of an in vivo physiological role for APP Cu binding came from $APP^{-/-}$ knockout mice studies. The absence of APP resulted in increased brain and liver Cu levels but no change in zinc or iron [20]. Moreover, the $APLP2^{-/-}$ knockout mice also had an increase in brain Cu levels but to a lesser extent than the $APP^{-/-}$ mice.

Subsequent studies in three different APP-transgenic mice models have confirmed a role for APP in modulating Cu homeostasis. The transgenic mice all displayed a decrease in brain Cu levels due to APP overexpression [21–23]. The knockout and transgenic data firmly establish the function of APP as a modulator of Cu homeostasis. This activity is particularly important as Cu homeostasis is a tightly regulated process in order to control copper's redox generating properties from causing damage. This results in the absence of free unbound intracellular copper and all the Cu is bound to proteins [166].

Cell-based studies have further elucidated the ability of APP to regulate Cu levels. Cultured primary cortical neurons or embryonic fibroblast cell lines from either APP and APLP2 knock out mice displayed increased Cu accumulation as the redundancy in the APP gene family was reduced. Conversely, primary cortical neurons from APP transgenic mice had lower levels of Cu [167]. Therefore, APP and to a lesser extent APLP2 are Cu-sensing proteins that regulate intracellular Cu levels. The overexpression of APP in the yeast *Pichia pastoris* has confirmed the histidines residues are important for APP and APLP2 mediated Cu transport [168]. It also established that APLP1 is inactive as a Cu transporter consistent with the sequence differences in its CuBD [25, 26] where it has a serine for histidine substitution at position 147.

The consequence of Cu binding to APP is to increase α-secretase cleavage of APP and a corresponding decrease in Aβ levels. Elevated Cu concentrations will reduce Aβ production and increase secretion of APP in a cell line transfected with human APP cDNA [27]. The physiological relevance of this effect was replicated in vivo where increasing brain Cu levels caused a decrease in Aβ production in APP transgenic mice [22, 23]. Moreover, modulating Cu levels in the APP23 mouse dramatically increased their survival.

The APP CuBD can also modulate Cu-mediated neurotoxicity in a species-dependent manner. $APP^{-/-}$ knockout primary cortical neurons are less susceptible to Cu-mediated toxicity as compared with wild-type neurons [24]. This correlated with APP reducing Cu(II) to Cu(I) as cell toxicity was associated with increased lipid peroxidation. This activity was localized to the APP N-terminal CuBD site as recombinant CuBD could potentiate Cu-mediated toxicity. This activity varies among the various APP paralogues and orthologues [25].

Although the CuBD sequence is similar among the paralogues and orthologues, there are sequence-dependent differences that profoundly affect the activity of this site. Conservation of the histidine residues corresponding with residues 147 and 151 of APP promoted Cu-mediated toxicity. The *Xenopus* APP and the human APLP2 CuBDs fell into this class. However, the *C. elegans* APL-1 CuBD has a tyrosine and lysine residues at positions 147 and 151, respectively, and it strongly protected against Cu-mediated neurotoxicity. Replacement of the histidines 147 and 151 with tyrosine and lysine residues conferred this neuroprotective Cu phenotype to human APP, APLP2, and *Xenopus* APP CuBD peptides.

Conversely, replacing the *C. elegans* tyrosine and lysine residues with histidines coverted it to a Cu-toxicity promoting sequence. Moreover, the toxic and protective CuBD phenotypes are associated with differences in Cu binding and reduction [26]. These studies identify a significant evolutionary change in the function of the CuBD in modulating Cu metabolism.

A possible in vivo molecular target for the APP:Cu or APLP2:Cu complexes has been identified as glypican-1 [169]. The glypican-1 molecule is a cell-surface proteoglycan that undergoes Cu-mediated degradation of its heparan sulfate chains. APP can bind glypican-1 with low nanomolar affinity, and this interaction inhibits APP-induced neurite outgrowth [170]. In a cell-free system, APP, but not APLP2, stimulates glypican-1 autodegradation in the presence of both Cu(II) and Zn(II), whereas the Cu(I) form of APP and the Cu(II) and Cu(I) forms of APLP2 inhibit autodegradation [169]. Primary cortical neurons and brain tissue from APP and APLP2 knockout mice had an increase in nitric oxide–catalyzed degradation of heparan sulfate compared with brain tissue and neurons from wild-type mice. Therefore, the rate of autoprocessing of glypican-1 is modulated by APP and APLP2 in neurons.

Importantly, these observations identified a functional relationship between the heparin/HS and copper-binding activities of the cysteine-rich region in APP and APLP2 in their modulation of the nitroxyl anion–catalyzed HS degradation in Gpc-1. Structural studies indicate this region is composed of a separate heparin-binding/growth factor domain [12] and a copper-binding domain [14] joined by a linker. The former domain should connect APP to the heparin sulfate in Gpc-1, and the second domain should be involved in modulating the Cu-dependent redox reactions required for NO-catalyzed HS degradation.

The three-dimensional structure of the human APP copper-binding domain (APP residues 124 to 189) has been determined by NMR spectroscopy [14]. It showed structural homology to copper chaperones, thus strongly supporting the in vivo data and suggesting that the APP copper-binding domain functions as a neuronal metal-transporter and/or metal-chaperone to modulate copper homeostasis. The Cu binding site had a distorted square planar arrangement toward a tetrahedral arrangement, which would favor Cu(I) binding. This is consistent with Cu(II) binding to the APP CuBD and promoting its reduction to Cu(I). However, the mechanism by which the APP and APLP2 function together in cellular copper homeostasis is unknown.

3.9 Conclusions

Significant progress has been made in elucidating both the structure and neuronal function of the APP and APLPs. The data points toward the APP-family acting upon the development of the synapse and presumably synaptic activity. The challenge is to ratify this with the other main effects associated with APP, in particular its upregulation after axonal injury and its copper-binding activity. Moreover, there is a need to understand how the alternative splicing and its processing via the α- and β-secretase pathways and the release of the AICD relates to its function. Finally, what is the relationship between the redundant and unique roles played by the different APP-family members? Determining the function of APP and relating this to its structure will provide a more complete understanding of the role of APP in AD and should provide information necessary for the development of therapeutic strategies.

References

1. Glenner GG, Wong CW. Alzheimer's disease: initial report of the purification and characterization of a novel cerebrovascular amyloid protein. Biochem Biophys Res Commun 1984;120:885-90.
2. Masters CL, Multhaup G, Simms G, et al. Neuronal origin of a cerebral amyloid: neurofibrillary tangles of Alzheimer's disease contain the same protein as the amyloid of plaque cores and blood vessels. EMBO J 1985;4:2757-63.
3. Masters CL, Simms G, Weinman NA, et al. Amyloid plaque core protein in Alzheimer's disease and Down syndrome. Proc Natl Acad Sci U S A 1985;82:4245-9.
4. Kang J, Lemaire H, Unterbeck A, et al. The precursor of Alzheimer's disease amyloid A4 protein resembles a cell-surface receptor. Nature 1987;325:733-6.
5. Goldgaber D, Lerman MI, McBride OW, et al. Characterization and chromosomal localization of a cDNA encoding brain amyloid of Alzheimer's disease. Science 1987;235:877-80.
6. Tanzi RE, Gusella JF, Watkins PC, et al. Amyloid β protein gene: cDNA, mRNA distribution and genetic linkage near the Alzheimer locus. Science 1987;235:880-4.

7. Yoshikai S, Sasaki H, Doh-ura K, et al. Genomic organization of the human-amyloid beta-protein precursor gene. Gene 1991;102:291-2.
8. Panegyres PK. The functions of the amyloid precursor protein gene. Rev Neurosci 2001;12:1-39.
9. Beyreuther K, Pollwein P, Multhaup G, et al. Regulation and expression of the Alzheimer's beta/A4 amyloid protein precursor in health, disease, and Down's syndrome. Ann N Y Acad Sci 1993;695:91-102.
10. Sandbrink R, Masters CL, Beyreuther K. βA4-amyloid protein precursor mRNA isoforms without exon 15 are ubiquitously expressed in rat tissues including brain, but not in neurons. J Biol Chem 1994;269:1510-7.
11. Shioi J, Pangalos MN, Ripellino JA, et al. The Alzheimer amyloid precursor proteoglycan (appican) is present in brain and is produced by astrocytes but not by neurons in primary neural cultures. J Biol Chem 1995;270:11839-44.
12. Rossjohn J, Cappai R, Feil SC, et al. Crystal structure of the N-terminal, growth factor-like domain of Alzheimer amyloid precursor protein. Nat Struct Biol 1999;6:327-31.
13. Dulubova I, Ho A, Huryeva I, et al. Three-dimensional structure of an independently folded extracellular domain of human amyloid-beta precursor protein. Biochemistry 2004;43:9583-8.
14. Barnham KJ, McKinstry WJ, Multhaup G, et al. Structure of the Alzheimer's disease amyloid precursor protein copper binding domain. A regulator of neuronal copper homeostasis. J Biol Chem 2003; 278:17401-7.
15. Wang Y, Ha Y. The x-ray structure of an antiparallel dimer of the human amyloid precursor protein E2 domain. Mol Cell 2004;15:343-53.
16. Small DH, Nurcombe V, Reed G, et al. A heparin-binding domain in the amyloid protein precursor of Alzheimer's disease is involved in the regulation of neurite outgrowth. J Neurosci 1994;14:2117-27.
17. Hesse L, Beher D, Masters CL, Multhaup G. The beta A4 amyloid precursor protein binding to copper. FEBS Lett 1994;349:109-16.
18. Multhaup G, Schlicksupp A, Hesse L, et al. The amyloid precursor protein of Alzheimer's disease in the reduction of copper(II) to copper(I). Science 1996;271:1406-9.
19. Bush AI, Multhaup G, Moir RD, et al. A novel zinc(II) binding site modulates the function of the beta A4 amyloid protein precursor of Alzheimer's disease. J Biol Chem 1993;268:16109-12.
20. White AR, Reyes R, Mercer JF, et al. Copper levels are increased in the cerebral cortex and liver of APP and APLP2 knockout mice. Brain Res 1999;842:439-44.
21. Maynard CJ, Cappai R, Volitakis I, et al. Overexpression of Alzheimer's disease amyloid-beta opposes the age-dependent elevations of brain copper and iron. J Biol Chem 2002;277:44670-6.
22. Bayer TA, Schafer S, Simons A, et al. Dietary Cu stabilizes brain superoxide dismutase 1 activity and reduces amyloid Abeta production in APP23 transgenic mice. Proc Natl Acad Sci U S A 2003;100: 14187-92.
23. Phinney AL, Drisaldi B, Schmidt SD, et al. In vivo reduction of amyloid-beta by a mutant copper transporter. Proc Natl Acad Sci U S A 2003;100:14193-8.
24. White AR, Multhaup G, Maher F, et al. The Alzheimer's disease amyloid precursor protein modulates copper-induced toxicity and oxidative stress in primary neuronal cultures. J Neurosci 1999;19:9170-9.
25. White AR, Multhaup G, Galatis D, et al. Contrasting, species-dependent modulation of copper-mediated neurotoxicity by the Alzheimer's disease amyloid precursor protein. J Neurosci 2002;22:365-76.
26. Simons A, Ruppert T, Schmidt C, et al. Evidence for a copper-binding superfamily of the amyloid precursor protein. Biochemistry 2002;41:9310-20.
27. Borchardt T, Camakaris J, Cappai R, et al. Copper inhibits beta-amyloid production and stimulates the non-amyloidogenic pathway of amyloid-precursor-protein secretion. Biochem J 1999;344 Pt 2:461-7.
28. Borchardt T, Schmidt C, Camarkis J, et al. Differential effects of zinc on amyloid precursor protein (APP) processing in copper-resistant variants of cultured Chinese hamster ovary cells. Cell Mol Biol 2000;46:785-95.
29. Donnelly RJ, Rasool CG, Bartus R, et al. Multiple forms of beta-amyloid peptide precursor RNAs in a single cell type. Neurobiol Aging 1988;9:333-8.
30. Weidemann A, König G, Bunke D, et al. Identification, biogenesis, and localization of precursors of Alzheimer's disease A4 amyloid protein. Cell 1989; 57:115-26.
31. Kitaguchi N, Takahashi Y, Tokushima Y, et al. Novel precursor of Alzheimer's disease amyloid protein shows protease inhibitory activity. Nature 1988;331:530-2.
32. Oltersdorf T, Fritz LC, Schenk DB, et al. The secreted form of the Alzheimer's amyloid precursor protein with the Kunitz domain is protease nexin-II. Nature 1989;341:144-7.
33. Ponte P, Gonzalez-DeWhitt P, Schilling J, et al. A new A4 amyloid mRNA contains a domain homologous to serine proteinase inhibitors. Nature 1988;331:525-7.
34. Tanzi RE, McClatchey AI, Lamperti ED, et al. Protease inhibitor domain encoded by an amyloid protein precursor mRNA associated with Alzheimer's disease. Nature 1988;331:528-30.

35. Dyrks T, Weidemann A, Multhaup G, et al. Identification, transmembrane orientation and biogenesis of the amyloid A4 precursor of Alzheimer's disease. EMBO J 1988;7:949-57.
36. Weidemann A, Eggert S, Reinhard FB, et al. A novel epsilon-cleavage within the transmembrane domain of the Alzheimer amyloid precursor protein demonstrates homology with Notch processing. Biochemistry 2002;41:2825-35.
37. Cao X, Sudhof TC. Dissection of amyloid-beta precursor protein-dependent transcriptional transactivation. J Biol Chem 2004;279:24601-11.
38. Cao X, Sudhof TC. A transcriptionally active complex of APP with Fe65 and histone acetyltransferase Tip60. Science 2001;293:115-20.
39. von Rotz RC, Kohli BM, Bosset J, et al. The APP intracellular domain forms nuclear multiprotein complexes and regulates the transcription of its own precursor. J Cell Sci 2004;117:4435-48.
40. Selkoe DJ, Podlisny MB, Joachim CL, et al. Beta-amyloid precursor protein of Alzheimer's disease occurs as 110- to 135-kilodalton membrane-associated proteins in neural and nonneural tissues. Proc Natl Acad Sci U S A 1988;85:7341-5.
41. Golde TE, Estus S, Usiak M, et al. Expression of beta amyloid protein precursor mRNAs: recognition of a novel alternatively spliced form and quantitation in Alzheimer's disease using PCR. Neuron 1990; 4:253-67.
42. Palmert MR, Podlisny MB, Witker DS, et al. The β-amyloid protein precursor of Alzheimer's disease has soluble derivatives found in human brain and cerebrospinal fluid. Proc Natl Acad Sci U S A 1989; 86:6338-42.
43. Olsson A, Csajbok L, Ost M, et al. Marked increase of beta-amyloid(1-42) and amyloid precursor protein in ventricular cerebrospinal fluid after severe traumatic brain injury. J Neurol 2004;251:870-6.
44. Xie Y, Yao Z, Chai H, et al. Potential roles of Alzheimer precursor protein A4 and beta-amyloid in survival and function of aged spinal motor neurons after axonal injury. J Neurosci Res 2003;73:557-64.
45. Abe K, Tanzi RE, Kogure K. Selective induction of Kunitz-type protease inhibitor domain-containing amyloid precursor protein mRNA after persistent focal ischemia in rat cerebral cortex. Neurosci Lett 1991;125:172-4.
46. Forloni G, Demicheli F, Giorgi S, et al. Expression of amyloid precursor protein mRNAs in endothelial, neuronal and glial cells: modulation by interleukin-1. Mol Brain Res 1992;16:128-34.
47. Monning U, Konig G, Banati RB, et al. Alzheimer beta A4-amyloid protein precursor in immunocompetent cells. J Biol Chem 1992;267:23950-6.
48. Li QX, Berndt MC, Bush AI, et al. Membrane-associated forms of the βA4 amyloid protein precursor of Alzheimer's disease in human platelet and brain: surface expression on the activated human platelet. Blood 1994;84:133-42.
49. Van Nostrand WE, Schmaier AH, Farrow JS, et al. Protease nexin-II (amyloid β-protein precursor): a platelet α-granule protein. Science 1990;248:745-8.
50. Moir RD, Lynch T, Bush AI, et al. Relative increase in Alzheimer's disease of soluble forms of cerebral a-β amyloid protein precursor containing the kunitz protease inhibitory domain. J Biol Chem 1998;273: 5013-9.
51. Ho L, Fukuchi K, Younkin SG. The alternatively spliced Kunitz protease inhibitor domain alters amyloid beta protein precursor processing and amyloid beta protein production in cultured cells. J Biol Chem 1996;271:30929-34.
52. Barrachina M, Dalfo E, Puig B, et al. Amyloid-beta deposition in the cerebral cortex in Dementia with Lewy bodies is accompanied by a relative increase in AbetaPP mRNA isoforms containing the Kunitz protease inhibitor. Neurochem Int 2005;46:253-60.
53. Konig G, Monning U, Czech C, et al. Identification and differential expression of a novel alternative splice isoform of the beta A4 amyloid precursor protein (APP) mRNA in leukocytes and brain microglial cells. J Biol Chem 1992;267:10804-9.
54. Pangalos MN, Efthimiopoulos S, Shioi J, et al. The chondroitin sulfate attachment site of appican is formed by splicing out exon 15 of the amyloid precursor gene. J Biol Chem 1995;270:10388-91.
55. Morin PJ, Medina M, Semenov M, et al. Wnt-1 expression in PC12 cells induces exon 15 deletion and expression of L-APP. Neurobiol Dis 2004;16:59-67.
56. De Strooper B, Annaert W. Where Notch and Wnt signaling meet. The presenilin hub. J Cell Biol 2001;152:F17-20.
57. Coulson EJ, Paliga K, Beyreuther K, Masters CL. What the evolution of the amyloid protein precursor supergene family tells us about its function. Neurochem Int 2000;36:175-84.
58. Daigle I, Li C. apl-1, a *Caenorhabditis elegans* gene encoding a protein related to the human β-amyloid protein precursor. Proc Natl Acad Sci U S A 1993; 90:12045-9.
59. Iijima K, Lee DS, Okutsu J, et al. cDNA isolation of Alzheimer's amyloid precursor protein from cholinergic nerve terminals of the electric organ of the electric ray. Biochem J 1998;330:29-33.
60. Okado H, Okamoto H. A *Xenopus* homologue of the human β-amyloid precursor protein: developmental regulation of its gene expression. Biochem Biophys Res Commun 1992;189:1561-8.

61. Rosen DR, Martin-Morris L, Luo LQ, White K. A Drosophila gene encoding a protein resembling the human beta-amyloid protein precursor. Proc Natl Acad Sci U S A 1989;86:2478-82.
62. De Strooper B, Annaert W. Proteolytic processing and cell biological functions of the amyloid precursor protein. J Cell Sci 2000;113:1857-70.
63. Anderson JP, Chen Y, Kim KS, Robakis NK. An alternative secretase cleavage produces soluble Alzheimer amyloid precursor protein containing a potentially amyloidogenic sequence. J Neurochem 1992;59:2328-31.
64. LeBlanc AC, Papadopoulos M, Bélair C, et al. Processing of amyloid precursor protein in human primary neuron and astrocyte cultures. J Neurochem 1997;68:1183-90.
65. Ikezu T, Trapp BD, Song KS, et al. Caveolae, plasma membrane microdomains for alpha-secretase-mediated processing of the amyloid precursor protein. J Biol Chem 1998;273:10485-95.
66. Buxbaum JD, Liu KN, Luo YX, et al. Evidence that tumor necrosis factor alpha converting enzyme is involved in regulated alpha-secretase cleavage of the Alzheimer amyloid protein precursor. J Biol Chem 1998;273:27765-7.
67. Lammich S, Kojro E, Postina R, et al. Constitutive and regulated α-secretase cleavage of Alzheimer's amyloid precursor protein by a disintegrin metalloprotease. Proc Natl Acad Sci U S A 1999;96:3922-7.
68. Asai M, Hattori C, Szabo B, et al. Putative function of ADAM9, ADAM10, and ADAM17 as APP alphasecretase. Biochem Biophys Res Commun 2003; 301:231-5.
69. Allinson TM, Parkin ET, Turner AJ, Hooper NM. ADAMs family members as amyloid precursor protein alpha-secretases. J Neurosci Res 2003;74:342-52.
70. De Strooper B, Umans L, Van Leuven F, Van Den Berghe H. Study of the synthesis and secretion of normal and artificial mutants of murine amyloid precursor protein (APP): cleavage of APP occurs in a late compartment of the default secretion pathway. J Cell Biol 1993;121:295-304.
71. Sambamurti K, Shioi J, Anderson JP, et al. Evidence for intracellular cleavage of the Alzheimer's amyloid precursor in PC12 cells. J Neurosci Res 1992;33:319-29.
72. Haass C, Hung AY, Schlossmacher MG, et al. β-amyloid peptide and a 3-kDa fragment are derived by distinct cellular mechanisms. J Biol Chem 1993;268:3021-4.
73. Hussain I, Powell D, Howlett DR, et al. Identification of a novel aspartic protease (Asp 2) as β-secretase. Mol Cell Neurosci 1999;14:419-27.
74. Vassar R, Bennett BD, Babu-Khan S, et al. Beta-secretase cleavage of Alzheimer's amyloid precursor protein by the transmembrane aspartic protease BACE. Science 1999;286:735-41.
75. Sinha S, Anderson JP, Barbour R, et al. Purification and cloning of amyloid precursor protein β-secretase from human brain. Nature 1999;402:537-40.
76. Yan R, Bienkowski MJ, Shuck ME, et al. Membrane-anchored aspartyl protease with Alzheimer's disease β-secretase activity. Nature 1999;402:533-7.
77. Busciglio J, Gabuzda DH, Matsudaira P, Yankner BA. Generation of beta-amyloid in the secretory pathway in neuronal and nonneuronal cells. Proc Natl Acad Sci U S A 1993;90:2092-6.
78. Seubert P, Vigo-Pelfrey C, Esch F, et al. Isolation and quantification of soluble Alzheimer's β-peptide from biological fluids. Nature 1992;359:325-7.
79. Estus S, Golde TE, Kunishita T, et al. Potentially amyloidogenic, carboxyl-terminal derivatives of the amyloid protein precursor. Science 1992;255:726-8.
80. Golde TE, Estus S, Younkin LH, et al. Processing of the amyloid protein precursor to potentially amyloidogenic derivatives. Science 1992;255:728-30.
81. Bennett BD, Babu-Khan S, Loeloff R, et al. Expression analysis of BACE2 in brain and peripheral tissues. J Biol Chem 2000;275:20647-51.
82. Wen Y, Onyewuchi O, Yang S, et al. Increased beta-secretase activity and expression in rats following transient cerebral ischemia. Brain Res 2004;1009:1-8.
83. Haass C. Take five—BACE and the gamma-secretase quartet conduct Alzheimer's amyloid b-peptide generation. EMBO J 2004;11:483-8.
84. Pollwein P, Masters CL, Beyreuther K. The expression of the amyloid precursor protein (APP) is regulated by two GC-elements in the promoter. Nucleic Acids Res 1992;20:63-8.
85. von Koch CS, Lahiri DK, Mammen AL, et al. The mouse *APLP2* gene -chromosomal localization and promoter characterization. J Biol Chem 1995;270: 25475-80.
86. Mattson MP, Cheng B, Culwell AR, et al. Evidence for excitoprotective and intraneuronal calcium-regulating roles for secreted forms of the beta-amyloid precursor protein. Neuron 1993;10:243-54.
87. Schubert D, Jin LW, Saitoh T, Cole G. The regulation of amyloid β protein precursor secretion and its modulatory role in cell adhesion. Neuron 1989;3:689-94.
88. Van Nostrand WE. Zinc (II) selectively enhances the inhibition of coagulation factor XIa by protease nexin-2/amyloid beta-protein precursor. Thromb Res 1995;78:43-53.
89. Smith RP, Higuchi DA, Broze GJ, Jr. Platelet coagulation factor XI_a-inhibitor, a form of Alzheimer amyloid precursor protein. Science 1990;248:1126-8.
90. Henry A, Li QX, Galatis D, et al. Inhibition of platelet activation by the Alzheimer's disease

amyloid precursor protein. Br J Haematol 1998;103:402-15.
91. Schubert D, Behl C. The expression of amyloid beta protein precursor protects nerve cells from beta-amyloid and glutamate toxicity and alters their interaction with the extracellular matrix. Brain Res 1993; 629:275-82.
92. Furukawa K, Sopher BL, Rydel RE, et al. Increased activity-regulating and neuroprotective efficacy of alpha-secretase-derived secreted amyloid precursor protein conferred by a C-terminal heparin-binding domain. J Neurochem 1996;67:1882-96.
93. Smith-Swintosky VL, Pettigrew LC, Craddock SD, et al. Secreted forms of beta-amyloid precursor protein protect against ischemic brain injury. J Neurochem 1994;63:781-4.
94. McKenzie JE, Gentleman SM, Roberts GW, et al. Increased numbers of βAPP-immunoreactive neurones in the entorhinal cortex after head injury. Neuroreport 1994;6:161-4.
95. Nakamura Y, Takeda M, Niigawa H, et al. Amyloid β-protein precursor deposition in rat hippocampus lesioned by ibotenic acid injection. Neurosci Lett 1992;136:95-8.
96. Ciallella JR, Ikonomovic MD, Paljug WR, et al. Changes in expression of amyloid precursor protein and interleukin-1beta after experimental traumatic brain injury in rats. J Neurotrauma 2002;19:1555-67.
97. Card JP, Meade RP, Davis LG. Immunocytochemical localization of the precursor protein for β-amyloid in the rat central nervous system. Neuron 1988;1:835-46.
98. Masliah E, Mallory M, Ge N, Saitoh T. Amyloid precursor protein is localized in growing neurites of neonatal rat brain. Brain Res 1992;593:323-8.
99. Storey E, Beyreuther K, Masters CL. Alzheimer's disease amyloid precursor protein on the surface of cortical neurons in primary culture co-localizes with adhesion patch components. Brain Res 1996;735:217-31.
100. Löffler J, Huber G. β-Amyloid precursor protein isoforms in various rat brain regions and during brain development. J Neurochem 1992;59:1316-24.
101. Clarris HJ, Key B, Beyreuther K, et al. Expression of the amyloid protein precursor of Alzheimer's disease in the developing rat olfactory system. Brain Res Dev Brain Res 1995;88:87-95.
102. Koo EH, Park L, Selkoe DJ. Amyloid β-protein as a substrate interacts with extracellular matrix to promote neurite outgrowth. Proc Natl Acad Sci U S A 1993;90:4748-52.
103. Milward E, Papadopoulos R, Fuller SJ, et al. The amyloid protein precursor of Alzheimer's disease is a mediator of the effects of nerve growth factor on neurite outgrowth. Neuron 1992;9:129-37.
104. Mok SS, Sberna G, Heffernan D, et al. Expression and analysis of heparin-binding regions of the amyloid precursor protein of Alzheimer's disease. FEBS Lett 1997;415:303-7.
105. Qiu WQ, Ferreira A, Miller C, et al. J. Cell-surface beta-amyloid precursor protein stimulates neurite outgrowth of hippocampal neurons in an isoform-dependent manner. J Neurosci 1995;15:2157-67.
106. Ribaut-Barassin C, Dupont JL, Haeberle AM, et al. Alzheimer's disease proteins in cerebellar and hippocampal synapses during postnatal development and aging of the rat. Neuroscience 2003;120: 405-23.
107. Bush AI, Martins RN, Rumble B, et al. The amyloid precursor protein of Alzheimer's disease is released by human platelets. J Biol Chem 1990;265:15977-83.
108. Evin G, Zhu A, Holsinger RM, et al. Proteolytic processing of the Alzheimer's disease amyloid precursor protein in brain and platelets. J Neurosci Res 2003;74:386-92.
109. Skovronsky DM, Lee VM, Pratico D. Amyloid precursor protein and amyloid beta peptide in human platelets. Role of cyclooxygenase and protein kinase C. J Biol Chem 2001;276:17036-43.
110. Storey E, Cappai R. The amyloid precursor protein of Alzheimer's disease and the Abeta peptide. Neuropathol Appl Neurobiol 1999;25:81-97.
111. Nishimoto I, Okamoto T, Matsuura Y, et al. Alzheimer amyloid protein precursor complexes with brain GTP-binding protein G. Nature 1993; 362:75-9.
112. Okamoto T, Takeda S, Murayama Y, et al. Ligand-dependent G protein coupling function of amyloid transmembrane precursor. J Biol Chem 1995;270: 4205-8.
113. Giambarella U, Yamatsuji T, Okamoto T, et al. G protein betagamma complex-mediated apoptosis by familial Alzheimer's disease mutant of APP. EMBO J 1997;16:4897-907.
114. Mahlapuu R, Viht K, Balaspiri L, et al. Amyloid precursor protein carboxy-terminal fragments modulate G-proteins and adenylate cyclase activity in Alzheimer's disease brain. Brain Res Mol Brain Res 2003;117:73-82.
115. Yamatsuji T, Matsui T, Okamoto T, et al. G protein-mediated neuronal DNA fragmentation induced by familial Alzheimer's disease-associated mutants of APP. Science 1996;272:1349-52.
116. Garcia-Jimenez A, Cowburn RF, Ohm TG, et al. Loss of stimulatory effect of guanosine triphosphate on [(35)S]GTPgammaS binding correlates with Alzheimer's disease neurofibrillary pathology in entorhinal cortex and CA1 hippocampal subfield. J Neurosci Res 2002;67:388-98.

117. Kerr ML, Small DH. Cytoplasmic domain of the beta-amyloid protein precursor of Alzheimer's disease: function, regulation of proteolysis, and implications for drug development. J Neurosci Res 2005;80:151-9.
118. Kimberly WT, Zheng JB, Guenette SY, Selkoe DJ. The intracellular domain of the beta-amyloid precursor protein is stabilized by Fe65 and translocates to the nucleus in a notch-like manner. J Biol Chem 2001;276:40288-92.
119. Perkinton MS, Standen CL, Lau KF, et al. The c-Abl tyrosine kinase phosphorylates the Fe65 adaptor protein to stimulate Fe65/amyloid precursor protein nuclear signaling. J Biol Chem 2004;279: 22084-91.
120. Pietrzik CU, Yoon IS, Jaeger S, et al. FE65 constitutes the functional link between the low-density lipoprotein receptor-related protein and the amyloid precursor protein. J Neurosci 2004;24:4259-65.
121. von der Kammer H, Loffler C, Hanes J, et al. The gene for the amyloid precursor-like protein APLP2 is assigned to human chromosome 11q23-q25. Genomics 1994;10:308-11.
122. Wasco W, Gurubhagavatula S, Paradis MD, et al. Isolation and characterization of APLP2 encoding a homologue of the Alzheimer's associated amyloid β protein precursor. Nat Genet 1993;5:95-100.
123. Sandbrink R, Masters CL, Beyreuther K. APP gene family. Alternative splicing generates functionally related isoforms. Ann N Y Acad Sci 1996;777:281-7.
124. Wasco W, Bupp K, Magendantz M, et al. Identification of a mouse brain cDNA that encodes a protein related to the Alzheimer's disease-associated amyloid β protein precursor. Proc Natl Acad Sci U S A 1992;89:10758-62.
125. Collin RWJ, van Strein D, Leunissen JAM, Marten GJM. Identification and expression of the first non-mammalian amyloid-β precursor-like protein APLP2 in the amphibian Xenopus laevis. Eur J Biochem 2004;271:1906–12.
126. Slunt HH, Thinakaran G, Von Koch C, et al. Expression of a ubiquitous, cross-reactive homologue of the mouse β-amyloid precursor protein (APP). J Biol Chem 1994;269:2637-44.
127. Thinakaran G, Kitt CA, Roskams AJ, et al. Distribution of an APP homolog, APLP2, in the mouse olfactory system: a potential role for APLP2 in axogenesis. J Neurosci 1995;15:6314-26.
128. Cappai R, Mok SS, Galatis D, et al. Recombinant human amyloid precursor-like protein 2 (APLP2) expressed in the yeast Pichia pastoris can stimulate neurite outgrowth. FEBS Lett 1999;442:95-8.
129. Holback S, Adlerz L, Iverfeldt K. Increased processing of APLP2 and APP with concomitant formation of APP intracellular domains in BDNF and retinoic acid-differentiated human neuroblastoma cells. J Neurochem 2005;95:1059-68.
130. Thinakaran G, Kitt CA, Roskams AJI, et al. Distribution of an APP homolog, APLP2, in the mouse olfactory system: a potential role for APLP2 in axogenesis. J Neurosci 1995;15:6314-26.
131. Guo J, Thinakaran G, Guo Y, et al. A role for amyloid precursor-like protein 2 in corneal epithelial wound healing. Invest Ophthalmol Vis Sci 1998; 39:292-300.
132. Kummer C, Wehner C, Quast T, et al. Expression and potential function of amyloid precursor proteins during cutaneous wound repair. Exp Cell Res 2002; 280:222–32.
133. Beckman M, Iverfeldt K. Increased gene expression of β-amyloid precursor protein and its homologues APLP1 and APLP2 in human neuroblastoma cells in response to retinoic acid. Neurosci Lett 1997; 221:73-6.
134. White AR, Zheng H, Galatis D, et al. Survival of cultured neurons from amyloid precursor protein knock-out mice against Alzheimer's amyloid-beta toxicity and oxidative stress. J Neurosci 1998; 18:6207-17.
135. White AR, Maher F, Brazier MW, et al. Diverse fibrillar peptides directly bind the Alzheimer's amyloid precursor protein and amyloid precursor-like protein 2 resulting in cellular accumulation. Brain Res 2003;966:231-44.
136. Hanes J, von der Kammer H, Kristjansson GI, Scheit KH. The complete cDNA coding sequence for the mouse CDEI binding protein. Biochim Biophys Acta 1993;1216:154-6.
137. Blangy A, Vidal F, Cuzin F, et al. CDEBP, a site-specific DNA-binding protein of the 'APP-like' family, is required during the early development of the mouse. J Cell Sci 1995;108:675-83.
138. Rassoulzadegan M, Yang YH, Cuzin F. APLP2, a member of the Alzheimer precursor protein family, is required for correct genomic segregation in dividing mouse cells. EMBO J 1998;17:4647-56.
139. Wasco W, Brook JD, Tanzi RE. The amyloid precursor-like protein (APLP) gene maps to the long arm of human chromosome 19. Genomics 1993; 15:237-9.
140. Lyckman AW, Confaloni AM, Thinakaran G, et al. Post-translational processing and turnover kinetics of presynaptically targeted amyloid precursor superfamily proteins in the central nervous system. J Biol Chem 1998;273:11100-6.
141. Suzuki T, Ando K, Isohara T, et al. Phosphorylation of Alzheimer β-amyloid precursor-like proteins. Biochemistry 1997;36:4643-9.

142. Scheinfeld MH, Ghersi E, Laky K, et al. Processing of β-amyloid precursor-like protein-1 and -2 by γ-secretase regulates transcription. J Biol Chem 2002;277:44195-201.
143. Paliga K, Peraus G, Kreger S, et al. Human amyloid precursor-like protein 1—cDNA cloning, ectopic expression in COS-7 cells and identification of soluble forms in the cerebrospinal fluid. Eur J Biochem 1997;250:354-63.
144. Lorent K, Overbergh L, Moechars D, et al. Expression in mouse embryos and in adult mouse brain of three members of the amyloid precursor protein family, of the alpha-2-macroglobulin receptor/low density lipoprotein receptor-related protein and of its ligands apolipoprotein E, lipoprotein lipase, alpha-2-macroglobulin and the 40,000 molecular weight receptor-associated protein. Neuroscience 1995;65:1009-25.
145. Kim TW, Wu K, Xu JL, et al. Selective localization of amyloid precursor-like protein 1 in the cerebral cortex postsynaptic density. Mol Brain Res 1995; 32:36-44.
146. Zheng H, Jiang M, Trumbauer ME, et al. beta-Amyloid precursor protein-deficient mice show reactive gliosis and decreased locomotor activity. Cell 1995;81:525-31.
147. Phinney AL, Calhoun ME, Wolfer DP, et al. No hippocampal neuron or synaptic bouton loss in learning-impaired aged beta-amyloid precursor protein-null mice. Neuroscience 1999;90:1207-16.
148. Dawson GR, Seabrook GR, Zheng H, et al. Age-related cognitive deficits, impaired long-term potentiation and reduction in synaptic marker density in mice lacking the β-Amyloid precursor protein. Neuroscience 1999;90:1-13.
149. Seabrook GR, Smith DW, Bowery BJ, et al. Mechanisms contributing to the deficits in hippocampal synaptic plasticity in mice lacking amyloid precursor protein. Neuropharmacology 1999;38:349-59.
150. Stephan A, Davis S, Salin H, et al. Age-dependent differential regulation of genes encoding APP and alpha-synuclein in hippocampal synaptic plasticity. Hippocampus 2002;12:55-62.
151. Muller U, Cristina N, Li ZW, et al. Behavioral and anatomical deficits in mice homozygous for a modified beta-amyloid precursor protein gene. Cell 1994;79:755-65.
152. Tremml P, Lipp HP, Müller U, et al. Neurobehavioral development, adult openfield exploration and swimming navigation learning in mice with a modified β-amyloid precursor protein gene. Behav Brain Res 1998;95:65-76.
153. Steinbach JP, Muller U, Leist M, et al. Hypersensitivity to seizures in beta-amyloid precursor protein deficient mice. Cell Death Differ 1998; 5:858-66.
154. Magara F, Müller U, Li ZW, et al. Genetic background changes the pattern of forebrain commissure defects in transgenic mice underexpressing the β-amyloid-precursor protein. Proc Natl Acad Sci U S A 1999;96:4656-61.
155. Perez RG, Zheng H, Van der Ploeg LHT, Koo EH. The β-amyloid precursor protein of Alzheimer's disease enhances neuron viability and modulates neuronal polarity. J Neurosci 1997;17:9407-14.
156. Harper SJ, Bilsland JG, Shearman MS, et al. Mouse cortical neurones lacking APP show normal neurite outgrowth and survival responses in vitro. Neuroreport 1998;9:3053-7.
157. Heber S, Herms J, Gajic V, et al. Mice with combined gene knock-outs reveal essential and partially redundant functions of amyloid precursor protein family members. J Neurosci 2000;20:7951-63.
158. Lorenzo A, Yuan M, Zhang Z, et al. Amyloid beta interacts with the amyloid precursor protein: a potential toxic mechanism in Alzheimer's disease. Nat Neurosci 2000;3:460-4.
159. McNamara MJ, Ruff CT, Wasco W, et al. Immunohistochemical and in situ analysis of amyloid precursor-like protein-1 and amyloid precursor-like protein-2 expression in Alzheimer's disease and aged control brains. Brain Res 1998;804:45-51.
160. von Koch CS, Zheng H, Chen H, et al. Generation of APLP2 KO mice and early postnatal lethality in APLP2/APP double KO mice. Neurobiol Aging 1997;18:661-9.
161. Wang P, Yang G, Mosier DR, et al. Defective neuromuscular synapses in mice lacking amyloid precursor protein (APP) and APP-Like protein 2. J Neurosci 2005;25:1219-25.
162. Torroja L, Luo L, White K. APPL, the Drosophila member of the APP-family, exhibits differential trafficking and processing in CNS neurons. J Neurosci 1996;16:4638-50.
163. Torroja L, Packard M, Gorczyca M, White K, Budnik V. The Drosophila beta-amyloid precursor protein homolog promotes synapse differentiation at the neuromuscular junction. J Neurosci 1999; 19:7793-803.
164. Huang X, Atwood CS, Hartshorn MA, et al. The Aβ peptide of Alzheimer's disease directly produces hydrogen peroxide through metal ion reduction. Biochemistry 1999;38:7609-16.
165. Huang X, Cuajungco MP, Atwood CS, et al. Cu(II) potentiation of Alzheimer abeta neurotoxicity. Correlation with cell-free hydrogen peroxide production and metal reduction. J Biol Chem 1999; 274:37111-6.

166. Rae TD, Schmidt PJ, Pufahl RA, et al. Undetectable intracellular free copper: the requirement of a copper chaperone for superoxide dismutase. Science 1999;284:805-8.
167. Bellingham SA, Ciccotosto GD, Needham BE, et al. Gene knockout of amyloid precursor protein and amyloid precursor-like protein-2 increases cellular copper levels in primary mouse cortical neurons. J Neurochem 2004;91:423-8.
168. Treiber C, Simons A, Strauss M, et al. Clioquinol mediates copper uptake and counteracts copper efflux activities of the amyloid precursor protein of Alzheimer's disease. J Biol Chem 2004;279:51958-64.
169. Cappai R, Cheng F, Ciccotosto GD, et al. The amyloid precursor protein (APP) of Alzheimer's disease and its paralog, APLP2, modulate the Cu/Zn-nitric oxide-catalyzed degradation of glypican-1 heparan sulfate in vivo. J Biol Chem 2005;280:13913-20.
170. Williamson TG, Mok SS, Henry A, et al. Secreted glypican binds to the amyloid precursor protein of Alzheimer's disease (APP) and inhibits APP-induced neurite outgrowth. J Biol Chem 1996;271:31215-21.

4 The Involvement of Aβ in the Neuroinflammatory Response

Piet Eikelenboom, Willem A. van Gool, Annemieke J.M. Rozemuller, Wiep Scheper, Rob Veerhuis, and Jeroen J.M. Hoozemans

4.1 Introduction

In the same year as Alzheimer described the case of Auguste D. as a peculiar disease of the cerebral cortex, Fischer published his classic paper about miliary plaque formation in a large number of brains from patients with senile dementia [1]. In this paper and a following one from 1910, Fischer stated that plaque formation is the result of the deposition of a peculiar foreign substance in the cortex that induces a regenerative response of the surrounding nerve fibers [2]. He described spindle-shaped thickening of nerve fibers terminating with club forms in the corona of plaques (Fig. 4.1). These altered nerve fibers were considered as axonal sprouting, and the terminal club forms showed a strong similarity with the club-shaped buddings of axons found in developing nerve fibers and after transections of peripheral nerves as described by Cajal some years earlier. According to Fischer, the crucial step of the plaque formation is the deposition of a foreign substance that provokes a local inflammatory response step followed by a regenerative response of the surrounding nerve fibers. However, Fischer could not find morphological characteristics of an inflammatory process around the plaques after extensive histopathological observations including complement binding studies. The only tissue reaction appeared to be an overgrowth of club-formed neurites.

In the last quarter of the 20th century, the composition of plaques was elucidated on a molecular level using various new techniques. In 1984, Glenner and Wong identified the major component of the amyloid deposits in Alzheimer's disease (AD) brains, the so-called β-amyloid peptide (Aβ) [3]. The following years it was found that this 40- to 42-amino-acid peptide was a cleavage product of a much larger membrane spanning protein, the β-amyloid precursor protein (βAPP) [4–6]. Studies in familial AD presented evidence that an altered metabolism of βAPP with progressive deposition of the Aβ fragment is a crucial event in the pathogenesis of AD. This work led to the controversial concept that AD may be a primarily amyloid-driven process, with the neuritic tau-pathology (neurofibrillary tangles and neuropil threads) being an important secondary phenomenon that is closely correlated with the syndrome of dementia [7, 8].

Although the formation of fibrillar forms of Aβ plays a crucial role in the pathogenesis of AD, the presence of diffuse deposits of Aβ in the cerebral cortex of nondemented elderly and in brain regions of AD patients not associated with clinical symptoms, such as the cerebellum, suggests that the deposition of Aβ by itself is not sufficient to produce the AD clinical symptoms [9, 10]. AD most likely results from a complex sequence of steps involving multiple factors beyond the production and deposition of Aβ alone. During the past 20 years, a variety of inflammatory proteins have been reported to be associated with the amyloid plaques. The idea that inflammation is implicated in AD pathology has received support from epidemiological studies indicating that the use of anti-inflammatory drugs can prevent or retard the process of AD [11–13]. In this chapter, we will review the evidence of the original assumption of Fischer that a peculiar substance in AD can induce a local chronic inflammatory response with a reactive aberrant regenerative response of neurons, which is highly topical in

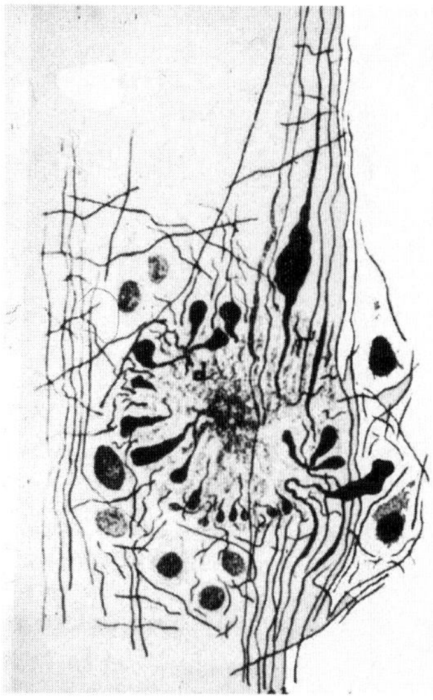

FIGURE 4.1. Drawing of a senile plaque by O. Fischer [1].

current AD research. The clinical and therapeutical implications of this view will be discussed.

4.2 Senile Plaques: The Nidus of a Chronic Inflammatory Response

In 1982, we demonstrated in an immunohistochemical study that senile plaques contain the early complement factors C1q, C3, and C4, and these findings were confirmed by others [14–16]. With a panel of specific monoclonals directed against neoepitopes, which are specific for activated complement products and not present on native proteins, it could be demonstrated that the complement proteins in plaques were not the result of passive absorption but the result of complement activation [17]. McGeer and co-workers extended these findings by the demonstration of the presence of the terminal membrane attack complex, indicating that a full-blown activation of the complement cascade occurs in senile plaques [18].

At the end of the 1980s, several groups demonstrated with monoclonals, directed against cells of the monocyte-macrophage cell lineage, immunohistochemically an association of clusters of activated microglia (brain macrophages) with senile plaques (Fig. 4.2A) [19–22]. The association of amyloid plaques with complement proteins (Fig. 4.2B), as well as clusters of activated microglia, strongly suggest some form of an inflammatory process. In contrast, the absence of immunoglobulins and T-cell subsets within or around plaques indicates that humoral or classical cellular immune-mediated responses are not involved in cerebral β-amyloid plaque formation [23]. Also the recruitment of leukocytes from the blood into the inflammatory foci in the neuropil would require adhesive interactions between leukocytes and endothelial cells of brain capillaries. However, no (increased) expression of the most relevant intercellular adhesion molecules (ICAM-1, VCAM-1, E-selectin) has been found on endothelial cells of capillaries in AD brains [24]. Thus, unlike other brain disorders such as multiple sclerosis [25] and HIV-dementia [26] in which the expression of E-selectin and VCAM-1 coincides with monocyte/macrophage infiltration, the influx of blood-borne cells is not likely to occur in AD brains. Taken together, these data support the view that the (fibrillar) Aβ plaques in AD brains are closely associated with a locally induced, nonimmune mediated, chronic inflammatory type of response without any apparent influx of leukocytes from the blood.

A wealth of data indicate now the extracellular deposition of Aβ in AD brains as one of the triggers of inflammation [27]. For example, Aβ activates microglia by binding to the receptor for advanced glycation end products (RAGE) [28] and to other scavenger receptors [29, 30]. Furthermore, the LPS receptor, CD14, interacts with fibrillar Aβ [31], and microglia kill Aβ1-42 damaged neurons by a CD14-dependent process [32]. The involvement of CD14 in Aβ-induced microglia activation strongly suggests that innate immunity is linked with AD pathology. The concept that Aβ peptide can induce a local inflammatory-type response received impetus from the in vitro findings that fibrillar Aβ can bind C1 and hence potentially activate the classical complement pathway in an antibody-independent fashion [33]. Such activated early complement factors could play an important role in the local recruitment and activation of microglial cells expressing the complement receptors CR3 and CR4 [22].

In vitro studies indicate that a certain degree of Aβ formation is required for the initiation of the complement system [34]. This in vitro finding is consistent with the immunohistochemical data in AD brains showing no or a weak immunostaining for early

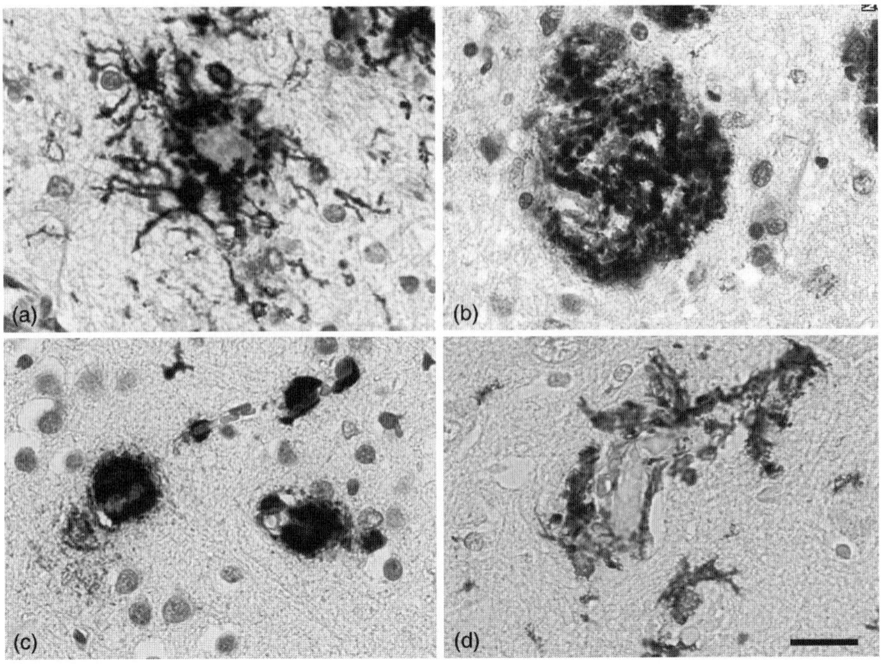

FIGURE 4.2. (A) Asociation of clusters of activated microglia (brain macrophages) immunostained for HLA DP/DQ/DR (CR3/43) with a congophilic plaque in an Alzheimer's disease case. (B) The association of complement protein C3d with a classical amyloid plaque. (C) Immunostaining of Aβ deposits around small blood vessels in a "vascular variant of Alzheimer's disease" case. (D) Clusters of activated microglia (immunostained for HLA DP/DQ/DR [CR3/43]) localized with small congophilic blood vessel in a "vascular variant of Alzheimer's disease" case. Bar represents 20 μm.

complement components in diffuse plaques composed of non- or low-grade fibrillar Aβ peptide [24]. The diffuse plaques are not associated with activated microglia and altered neurites, in contrast with the so-called classical and neuritic plaques, which are characterized by congophilic fibrillar Aβ deposits. So, the nidus for the chronic inflammatory response in AD brains is the plaque containing fibrillar Aβ deposits but not the diffuse plaque with the non-congophilic low-fibrillar Aβ depositions [10, 35].

After the initial reports on complement proteins and activated microglia in senile plaques, a long list of inflammation-related proteins, such as complement factors, acute-phase proteins, and pro-inflammatory cytokines, were found to be localized in senile plaques (for a review, see Ref. 36). Activated microglia, particularly in the vicinity of senile plaques, has been shown to be immunoreactive with antibodies for interleukin-1 [37], interleukin-6 [38, 39], and tumor necrosis factor-α [38]. The so-called Aβ-associated proteins (most of them are acute-phase proteins) include, apart from the complement factors, α1-antichymotrypsin, ICAM-1, α2-macroglobulin, clusterin, apolipoprotein E (ApoE), serum amyloid P component (SAP), and heparan sulfate proteoglycans. In vitro studies showed that most of these Aβ-associated proteins are involved in the amyloidogenic process. For example, ApoE and complement factor C1q can accelerate Aβ fibrillogenesis [40].

One of the biological functions of ApoE is to transport Aβ. The ε4 allele of ApoE is a risk factor for AD and cerebral amyloid angiopathy [41, 42]. There is strong evidence to suggest that the major mechanism underlying the link between ApoE and both AD and congophilic angiopathy is related to the ability of ApoE to interact with the Aβ peptide and influence its clearance, aggregation, and conformation [42]. Clusterin may prevent Aβ aggregation because in cerebrospinal fluid, clusterin is found to be complexed with Aβ thereby maintaining the solubility of Aβ in biological fluid [43].

SAP and heparan sulfate proteoglycans are thought to be essential for amyloid formation and persistence [44, 45]. SAP may protect amyloid deposits against proteolytic breakdown and prevent Aβ phagocytosis by microglia. In the presence of chondroitin sulfate proteoglycans or astrocyte conditioned medium that contains this proteoglycan, microglial capacity to remove deposits of Aβ in culture dishes is inhibited [46]. This indicates that astrocyte-derived factors may downregulate the actions of microglia. In contrast with the other Aβ-associated proteins, α1-antichymotrypsin specifically accumulates in plaques containing the Aβ-peptide but not in other types of amyloid [47]. Hence, α1-antichymotrypsin seems to be involved primarily in the process of Aβ production and deposition and the other plaque-associated proteins more generally in the process of amyloidogenesis independent of the specific chemical structure of the amyloid peptide.

The lack of evidence for blood-brain barrier dysfunction in AD suggests that these Aβ-associated proteins are produced locally [48]. Indeed, with possible exception of the amyloid P component, the messenger RNA for these proteins can be found in brain tissue [49]. Astrocytes are known to synthesize a variety of Aβ-associated proteins including complement factors, α1-antichymotrypsin, and lipid transporters like ApoE and clusterin (apolipoprotein J) [36]. Surprisingly, the major mRNA signal for complement factors and the complementary regulatory proteins is found in neurons and not in glial cells [50, 51].

The involvement of neurons as a source of inflammatory mediators in response to brain injury in AD and lesion studies in animal models was first suggested by Finch and co-workers [52–54]. Neurons in AD brain were found to express mRNA for C1q, C4, and clusterin as detected by combined RNA *in situ* hybridization and immunocytochemistry. In vitro, the production of early complement proteins by neuronal cell cultures increased in response to the cytokines Il-1, Il-6, and TNF-α, which are found in amyloid plaques [55]. These findings implicate that neurons are active players in the inflammatory response in AD brains.

4.3 Concept of Neuroinflammation

Although inflammation is a well-recognized pathological phenomenon, the precise definition of inflammation remains obscure [56, 57]. Consequently, inflammation can be defined in clinical, pathological, and molecular terms. Clinically, the brain of AD patients does not show the cardinal symptoms of Celsus: dolor, tumor, calor, and rubor (i.e., pain, swelling, heat, redness). At the histopathological level, while cells associated with a classical acute inflammatory response (neutrophils) are absent, AD brains show miliary foci with clusters of activated microglia (brain macrophages) indicating a process of focal recruitment and activation of mononuclear phagocytes.

At the molecular level, amyloid plaques in AD brains contain numerous proteins associated with an inflammatory response, including activated complement factors, acute-phase proteins, and proinflammatory cytokines. However, most of these proteins have pleiotrophic effects dependent on their concentrations, and so the precise role of most of these molecules in the amyloid formation is largely unknown. At the present time, the most convincing argument to support the concept of chronic inflammation is related to the histopathological and immunohistochemical observations of recruitment and focal accumulation of phagocytic cells to meet the classical criteria for an inflammatory process as suggested by Metchnikoff [58].

Inflammation is often regarded as a stereotypical nonspecific response to destructive stimuli, and chronic inflammation occurs when there is a failure to eliminate the initiating targets. In most tissues, acute injury is followed by a release of histamine with vascular changes as a consequence. This results in exudation of fluid into the injured tissue and migration of neutrophils. Such a response resulting in fluid exudation with raised intracranial pressure and tissue destruction by neutrophils would be detrimental with respect to the requirement for tight homeostatic control of the neuronal environment to permit efficient neuronal transmission and to maintain a postmitotic neuronal population. Thus, it is conceptually possible that the brain, and the endothelial interface with the bloodstream, has become adapted in such a way as to prevent "bystander" tissue damage after injury. Therefore, in this regard, microglia activation could be considered as a specialized CNS response to injury.

In the normal CNS, most microglial populations are more downregulated than resident tissue macrophages in other organs, and the extent to which they become activated and upregulate a range of factors, including proinflammatory cytokines, complement receptors, and MHC class II receptors, would be a graded response dependent on the nature, severity, and extent of the stimulus. In this scheme, the presence of clusters of activated microglia in senile plaques in AD would be consistent with a neurological form of low-grade chronic inflammation [57].

It is unlikely that neurons are merely passive passengers in the sequence of inflammation that leads to neuronal loss. Recent findings indicate that the neurons themselves appear to be active players in the neuroinflammatory process in AD brains. Increased expression of complement factors and the inducible cyclooxygenase-2 [COX-2] is mainly found in neurons and not or to a lesser extent in glial cells in AD brains [50, 59–61]. Whether the increased levels of inflammation-related proteins within neurons reflect a protective reaction preventing neuronal damage, or stimulate degeneration, remains unknown. Microglial inflammatory mediators have neuropathic as well as neuroprotective actions. Thus, whereas excess levels of reactive oxygen species or TNF-α might cause neurotoxicity, mild oxidative stress and low-dose TNF-α could, alternatively, trigger the neuroprotective and/or anti-apoptotic genes [36, 62].

The role of glial cells is to support and sustain proper neuronal function, and microglia are no exception to this. Kreutzberg, Streit, and co-workers have studied the neuroprotective and proregenerative role of microglia in acute injured CNS [63, 64]. The primary mode of action of microglia may be CNS protection. However, upon excessive or sustained activation, microglia could significantly contribute to chronic neuropathologies. Dysregulation of microglial cytokine production could result in harmful actions of the defense mechanisms, leading to neurotoxicity, as well as disturbances in neural function as neurons are sensitive to cytokine signaling [65].

4.4 Brain Changes in an Early-Stage AD

At the neuropathological level, AD brains are characterized by plaques and tangles. There is a long-lasting and still ongoing debate about the question which lesion comes first: the plaque or the tangle [66, 67]. It has been repeatedly shown that in many cases, entorhinal tangles are the first morphological lesions to be detected in the brain of aging patients. However, these findings in the entorhinal system may not be generalized to the whole brain. In the isocortex, the plaques precede the tangles. In psychometrically well-evaluated subjects, it seems that in the aging process plaques and tangles develop independently. A majority of normal subjects have tangles in the entorhinal-hippocampal areas, but diffuse Aβ deposits are first detected in old subjects above 80 years of age [68, 69]. In subjects at the threshold of detectable dementia, high densities of senile plaques (predominately of the diffuse subtype) are observed [70]. These results suggest that senile plaques in the neocortical regions may not be part of normal aging but instead represent presymptomatic or unrecognized early symptomatic phenomena in AD. Duyckaert and co-workers proposed the following chronological sequence of neuropathological events in the neocortical regions: diffuse fibrillar Aβ deposits, fibrillar Aβ deposits (classical plaques), neurofibrillar tangles. We and others have studied the presence of some inflammation related events in relation to the proposed sequence of occurrence of neuropathological lesions in neocortical areas.

4.4.1 Microglia

In a clinicopathological study of a sample of clinically well-evaluated patients, the volume of tissue occupied by activated microglia, congophilic amyloid, Aβ and tau deposits were studied in neocortical areas [71]. The volume density of activated microglia cells (with CD-68 as marker) correlated highly with the volume density of congophilic deposits, but not with the volume density of Aβ or tau. If cases were ranked in increasing order of severity of clinical dementia, the peak volume densities of activated microglia and congophilic deposits occurred in moderately affected cases, whereas Aβ and tau steadily accumulated with progression of the disease. A decrease of congophilic deposits in the neuropil in the most severe AD cases was already reported [72].

The finding that formation of the congophilic amyloid/microglia complex is a relatively early event in the AD pathogenesis is in agreement with another recently published clinicopathological study. In this study, the CERAD classification was used to show that the prevalence of activated microglia were significantly increased in early stages, while the significant association between astrocytic reaction and clinically manifest dementia suggests that the occurrence of activated astrocytes reflects later stages of the disease, when dementia develops. Tau immunoreactivity in the cerebral neocortex was observed only in the neuropil of definite cases [73].

Studies using positron emission tomography (PET) and the peripheral benzodiazepine ligand PK11195 as marker for activated microglial cells, indicate that activation of microglia precedes cerebral atrophy in AD [74]. Thus, neuropathological and neuroradiological studies indicate that the activation of microglia is a relatively early pathogenic event that precedes the process of neuropil destruction in AD patients. Similarly, in prion disease, the onset of microglial activation was found to coincide with the earliest changes in cerebral morphology. In scrapie-infected mice, microglial activation occurs many weeks before neuronal loss and subsequent clinical signs of disease become apparent [75, 76].

4.4.2 Aβ-Associated Proteins

Intracerebral deposits of Aβ amyloid plaques are invariably associated with a number of proteins, including complement factors, α1-antichymotrypsin, ApoE, clusterin, SAP, and proteoglycans (Fig. 4.3). Strong immunostaining for C1q and SAP is observed in the dense-core and primitive plaques in the cerebral cortex of AD patients. Weak to moderate immunostaining is observed in a variable number of circumscript diffuse plaques in AD and in nondemented controls with plaques, but not in irregular-shaped diffuse Aβ plaques in nondemented controls [77, 78]. α1-Antichymotrypsin and ApoE are present

	PLAQUE TYPE				
Immunostaining	NON-FIBRILLAR		FIBRILLAR		
	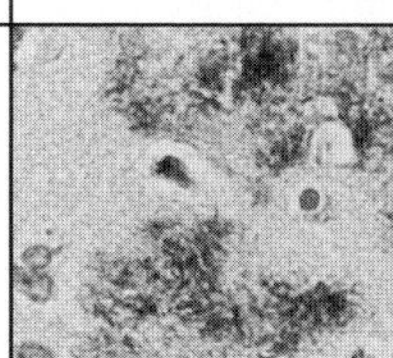	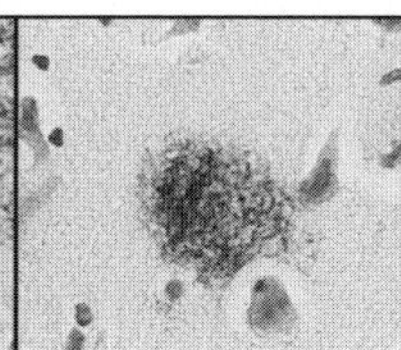	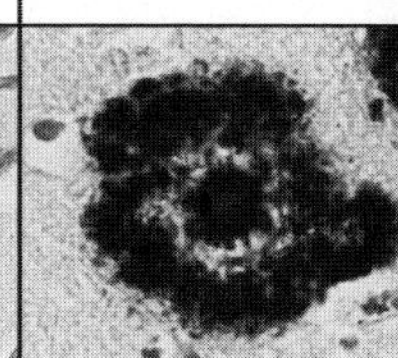		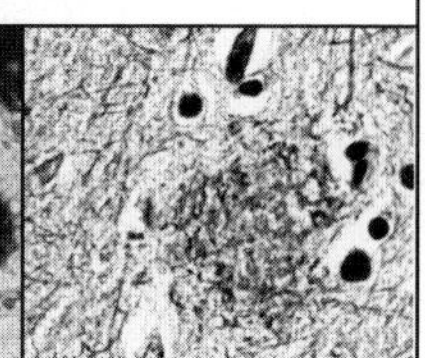
	Irregular shaped, diffuse	Circumscript (well demarcated)	Classic with dense core		Neuritic plaque
			core	corona	
SAP	–	±	++	+	+
C1q	–	±	++	+	+
C4d	±	±	++	+	+
C3d	±	±	++	+	+
ACT	+	+	++	+	+
ApoE	±	+	+	+	+
Tau (AT8)	–	–	+		+
Clustered microglia	–	–	++		±

FIGURE 4.3. Immunohistochemical distribution of SAP, C1q, C4d, C3d, ACT, ApoE, AT8, and activated microglia in morphologically distinguished cerebral Aβ plaque types; -, none; ±, maximally 50% of total; +, >75% of total; ++, all plaques (SAP, serum amyloid P component). Adapted from Ref. 78.

in all forms of plaques including the diffuse type. Accumulation of most of the Aβ-associated proteins is dependent on the degree of fibril density of the Aβ deposits and precede the appearance of clusters of activated microglia and neuronal tau-related changes, suggesting that the associated factors have a modulatory role in early stages of the amyloid-driven pathology cascade.

Only in those Aβ plaques that have accumulated SAP and C1q can clusters of activated microglia be observed in AD neocortex [78]. This suggests that microglia may be attracted to and activated by Aβ deposits of certain fibril density that, in addition, have fixed SAP and C1q. When exposed to a mixture of Aβ1-42, SAP, and C1q, a combination that is relevant to the in vivo situation, adult human microglia secrete significantly higher levels of proinflammatory cytokines in vitro than cells treated with Aβ1-42 alone [78]. Although fibril formation was enhanced in the presence of SAP and C1q, as judged by electron microscopy, cellular effects of the Aβ-SAP-C1q mixture may also be due to interactions of SAP and C1q with microglial acceptors sites, which include receptors for C1q [79, 80]. Taken together, these findings indicate a role of Aβ-associated proteins in Aβ deposition and removal and in microglial activation, and that both events are relatively early steps in the pathological cascade of AD [81].

4.4.3 Adhesion Molecules

Early on, investigators noted that the brain in AD is not only undergoing degeneration but also signs of regeneration and sprouting in and outside the plaques [1, 82, 83]. Regulation of tissue degradation and remodeling involves a complex network including proteases and protease inhibitors, cytokines, integrins, and adhesion molecules [84]. Some growth-promoting factors, such as GAP43, APP, laminin, and collagen IV, have been found in dystrophic neurites but not in neuropil threads outside the plaques [85–87]. Cell adhesion to the extracellular matrix is mediated by integrins, a set of heterodimeric cell-surface receptors that integrate the extracellular matrix or other cells with the intracellular network.

There are different subfamilies of integrins, each defined by a common β-subunit with multiple, distinct α-subunits. The dystrophic neurites associated with the fibrillar Aβ deposits in classical plaques are next to laminin and collagen IV also outlined by different β1 integrins including the laminin-receptor (α6/β1). Interactions between APP and laminin [88] or collagen IV [89] have been described in studies in vitro. The presence of low amounts of extracellular matrix components promotes neurite outgrowth in a dose-dependent manner [90]. The expression of cellular matrix adhesion molecules is regulated by transforming growth factor β1 (TGF-β1), which is present in amyloid plaques [91]. βAPP is in the strict definition of the term a cell adhesion molecule. βAPP can bind heparin and laminin and it appears capable of mediating cell-cell or cell-matrix adhesion.

In antisense βAPP transfected cells, adhesion is reduced and this can be repaired by addition of βAPP [92]. Furthermore, APP has neurite-growth promoting activity and in its secreted form appears to protect against neuronal excitotoxicity [93]. Thus, the plaque actually seems to form a local abnormal microenvironment that employs some of the same principles that are used during normal growth and development [94]. Aβ fibrils appear to have the ability to serve as pseudo "cell adhesion molecules." Aβ assembled into fibrils develops a β-sheet conformation and induces neurites in and around plaques to express the morphological features of dystrophic neurites.

Findings of Cotman and colleagues suggested that β-amyloid activates signal transduction via adhesion molecules and their cross-linking [95, 96]. Fibrillar Aβ could promote dystrophy through aberrant activation of signal transduction cascades, which leads to cytoskeletal changes [97]. Aβ binds to integrins and activates the focal adhesion proteins paxillin and focal adhesion kinase, which are downstream of integrin receptors, suggesting that focal adhesion signaling cascades might be involved in Aβ-induced neuronal dystrophy [98, 99].

Recent experiments indicate that fibrillar Aβ treatment induced integrin receptor clustering, paxillin tyrosine phosphorylation, and translocation to the cytoskeleton and promoted the formation of aberrant focal adhesion-like structures, suggesting the activation of focal adhesion signaling cascades [100]. Focal adhesion signaling induced by fibrillar Aβ may lead to deregulation of kinase and phosphatase activities responsible for tau hyperphosphorylation. Focal adhesion signaling leads to activation of cyclin-dependent kinase 5 (CDK5) and glycogen synthase 3β (GSK-3β), two kinases that phosphorylate tau at epitopes corresponding with those found in neurofibrillary tangles [101].

In summary, the aberrant activation of focal adhesion pathways appears to be critically involved in fibrillar Aβ-induced neuronal dystrophy. The ability of the neuron to respond dynamically to extracellular cues is reminiscent of plasticity mechanisms. In this regard, maladaptive neuronal plasticity may play a major role in AD [95, 100, 102].

4.4.4 Early Neuronal Changes

Cyclooxygenase-2 (COX-2) is involved in the production of prostaglandins and is upregulated at sites of inflammation [103]. It is an enzyme that gathered great interest in AD scientists because of its therapeutic potentials. While it was expected that activated microglia and astrocytes would show increased expression of COX-2 in AD, it was eventually found by immunohistochemistry that mainly neurons express COX-2, whereas astrocytes and microglia are almost unlabeled [59–61,104]. It appears that the neuronal COX-2 is upregulated in early stages of AD, whereas its expression is diminished in advanced stages of AD [105, 106]. Interestingly, this upregulation of COX-2 in early AD and downregulation in advanced AD correlate well with the prostaglandin E2 levels in the CSF, which are elevated in probable AD patients and which decline with increasing severity of dementia [107].

The role of COX-2 in early AD pathogenesis is still elusive. The expression of COX-2 in numerous types of cancers and the effect of selective COX-2 inhibitors on tumor growth suggest a role for COX-2 in cell-cycle control. A dysregulation of cyclins, cyclin-dependent kinases (CDKs), and their inhibitors has been observed in postmitotic neurons in AD [108, 109] and also in other neurodegenerative disorders like Parkinson disease (PD) [110, 111] and amyotrophic lateral sclerosis (ALS) [112, 113]. This suggests that proteins that normally function to control cell-cycle progression in dividing cells may play a role in the death of terminally differentiated postmitotic neurons. During our studies into the expression and role of neuronal COX-2 in AD, the question was raised whether neuronal COX-2 could also be involved in mediating cell-cycle changes in neurons during disease. Indeed, recent studies have shown that COX-2 expression in AD neurons parallels neuronal cell-cycle changes (Fig. 4.4) [106, 114]. It is possible that the increase in neuronal COX-2 expression leads to increased expression of cell-cycle mediators in postmitotic neurons, as shown using a transgenic mouse model with increased neuronal COX-2 expression [115]. Whether COX-2 can be used as a therapeutic target to modulate neuronal cell-cycle changes remains elusive.

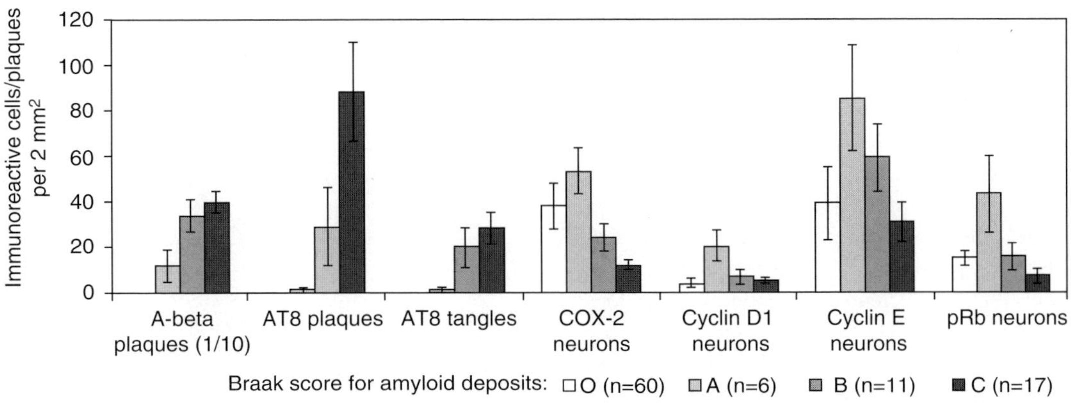

FIGURE 4.4. Shown are the mean immunoreactive scores of patients grouped according to the Braak score for Aβ deposits. COX-2, cyclin D1, cyclin E, and pRb are increased in neurons at Braak stage A, in which already a small number of Aβ plaques is present and almost no neurofibrillary changes are visible in the temporal cortex. At later Braak stages and with the increase of plaques and tangles, the number of neurons immunoreactive for cell-cycle proteins decreases. Data adapted from Refs. 106 and 126.

Although COX-2 may play a role, it is still elusive how and why terminally differentiated neurons in neurodegenerative disorders attempt to reenter the cell cycle. In AD, the presence of growth-associated and growth promoting factors as well as growth factor receptors around the plaques might be an indication of an increased mitogenic force [116]. In addition, conditioned medium from Aβ-stimulated microglia can also trigger neuronal cell division followed by cell death [117]. Aβ protein itself has mitogenic properties and can induce cell cycle–mediated cell death in cultured neurons [118]. Initial studies implicating cell-cycle events in degenerating neurons in AD showed induction and activation of CDC2 and its partner cyclin B1 in postmitotic neurons [119, 120]. CDC2 activity has been proposed to play a major role in the hyperphosphorylation of the tau protein and the subsequent formation of neurofibrillary tangles [119], which suggests a direct link between the reactivation of the cell cycle and the pathogenesis of AD. The reexpression of cell cycle proteins is also closely associated with apoptosis in neurons [118, 121]. These findings led to the suggestion that uncoordinated expression of cell-cycle molecules and the resulting breach of cell-cycle checkpoints is one of the primary mechanisms by which postmitotic neurons undergo apoptotic death [102, 122].

Cell-cycle changes can be detected in neurons that are vulnerable to neurodegenerative changes that are associated with AD [120, 123–125]. This implies that neuronal cell-cycle changes are involved in the early steps of AD neurodegeneration. Cell-cycle proteins cyclin D1, cyclin E, and phosphorylated retinoblastoma protein (ppRb) are found to be increased in cases with Braak stage A for amyloid deposits [106, 126]. These cases already show some Aβ deposits but lack neurofibrillary changes (Fig. 4.4). In later Braak stages, these neuronal cell changes become less apparent. Double-immunohistochemistry for ppRb and the neurofibrillary marker AT8 shows that the nuclear expression of ppRb does not coincide with the occurrence of neurofibrillary changes inside the neuron [126]. These data support the view that the increase of cell-cycle proteins is an early event in the pathogenesis that occurs before the formation of neurofibrillary tangles.

In general, an aberrant cell-cycle reentry has been implicated in neuronal death during the pathogenesis of AD as well as other neurodegenerative disorders. Interestingly, neurodegenerative diseases like AD, ALS, and PD do not only show neuronal cell-cycle abnormalities [110, 113, 127] but also have aggregation of abnormal or misfolded proteins in common [128]. The accumulation of misfolded or aggregated proteins in the endoplasmic reticulum (ER) activates a homeostatic pathway: the "unfolded protein response" (UPR) [129, 130]. The activation of the UPR results in an overall decrease in translation, increased protein degradation, and in increased levels of ER chaperones like BiP/GRP78 to increase the protein folding capacity of the ER. In vitro data show that activation of the UPR induces a G1 phase arrest, linking the occurrence of unfolded proteins in the ER to altered control of cell-cycle regulation [131, 132]. The occurrence of misfolded proteins in the ER and the resulting UPR could directly mediate the regulation of cell-cycle proteins in postmitotic neurons. In a recent study, we investigated the role of the UPR in cell-cycle regulation during AD pathogenesis [133]. Activation of the UPR, as measured by the levels of BiP/GRP78, is progressively occuring in AD as compared with nondemented control cases. Furthermore, activation of the UPR also negatively correlates with the expression of cell-cycle proteins (Fig. 4.5).

Activation of the UPR in a neuronal cell model inhibits cell-cycle progression showing a direct

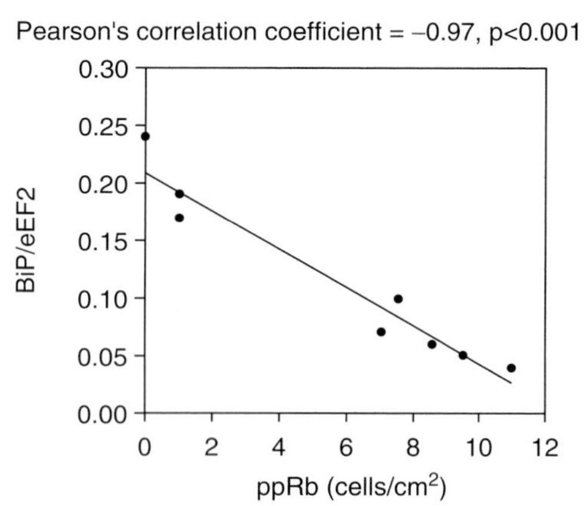

FIGURE 4.5. Correlation between neuronal ppRb immunoreactivity and relative BiP/GRP78 expression levels in AD and nondemented temporal cortex. Relative expression levels of BiP/GRP78 as determined by Western blot analysis were correlated with the occurrence of nuclear ppRb immunoreactivity in neurons in the temporal cortex.

link between UPR activation and cell-cycle regulation in neurons. This interaction between the UPR and an aberrant cell cycle in postmitotic neurons might eventually determine the fate of a neuron during the progression of AD. On the other hand, these data suggest that there are two phases in AD pathogenesis: an early neuronal response involving COX-2 and cell-cycle changes followed by a second phase involving an advanced stage of protein aggregation and neurofibrillary changes. The first phase could be a response of the neurons to extracellular (inflammatory) cues activating mechanisms that induce plasticity. The second phase reflects the inability of the neurons to regenerate, resulting in widespread neurodegeneration [134].

4.4.5 Convergence of the Immunohistochemical Data and Gene Findings

Recently, new tools have been developed that can address the complexity of the pathogenesis of Alzheimer's disease. Gene microarrays simultaneously allow the study of the activity of multiple cellular pathways. Although microarray data interpretation is hindered by low statistical power and high false positives and negatives, recent microarray studies have confirmed the involvement of several cellular pathways in AD pathogenesis.

An earlier study, comparing gene expression in the CA1 of the hippocampus between AD and control subjects already indicated the involvement of apoptotic and neuroinflammatory signaling [135]. More recently, Blalock and colleagues performed an analyses of the correlation between hippocampal gene expression with Mini-Mental State Examination (MMSE) and Neurofibrillary Tangles (NFT) scores [136]. Upregulation of biological process categories included genes regulating cell proliferation and differentiation and genes encoding cell adhesion and complement factors. Most interestingly, proliferation and prostaglandin synthesis pathways were among the main categories of upregulated genes in incipient AD cases.

It has been recognized by immunohistochemical studies that inflammation, synaptic dysfunction, glial reactivity, protein misfolding, lipogenesis, and cell-cycle disturbances are involved in AD. Although cDNA microarray is a relatively new and emerging technique, it confirms the immunohistochemical findings for the early involvement of inflammatory and regenerative pathways in AD pathogenesis.

4.5 Inflammation in Transgenic Models

Familial autosomal dominant mutations identified in AD patients have been introduced in transgenic mice to establish models that reconstitute the pathogenic process associated with Aβ amyloidosis [137–139]. These models display several pathological characteristics of AD such as Aβ-immunoreactive plaques that are accompanied by dystrophic neurites and reactive gliosis. The different transgenic models display various types of plaques early in the amyloidogenic process. In some models, diffuse and compact fibrillar plaques accumulate concomitantly even at the earliest stages of deposition, in contrast with other models in which exclusively fibrillar compact plaques are seen. In these models, the amyloid deposits are associated with an inflammatory response characterized by clustering of activated microglia, complement factors, and glial expression of both pro- and anti-inflammatory cytokines [140–144]. In some of these models, microglia are associated with compact deposits only. The TgCRND8 mouse model exhibits neuropathological changes with a robust increase in cerebral Aβ level and formation of diffuse and compact plaques as early as 9–10 weeks of age. The formation of plaques was concurrent with the appearance of activated microglia and followed by the clustering of activated astrocytes around plaques at 13–14 weeks of age [145]. The simultaneous deposition of plaques and the activation of the inflammatory processes underline the relationship between both events in the initial stage of neuropathological brain changes. Although the fibrillar Aβ-induced inflammatory response is a relatively early event in transgenic mice, the earliest cognitive impairment is correlated with the accumulation of intraneuronal Aβ in the hippocampus and amygdala before plaque pathology become apparent [146].

Transgenic mouse cell lines expressing human βAPP harboring the vasculotropic Dutch and/or Iowa mutations exhibit an early and robust cerebral microvascular accumulation of fibrillar Aβ amyloid exhibiting strong thioflavin S staining and numerous largely diffuse plaque-like structures in the parenchyma [147, 148]. The distribution of Aβ in these transgenic cell lines is consistent with the cerebral Aβ distribution that is seen in patients with the Dutch and Iowa disorders. The depostion of cerebral microvascular amyloid in the transgenic mice harboring the vasculotropic mutation is accompanied by large increases in the numbers of neuroinflammatory reactive astrocytes and activated microglia as well as elevated cerebral levels of the proinflammatory cytokines Il-1β and Il-6 [149].

Transgenic models seem also a promising model to elucidate the role of the Aβ-associated factors in amyloidogenesis. Studies in these models have already established the role of ApoE, α1-antichymotrypsin (ACT), complement factors, and clusterin in amyloid formation. The transgenic hAPP mouse studies show that increased expression of some Aβ-associated proteins (ApoE, ACT) leads to higher amyloid load, whereas inhibition of complement factors results in low amyloid load. On the other hand, the amyloid formation is strongly hampered in mouse strains that expressed mutant hAPP and are "null" for ApoE [150]. When ACT transgenic mice are crossed to transgenic hAPP mice, the ACT/APP mice have twice the amyloid load and plaque density compared with the mice carrying mutant hAPP alone [151]. Inhibition of complement activation in the brain of hAPP mice by expressing soluble complement receptor-related protein (sCrry), a complement inhibitor, lead to a two- to threefold higher amyloid load and more neuronal loss than in age-matched hAPP mice [152]. In transgenic hAPP mice crossed with clusterin [-/-] mice, the levels of Aβ deposits are similar to these in hAPP mice expressing clusterin, but there are significantly fewer fibrillar Aβ deposits. In the absence of clusterin, neuritic dystrophy associated with the amyloid deposits is markedly reduced, resulting in dissociation between amyloid formation and neuritic dystrophy [153]. All these observations in transgenic mice models support the idea that Aβ-associated proteins play an important role in the dynamic balance between Aβ deposition and removal.

The fundamental discussion about the beneficial or detrimental aspects of inflammation in amyloid deposition and its therapeutical consequences are well illustrated by the findings from inflammation-based treatment strategies in transgenic mice models. Recent work in transgenic models has revealed that either intercranial lipopolysaccharide (LPS) injection or treatment with the nitric oxide–releasing nonsteroidal anti-inflammatory drug NCX-2216 potentiates microglial activation and leads to reduction in Aβ plaque load [154, 155]. Another inflammation-based treatment strategy is immunization with Aβ [156]. Immunization of the young animals prevents the development of amyloid plaque formation, and in older animals it markedly reduces the extent and progression of amyloid pathology. Injections with anti-Aβ antibodies cleared the plaques in the cortex of transgenic mice and activated the microglia [157, 158]. The therapeutic option for vaccination in AD patients is hampered by severe side effects [159]. These side effects reflect most probably the double-edged sword role of the inflammatory response in AD pathogenesis.

4.6 Inflammation and the Pathological Cascade

Although the role of inflammatory molecules in the pathological process of AD is not fully understood, current findings indicate that these molecules may be involved in a number of key steps in the proposed amyloid-driven cascade (Fig. 4.6) [160].

1. The brain concentration of Aβ is the result of the equilibrium between the Aβ-producing enzymes and the catabolic enzymes involved in Aβ degradation. During the past few years, a growing list of candidate enzymes for Aβ degradation has been described, including the metalloproteases, for example, insulin-degrading enzyme, neprilysin, angiotensin converting enzyme, and serine proteases such as plasmin [161]. It has been shown that Il-1 (possibly together with other cytokines) can regulate βAPP synthesis and Aβ production in vitro [162–164]. Such a cytokine-induced production in vivo may initiate a vicious circle whereby Aβ deposits stimulate further cytokine production by activated microglia to even higher synthesis rates of βAPP and its Aβ fragments. There is a lack of information about the effect of inflammatory mediators on the enzymes involved in Aβ degradation.

2. The Aβ-associated proteins (most of which are acute-phase proteins) are involved in regulation of the Aβ amyloidogenic process. These proteins

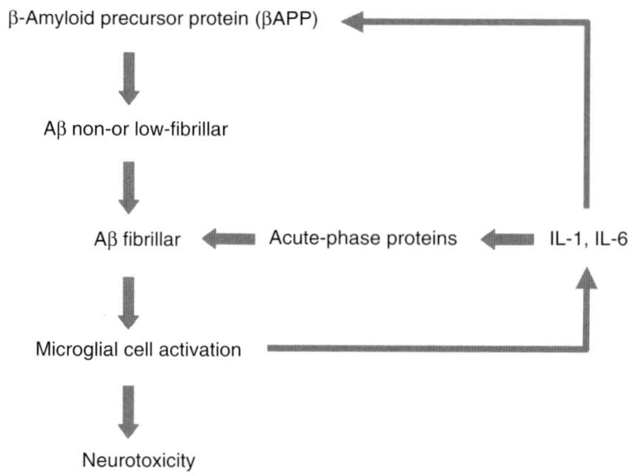

FIGURE 4.6. Mismetabolism of the β-amyloid precursor protein (βAPP) with progressive deposition of its Aβ fragment is a crucial event in the pathogenesis of AD. Once aggregated, Aβ is able to activate the classical complement pathway, resulting in the attraction and activation of microglial cells. In turn, these microglial cells produce multiple proinflammatory and neurotoxic factors. Factors such as interleukin-1 (Il-1) and -6 (Il-6) can reinforce the pathological amyloid cascade by a positive feedback loop. Modified from Ref. 160.

are involved in the fibrillization, deposition, and removal of the Aβ peptide as discussed earlier.

3. Once fibrillar, Aβ can induce a microglia-mediated chronic inflammatory response. Activated microglial cells produce and release potentially toxic products, including reactive oxygen species, proinflammatory cytokines, excitotoxins, and proteases, which could damage the neighboring neurons. Recent studies suggest that the oligomeric forms of Aβ are more toxic for neurons than the high-fibrillar forms. However, the high-fibrillar forms of Aβ that are in vivo associated with activated complement fragments induce the inflammatory response leading to gliosis and destruction of functional nervous tissue architecture.

It is important to keep in mind that the involvement of numerous inflammatory proteins in the pathological cascade is not related to a single pathogenic event but to a number of subsequent steps. Most of these proteins have pleiotrophic effects depending on their concentrations, and thus the precise role of these molecules in the different steps of the pathological cascade is largely unknown. In addition to the production of proinflammatory cytokines, microglial cells can also produce anti-inflammatory cytokines such as interleukin-10 (Il-10) [165]. The neuroinflammatory response includes both beneficial and deleterious effects on the progression of the disease process. On the one hand, inflammatory activation by Aβ could be viewed as a potential contributor to AD neurodegenerative processes, however, inflammatory proteins, particularly complement proteins, may also play a role in microglial-mediated Aβ removal [166].

The role of inflammation as a double-edge sword in neurodegenerative disorders attracts much interest in current AD research [167]. This is not surprising because eliminating pathogenic stimuli, such as the removal of fibrillar Aβ deposits, and tissue repair with scar formation are essential characteristics of inflammatory processes. In this context, it is interesting to recall the suggestion that there are two phases in AD neurodegeneration: a first phase, involving increased neuronal COX-2 and cell-cycle protein expression, as a response to induce neuronal plasticity, and a second phase in which neurons fail to cope with the increasing presence of unfolded proteins and eventually undergo neurofibrillary degeneration. Aβ deposition, inflammation, and neuroregenerative mechanisms are related and early pathogenic events in AD that can be also seen in the transgenic mouse AD model, while "later" neurodegenerative characteristics are not seen in these models. The precise relation between the neuroregenerative and neurodegenerative events in AD pathology remains elusive.

4.6 Inflammation-Related Systemic Changes in AD Patients

A systemic consequence of a local inflammatory response is the acute-phase response. This response is characterized by a change in plasma concentrations of proteins, collectively known as acute-phase reactants. In serum of AD patients, a significant increase of the levels of several acute-phase proteins has been found [168]. Most notably, an increase in serum levels of the acute-phase reactant α1-antichymotrypsin has been reported in several studies [169–173]. Moreover, increased serum levels of Il-6 and TNF-α and decreased levels of albumin have been reported in some studies

[174–176]. The acute-phase response can be considered as part of a complex generalized stress reaction in which the activation of the sympathic nervous system coincides with endocrine changes, including the activation of the hypothalamic-pituitary-adrenal (HPA) axis. Abnormalities of the HPA system linked to AD include both basal cortisol hypersecretion and insufficient cortisol suppression after dexamethasone administration [177–180]. Another sign of activation of the HPA axis in AD patients is the increased neuronal expression of mRNA for corticotropin-releasing hormone in the hypothalamic paraventricular nucleus [181].

With respect to the activation of the sympathetic system, it has been reported that the basal 3-methoxy-4-hydroxyphenylglycol levels were positively associated with the degree of cognitive impairment in AD patients [182]. Although plasma 3-methoxy-4-hydrophenylglycol is a much better indicator of peripheral rather than central noradrenalin metabolism, these findings could reflect alterations in the central noradrenergic activity. In AD patients, the cerebrospinal fluid levels of 3-methoxy-4-hydrophenylglycol correlated positively with post-dexamethasone cortisol levels and with rating of dementia severity [180]. A strong activation of the remaining noradrenergic neurons in the locus coeruleus has been reported in AD brains [183]. The findings concerning the activation of the HPA axis and the sympathic system, together with changes in the levels of some acute-phase reactants, indicate that a systemic acute-phase response can be found in AD patients [184].

4.7 Inflammation and the Epidemiological Findings

Recent epidemiological and genetic studies favored the idea that the acute-phase response in AD patients can be a crucial part of the pathophysiology. In four different prospective case-cohort studies, it has been shown that high serum levels of the acute-phase proteins α1-antichymotrypsin, C-reactive protein, and Il-6 and low serum levels of albumin were each associated with an increased risk of cognitive decline/AD [185–188]. In a recent study, Yaffe and colleagues reported that elderly subjects, with a metabolic syndrome and a high serum level of Il-6 and C-reactive protein, were more likely to experience cognitive decline in the next 4 years, compared with those with a metabolic syndrome and low levels of these inflammatory markers [189].

These epidemiological findings from several case cohort studies indicate that nondemented subjects with an acute-phase response profile in serum are at risk of developing AD. The acute-phase response is initiated and orchestrated by cytokines, most notably Il-1. Several studies have shown that an Il-1α -899 C/T gene polymorphism is associated with AD. A strong association between the Il-1α T/T genotype and AD onset before 65 years was found, with carriers of this genotype showing an onset 9 years earlier than Il-1α C/C carriers [190]. This study also reported a weaker association with age of onset for the Il-1β and Il-1 receptor agonist genes. In neuropathologically confirmed AD patients, the prevalence of the Il-1α T/T genotype was higher than in controls (odds ratio 3.0 controlled for age and ApoE status) [191]. Other authors also found an increased risk for AD for the heterogeneous carriers of the C/T genotype and much stronger for the homogeneous carriers of the T/T genotype [192]. These findings were further confirmed in a study reporting the association of Il-1 T/T genotype with increased risk of early onset of AD. Clinically, this genotype was associated with earlier age of onset but not with a change in the rate of progression of AD [193].

Others reported that the risk of this Il-1α allele polymorphism is not restricted to AD patients of a particular age and found the association in both early-onset and late-onset AD patients [194]. However, the association between this Il-1α polymorphism and (late-onset) AD could not be confirmed in other studies [195–198]. In a meta-analysis on the association between the Il-1α genotypes and AD, the data showed a significant but modest association in patients with an early-onset AD but not in late-onset AD [199]. In a recent study, it was found that the polymorphism association in the Il-1α gene influences the microglial load (volumetric percentage of the brain occupied by microglia) in AD brains. It was 31% greater in patients with one T allele and 62% in patients with the TT genotype but no effects on microglial load occurred with polymorphisms in Il-1β [200]. Results of studies on polymorphisms of Il-1β, Il-1Ra, Il-6, and TNF-α as risk factors for AD show contradictory findings, which makes it difficult to draw conclusions [189, 201–211].

A potential role of polymorphisms of Aβ-associated proteins as genetic risk factor of Alzheimer's disease is strongly suggested by genetic association of the apolipoprotein E4 (ApoE4) allele as a susceptibility gene increasing the risk and lowering the age of onset distribution of AD [41]. It has been reported that the ApoE4-associated risk is modified by α1-antichymotrypsin (ACT) polymorphism [212]. A high frequency of a combined ACT A/A and ApoE4 genotype was found in patients with a familial late-onset AD [213]. Others reported that the ACT T/T genotype was overrepresented in patients with early onset of sporadic AD but no relationship with ApoE genotype was found suggesting ACT T/T genotype is an independent risk factor of early-onset AD [214]. The concomitant ACT T/T and Il-1β T/T strongly increased the risk of AD and the age of onset of the disease. Patients with these genotypes showed the highest levels of plasma ACT and Il-1β [215]. However, several studies from China, Germany, and Japan could not confirm an association between ACT polymorphism and AD [216–223]. In respect to these inconsistent data between ACT genetic variation and AD risk, Kamboh and co-workers have recently studied the relationship between ACT polymorphism with age of onset and disease duration [224]. They found in male AD patients that the mean of age-of-onset and the disease duration among ACT/AA homozygotes were significantly lower than that in the combined AT+TT genotype group.

A genetic association analysis for AD and α2-macroglobulin (A2M) has also been controversial. Initially, an association between AD and an intronic deletion polymorphism in the spliced site of exon 18 of A2M was reported in a sample of discordant sibships [225]. While this initial finding was later replicated in independent family-based AD samples, case-control association studies of AD and A2M18i deletion polymorphism have been largely negative (for a review, see Ref. 226). The discrepancy between the generally positive association findings in family-based samples and the generally negative association findings in case-control samples suggests that A2M may be a risk factor primarily in individuals with a family history of AD. For the complement factors, an association for C3 [227] and C4 [228] phenotypes and AD has been reported but these findings could not be replicated [229, 230].

In conclusion, several epidemiological studies have shown consistently in prospective case-cohort studies that a higher serum level of certain acute-phase reactant is a risk factor for AD. With respect to the association of polymorphisms of cytokines and Aβ-associated proteins, the role of ApoE4 as risk factor is firmly established. Il-1α polymorphism could be a risk factor in early-onset AD but probably not in late-onset AD. For the other cytokines and acute-phase proteins, the findings about an association between polymorphisms and AD are too inconclusive to consider them at this moment as genetic risk factors for AD.

4.8 Inflammation and the Etiology of AD Subtypes

In the past decade, the research agenda for unraveling the pathogenesis of AD was strongly dominated by the findings in rare autosomal dominant variants of AD. The finding that most studied causal mutations in familial AD lead to higher production of Aβ1-42 has stimulated the concept that mismetabolism of βAPP with increased production of its Aβ fragment must be considered as the crucial pathogenic event in all forms of AD. However, it is becoming increasingly clear that factors other than mismetabolism of βAPP can initiate or stimulate the pathological cascade. In this chapter, we have reviewed the evidence from genetic, epidemiological, pathological, and experimental transgenic animal studies that inflammation-related mechanisms are most likely involved in the early stages of the pathological process. The involvement of cytokines and acute-phase proteins in Aβ production, fibrillization, deposition, and removal indicate that inflammatory molecules are involved in early key events in the pathological cascade. In this respect, the findings in transgenic AD models are illustrative. On one hand, these models convincingly document the important effect of βAPP or presenilin mutants, but, on the other hand, these models show also that cross-breeding of mice with variation in the expression of Aβ-associated proteins strongly influence the rate and load of cerebral amyloid deposition. In addition, immunization studies in the transgenic mouse models illustrate the importance of Aβ removal for the process of amyloid deposition. These findings indicate the involvement of multiple factors in the initial steps of the pathological cascade and could explain the heterogeneity of AD.

In the autosomal dominant forms, the initial event is increased Aβ1-42 deposition that elicits a brain inflammatory response. An example where inflammatory mechanisms could play a role in initiating AD is the development of AD after head trauma. It has been proposed that in these cases, the βAPP overexpression and increased Aβ production is a direct consequence of the Il-1–driven acute-phase response [231].

Most Down syndrome patients develop AD pathology after the age of 50. With respect to the role of inflammatory mechanisms in AD, it is noteworthy that in earlier days, chronic inability to resist infection was a major cause of death in patients with AD. The most likely reason for susceptibility to infection in Down syndrome is that gene dosage results in altered expression of a gene on chromosome 21 that is crucial for an adequate immune response. The observation that the deposition of diffuse Aβ plaques precedes other Alzheimer-related brain lesions by many years, together with the discovery, that the βAPP gene is localized on chromosome 21, which is overexpressed in Down syndrome, suggest that the increased expression of βAPP and consequent deposition of Aβ is the prime cause of AD in Down patients [232].

However, it is important to realize that in addition to βAPP, several other proteins that are implicated in the regulation of inflammation and oxidative stress (e.g., superoxide dismutase and carbonyl reductase) are encoded on chromosome 21 [233]. Taylor and co-workers have demonstrated an altered expression of the leukocyte adhesion molecules belonging to the β2 integrin subfamily in patients with Down syndrome [234]. Their members constitute a family of three noncovalently associated αβ-heterodimers with homologous α-subunits and a common β-subunit that is encoded on chromosome 21. The most important ligands for β2 integrins are ICAM-1 and the activated fragments of complement factor C3. As mentioned earlier, the amyloid plaques in AD are characterized by the presence of activated complement fragments, ICAM-1, and clusters of activated microglia that strongly express the leukocyte adhesion molecules of the β2 integrin family. The activated microglia with the complement receptors CR3 and CR4 (members of the β2 integrin family) can play an essential role in the phagocytosis of complement-opsonized Aβ fibrils [22, 24]. As the amyloid burden in AD brains is most likely determined by a dynamic balance between amyloid deposition and resolution [235], it is important to note that both βAPP and β2 integrins, which are involved in amyloid production and removal, respectively, are encoded on chromosome 21. Therefore, the high amyloid burden of amyloid found in Down syndrome patients with AD could be the net result of high Aβ production and impaired complement-mediated phagocytosis of Aβ.

Another example for the involvement of inflammation in the etiology of certain subtypes could be the role of vascular factors in the etiology of AD [236]. Accumulating evidence suggests inflammation as a secondary injury mechanism after ischemia and stroke [237]. So, head trauma and ischemia do not only cause acute brain damage but also induce brain inflammatory responses that could contribute to the development and/or aggravation of AD pathology. In relatively older patients with a clinical dementia syndrome, the neuropathological findings show frequently both vascular and Alzheimer changes. This form of dementia is described as a mixed type dementia, a combination of two different pathologies that are both common in the elderly [238]. From a pathogenic view, it can be hypothesized that this clinical syndrome is not simply the result of summation of two different, independent disease processes but rather the outcome of synergistic interactions between the vascular and Alzheimer components that are both mediated by neuroinflammatory processes [239]. After the proposal of Blennow and Wallin [240] and Hoyer [241] to distinguish AD in type I (early onset) and type II (sporadic late onset), we would suggest that in type I AD mismetabolism of βAPP with increased Aβ deposition is frequently the initial and crucial pathogenic event that is followed by a fibrillar Aβ-induced neuroinflammatory response.

In contrast, in type II AD a broad variety of inflammatory molecules, including cytokines and acute-phase reactants, seem to play a major role in the initiation of the pathological cascade (Fig. 4.7) [242]. Although both forms of AD do not form a single, homogeneous nosological entity, the clinical picture and neuropathological end stage characteristics are strikingly uniform. The very same

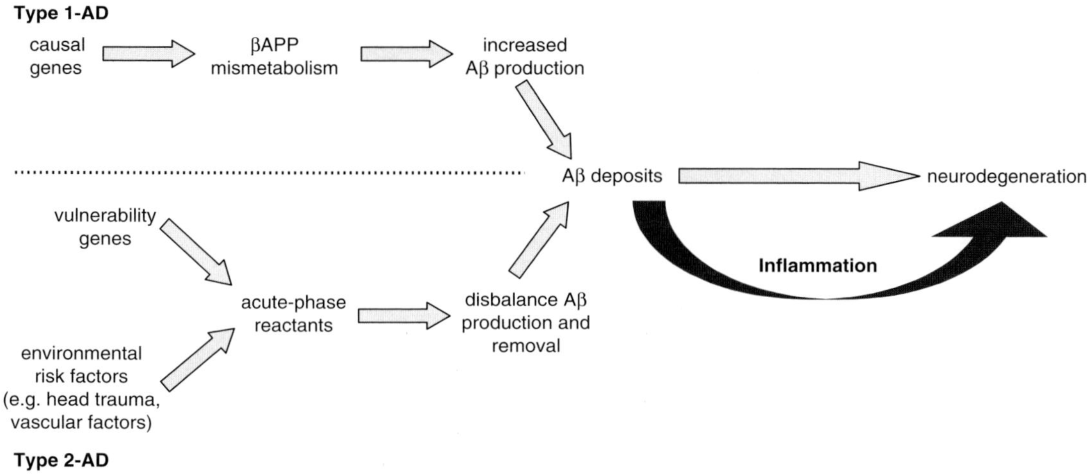

FIGURE 4.7. Illustration of the differences in etiology between type I and type II AD.

combination of pathogenic heterogeneity with homogeneity in clinical appearance is not uncommon in medicine and known for diabetes and arthrosclerosis. In the early-onset form of diabetes mellitus, type I DM, the insulin production is deficient, and in type II the function of the insulin receptor is damaged. There is increasing evidence that the inflammatory response is associated with the presence of insulin resistance. Experimental studies in humans and animals show that treatment with proinflammatory cytokines produce hypertriglyceridemia and insulin resistance. TNF-α downregulates the tyrosine kinase activity of the insulin receptor, thereby increasing insulin resistance [243].

Recent research suggests that atherosclerosis is a lipid-driven macrophage-dependent process [244, 245]. Inflammatory processes mark all stages of atherogenesis; from endothelial activation to eventual plaque rupture. It is well-known that a high plasma concentration of cholesterol, in particular those of low-density lipoprotein cholesterol, is one of the principal risk factors for atherosclerosis. Although hypercholesterolemia is important in approximately 50% of patients with cardiovascular disease, other factors need to be taken into consideration. Over the past decade, it was found that inflammatory mechanisms couple dyslipidemia to atheroma formation. Arteriosclerosis, diabetes mellitus, and AD have in common that their etiology is heterogeneous and that the late-onset variant is multifactorial. The etiological event can be a disturbance of (altered) cholesterol production, insulin, or Aβ production but also inflammation. The inflammatory processes can lead to the development of insulin resistance and disturbances in the removal of lipoproteins and Aβ with as consequence the development of (late-onset) diabetes, arteriosclerosis and AD, respectively. The common etiological role of inflammation in the late-onset variants of these disorders could explain that diabetes mellitus and arteriosclerosis are considered as risk factors for the late-onset form of AD.

4.9 Inflammation and the Clinical Symptoms

In the cerebral cortex of elderly nondemented subjects with high numbers of diffuse Aβ deposits, immunohistochemical signs for an inflammatory process are absent. In brain areas of AD patients not linked to Alzheimer symptomatology but with widespread deposition of Aβ (such as the cerebellum), the levels of acute-phase proteins and early complement factors, as well as the numbers of activated microglia, are low (10, 246). In a clinicopathological study including demented and nondemented cases, Lue and co-workers [247] found correlations between inflammatory markers, such as complement activation and activated microglia, and synapse loss, a major correlate of cognitive decline in AD patients.

The idea that the site of inflammation is related to the clinical manifestations can be illustrated by neuropathological findings in brains from patients with hereditary cerebral hemorrhage with amyloidosis–Dutch type (HCHWA-D). This disorder belongs to the cerebral Aβ diseases and is a rare autosomal dominant characterized by a single base mutation at codon 693 of the APP gene [248]. Clinically, HCHWA-D is characterized by recurrent hemorrhagic strokes and at the neuropathological level by extensive vascular amyloid and diffuse plaques in the absence of neuritic plaques and neurofibrillary degeneration. The congophilic angiopathy in HCHWA-D is associated with strong monocyte/macrophage reactivity, but there is no evidence for microglia activation seen in the cerebral cortex [249, 250].

However, in AD brains the congophilic angiopathy (with exception of "drüsige Entartung," see below" is not associated with an increased number of cells expressing monocyte/macrophage markers in contrast with the fibrillar Aβ plaques in the neuropil [251–254]. Therefore, in AD brains the inflammatory response is associated with fibrillar Aβ deposits in the neuropil, whereas in HCHWA-D brains, the inflammatory response is associated exclusively with the fibrillar Aβ deposits in the vascular walls. The most characteristic clinical features of AD and HCHWA-D are dementia and recurrent strokes, respectively. Hence, these findings indicate that in both AD and HCHWA-D, the clinical symptoms are associated to a great extent with the site of inflammation [184].

In some cases with AD, the pathological process differed from that typically seen in AD. These patients show a severe amyloid angiopathy associated with perivascular tau neurofibrillary pathology in absence of neuritic plaques unrelated to blood vessels [255]. The vascular plaques are related to capillaries, and the amyloid deposits, radiating from the vessel wall into the surrounding neuroparenchyma, are associated with a crown of tau-positive neuritis and astrogliosis. This phenomenon is also called dyshoric angiopathy or microcapillary amyloid angiopathy. The neuropathology of these atypical AD cases is different from that observed in HCHWA-D because of the presence of dementia with neurofibrillary lesions and the absence of deadly cerebral hemorrhages and from the pathology found in typical AD because of the absence of senile plaques in the neuroparenchyma. This atypical form of AD has been described and diagnosed as "vascular variant of Alzheimer's disease" [256]. In our cases with this AD variant, we found that the vascular amyloid plaques were immunolabeled for the complement proteins and always associated with clusters of activated microglia (Fig. 4.2). These findings indicate that the site of the chronic inflammatory response in these cases is related to the microcapillary amyloid angiopathy [257].

Cognitive symptoms are the cardinal clinical signs of a dementia syndrome. These symptoms are related to destruction of neuronal circuits in hippocampal and neocortical brain regions, and these cognitive deficits are seen in relatively late stages of the underlying disease process. Epidemiological studies indicate that depressive-like symptoms, such as loss of interest and energy, as well as mental slowing (subjective bradyphrenia), can be present at a preclinical stage of AD [258, 259].

In human prion disease, it is also reported that psychiatric symptoms can precede the neurological symptoms. Thus, in the new BSE-related variant of Creutzfeldt-Jakob disease (nvCJD), psychiatric symptoms (especially depression) are an early and prominent clinical feature preceding other neurological symptoms in many cases [260]. Experimental animal and human studies have shown that proinflammatory cytokines produced as a part of the "stereotypical" macrophage/microglia response to injury can induce behavioral changes such as a "depressive-like" syndrome [261–264]. As discussed earlier, the fibrillar Aβ-induced activation of microglia is a relatively early pathogenic event in AD brains and precedes the process of severe neuropil destruction. The effect of proinflammatory cytokines derived from activated microglia may cause disturbances in the neurotransmission leading to behavioral changes at an early stage of the disease with no or little structural brain tissue damage. The characteristic cognitive symptoms in more advanced stages of AD are the result of the inflammation-related events that lead to neuropil destruction. Thus, distinct inflammatory mechanisms seem to be involved in a broad spectrum of behavioral and cognitive symptoms in several stages of AD [265].

4.10 Inflammation and Therapeutic Aspects

Based on observations from neuropathology, genetics, epidemiology, as well as from in vitro and animal experiments, the inflammatory component of AD has been considered a compelling target for therapeutic intervention. The idea that the neuroinflammatory response is an interesting therapeutic target was strongly stimulated by epidemiological studies that the use of classical NSAIDs could prevent or retard AD [11, 12]. The first clinical trials fostered the hopes for anti-inflammatory treatments of AD patients. In two small studies, the effects of indomethacin and diclofenac, both classical NSAIDs, were studied.

The therapeutic activity of indomethacin, a NSAID that crosses the blood-brain barrier, was investigated in a double-blind, placebo-controlled pilot study [266]. A small positive effect on the cognitive outcome measurement was reported. However, during the 6-month treatment period, the dropout rate in the indomethacin group was approximately 40%, mostly owing to drug-related gastrointestinal adverse events. The second trial suggested disease stabilization to some degree in patients treated with diclofenac in combination with misoprostol [267]. However, the observed differences in this small study failed to reach significance on intention-to-treat analysis of standard outcome measures. A consistent picture emerged from four larger randomized controlled trials with longer treatment periods. Studies on the effect of prednisone, hydroxychloroquine, naproxen (a traditional nonselective NSAID), celecoxib, and rofecoxib (both selective COX-2 inhibitors) in patients with early AD, with relatively few dropouts, all failed to document a benefit in favor of patients that were treated with the specific anti-inflammatory drug under study [266–273] (Fig. 4.8). When these data are put together, it is clear that the best available evidence to date does not support the idea that AD patients benefit from treatment with anti-inflammatory drugs [274]. How can this finding be explained in the light of the widespread support

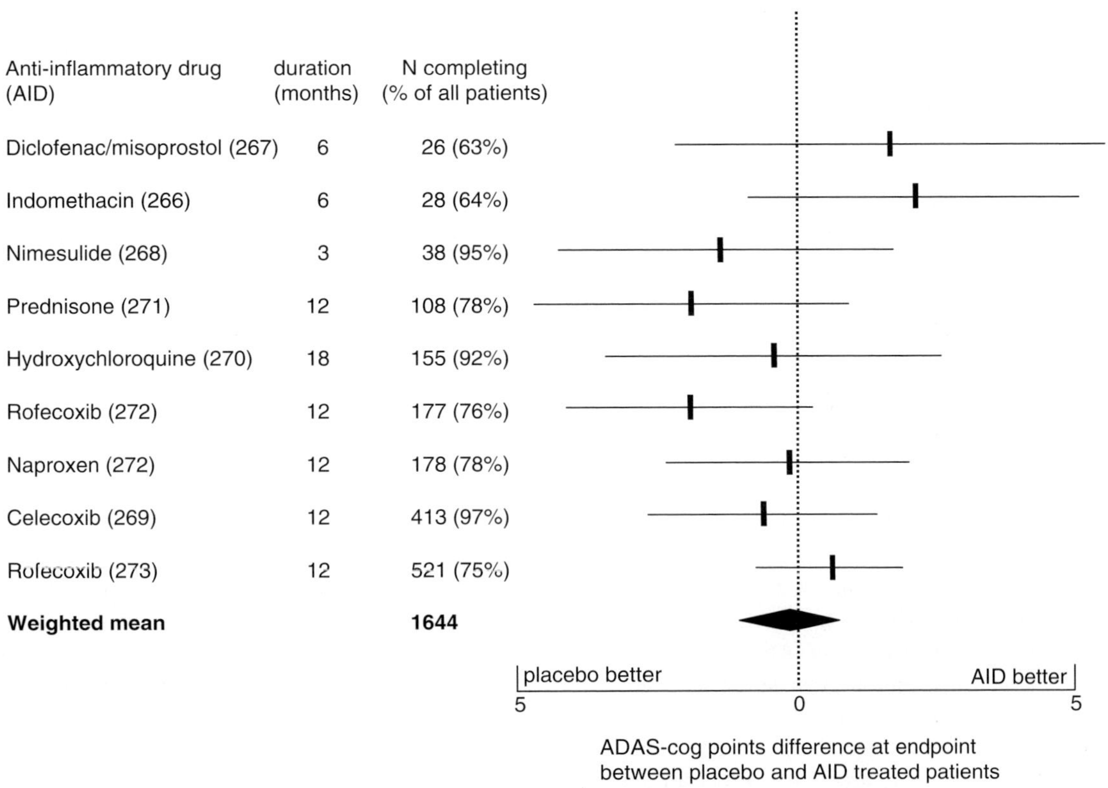

FIGURE 4.8. Overview of randomized clinical trials on the effects of anti-inflammatory drugs (AID) on the course of Alzheimer's disease. The trials are identified by study drug, the name of the first author, and the year of publication (see References), and they are listed according to their size indicated by the number of patients completing the study. For each trial, the difference between AID and placebo-treated patients in change of the Alzheimer's Disease Assessment Scale (ADAS-cog) scores is shown. The diamond represents the weighted mean, and its size reflects the 95% confidence interval of this measure.

for the inflammatory hypothesis of AD? What was wrong with the studies of anti-inflammatory treatment in AD? Was it the concept of neuroinflammation, the class of drugs that were used in the trials, or was it the timing of anti-inflammatory treatment?

In respect to the concept of the role of neuroinflammation in AD, it is important to keep in mind that inflammation is not linked to a single pathogenic event but that inflammatory mediators are involved in a number of key steps of the pathological cascade. As discussed before, there is a lack of knowledge on the detrimental or protective role of each of the inflammatory molecules involved in pathological cascade in AD. Several studies in transgenic mice encoding the familial AD mutations have shown that immunization with Aβ peptide reduces deposition of cerebral fibrillar Aβ deposits and that this is associated with the beneficial behavioral effects [154, 275].

The idea of treatment with anti-inflammatory drugs is based on the reduction of the inflammatory reaction, whereas immunization leads to stimulation of the inflammatory response that may be beneficial for Aβ removal. Treatments with either anti-inflammatory drugs are based on reduction of the inflammatory process, whereas immunization stimulates a more efficient phagocytic activity of microglia. The immunization story in transgenic mice suggests that inflammatory mechanisms play a beneficial role in the removal of Aβ [166]. When there is failure of Aβ removal, microglia become prolonged highly activated, and they produce potential neurotoxic factors. However, it would be possible that both therapeutic options are not mutually exclusive and that the effects of immunization and anti-inflammatory drug therapy may act on different inflammation-mediated events in the pathological cascade.

The second explanation for the failure of anti-inflammatory drug treatment of AD patients could be the timing of the treatment. An important difference between epidemiological studies suggesting protective effects of inhibition of inflammatory processes and the clinical trials is that both deal with entirely different parts of the time frame of the disease. In a long-term prospective population study of the incidence of AD, it was found that for those whose cumulative use of NSAIDs was 2 years or more, the relative risk of developing AD was reduced by 80% [12]. This 2-year lag-time may explain some of the negative findings in previous epidemiological studies because most studies relied on brief periods of follow-up after classifying patients according to NSAIDs use [276]. The 2-year lag-time seems also biologically plausible because neuropathological and neuroradiological studies indicate that neuroinflammation is an early pathogenic event that precedes the process of severe neuropil destruction in AD patients. Similarly, elevation of neuronal COX-2 activity is an early event in the pathogenesis of AD. The cognitive deficits are the cardinal clinical signs of a dementia syndrome, and these symptoms are related to destruction of hippocampal and neocortical brain regions. Therefore, the cognitive deficits are generally assumed to reflect relatively late stages of the underlying process. Inhibition of the neuroinflammatory process even at the time that the first symptoms of dementia exceed clinical detection thresholds might be simply too late to attenuate the alleged detrimental effects of inflammatory processes. This view implicates that intervention with anti-inflammatory drugs should take place in the earliest stage of the pathogenesis.

The third explanation for the failure of the anti-inflammatory drug treatment in AD patients could be the choice of the studied drugs. Theoretically speaking, it could be possible that the positive effects of drugs with a broad range of anti-inflammatory actions, such as prednisone and hydrochloroquine, on the harmful component of inflammation can be neutralized by a negative effect of these drugs on the beneficial components of the inflammatory response. Therefore, potential positive clinical effects of "broad" anti-inflammatory drugs on certain components of the inflammatory response can remain unrecognized in clinical trials. The positive epidemiological findings with anti-inflammatory drugs to prevent or retard AD are reported for the classical NSAIDs, which are known to inhibit both COX-1 and COX-2. Both COX isoenzymes have high structural identity but differ in substrate and inhibitor activity and are involved in the first steps of the synthesis of prostaglandins from the substrate arachidonic acid [103]. COX-1 is normally expressed constitutively and is involved in the production of prostaglandins and effective housekeeping functions. Under normal condition, COX-2 has a low expression in most human tissues, but it can be induced by inflammatory stimuli such as Il-1.

With respect to the adverse effect of the classic NSAIDs, the novel class of gastrointestinal-sparing

COX-2 selective NSAIDs seemed to be promising in the treatment of AD patients. This therapeutic perspective has stimulated investigations into the role and distribution of both COX enzymes in normal and AD brains. Surprisingly, COX-2 has been immunohistochemically detected in neurons in normal and AD brains, whereas astrocytes and microglia are almost unlabeled [59, 60, 61, 104]. In contrast, the immunoreactivity for COX-1 is found particularly in the activated microglia cells associated with plaques [60, 61]. In vitro studies with adult human microglia cells show that neither the proinflammatory cytokines that are increased at sites of Aβ plaques nor Aβ1-42 induces COX-2 expression in these cells [277]. Therefore, the distribution patterns are strikingly different for COX-1 and COX-2 in AD brains.

The initial idea for COX-2 inhibitors as a drug for AD was based on the idea of limitation of side effects of the classical NSAIDs. The current findings implicate distinct cellular expression of COX-1 and COX-2. As discussed before, the neuronal upregulation of COX-2 is found in early stages of AD and diminished neuronal COX-2 expression in advanced stages of AD [105, 106]. For treatment of AD patients with COX-2 inhibitors, it is important to realize that it is yet unclear in which respect the neuronal COX-2 upregulation in early stages and downregulation in advanced stages are involved in protective or damaging mechanisms. Irrespective of the issue of the selectivity of NSAIDs in COX-1 or COX-2 inhibition during the past few years, several studies suggest that classical NSAIDs have modes of action that are independent of COX activity [27]. Some of the widely used classical NSAIDs, such as indomethacin and ibuprofen, can activate the nuclear receptor peroxisome proliferator receptor gamma (PPARγ), which has been shown to inhibit the production of proinflammatory cytokines [278]. Recent findings indicate that PPARγ can induce a clearance mechanism for the Aβ peptide [279].

A variety of experimental studies indicate that a subset of classical NSAIDs such as ibuprofen, flurbiprofen, indomethacin, and sulindac also possess Aβ42-lowering properties in both AD transgenic mice and cell cultures of peripheral, glial, and neuronal origin [280–282]. While COX inhibition occurs at low concentrations in vitro (nM to low μM range) the Aβ-lowering activity is observed at high concentrations (50 μM) [27]. The inhibition of Aβ1-42 levels by a subgroup of NSAIDs is based on direct modulation of γ-secretase activity [283]. Recently, it was demonstrated that these NSAIDs have an allosteric effect on γ-secretase by which these drugs selectively reduce Aβ1-42 but do not affect processing of other γ-secretase targets [284, 285]. These findings illustrate the possibility to develop drugs that lower the amyloid burden without affecting other important physiological pathways (e.g., Notch cleavage). In a recent published paper, the role of downstream prostaglandin pathways in COX-mediated inflammation and Aβ production was investigated [286].

Aged transgenic APPSwe-PS1 mice crossed to mice with deletion of the prostaglandin E2 E prostanoid subtype 2 (EP2) receptor show a marked reduction in lipid peroxidation and a significant decrease in Aβ levels. The current findings indicate that PGE2 signaling via the EP2 receptor promotes age-dependent oxidative damage and increased Aβ peptide burden in this model, possibly via effects of increased oxidative stress on BACE1 activity in processing APP. Flurbiprofen has been proposed as a candidate drug for the treatment of AD [282]. To avoid the gastrointestinal side effects of classical NSAIDs, which limit their chronic use, two different strategies have been identified [27]. One is the use of the *R* enantiomer of flurbiprofen, which maintains the Aβ-lowering properties of the racemate but does not cause gastric damage due to a lack of COX inhibitory activity [281]. The other strategy is based on the use of NO-releasing derivates of flurbiprofen, which have been shown in animal studies to reduce brain inflammation and Aβ burden [155, 287].

The discovery that a subset of NSAIDs such as ibuprofen, indomethacin, and fluriprofen may have direct Aβ-loweing properties in cell cultures as well as in transgenic models of AD amyloidosis suggest new pharmacological properties of these drugs with novel therapeutic implications for the treatment of AD [288]. The clinical trials with naproxen, rofecoxib, celecoxib, all with negative results, are performed with NSAIDs that have the least potency in modulating Aβ in experimental models.

4.11 Conclusions and Future Directions

Studies performed over past 20 years to elucidate the molecular composition of plaques have shown that the original assumption of Fischer dating from

1907 that an inflammatory process occurs in AD brain was indeed correct. The challenge for this response is the extracellular deposition of fibrillar Aβ that Fischer considered as a "foreign substance." The recent finding that removal of this substance by anti-Aβ antibodies leads to clearance of the Aβ deposits with subsequent reduction of the plaque-associated (neuroregenerative) dystrophic neurites in transgenic APP support this notion [289]. Immunohistochemical and gene profiling findings in the initial stages of AD pathology showing upregulation of genes involved in cell-cycle regulation, adhesion, and inflammation indicate the early involvement of inflammatory and regenerating pathways in AD pathogenesis. These brain changes precede the tau-relatred neurofibrillary pathology and the extensive process of neurodestruction and (astro)gliosis.

The role of inflammation in the pathological cascade is not restricted to a single event, but inflammatory mechanisms appear to be involved in nearly every pathogenic step of the pathological cascade. For the near future, the beneficial and detrimental aspects of the inflammatory mediators will have to be investigated in vitro in cell cultures that reflect the distinct steps of the pathological cascade. In neuronal cell cultures, the role of cytokines in the metabolism of βAPP and the production of its Aβ fragments can be studied. In in vitro models, the role of Aβ-associated proteins on Aβ aggregation can be evaluated with thioflavin assays and electron microscopy. In cell cultures, the effects of Aβ alone or complexed with Aβ-associated proteins on microglial activation and neuronal toxicity can be studied. Likewise, neuronal-glial interactions and the effects of Aβ whether or not complexed with certain Aβ-associated proteins on these interactions can be investigated in mixed neuronal and glial cell cultures. The advantage of this approach is that the role of inflammatory mediators can be studied on a mechanistic level for each of the distinct steps of the pathological cascade with the relevant human peptides and cell types.

Other promising avenues for the near future are (1) the recent neuropathological findings that inflammatory mediators are upregulated in early stages of the disease process, (2) the epidemiological findings that nondemented subjects with high serum levels of acute-phase proteins have a higher risk to develop AD, and (3) the observation that Il-1α polymorphisms seem to be a genetic risk factor. Taken together, these pathological and epidemiological findings suggest that inflammation-related mechanisms can play a role in the etiology of certain (sub)types of AD. Nearly 100 years after the assumption of Fischer that the senile plaque is a nidus of inflammation, a role of inflammatory mechanisms in amyloid plaque formation is well established. The research agenda for the near future will include the etiological, clinical, and therapeutic implications of the view that inflammatory mechanisms are involved in the pathological cascade of AD.

References

1. Fischer O. Miliare Nekrosen mit drusigen Wucherungen der Neurofibrillen, eine regelmässige Veränderung der Hirnrinde bei seniler Demenz. Monatsch f Psychiat u Neurol 1907; 22:361-372.
2. Fischer O. Die presbyophrene Demenz, deren anatomische Grundlage und klinische Abgrenzung. Z Ges Neurol u Psychiat 1910; 3:371-471.
3. Glenner GG, Wong CW. Alzheimer's disease: initial report of the purification and characterization of a novel cerebrovascular protein. Biochem Biophys Res Commun 1984; 16:885-890.
4. Goldgaber D, Lerman MI, McBride OW, et al. Characterization and chromosomal localization of a cDNA encoding brain amyloid of Alzheimer's disease. Science 1987; 235:877-880.
5. Kang J, Lemaire HG, Unterbeck A, et al. The precursor of Alzheimer's disease A4 protein resembles a cell-surface receptor. Nature 1987; 325:733-736.
6. Tanzi RE, Gusella JF, Watkins PC, et al. Amyloid β protein gene: cDNA, mRNA distribution, and genetic linkage near the Alzheimer locus. Science 1987; 235: 880-884.
7. Hardy J, Allsop D. Amyloid deposition as the central event in the etiology of Alzheimer's disease. Trends Pharmacol Sci 1991; 12:383-388.
8. Selkoe DJ. The molecular pathology of Alzheimer's disease. Neuron 1991; 6:487-498.
9. Rozemuller JM, Eikelenboom P, Stam FC, et al. A4 protein in Alzheimer's disease: primary and sevondary cellular events in extracellular amyloid depostion. J Neuropathol Exp Neurol 1989; 48:647-663.
10. Rozemuller JM, Stam FC, Eikelenboom P. Acute phase proteins are present in amorphous plaques in the cerebral but not in the cerebellar cortex in patients with Alzheimer's disease. Neurosci Lett 1990; 119:75-78.

11. McGeer PL, Schulzer M, McGeer EG. Arthritis and anti-inflammatory agents as possible protective factors for Alzheimer's disease: a review of 17 epidemiological studies. Neurology 1996; 47:425-432.
12. In 't Veld BA, Ruitenberg A, Hofman A, et al. Nonsteroidal antiinflammatory drugs and the risk of Alzheimer's disease. N Engl J Med 2001; 345: 1515-1521.
13. Szekely CA, Thorne JE, Zandi PP, et al. Nonsteroidal anti-inflammatory drugs for the prevention of Alzheimer's disease: a systematic review. Neuroepidemiology 2004; 23:159-169.
14. Eikelenboom P, Stam FC. Immunoglobulins and complements factors in senile plaques. Acta Neuropathol (Berl) 1982; 57:239-242.
15. Ishiii T, Haga S. Immuno-electron-microscopic localization of complement in amyloid fibrils of senile plaques. Acta Neuropathol 1984; 63:296-300.
16. Pouplard A, Emile J. New immunological findings in senile dementia. Interdiscipl Topics Gerontol 1985; 19:62-71.
17. Eikelenboom P, Hack CE, Rozemuller JM, et al. Complement activation in amyloid plaques in Alzheimer's disease. Virchows Arch B (Cell Pathol) 1989; 56:259-262.
18. McGeer PL, Akiyama H, Itgaki S, et al. Immune system response in Alzheimer's disease. Can J Neurol Sci 1989; 16:516-527.
19. McGeer PL, Itagaki S, Tago H, et al. Reactive microglia in patients with senile dementia of Alzheimer-type are positive for the histocompatibility glycoprotein HLA-DR. Neurosci Lett 1987; 79: 195-200.
20. Poulard-Barthelaix A, Dubas F, Jabbour W, et al. An immunological view on the etiology and pathogenesis of Alzheimer's disease. In: Bes A, editor. Senile Dementia Early Detection. Paris: John Libbey, Eurotext, 1986:216-222.
21. Rogers J, Luber-Narod J, Styren SD, et al. Expression of the immune system-associated antigen by cells of the human central nervous system. Relationship to the pathology of Alzheimer's disease. Neurobiol Aging 1988; 9:339-349.
22. Rozemuller JM, Eikelenboom P, Pals ST, et al. Microglial cells around plaques in Alzheimer's disease express leucocyte adhesion of the LFA/1 family. Neurosci Lett 1989; 101:288-292.
23. Eikelenboom P, Rozemuller JM, Kraal G, et al. Cerebral amyloid plaques in Alzheimer's disease but not in scrapie-affected mice are closely associated with an chronic inflammatory process. Virchows Arch B (Cell Pathol) 1991; 60:329-336.
24. Eikelenboom P, Veerhuis R. The role of complement and activated microglia in the pathogenesis of Alzheimer's disease. Neurobiol Aging 1996; 17: 673-680.
25. Cannella B, Raine CS. The adhesion molecule and cytokine profile of multiple sclerosis lesions. Ann Neurol 1995; 37:424-435.
26. Nottet HS, Persidsky Y, Sasseville VG, et al. Mechanisms for the transendothelial migration of HIV-1-infected monocytes into the brain. J Immunol 1996; 156:1284-1295.
27. Gasparini L, Ongini E, Wenk G. Non-steroidal anti-inflammatory drugs (NSAIDs) in Alzheimer's disease: old and new mechanisms of action. J Neurochem 2004; 91:521-536.
28. Yan SD, Chen X, Fu J, et al. RAGE and amyloid-β peptide neurotoxicity in Alzheimer's disease. Nature 1996; 382:685-691.
29. El Khoury J, Hickman SE, Thomas CA, et al. Scavenger receptor mediated adhesion of micoglia to amyloid-β fibrils. Nature 1996; 382:716-719.
30. Paresce DM, Ghosh RN, Maxfield FR. Microglial cells internalize aggregates of the Alzheimer's disease amyloid-β-protein via a scavenger receptor. Neuron 1996; 17:553-565.
31. Fassbender K, Walter S, Kuhl S, et al. The LPS receptor (CD14) links innate immunity with Alzheimer's disease. FASEB J 2004; 18:203-205.
32. Bate C, Veerhuis R, Eikelenboom P, Williams A. Microglia kill Aβ1-42 damaged neurons by a CD14 dependent process. Neuroreport 2004; 28:1427-1430.
33. Rogers J, Cooper NR, Webster S, et al. Complement activation by β-amyloid in Alzheimer's disease. Proc Natl Acad Sci U S A 1992; 89:10016-10020.
34. Snyder SW, Wang Gt, Barrett L, et al. Complement C1q does not bind monomeric β-amyloid. Exp Neurol 1994; 128:136-142
35. Itagaki S, McGeer PL, Akiyama H, et al. Relationship of microglia and astrocytes to amyloid deposits of Alzheimer's disease. J Neuroimmunol 1989; 24:173-182.
36. Akiyama H, Barger S, Barnum S, et al. Inflammation and Alzheimer's disease. Neurobiol Aging 2000; 21: 383-421.
37. Griffin WST, Stanley LC, Ling C, et al. Brain interleukin-1 and S-100 immunoreactivity are elevated in Down syndrome and Alzheimer's disease. Proc Natl Acad Sci U S A 1989; 86:7611-7615.
38. Dickson DW, Lee SC, Mattiace LA, et al. Microglia and cytokines in neurological diseases, with special reference to AIDS and Alzheimer's disease. Glia 1993; 7:75-83.
39. Huell M, Straus S, Volk B, et al. Interleukin-6 is present in early stages of plaque formation and is restricted to the brains of Alzheimer's disease. Acta Neuropathol 1995; 89:544-551.

40. Webster S, O'Barr S, Rogers J. Enhanced aggregation of β structure of the amyloid β pepotide after incubation with C1q. J Neurosci Res 1994; 39:448-456.
41. Corder EH, Saunders EM, Strittmatter WJ, et al. Gene dose of apolipoprotein E type 4 llele and the risk of Alzheimer's disease in late onset families. Science 1993; 261:921-923.
42. Holtzman DM. In vivo effects of ApoE and clusterin on amyloid-β metabolism and neuropathology. J Mol Neurosci 2004; 23:247-254.
43. Ghiso J, Matsubara E, Koudinov A, et al. The cerebrospinal-fluid soluble form of Alzheimer's amyloid β is complexed to SP-40,40 (apolipoproten J), an inhibitor of the complement membrane-attack-complex. Biochem J 1993; 293:27-30.
44. Emsley J, White HE, O'Hara BP, et al. Structure of the pentameric human serum amyloid P component. Nature 1994; 367:338-345.
45. Snow AD, Sekiguchi R, Nochlin D, et al. An important role of heparan sulphate proteoglycan (Perlecan) in a model system for the deposition and persistence of fibrillar β-amyloid in the rat brain. Neuron 1994; 12:219-234.
46. Shaffer LM, Dority MD, Gupta-Bansal R, et al. Amyloid β protein removal by neuroglial cells in culture. Neurobiol Aging 1995; 16:737-745.
47. Abraham CR, Shirahama T, Potter H. α1-Antichymotrypsin is associated solely with amyloid deposits containining the β-protein. Amyloid and cell localization of α1-antichymotrypsin. Neurobiol Aging 1990; 11:123-129.
48. Rozemuller JM, Eikelenboom P, Kamphorst W, et al. Lack of evidence for dysfunction of the blood-brain barrier in Alzheimer's disease: an immunohistochemical study. Neurobiol Aging 1988; 9:383-391.
49. Kalaria RN. The immunopathology of Alzheimer's disease and related pathology. Brain Pathol 1993; 3:333-347.
50. Shen Y, Li R, McGeer EG, McGeer PL. Neuronal expression of MRNAs for complement proteins of the classical pathway in Alzheimer's disease. Brain Res 1997; 769:391-395.
51. Veerhuis R, Janssen I, Hoozemans JJM, et al. Complement C1-inhibitor expression in Alzheimer's disease. Acta Neuropathol 1998; 96:628-636.
52. Lampert-Etchels M, McNeill TH, et al. Sulfated glycoprotein-2 is increased in rat hippocampus following entorhinal cortex lesioning. Brain Res 1991; 563: 101-106.
53. Johnson SA, Lampert-Etchells M, Pasinetti GM, et al. Complement mRNA in the brain: responses to Alzheimer's disease and experimental brain lesioning. Neurobiol Aging 1992; 13:641-648.
54. Rozovsky I, Morgan TE, Willoughby DA, et al. Selective expression of clusterin (SGP-2) and complement C1qB and C4 during responses to neurotoxins in vivo and in vitro. Neuroscience 1994; 62:741-758.
55. Veerhuis R, Janssen I, De Groot CJA, et al. Cytokines associated with amyloid plaques in Alzheimer's disease brain stimulate human glial and neuronal cultures to secrete early complement factors, but not C1-inhibitor. Exp Neurol 1999; 160:289-299.
56. McGeer PL, McGeer EG. Inflammation, autotoxicity and Alzheimer's disease. Neurobiol Aging 2001 22: 799-809.
57. Eikelenboom P, Bate C, van Gool WA, et al. Neuroinflammation in Alzheimer's disease and prion disease. Glia 2002; 40:232-239.
58. Metchnikoff E. Leçons sur la pathologie comparée de l'inflammation. Paris: Masson, 1892.
59. Oka A, Takashima S. Induction of cyclo-oxygenase 2 in brains of patients with Down's syndrome and dementia of Alzheimer type: specific localization in affected neurons and axons. Neuroreport 1997; 8: 161-164.
60. Yermakova AV, Rollins J, Callahan LM, et al. Cyclooxygenase-1 in human Alzheimer and control brain: quantitative analysis of expression by microglia and CA3 hippocampal neurons. J Neuropathol Exp Neurol 1999; 8:1135-1146.
61. Hoozemans JJM, Rozemuller JM, Janssen I, et al. Cyclooxygenase expression in microglia and neurons in Alzheimer's disease and control brains. Acta Neuropathol 2001; 101:2-8.
62. Rozemuller JM, van Muiswinkel FL. Microglia and neurodegeneration. Eur J Clin Invest 2000; 30:469-470.
63. Kreutzberg GW. Microglia: a sensor for pathological events in the CNS. Trends Neurosci 1996; 45:239-247.
64. Streit WJ. Microglia as neuroprotective, immunocomptent cells of the CNS. Glia 2002; 40:133-139.
65. Hanisch UK. Microglia as a source and target of cytokines. Glia 2002; 40:140-155.
66. Duyckaerts C. Looking for the link between plaques and tangles. Neurobiol Aging 2004; 25:735-739.
67. Schönheit B, Zarski R, Ohm TG. Spatial and temporal relationships between plaques and tangles in Alzheimer pathology. Neurobiol Aging 2004; 25: 697-711.
68. Price JL, Davis PB, Morris JC, et al. The distribution of tangles, plaques and related immunohistochemical markers in healthy aging and Alzheimer's disease. Neurobiol Aging 1991; 12:295-312.
69. Price JL, Morris JC. Tangles and plaques in nondemented aging and 'preclinical' Alzheimer's disease. Ann Neurol 1999; 45:358-368.

70. Morris JC, Storandt M, McKeel DW Jr, et al. Cerebral amyloid deposition and diffuse plaques in 'normal' aging. Evidenve for presymptomatic and very mild Alzheimer's disease. Neurology 1996; 46: 707-719.
71. Arends YM, Duyckaerts C, Rozemuller JM, et al. Microglia, amyloid and dementia in Alzheimer's disease. A correlative study. Neurobiol Aging 2000; 21: 39-47.
72. Brun A, Englund E. Regional pattern of degeneration in Alzheimer's disease: neuronal loss and histopathological grading. Histopathology 1981; 5:549-564.
73. Vehmas AK, Kawas CH, Stewart WF, et al. Immunoreactive cells and cognitive decline in Alzheimer's disease. Neurobiol Aging 2003; 24: 321-331.
74. Cagnin A, Brooks DJ, Kennedy AM, et al. In-vivo measurement of microglia in dementia. Lancet 2001; 358:461-467.
75. Williams AE, van Dam A-M, Ritchie D, et al. Immunohistochemical appearance of cytokines, prostaglandin E2, and lipocortin-1 in the CNS during the incubation period of murine scrapie correlates with progressive PrP accumulations. Brain Res 1997; 754:171-180.
76. Jeffrey M, Halliday WG, Bell J, et al. Synaps loss associated with abnormal PrP precedesneuronal degeneration in the scrapie-infected murine hippocampus. Neuropathol Appl Neurobiol 2000; 26: 41-54.
77. Zhan SS, Veerhuis R, Kamphorst W, et al. Distribution of β-amyloid associated proteins in plaques in Alzheimer's disease and in non-demented elderly. Neurodegeneration 1995; 4:291-297.
78. Veerhuis R, van Breemen MJ, Hoozemans JJM, et al. Amyloid β-associated proteins C1q and SAP enhance the Aβ1-42 peptide-induced cytokine secretion by adult human microglia in vitro. Acta Neuropathol 2003; 105:135-144.
79. Gasque P, Dean YD, McGreal EP, et al. Complement components of the innate system in health and disease in the CNS. Immunopharmacology 2000; 49: 171-186.
80. Webster SD, Yang AJ, Margol L, et al. Complement component C1q modulates the phagocytosis of Aβ by microglia. Exp Neurol 2000; 161:127-138.
81. Veerhuis R, Boshuizen RS, Familian A. Amyloid associated poteins in Alzheimer's and prion disease. Curr Drug Targets CNS Neurol Disord 2005; 4:325-348.
82. Bouman L. Senile plaques. Brain 1934; 57:128-142.
83. Geddes JW, Monaghan DT, Cotman CW, et al. Plasticity of hippocampal circuitry in Alzheimer's disease. Science 1995; 230:1179-1181.
84. Masure S, Opdenakker G. Cytokine-mediated proteolysis in tissue remodelling. Experientia 1989; 45: 542-549.
85. Cras P, Kawai M, Lowery D, et al. Senile plaque neurites accumulate amyloid precursor protein. Proc Natl Acad Sci U S A 1991; 88:7552-7556.
86. Eikelenboom P, Zhan SS, Kamphorst W, et al. Cellular and substrate adhesion molecules (integrins) and their ligands in cerebral amyloid plaques in Alzheimer's disease. Virchows Arch 1994; 424:421-427.
87. Zhan SS, Kamphorst W, VanNostrand WE, et al. Distribution of neuronal growth-promoting factors and cytoskeletal proteins in altered neurites in Alzheimer's disease and non-demented elderly. Acta Neuropathol (Berl) 1995; 89:365-362.
88. Narindrasorasak S, Lowery DE, Altman RA, et al. Characterization of high affinity binding between laminin and Alzheimer's amyloid precursor proteins. Lab Invest 1992; 67:643-652.
89. Breen KC. APP-collagen interaction is mediated by a heparin bridge mechanism. Mol Chem Neuropathol 1992; 16:109-121.
90. Koo EH, Park L, Selkoe DJ. Amyloid-β protein as substrate interacts with extracellular matrix to promote neurite outgrowth. Proc Natl Acad Sci U S A 1993; 90:1564-1568.
91. Van der Wal E, Gomez-Pinilla F, Cotman CW. Transforming growth factor-β in plaques in Alzheimer and Down pathologies. Neuroreport 1993; 4:69-72.
92. LeBlanc AC, Kovacs DM, Chen HY, et al. Role of amyloid precursor protein (APP): study with antsense transfection of human neuroblastoma cells. J Neurosci Res 1992; 31:635-645.
93. Mattson MP, Cheng B, Culwell AR, et al. Evidence for excitoprotective and intraneuronal calcium-regulating roles for secreted forms of the β-amyloid precursor protein. Neuron 1993; 10:243-254.
94. Cotman CW, Hailer NP, Pfister KK, et al. Cell adhesion molecules in neural plasticity and pathology: similar mechanisms, distinct organisation. Progr Neurobiol 1998; 55:659-669.
95. Cotman CW. Beta-amyloid peptide, peptide selfassembly, and the emergence of biological activities. A new principle in peptide function and the induction of neuropathology. Ann N Y Acad Sci 1997; 814:1-16.
96. Cribbs DH, Kreng VM, Anderson AJ, et al. Crosslinking of concavalin A receptors on cortical neurons induces programmed cell death. Neuroscience 1996; 75:173-185.
97. Saitoh T Horsburg H, Masliah E. Hyperactivation of signal transduction systems in Alzheimer's disease. Ann N Y Acad Sci 1993; 695:34-41.

98. Kowalska MA, Badellino K. β-Amyloid protein induces platelet aggregation and supports platelet adhesion. Biochem Biophys Res Commun 1994; 205:1829-1835.
99. Zhang Y, Hayes A, Pritchard A, et al. Interleukin-6 promoter polymorphism: risk and pathology of Alzheimer's disease. Neurosci Lett 2004; 362:99-102.
100. Grace EA, Busciglio J. Aberrant activation of focal adhesion proteins mediates fibrillar amyloid-β-induced neuronal dystrophy. J Neurosci 2003; 23: 493-502.
101. Flaherty DB, Soria JP, Tomasiewicz HG, et al. Phosphorylation of human tau protein by microtubule-associated kinases: GSK3β and cdk5 are key participants. J Neurosci Res 2000; 62:463-472.
102. Arendt T. Alzheimer's disease as a disorder of mechanisms underlying structural brain self-organization. Neuroscience 2001; 102:723-765.
103. Vane JR, Bakhle YS, Botting RM. Cycloocygenase 1 and 2. Annu Rev Pharmacol Toxicol 1998; 38:97-120.
104. Pasinetti GM, Aisen PS. Cyclooygenase-2 expression is increased in frontal cortex of Alzheimer's disease. Neuroscience 1998; 87:319-324.
105. Yermakova AV, O'Banion MK. Downregulation of neuronal cyclooxygenase-2 expression in end stage Alzheimer's disease. Neurobiol Aging 2001; 22 823-826.
106. Hoozemans JJM, Bruckner MK, Rozemuller AJM, et al. Cyclin D1 and cyclin E are co-localized with cyclo-oxygenase 2 (COX-2) in pyramidal neurons in Alzheimer's disease temporal cortex. J Neuropathol Exp Neurol 2002; 61:678-688.
107. Combrinck M, Williams J, De Berardinis MA, et al. Levels of CSF prostaglandin E2, cognitive decline and survival in Alzheimer's disease. J Neurol Neurosurg Psychiatry 2006; 77:85-88.
108. Arendt T, Rodel L, Gartner U, et al. Expression of the cyclin-dependent kinase inhibitor p16 in Alzheimer's disease. Neuroreport 1996; 7:3047-3049.
109. Nagy Z, Esiri MM, Cato AM, et al. Cell cycle markers in the hippocampus in Alzheimer's disease. Acta Neuropathol (Berl) 1997; 94:6-15.
110. Jordan-Sciutto KL, Dorsey R, Chalovich EM, et al. Expression patterns of retinoblastoma protein in Parkinson disease. J Neuropathol Exp Neurol 2003; 62:68-74.
111. Lee SS, Kim YM, Junn E, et al. Cell cycle aberrations by alpha-synuclein over-expression and cyclin B immunoreactivity in Lewy bodies. Neurobiol Aging 2003; 24:687-696.
112. Nguyen MD, Boudreau M, Kriz J, et al. Cell cycle regulators in the neuronal death pathway of amyotrophic lateral sclerosis caused by mutant superoxide dismutase 1. J Neurosci 2003; 23:2131-2140.
113. Ranganathan S, Bowser R. Alterations in G(1) to S phase cell-cycle regulators during amyotrophic lateral sclerosis. Am J Pathol 2003; 162:823-835.
114. Mirjany M, Ho L, Pasinetti GM. Role of cyclooxygenase-2 in neuronal cell cycle activity and glutamate-mediated excitotoxicity. J Pharmacol Exp Ther 2002; 301:494-500.
115. Xiang Z, Ho L, Valdellon J, et al. Cyclooxygenase (COX)-2 and cell cycle activity in a transgenic mouse model of Alzheimer's disease neuropathology. Neurobiol Aging 2002; 23:327-334.
116. Arendt T. Synaptic plasticity and cell cycle activation in neurons are alternative effector pathways: the 'Dr. Jekyll and Mr. Hyde concept' of Alzheimer's disease or the yin and yang of neuroplasticity. Prog Neurobiol 2003; 71:83-248.
117. Wu Q, Combs C, Cannady SB, et al. Beta-amyloid activated microglia induce cell cycling and cell death in cultured cortical neurons. Neurobiol Aging 2000; 21:797-806.
118. Copani A, Condorelli F, Caruso A, et al. Mitotic signaling by beta-amyloid causes neuronal death. FASEB J 1999; 13:2225-2234.
119. Vincent I, Jicha G, Rosado M, et al. Aberrant expression of mitotic cdc2/cyclin B1 kinase in degenerating neurons of Alzheimer's disease brain. J Neurosci 1997; 17:3588-3598.
120. Busser J, Geldmacher DS, Herrup K. Ectopic cell cycle proteins predict the sites of neuronal cell death in Alzheimer's disease brain. J Neurosci 1998; 18:2801-2807.
121. Kranenburg O, Van der Eb AJ, Zantema A. Cyclin D1 is an essential mediator of apoptotic neuronal cell death. EMBO J 1996; 15:46-54.
122. Liu DX, Greene LA. Neuronal apoptosis at the G1/S cell cycle checkpoint. Cell Tiss Res 2001; 305:217-228.
123. Gartner U, Holzer M, Arendt T. Elevated expression of p21ras is an early event in Alzheimer's disease and precedes neurofibrillary degeneration. Neuroscience 1999; 91:1-5.
124. Hoozemans JJM, Veerhuis R, Rozemuller AJM, et al. Non-steroidal anti-inflammatory drugs and cyclooxygenase in Alzheimer's disease. Curr Drug Targets 2003; 4:461-468.
125. Yang Y, Mufson EJ, Herrup K. Neuronal cell death is preceded by cell cycle events at all stages of Alzheimer's disease. J Neurosci 2003; 23:2557-2563.
126. Hoozemans JJM, Veerhuis R, Rozemuller AJM, et al. Neuronal COX-2 expression and phosphorylation of pRb precede p38 MAPK activation and neurofibrillary changes in AD temporal cortex. Neurobiol Dis 2004; 15:492-499.

127. Husseman JW, Nochlin D, Vincent I. Mitotic activation: a convergent mechanism for a cohort of neurodegenerative diseases. Neurobiol Aging 2000; 21: 815-828.
128. Taylor JP, Hardy J, Fischbeck KH. Toxic proteins in neurodegenerative disease. Science 2002; 296: 1991-1995.
129. Forman MS, Lee VM, Trojanowski JQ. 'Unfolding' pathways in neurodegenerative disease. Trends Neurosci 2003; 26:407-410.
130. Rutkowski DT, Kaufman RJ. A trip to the ER: coping with stress. Trends Cell Biol 2004; 14:20-28.
131. Brewer JW, Hendershot LM, Sherr CJ, et al. Mammalian unfolded protein response inhibits cyclin D1 translation and cell-cycle progression. Proc Natl Acad Sci U S A 1999; 96:8505-8510.
132. Brewer JW, Diehl JA. PERK mediates cell-cycle exit during the mammalian unfolded protein response. Proc Natl Acad Sci U S A 2000; 97: 12625-12630.
133. Hoozemans JJM, Veerhuis R, Rozemuller JM, et al. The unfolded protein is activated in Alzheimer's disease. Acta Neuropathol 2005; 110:165-172.
134. Hoozemans JJM, Veerhuis R, Rozemuller JM, et al. Neuroinflammation and regeneration in the early stages of Alzheimer's disease pathology. Int J Devl Neurosci 2006; 24:157-165.
135. Colangelo V, Schurr J, Ball MJ, et al. Gene expressing profiling of 12633 genes in Alzheimer hippocampal CA1: transcription and neurotrophic factor down-regulation and up-regulation of apoptotic and pro-inflammatory signaling. J Neurosci Res 2002; 249:1242-1245.
136. Blalock EM, Geddes JW, Chen KC, et al. Incipient Alzheimer's disease: microarray correlation analyses reveal major transcriptional and tumor suppressor responses. Proc Natl Acad Sci U S A 2004; 101: 2173-2178.
137. Games D, Adams D, Alessandrini R, et al. Alzheimer-type neuropathology in transgenic mice overexpressing V717F β-amyloid precursor protein. Nature 1995; 373:523-527.
138. Duff K, Eckman C, Zehr C, et al. Increased Aβ42(43) in brains of mice expressing mutant presenilin 1. Nature 1996; 383:710-713.
139. Hsiao K, Chapman P, Nilsen S, et al. Correlative memory deficits, Aβ elevation, and amyloid plaques in transgenic mice. Science 1996; 274:99-102.
140. Apelt J Schliebs R. β-amyloid–induced glial expression of both pro- and anti-inflammatory cytokines in cerebral cortex of transgenic Tg2576 mice with Alzheimer plaque pathology. Brain Res 2001; 894:21-30.
141. Stalder M, Phinney A, Probst A, et al. Association of microglia with amyloid plaques in brains of APP23 transgenic mice. J Am Pathol 1999; 154:1673-1684.
142. Benzing WC, Wujek JR, Ward EK, et al. Evidence for glial mediated inflammation in aged APP(SW) transgenic mice. Neurobiol Aging 1999; 20:581-589.
143. Matsuoka Y, Picciano M, Malester B, et al. Inflammatory responses to amyloidosis in a transgenic mouse model of Alzheimer's disease. Am J Pathol 2001; 158:1345-1354.
144. Wegiel J, Wang H, Imaki H, et al. The role of microglial cells and astrocytes in fibrillar plaque evolution in transgenic (sw) mice. Neurobiol Aging 2001; 22:49-61.
145. Dudal S, Krzywkowski P, Paquette J, et al. Inflammation occurs early during the Aβ deposition process in TGCRND8 mice. Neurobiol Aging 2004; 25:861-971.
146. Billings LM, Odo S, Green KN, et al. Intraneuronal Aβ causes the onset of early Alzheimer's disease-related cognitive deficits in transgenic mice. Neuron 2005; 45:675-688.
147. Herzig MC, Winkler DT, Burgermeister P, et al. Aβ is targeted to the vasculature in a mouse model of hereditary cerebral hemorrhage with amyloidosis. Nat Neurosci 2004; 7:954-960
148. Davis J, Xu F, Deane R, et al. Early-onset and robust cerebral microvascular accumulation of amyloid-β protein in transgenic mice expressing low levels of a vascular tropic Dutch/Iowa mutant form of amyloid β-protein precursor. J Biol Chem 2004; 279:20296-20306.
149. Miao J, Xu F, Davis J, Otte-Höller, et al. Cerebral microvascular amyloid β protein deposition induces vascular degeneration and Neuroinflammation in transgenic mice expressing human vasculotropic mutant amyloid β precursor protein. Am J Pathol 2005; 167:505-515.
150. Bales KR, Verina T, Dodel RC, et al. Lack of apoliprotein E dramatically reduces amyloid β-peptide deposition. Nat Genet 1997; 17:263-264.
151. Nilsson LNG, Bales KR, DiCarlo G, et al. α1-Antichymotrypsin promotes β-sheet amyloid plaque formation in a transgenic mouse model of Alzheimer's disease. J Neurochem 2001; 21:1444-1451.
152. Wyss-Coray T, Yan F, Lin AH, Lambris JD, et al. Prominent neurodegeneration and increased plaque formation in complement inhibited Alzheimer's mice. Proc Natl Acad Sci U S A 2002; 99:10837-10842.
153. DeMattos RB, O'dell MA, Parsadanian M, et al. Clusterin promotes amyloid plaque formation and it is critical for neuritic toxicity in a mouse model of Alzheimer's disease. Proc Natl Acad Sci U S A 2002; 99:10843-10848.

154. DiCarlo G, Wilcock D, Henderson D, et al. Intrahippocampal LPS injections reduce Aβ load in APP+PS1 transgenic mice. Neurobiol Aging 2001; 22:1007-1012.
155. Jantzen PT, Connor KE, DiCarlo G, et al. Microglial activation and β-amyloid deposits reduction caused by a nitric oxide-relasing nonsteroidal anti-inflammatory drug in amyloid precursor protein plus presenilin transgenic mice. J Neurosci 2002; 22:2246-2254.
156. Schenk D, Barbour R, Dunn W, et al. Immunization with Aβ attenuates Alzheimer-disease-like pathology in PDAPP mouse. Nature 199; 400:173-177.
157. Bard F, Cannon C, Barbour R, et al. Peripherally administered antibodies against amyloid-β enter the central nervous system and reduce pathology in a mouse model of Alzheimer's diseases. Nat Med 2000; 6:916-919.
158. Bacskai BJ, Kajdasz ST, Christie RH, et al. Imaging of amyloid-β deposits in brains of living mice permits direct observation of clearance of plaques with immunotherapy. Nat Med 2001; 7:369-372.
159. Nicoll JAR, Wilkinson D, Holmes C, et al. Neuropathology of human Alzheimer's disease after immunization with amyloid-β peptide: a case report. Nat Med 2003; 9:448-452.
160. Eikelenboom P, Zhan SS, van Gool WA, et al. Inflammatory mechanisms in Alzheimer's disease. Trends Pharmacol Sci 1994; 15:447-450.
161. Carson JA, Turner AJ. β-amyloid catabolism, roles for neprilysin (NEP) and other metallopeptidases? J Neurochem 2000; 81:1-8.
162. Goldgaber D, Harris HW, Hla T, et al. Interleukin 1 regulates synthesis of amyloid β-protein precursor mRNA in human endothelial cells. Proc Natl Acad Sci U S A 1989; 86:7606-7610.
163. Blasko I, Marx F, Steiner E, et al. TNFα plus IFNγ induce the production of Alzheimer β-amyloid peptides and decrease the secretion of APPs. FASEB J 1999; 13:63-68.
164. Rogers JT, Leiter LM, McPhee J, et al. Translation of the Alzheimer amyloid precursor protein mRNA is upregulated by interleukin-1 through 5′-untranslated regions sequences. J Biol Chem 1999; 274:6421-6431.
165. De Groot CJ, Hulshof S, Hoozemans JJM, et al. Establishment of microglial cell cultures derived from postmortem human adult brain tissue: immunophenotypical and functional characterization. Microsc Res Tech 2001; 54:34-39.
166. Rogers J, Strohmeyer R, Kovelowski CJ, et al. Microglia and inflammatory mechanisms in the clearance of amyloid β peptide. Glia 2002; 40:260-269.
167. Wyss-Coray T, Mucke L. Inflammation in neurodegenerative disease: a double-edged sword. Neuron 2002; 35:419-432.
168. Giometto B, Argentiero V, Sanson F, et al. Acute-phase proteins in Alzheimer's disease. Eur Neurol 1988; 28; 30-33.
169. Hinds TR, Kukull WA, Van Belle, et al. Relationship between α1-antichymotrypsin and Alzheimer's disease. Neurobiol Aging 1994; 15:21-27.
170. Licastro F, Parnetti L, Morini MC, et al. Acute phase reactant α1-antichymotrypsin is increased in cerebrospinal fluid and serum of patients with probable Alzheimer's disease. Alzheimer Dis Assoc Disord 1995; 9:112-118.
171. Matsubara E, Hirai S, Amari M, et al. α1-antichymotrypsin as a possible biomedical marker for Alzheimer-type dementia. Ann Neurol 1990; 28: 561-567.
172. Sun YX, Minthon L, Wallmark A, et al. Inflammatory markers in matched plasma and cerebrospinal fluid from patients with Alzheimer's disease. Dement Geriatr Cogn Diord 2003; 16:136-144.
173. DeKosky ST, Ikonomovic MD, Wang X, et al. Plasma and cerebrospinal fluid α1-antichymotrypsin levels in Alzheimer's disease: correlation with cognitive impairment. Ann Neurol 2003; 53:81-90.
174. Fillit H, Ding WH, Buee L, et al. Elevated circulating tumor necrosis factor levels in Alzheimer's disease. Neurosci Lett 1991; 129:318-320.
175. Singh VK, Guthinkonda P. Circulating cytokines in Alzheimer's disease. J Psychiatr Res 1997; 31:657-660.
176. Maes M, DeVos N, Wauters A, et al. Inflammatory markers in younger vs elderly normal volunteers and in patients with Alzheimer's disease. J Psychiatr Res 1999; 33:397-405.
177. Raskind MA, Peskind E, Rivard MF, et al. The dexamethasone suppression test and cortisone circadiane rhytm in primary degenerative dementia. Am J Psychiatry 1982; 139:1468-1471.
178. Greenwald BS, Mathe AA, Mohs RC, et al. Cortisol and Alzheimer's disease. II. Dexamehasone suppression, dementia severity, and affective symptoms. Am J Psychiatry 1996; 143:442-446.
179. Ferrier IN, Pascual J, Charlton BG, et al. Cortisol, ACTH, and dexamethosone concentrations in a psychogeriatric population. Biol Psychiatry 1988; 23: 252-260.
180. Molchan SE, Hill JL, Mellow AM, et al. The dexamethasone suppression test in Alzheimer's disease and major depression: relationship between dementia severity, depression, and CSF monoamines. Int Psychogeriatr 1990; 2:99-122.
181. Raadsheer FC, Van Heerikhuize JJ. Lucassen PJ, et al. Corticotropin-releasing hormone mRNA levels in the paraventricular nucleus of patients with Alzheimer's disease and depression. Am J Psychiatry 1995; 152:1372-1376.

182. Lawlor BA, Bierer LM, Ryan RM, et al. Plasma 3-methoxy-4-hydrophenylglycol (MHPG) and clinical symptoms in Alzheimer's disease. Biol Psychiatry 1995; 95:185-188.
183. Hoogendijk WJG, Feenstra MGP, Botterblom MHA, et al. Increased activity of surviving neurons in Alzheimer's disease. Ann Neurol 1999; 45:82-91.
184. Eikelenboom P, Rozemuller JM, Van Muiswinkel FL. Inflammation and Alzheimer's disease: relationships between pathogenic mechanisms and clinical expression. Exp Neurol 1998; 154:89-98.
185. Schmidt R, Schmidt H, Curb JD, et al. Early inflammation and dementia: a 25-years follow-up of the Honolulo-Asia Aging Study. Ann Neurol 2002; 52: 168-174.
186. Yaffe K, Lindquist K, Penninx BW, et al. Inflammation and cognition in well-functioning African-American and white elders. Neurology 2003; 61:76-80.
187. Engelhart MJ, Geerlings MI, Meijer J, et al. Inflammatory proteins in plasma and the risk of dementia: the Rotterdam study. Arch Neurol 2004; 61:668-672.
188. Dik MG, Jonker C, Hack CE, et al. Serum inflammatory proteins and cognitive decline in older patients. Neurology 2005; 64:1371-1377.
189. Yaffe K, Kanaya A, Lindquist K, et al. The metabolic syndrome, inflammation, and risk of cognitive decline. JAMA 2004; 292:2237-2242.
190. Grimaldi LM, Casadei VM, Ferri C, et al. Association of early-onset Alzheimer's disease with an interleukin-1α gene polymorphism. Ann Neurol 2000; 47:361-365.
191. Nicoll JAR, Mrak RE, Graham DI, et al. Association of interleukin-1 gene polymorphisms with Alzheimer's disease. Ann Neurol 2000; 47: 365-368.
192. Du Y, Dodel RC, Eastwood BJ, et al. Association of an interleukin 1α polymorphism with Alzheimer's disease. Neurology 2000; 55:480-483.
193. Rebeck GW. Conformation of the genetic association of interleukin-1A with early onset sporadic Alzheimer's disease. Neurosci Lett 2000; 293:75-77.
194. Combarros O, Sanchez-Guerra M, Infante J, et al. Gene dose-dependent association of interleukin-1A allele polymorphism with Alzheimer's disease. J Neurol 2002; 249:1242-1245.
195. Fidani L, Goulas A, Mittsou V, et al. Interleukin-1A polymorphism is not associated with late onset Alzheimer's disease. Neurosci Lett 2002; 323:81-83.
196. Kuo YM, Liao PC, Lin C, et al. Lack of association interleukin-1α polymorphism and Alzheimer's disease or vascular dementia. Alzheimer Dis Assoc Disord 2003; 17:94-97.
197. Li XQ, Zhang JW, Zhang ZX, et al. Interleukin-1 gene cluster polymorphisms and risk of Alzheimer's disease in Chinese Han population. J Neural Transm 2004; 111:1183-1190.
198. Tsai SJ, Liu HC, Liu TY, et al. Lack of association between the interleukin-1α gene (C-889)T polymorphism and Alzheimer's disease in a Chinese poulation. Neurosci Lett 2003; 343:93-96.
199. Rainero I, Bo M, Ferrero M, et al. Association between the interleukin-1α gene and Alzheimer's disease: a meta-analysis. Neurobiol Aging 2004; 25:1293-1298.
200. Hayes A, Green EK, Pritchard A, et al. A polymorphic variation in the interleukin 1A gene increases brain microglial cell activity in Alzheimer's disease. J Neurol Neurosurg Psychiatry 2001; 75:1475-1477.
201. Bagli M, Papassotiropoulos A, Knapp M, et al. Association between an interleukin-6 promotor and 3′flanking region haplotype and reduced Alzheimer's risk in a German population. Neurosci Lett 2000; 283:109-112.
292. Bhojak TJ,DeKosky ST, Ganguli M, et al. Genetic polymorphisms in the cathepsin G and interleukin-6 genes and the risk of Alzheimer's disease. Neurosci Lett 2000; 288:21-24.
203. McCusker SM, Curran MD, Dynan KB, et al. Association between polymorphism in regulatory regions of gene encoding tumour necrosis factor α and risk of Alzheimer's disease and vascular dementia: a case-control study. Lancet 2001; 357: 436-439.
204. Culpan D, MacGowan SH, Ford JM, et al. Tumour necrosis factor-α gene polymorphisms and Alzheimer's disease. Neurosci Lett 2003; 350:61-65.
205. Ehl C, Kolsch H, Ptok U, et al. Association of an interleukin-1β gene polymorphism at position –511 with Alzheimer's disease. Int J Mol Med 2003; 11: 235-238.
206. Licastro F, Grimaldi LME, Bonafè M, et al. Interleukin-6 gene alleles affect the risk of Alzheimer's disease and the levels of the cxytokine in blood and brain. Neurobiol Aging 2003; 24:921-926.
207. Ma SL, Tang NLS, Lam LCW, et al. Lack of association of the interleukin-1β gene polymorphism with Alzheimer's disease in a Chinese population. Dement Geriatr Cogn Disord 2003; 16:265-268.
208. Rosenmann H, Meiner Z, Dressner-Pollak R, et al. Lack of association of interleukin-1β polymorphism with Alzheimer's disease in the Jewish population. Neurosci Lett 2004; 363:131-133.
209. Sciacca FL, Ferri C, Licastro F, et al. Interleukin-1B polymorhism is associated with age at onset of Alzheimer's disease. Neurobiol Aging 2003; 24: 927-931.

210. Zhang C, Lampert MP, Bunch C, et al. Focal adhesion kinase expressed by nerve cell lines shows increased tyrosine phosphorylation in response to Alzheimer's Aβ peptide. J Biol Chem 1994; 269: 2547-2550.
211. Seripa D, Matera MG, Dal Forno G, et al. Genotypes and haplotypes in the Il-1 gene cluster: analysis of two genetically and diagnostically distinct groups of Alzheimer's patients. Neurobiol Aging 2005; 26:455-464.
212. Kamboh MI, Sanghera DK, Ferrell RE, et al. APOE*4-associated Alzheimer's disease risk is modified by α1-antichymotrypsin polymorphism. Nat Genet 1995; 10:486-488.
213. Nacmias B, Marcon G, Tedde A, et al. Implication of α1-antichymotrypsin polymorphism in familial Alzheimer's disease. Neurosci Lett 1998; 244:85-88.
214. Licastro F, Pedrini S, Giovoni M, et al. Apolipoprotein E and α1-antichymotrypsin allele polymorphism in sporadic and familial Alzheimer's disease. Neurosci Lett 1999; 270:129-132.
215. Licastro F, Pedrini S, Ferri C, et al. Gene polymorhism affecting α1-antichymotrypsin and interleukin 1 plasma levels increases Alzheimer's disease risk. Ann Neurol 2000; 48:388-391.
216. Muramatsu T, Matsushita S, Arai H, et al. Alpha1-antichymotrypsin gene poymorphism and risk for Alzheimer's disease. J Neural Transm 1996; 103: 1205-1210.
217. Yoshizawa T, Yamakawa-Kobayashi K, Hamaguchi H, et al. Alpha1-antichymotrypsin polymorphism in Japanese cases of Alzheimer's disease. J Neurol Sci 1997; 152:136-139.
218. Itabashi S, Aria H, Matsui T, et al. Absence of association of α1-antichymotrypsin polymorphisms with Alzheimer's disease: a report on autopsied confirmed cases. Exp Neurol 1998; 151:237-240.
219. Schwab SG, Bagli M, Papassotiropoulos A, et al. Alpha1-1-antichymotrypsin gene polymorphism and risk for sporadic Alzheimer's disease in a german population. Dement Geriatr Cogn Disord 1999;10:469-472.
220. Sodeyama N, Yamada M, Itoh Y, et al. Lack of genetic associations of α1-antichymotrypsin polymorphism with alzheimer-type neuropathological changes or sporadic Alzheimer's disease. Dement Geriatr Cogn Disord 1999; 10:221-225.
221. Wang YC, Liu TY, Chi CW, et al. No association between α1-antichymotrypsin polymorphism and Alzheimer's disease. Neuropsychobiology 1999; 40:67-70.
222. Kim KW, Jhoo JH, Lee KU, et al. No association between α1-antichymotrypsin polymorphism and Alzheimer's disease in Koreans. Am J Med Genet 2000; 91:355-358.
223. Ki CS, Na DL, Kim JW. Alpha-1 antichymotrypsin and alpha-2 macroglobulin gene polymorphisms are not associated with Korean late-onset Alzheimer's disease. Neurosci Lett 2001; 302:69-72.
224. Kamboh MI, Minster RL, Kenney M, et al. Alpha-1-antichymotrypsin (ACT or SERPINA3) polymorphism may affect age-at-onset and disease duration of Alzheimer's disease. Neurobiol Aging 2005 Aug 29 [Epub ahead of print].
225. Blacker D, Wilcox MA, Laird NM, et al. Alpha-2 macroglobulin is genetically associated with Alzheimer's disease. Nat Genet 1998; 19:357-360.
226. Saunders AJ, Bertram L, Mullin K, et al. Genetic association of Alzheimer's disease with multiple polymorphisms in alpha-2-macroglobulin. Hum Mol Genet 2003; 12:2765-2776.
227. Mehne P, Grünwald P, Gerner-Beuerle E. Ein serogenetischer Beitrag zur Ätiopathogenese der Alzheimerschen Erkrankung. Akt Gerontol 1976; 6: 259-264.
228. Nerl C, Mayeux R, O'Neill GJ. HLA-linked complement markers in Alzheimer's and Parkinson's disease C4 variant (C4B2) – a possible marker for senile dementia of the Alzheimer type. Neurology 1984; 34:310-314.
229. Eikelenboom P, Vink-Starreveld ML, Jansen W, et al. C3 and haptoglobin polymorphism in dementia of the Alzheimer type. Acta Psychiatr Scand 1984; 69:140-142.
230. Eikelenboom P, Goetz J, Pronk JC, et al. Complement C4 phenotypes in dementia of the Alzheimer type. Hum Hered 1988; 38:48-51.
231. Gentleman SM, Graham DI, Roberts GW. Molecular pathology of head trauma: altered βAPP and the etiology of Alzheimer's disease. Prog Brain Res 1993; 96:237-246.
232. Rumble B, Retallack R, Hilbich C, et al. Amyloid A4 protein and its precursor in Down's syndrome and Alzheimer's disease. N Engl J Med 1999; 320: 1446-1452.
233. Lemieux N, Malfoy B, Forrest GL. Human carbonyl reductase (CBR) localized to band 21q22 by high resolution fluorescence in situ hybridisation displays gene dose effect in trisomy 21 cells. Genomics 1993; 15:169-172.
234. Taylor GM. Altered expression of lymphocyte functional antigen in Down syndrome. Immunol Today 1987; 8:366-369.
235. Hyman BT, Marzloff K, Arrigada V. The lack of accumulation of senile plaques or amyloid burden in Alzheimer's disease suggests a dynamic balance between amyloid deposition and removal. J Neuropathol Exp Neurol 1993; 52:594-600.
236. Hofman A, Ott A, Breteler MMB, et al. Atherosclerosis, apolipoprotein E, and the preva-

lence of dementia and Alzheimer's disease in the Rotterdam Study. Lancet 1997; 349:151-154.
237. Danton GH, Dietrich WD. Inflammatory mechanisms after ischemia and stroke. J Neuropathol Exp Neurol 2003; 62:17-136.
238. Van Gool WA, Eikelenboom P. The two faces of Alzheimer's disease. J Neurol 2000; 247:500-505.
239. Koistinaho M, Koistinaho J. Interactions between Alzheimer's disease and cerebral ischemia-focus on inflammation. Brain Res Rev 20005; 48:240-250.
240. Blennow K, Wallin A. Clinical heterogenity of probable Alzheimer's disease. J Geriatr Psychiatry Neurol 1992; 5:106-113.
241. Hoyer S. The brain insulin signal transduction system and sporadic (type II) Alzheimer's disease: an update. J Neural Transm 2002; 109:341-360.
242. Eikelenboom P, van Gool WA. Neuroinflammatory perspectives on the two faces of Alzheimer's disease. J Neural Transm 2004; 111:281-294.
243. Hotamisligil GS, Budavari A, Murray D, et al. Reduced tyrosine kinase activity of the insulin receptor in obesitas-diabetes, Central role of tumour necrosis factor-α. J Clin Invest 1994; 94:1543-1549.
244. Ross R. Arterosclerosis—an inflammatory disease. N Engl J Med 1999; 340:115-126.
245. Libby P. Inflammation in arterosclerosis. Nature 2002; 420:868-874.
246. Brachova L, Lue LF, Schultz J, et al. Association cortex, cerebellum, and serum concentrations of C1q and factor B in Alzheimer's disease. Mol Brain Res 1993; 18:329-334.
247. Lue LF, Brachova L. Civin WH, et al. Inflammation, Aβ deposition, neurofibrillary tangles as correlates of Alzheimer's disease neurodegeneration. J Neuropathol Exp Neurol 1996; 55:1083-1088.
248. Bornebroek M, Haan J, Maat Schieman MLC, et al. Hereditary cerebral hemorrhage with amyloidosis-Dutch type. I—A review of clinical, radiological and genetic aspects. Brain Pathol 1996; 6:111-114.
249. Rozemuller JM, Bots GT, Roos RAC, et al. Acute phase proteins but not activated microglia are present in parenchymal β/A4 deposits in the brains of patients with hereditary cerebral hemorrhage with amyloidosis-Dutch type. Neurosci Lett 1992; 140: 137-140.
250. Maat-Schieman MLC, van Duinen SG, Rozemuller AJM, et al. Association of vascular amyloid-β and cells of the mononuclear phagocyte system in hereditay cerebral hemorrhage with amyloidosis (Dutch) and Alzheimer's disease. J Neuropathol Exp Neurol 1997; 56:273-284.
251. Eikelenboom P, Rozemuller JM, Fraser H, et al. Neuroimmunological mechanisms in cerebral amyloid deposition in Alzheimer's disease. In: Ishii T, Allsop D, Selkoe DJ, editors. Frontiers of Alzheimer Research. Amsterdam, New York, Oxford: Excerpta Medica, 1999:259-271.
252. Maat-Schieman MLC, Rozemuller AJM, van Duinen SG, et al. Microglia in diffuse plaques in hereditary cerebral hemorrhage with amyloidosis (Dutch). An immunohistochemical study. J Neuropathol Exp Neurol 1994:483-491.
253. Verbeek MM, Otte-Holler I, Westphal JR, et al. Differential expressing of intercellular adhesion molecule-1 (ICAM-1) in the Aβ containing lesions in brains of patients with dementia of the Alzheimer type. Acta Neuropathol 1996; 91; 608-615.
254. Verbeek MM, Eikelenboom P, de Waal RMW. Differences between the pathogenesis of senile plaques and congophilic angiopathy in Alzheimer's disease. J Neuropathol Exp Neurol 1997; 56:751-761.
255. Vidal R, Calero M, Piccardo P, et al. Senile dementia associated with amyloid β protein angiopathy and tau perivascular pathology but not neuritic plaques in patients homozygous for the APOE-ε4 allele. Acta Neuropathol 2000; 100:1-12.
256. Yamada M, Itoh Y, Suematsu N, et al. Vascular variant of Alzheimer's disease characterized by severe plaque like β protein angiopathy. Dement Geriatr Cogn Disord 1997; 8:163-168.
257. Rozemuller AJM, van Gool WA, Eikelenboom P. The neuroinflammatory response in plaques and amyloid angiopathy in Alzheimer's disease: therapeutic implications. Curr Drug Targets CNS Neurol Disord 2005; 4:223-233.
258. Devenand DP, Sano M, Tang M-X, et al. Depressed mood and the incidence of Alzheimer's disease living in the community. Arch Gen Psychiatry 1996; 53; 175-182.
259. Geerlings MI, Schmand B, Braam AW, et al. Depressive symptoms and risk of Alzheimer's disease in more highly educated older people. J Am Geriatr Soc 2000; 48:1092-1097.
260. Zeidler M, Johnstone EC, Bamber RW, et al. New variant Creutzfeldt-Jakob disease: psychiatric features. Lancet 1997; 350:908-910.
261. Tilders FJH, Schmidt ED, Hoogendijk WJG, et al. Delayed effect of stress and immune activation. Ballieres Best Pract Res Clin Endocrinol Metab 1999; 13:523-540.
262. Dantzer R, Aubert A, Bluthe RM, et al. Sickness behavior: a neuroimmune-based response to infectious disease. In: Patterson P, Kordon C, Christen Y, editors. Neuro-immune Interactions in Neurologic and Psychiatric Disorders. Berlin, Heidelberg: Springer, 2000:169-184.
263. Yirmiya R, Pollak Y, Morag M, et al. Illness, cytokines, and depression. Ann N Y Acad Sci 2001; 917:488-499.

264. Reichenberg A, Yirmiya R, Schuld AM, et al. Cytokine-associated emotional and cognitive disturbances in humans. Arch Gen Psychiatry 2001; 56:445-452.
265. Eikelenboom P, Hoogendijk WJG, Jonker C, et al. Immunological mechanisms and the spectrum of psychiatric syndromes in Alzheimer's disease. J Psychiatr Res 2002; 36:269-280.
266. Rogers J, Kirby LC, Hempelman SR, et al. Clinical trial of indomethacin in Alzheimer's disease. Neurology 1993; 43:1609-1611.
267. Scharf S, Mander A, Ugoni A, et al. A double-blind, placebo-controlled trial of diclofenac/misoprostol in Alzhemer's disease. Neurology 1999; 53:197-201.
268. Aisen PS, Schmeidler J, Pasinetti GM. Randomized pilot study of nimesulfide treatment in Alzheimer's disease. Neurology 2002; 58:327-334.
269. Sainati SM, Ingram DM, Talwalker S, et al. Results of a double-blind, randomized, placebo-controlled study of celecoxib in the treatment of progression of Alzheimer's disease. Proceedings of the Sixth International Stockholm/Springfield Symposium on Advances in Alzheimer therapy, 2000:180.
270. Van Gool WA, Weinstein HC, Scheltens Ph, et al. Effect of hydrocholoroquine on the progression of dementia in early Alzheimer's disease: an 18-month randomised, double-blind, placebo-controlled study. Lancet 2001; 358:455-460.
271. Aisen PS, Davis KL, Berger JD, et al. A randomized controlled trial of predinsone in Alzheimer's diseae. Alzheimer's disease Cooperative Study. Neurology 2000; 54:588-593.
272. Aisen PS, Schafer KA, Grundman M, et al. Effects of rofecoxib or naproxen vs placebo on Alzheimer's disease progression: a randomized controlled trial. JAMA 2003; 289:2819-282.
273. Reines SA, Block GA, et al. Rofecoxib: no effect of Alzheimer's disease in a 1-year, randomized, blinded, controlled study. Neurology 2004; 62:66-71.
274. Van Gool WA, Aisen PS, Eikelenboom P. Anti-inflammatory therapy in Alzheimer's disease: is hope still alive. J Neurol 2003; 250:788-792.
275. Morgan D, Diamond DM, Gottschall PE, et al. Aβ peptide vaccination prevents memory loss in an anmal model of Alzheimer's disease. Nature 2000; 408:982-985.
276. Breitner JC, Zandi PP. Do nonsteroidal antiinflammatory drugs reduce the risk of Alzheimer's disease. N Engl J Med 2001; 345:1567-1568.
277. Hoozemans JJM, Veerhuis R, Janssen I, et al. The role of cyclooxygenase-1 and –2 activity in prostaglandin E2 secretion by cultured human and microglia: Implications for Alzheimer's disease. Brain Res 2002; 951:218-226.
278. Landreth GE, Heneka MT. Anti-inflammatory actions of perixisome proliferator-activated receptor γ agonist in Alzheimer's disease. Neurobiol Aging 2001; 22:937-944.
279. Camacho IE, Serneels L, Spittaels K, et al. Peroxisome proliferator-activated receptor γ induces a clearance mechanism for the amyloid-β peptide. J Neurosci 2004; 24:10908-10917.
280. Weggen S, Eriksen JL, Das P, et al. A subset of NSAIDs lower amyloidogenic Aβ42 independently of cyclooxygenase activity. Nature 2001; 414:212-216.
281. Morihara T, Chu T, Ubeda O, et al. Selective inhibition of Aβ42 production by NSAID R enantiomers. J Neurochem 2002; 83:1009-1012.
282. Eriksen JL, Sagi SL, Smith TE, et al. NSAIDs and enantiomers of fluriprofen target γ-secretase activity and lower Aβ in vivo. J Clin Invest 2003; 112: 440-449.
283. Weggen S, Eriksen JL, Sagi SA, et al. Evidence that nonsteroidal anti-inflammatory drugs decrease amyloid beta 42 production by direct modulation of gamma-secretase activity. J Biol Chem 2003; 278: 31831-31837.
284. Beher D, Clarke EE, Wrigley JD, et al. Selected non-steroidal anti-inflammatory drugs and their derivatives target γ-secretase at a novel site. Evidence for an allosteric mechanism. J Biol Chem 2004; 279:43419-43426.
285. Lleo A, Berezovska O, Herl L, et al. Nonsteroidal antiinflammatory drugs lower Aβ42 and change presenilin conformation. Nat Med 2004; 10:1065-1066.
286. Liang X, Wang Q, Hand T, et al. Deletion of the prostaglandin E2, EP2 receptor reduces oxidative damage and amyloid burden in a model of Alzheimer's disease model. J Neurosci 2005; 25: 10180-10187.
287. Wenk GL, Rosi S, McGann K, et al. A nitric oxide-donating flurbiprofen derivative reduces neuroinflammation without interaction with galantamine in the rat. Eur J Pharmacol 2002; 453:319-324.
288. Townsend KP, Patrico D. Novel therapeutic opportunities for Alzheimer's disease: focus on nonsteroidal anti-inflammatory drugs. FASEB J 2005; 19:1592-1601.
289. Brendza RP, Bacskai BJ, Cirrito JR, et al. Anti-Aβ antiboy treatment promotes the rapid recovery of the amyloid associated neuritic dystrophy in PDAPP transgenic mice. J Clin Invest 2005; 115: 428-433.

5
Amyloid β-Peptide(1-42), Oxidative Stress, and Alzheimer's Disease

D. Allan Butterfield

5.1 Introduction

Alzheimer's disease (AD) a progressive, age-related neurodegenerative disorder that affects memory, cognition, and speech, is present in more than 4 million persons. The number of cases of AD will significantly elevate, because the mean population in the United States is increasing [1]. AD is characterized pathologically by the presence of extracellular senile plaques, intracellular neurofibrillary tangles, and synapse loss. Senile plaques are composed of an amyloid beta-peptide (Aβ) core surrounded by dystrophic neurites.

Amyloid precursor protein (APP) is a transmembrane glycoprotein of unknown function that is present in many cells. The protease α-secretase cleaves APP between residues 16 and 17 of Aβ(1-42) to release soluble APP and form a C-terminal fragment of APP. β-secretase proteolytically cleaves APP at the N-terminal side of Aβ(1-42), while γ-secretase cleaves APP on the carboxy-terminus of this sequence. γ-Secretase cleavage takes place at different residues near the carboxy terminus of Aβ resulting principally in the 40-mer and 42-mer, Aβ(1-40) and Aβ(1-42), respectively. These two peptides comprise most of the brain-resident peptide. The more toxic of the two peptides, Aβ(1-42), aggregates more quickly than Aβ(1-40). Aβ(1-42) plays a central role in the pathogenesis of AD, mostly evidenced by the observation of mutations in the genes for APP or presenilin-1 and presenilin-2, all of which result in familial AD and increased production of Aβ(1-42) [2]. We and others have also demonstrated that the AD brain is under extensive oxidative stress as indexed by protein oxidation and lipid peroxidation [3–7]. Moreover, Aβ(1-42) induces protein oxidation and lipid peroxidation both in vitro and in vivo [3–6, 8–11]. Thus, Aβ(1-42), central to the pathogenesis of AD, is likely also to be central to the oxidative stress under which the AD brain exists.

We developed a unifying model for the pathogenesis of AD based on the central role of Aβ(1-42) as a mediator of free radical–induced oxidative stress in AD brain [4, 12–14]. In this model, Aβ(1-42) inserts into the lipid bilayer as a small aggregate resulting in lipid peroxidation and oxidative modification of proteins [3, 15], both of which are inhibited by vitamin E [16]. In addition, the AD-related peptide Aβ(1-42) causes an influx of Ca^{2+} into the neuron, resulting in loss of intracellular Ca^{2+} homeostasis, mitochondrial dysfunction, and ultimately cell death [17, 18].

In this review, the role of Aβ(1-42)-induced lipid peroxidation and protein oxidation in the pathogenesis of AD is discussed. Additionally, we point out the importance of the single methionine of Aβ(1-42) (residue 35 of this 42-mer) to the oxidative stress and neurotoxic properties of Aβ(1-42).

5.2 Aβ(1-42)-Mediated Lipid Peroxidation and Protein Oxidation

The 1300-g normal brain, though small, consumes more than 30% of inspired oxygen. Unfortunately, the brain is especially vulnerable to lipid peroxidation due to the relatively high abundance of polyunsaturated fatty acids (PUFAs), such as arachidonic acid and docosohexenoic acid, the presence of redox metal ions that can take part in free-radical reactions, and the relatively low abundance of brain-resident antioxidants. These factors, coupled to the high rate of oxygen respiration in the brain, lead to lipid peroxiation, which is initiated by a free radical–mediated hydrogen atom abstraction from an unsaturated carbon on a lipid-resident acyl chain, resulting in the formation of a carbon-centered lipid radical (L·). Because oxygen is both paramagnetic and of zero dipole moment, the lipid radical can readily react with lipid-soluble molecular oxygen to form a peroxyl radical (LOO·). This latter reactive free radical subsequently expropriates a hydrogen atom from a neighboring unsaturated lipid acyl chain, forming a lipid hydroperoxide (LOOH) and another carbon-centered lipid radical (L·), Thus, the free-radical chain reaction is propagated. If chain-breaking antioxidants, such as vitamin E are present, the chain reaction is terminated (Fig. 5.1).

Lipid peroxidation leads to the production of reactive alkenals such as 4-hydroxy-2-nonenal (HNE) and 2-propen-1-al (acrolein), both of which are increased in AD brain [15, 19, 20]. These electrophilic α,β-unsaturated aldehydes easily react with protein-bound cysteine, lysine, and histidine residues by Michael addition to form covalently bound adducts that change protein conformation and structure [21],

$$LH + X\cdot \rightarrow L\cdot + XH \quad (1)$$

$$L\cdot + O_2 \rightarrow LOO\cdot \quad (2)$$

$$LOO\cdot + LH \rightarrow LOOH + L\cdot \quad (3)$$

$$LOO\cdot + LOO\cdot \rightarrow \text{nonradical} + O_2 \quad (4)$$

FIGURE 5.1. Mechanism of lipid peroxidation. The free radical X· abstracts a H atom from unsaturated sites on the fatty acid chains of phospholipids (LH) to produce a carbon-centered free radical (L·) (1). The latter in turn is immediately bound by paramagnetic oxygen to form lipid peroxyl free radicals (LOO·) (2). The chain reaction is propagated by attack of LOO· on another fatty acid chain to form the lipid hydroperoxide and L· again (3). The chain reaction is terminated by radical-radical recombination (4).

resulting in loss of protein function and initiation of cell death (Fig. 5.2).

Aβ(1-42) leads to oxidative stress in vivo [8, 10, 21]. Increased protein oxidation (and where measured, lipid peroxidation as well) was found in *C. elegans* that express human Aβ(1-42) [8, 10] and in brains from knock-in mice with the mutated human gene for APP, PS-1, or the double mutant APP/PS-1 [11, 22, 23].

The mitochondrial enzymes pyruvate dehydrogenase and α-ketoglutarate dehydrogenase are inactivated by HNE or acrolein, presumably by covalent modification of the lipoic acid cofactors of each enzyme via Michael addition [22]. Acrolein and HNE, as well as Aβ(1-42), apparently covalently modify the transmembrane aminophospholipid-translocase (flippase), an ATP-requiring enzyme that maintains phospholipid asymmetry [17, 18]. Appearance of phosphatidylserine (PS) on the outer leaflet of the lipid bilayer is an early signal of apoptosis. Flippase activity is inhibited if a critical cysteine residue in the active site is not free. Consequently, oxidative modification of flippase by HNE or acrolein at this Cys residue could result in exposure of PS on the outer leaflet of the cell membrane leading to neuronal loss [17, 18].

As noted above, the AD brain is under extensive oxidative stress, manifested by, among other indices, increased oxidation of DNA [25]. We hypothesized that one means by which DNA would be oxidized in AD brain is if the protective function of the surrounding histone proteins were altered due to their oxidative modification. To test this hypothesis, we added HNE to histones and showed that (a) the conformation of histones was markedly altered as determined by magnetic resonance methods; (b) the resulting interactions of oxidatively modified histones with DNA were significantly changed from control, consistent with the notion that the protective functions of histones would be compromised in AD brain; and (c) acetylated histones seemed even more vulnerable to oxidative modification by HNE than nonacetylated histones [26]. Thus, we found evidence to support the hypothesis that the lipid peroxidation product, HNE, known to be elevated in AD brain [15, 20], may contribute to the vulnerability of DNA to oxidation in the AD brain.

Addition of Aβ(1-42) to neurons or synaptosomes resulted in increased HNE, with consequent

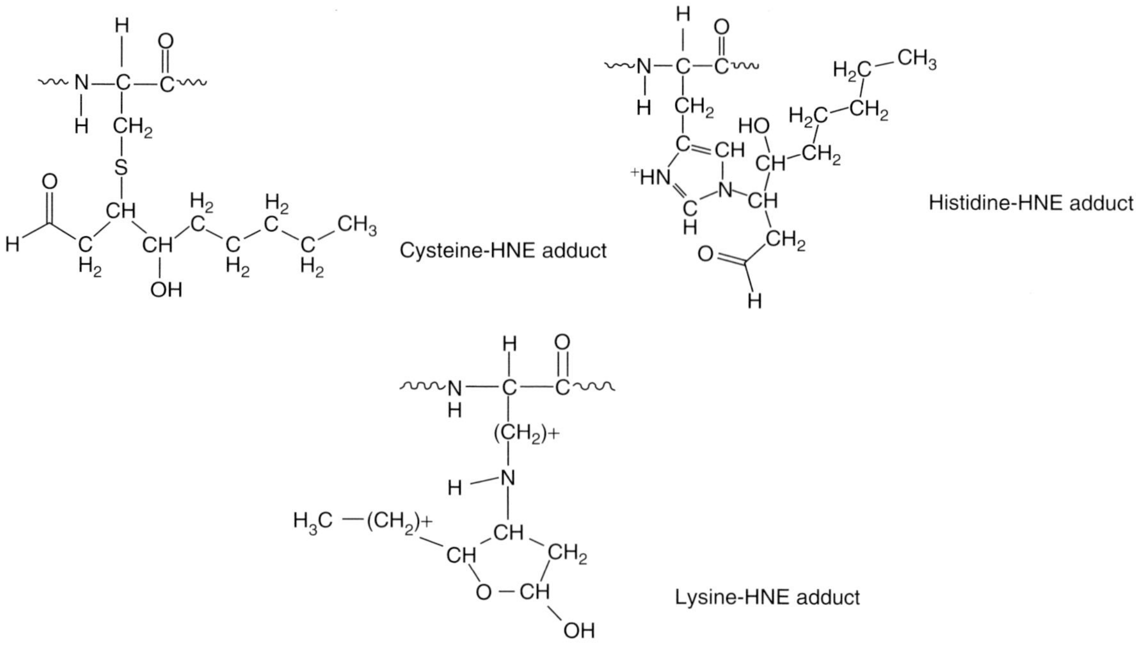

FIGURE 5.2. HNE adducts of cysteine and histidine formed by Michael addition, and the hemiacetal formed by HNE reaction with lysine.

covalent modification of key proteins [3, 15, 27]. Additionally, treatment of synaptsomes with Aβ(1-42) resulted in an increase in HNE bound to choline acetyltransferase and the glutamate transporter GLT-1 (EAAT2) [3, 15]. An increase in HNE bound to glutathione-*S*-transferase (GST), the multidrug resistance protein-1 (MRP1), and EAAT2 in AD brain also was found [15, 28]. The activities of GST and EAAT2 are decreased in AD brain [29, 30]. Thus, removal of HNE from neurons by the action of GST and MRP1 likely is compromised, resulting in accumulation of this harmful alkenal [28]. These findings are consistent with the notion that Aβ(1-42)-induced lipid peroxidation leads to HNE modification of important enzymes and transporters in AD brain, resulting in loss of function. Similar considerations might explain in part the decreased activity of choline acetyltransferase in AD brain compared with control [31].

Protein oxidation, which generally results in loss of function, is also evident in AD brain [3, 5, 15, 32]. Protein carbonyls are a marker of protein oxidation [33]. Four processes cause carbonyl moieties to be introduced to proteins: (a) free radical–induced scission of the peptide backbone; (b) oxidation of specific amino acid side chains; (c) HNE or acrolein covalent modification of proteins by Michael addition; and (d) glycoxidation reactions [33]. Protein carbonyls are measured by derivatization of the carbonyl moiety by 2,4-dinitrophenylhydrazine to form a hydrazone product, which can be detected spectroscopically or immunochemically (Fig. 5.3). Additionally, protein oxidation can be indexed by measure of 3-nitrotyrosine (3-NT) (Fig. 5.3). Increased levels of 3-NT have been reported in AD brain [7, 34–36] and CSF [37], and Aβ(1-42) addition to neurons results in elevated 3-NT [38, 39]. RNS leads to 3-NT in synaptosomes, and novel antioxidants are able to prevent damage to these synaptosomes or synaptosome-resident mitochondria [38–42].

Oxidative modification of glutamine synthetase (GS) and creatine kinase (CK) are found in AD brain, and both GS and CK have significantly decreased activity in AD brain [5, 32, 43, 44]. We have used proteomics to identify brain proteins that are excessively oxidatively modified in AD brain relative to control brain (Fig. 5.4) [34, 36, 43, 45–50]. These include: CK (BB isoform), phosphoglycerate mutase, glyceraldehydes-3-phosphate

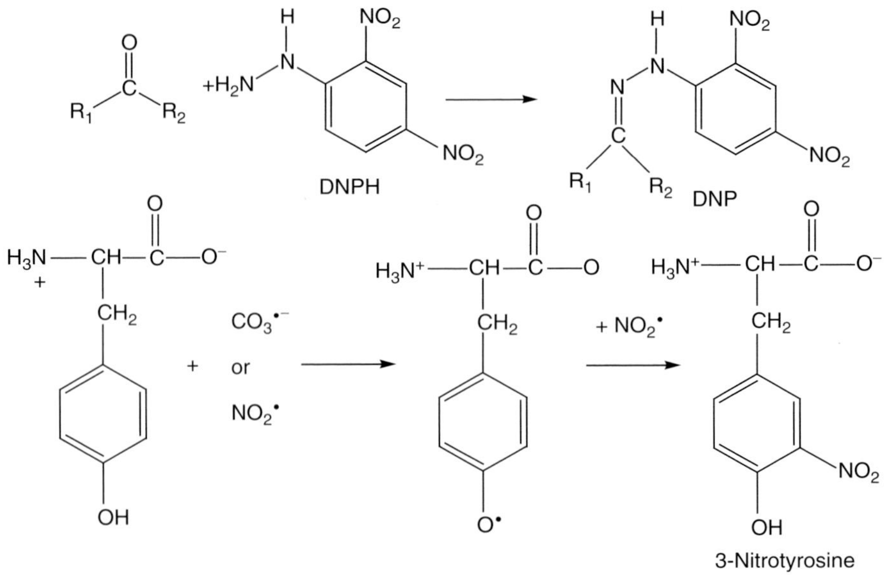

FIGURE 5.3. (Top) Derivatization of protein carbonyls by 2,4-dinitrophenylhydrazine. This reaction occurs as a consequence of the well-known Schiff base formation between a primary amine and a carbonyl functionality. (Bottom) Mechanism of formation of 3-NT. The 3-position of the aromatic ring of Tyr is attacked to form 3-NT as a consequence of the electronic structure around this aromatic site.

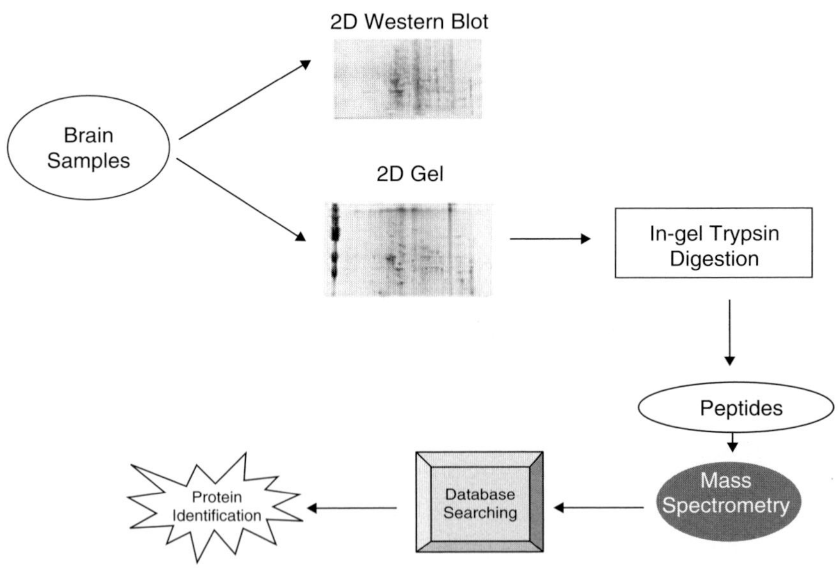

FIGURE 5.4. Schematic of proteomic identification of carbonylated proteins involving the parallel analysis for differences in protein expression and oxidative modification.

dehydrogenase, GS, ubiquitin carboxy-terminal hydrolyze L-1 (UCH L-1), α-enolase, triosphosphate isomerase, neuropolypeptide h3, and dihydropyrimidinase related protein-2 (DRP-2), among others. A wide spectrum of cellular functions including energy metabolism, glutamate uptake and excitotoxicity, proteosomal dysfunction, tau hyperphosphorylation, mitochondrial function, and neuronal communication are affected by these oxidized proteins. As noted above, oxidative modification of protein nearly always leads to loss of protein function. Thus, several plausible mechanisms of neurodegeneration can be proposed based on each of the oxidized proteins.

CK BB, α-enolase, phosphoglycerate mutase, glyceraldehydes-3-phosphate dehydrogenase, and triosphosphate isomerase are all directly or indirectly involved in the synthesis of ATP. Consistent with PET scanning findings that show decreased metabolism in AD brain [51, 52], CK and enolase activities are decreased in AD brain [5, 48]. Lack of ATP would cause dysfunction in ion pumps, electrochemical gradients, voltage-gated ion channels, and cell potential, all of which are needed to combat the oxidative stress of synaptic regions of neurons induced by Aβ(1-42).

The oxidative modification and dysfunction of EAAT2 in AD brain [15, 29] coupled to diminution of GS function as a result of its oxidation (as revealed by proteomics) would result in a decreased conversion of glutamate. This in turn would stimulate *N*-methyl-D-aspartate (NMDA) receptors leading to an increase in Ca^{2+} influx. Alterations in calcium homeostasis would lead to dysfunctional long-term potentiation (LTP), which, in turn, would affect learning and memory. Additionally, Ca^{2+}-mediated mitochondrial swelling, resulting in reactive oxygen species (ROS) and proapoptotic cytochrome c release, ER stress, and activation of calcium-sensitive proteases such as calpain and caspases, are downstream consequences of oxidative stress–related loss of Ca^{2+} homeostasis. These insults are known to lead to neuronal death, and we have hypothesized that such processes are important in AD brain [45, 49, 50].

Accumulation of damaged, misfolded, and aggregated proteins in AD brain may be due to proteasomal dysfunction [53, 54]. One protein involved in proteasome function is UCH L-1. Dysfunction of this protein is observed in AD brain [48]. UCH L-1 catalyzes removal of polyubiquitin from damaged proteins, and its dysfunction, as a result of its oxidation, would lead to excess protein ubiquitinylation, loss of activity of the proteasome, and accumulation of damaged or aggregated proteins, all of which are found in AD.

DRP-2, which has decreased expression in AD [55–57] and is oxidatively modified in AD brain [46], is involved in pathfinding and guidance for axonal outgrowth. Moreover, DRP-2 interacts with and modulates the function of collapsin, a protein involved in dendrite elongation and guidance to adjacent neurons. Therefore, DRP-2 is envolved in forming neuronal connections and maintaining neuronal communication. Consequently, the oxidation and diminished activity of DRP-2 could result in the reported shortened dendritic lengths in AD brain [58]. Neurons with shortened neurites are predicted to communicate less well with adjacent neurons, a process that could conceivably be important in a memory and cognitive disorder like AD.

Proteomics analysis has identified neuropolypeptide h3 as specifically nitrated in AD brain [34]. Neuropolypeptide h3 is also identified as phosphatidylethanolamine-binding protein (PEBP) and hippocampal cholinergic neurostimulating peptide (HCNP). A decrease in the function of PEBP could lead to loss of phospholipid asymmetry, resulting in the exposure of phosphatidylserine on the outer leaflet of the lipid bilayer, a signal of apoptosis. As noted, both Aβ(1-42) and the Aβ(1-42)-mediated lipid peroxidation product HNE lead to loss of lipid asymmetry, which may be relevant to oxidative stress–related AD [17, 18]. Upregulation of choline acetyltransferase (CAT) in cholinergic neurons after NMDA receptor activation is one function of HCNP [59]. CAT activity is known to be decreased in AD [31], and cholinergic deficits are prominent in AD brain [1, 60]. Aβ(1-42) leads to elevated HNE on CAT, possibly contributing to its loss of function in AD brain [3]. Nitration of neuropolypeptide h3 could lead to diminution of neurotrophic action on cholinergic neurons of the hippocampus and basal forebrain, which may be related to the observed decline in cognitive function in AD brain.

Proteomics studies in our laboratory are ongoing to identify proteins that are oxidatively modified by Aβ(1-42) in model systems relevant to AD [61–65]. The results of these studies show some common proteins that are oxidized by Aβ(1-42) in vivo and in AD brain, consistent with the notion that Aβ(1-42) significantly contributes to the oxidative stress of AD brain.

5.3 Methionine-35 of Aβ(1-42): Role in Aβ(1-42)-Induced Oxidative Stress and Neurotoxicity

Methionine 35 is a critical residue in Aβ(1-42)-mediated oxidative stress and neurotoxicity. Substitution of the sulfur atom of methionine 35 by a methylene group, $-CH_2-$ (norleucine), significantly modulates the oxidative stress and neurotoxicity of Aβ(1-42), but the fibrilar morphology of both peptides is similar [10]. Methionine 35 of Aβ(1-42) is also involved in the oxidative stress and neurotoxicity properties of this peptide in vivo. *C. elegans* expressing human Aβ(1-42) exhibited significantly increased protein oxidation, but replacement of the codon for Met by that for Cys in the DNA sequence for human Aβ(1-42) resulted in no increase in protein oxidation in the worm compared with *C. elegans* expressing native human Aβ(1-42) [10]. Additionally, studies involving a temperature inducible *C. elegans* model expressing human Aβ(1-42) revealed that protein oxidation preceeds the deposition of fibrilar aggregates [8]. This finding is consistent with increasing evidence that small soluble aggregates of Aβ(1-42) are the toxic species of this peptide [66–68]. Moreover, that Aβ(1-42) containing the norleucine deriviative of Aβ(1-42), which through producing fibrils, was not oxidative or neurotoxic supports our hypothesis that methionine is critically involved in the neurotoxic and oxidative properties of Aβ(1-42) [10, 69].

Lipid peroxidation is induced by Aβ(1-42) [15, 27] and is found in AD brain [15, 19, 20]. Because lipid peroxidation requires that the free radical involved must be located in the immediate vicinity of the labile H-atoms of unsaturated acyl-chains on phospholipids, this requirement suggests that the Met residue of Aβ(1-42) is located in the bilayer [70], a suggestion confirmed by others [71]. It has been proposed that, due to the hydrophobic carboxy terminus of Aβ(1-42), the peptide inserts into the lipid bilayer [70–72]. Aβ(1-42) adopts an α-helical conformation, similar to other proteins that insert into the lipid bilayer. A methionine sulfuranyl radical (MetS·) on Aβ(1-42) is formed by a one-electron oxidation [12–14, 69, 72–75]. This radical, in turn, can abstract a hydrogen atom from a neighboring unsaturated lipid resulting in the formation of a carbon-centered lipid radical (L·). Via mechanisms described above (Fig. 5.1), the carbon-centered radical on the lipid can readily react with molecular oxygen to form a peroxyl radical (LOO·). Hydrogen abstraction from a neighboring lipid results in the formation of a lipid hydroperoxide (LOOH) and another carbon-centered lipid radical (L·), thereby, propagating the free-radical chain reaction [69, 74, 75]. Both theoretical and experimental studies demonstrate that the α-helical secondary structure of the peptide provides stabilization of the sulfuranyl radical formed by a one-electron oxidation of methionine [72, 76]. Mutation of isoleucine 31 in Aβ(1-42) to proline, an α-helix breaker, attenuated the oxidative stress and neurotoxic properties of the native peptide, suggesting that the amide oxygen of isoleucine 31 in the α-helix conformation interacts with a lone pair of electrons on the sulfur atom of methionine 35, priming this atom for a one-electron oxidation [72]. Subsequently, the sulfuranyl radical of methionine can react with other moieties of methionine to form an α(alkylthio)alkyl radical of methionine ($-CH_2$-CH_2-S-CH_2 or $-CH_2$-CH-S-CH_3) [69, 72, 74, 76]. Such carbon-centered radicals provide potential substrates for reaction with molecular oxygen leading to the formation of peroxyl radicals, and consequently, potentiation of free-radical generation and HNE formation [69, 75, 77]. Recently, others have confirmed our hypothesis, directly demonstrating the existence of the sulfuranyl free radical in Aβ(1-40) [78]. Other researchers [79, 80] invoke Cu(II) reduction and subsequent H_2O_2 formation in the oxidative stress and neurotoxic properties of Aβ(1-42). Critical in this scenario are the three His residues at positions 6, 13, and 14 and the Tyr at position 10. The former are the likely binding sites for Cu(II) on Aβ(1-42), while Tyr 10 is proposed to be the source of the electron to reduce Cu(II) to Cu(I). However, substitution of the three His residues by asparagine (which has at least a 100-fold less binding affinity of Cu(II) than does His) or substitution of Tyr 10 by aromatic Phe (which, though still aromatic, is incapable of providing an election to Cu(II)) leads to peptides that are similarily toxic and oxidative as native Aβ(1-42) [81, 82]. In contrast, substitution of Met by norleucine, which still has the three His residues and Tyr 10 present, is no longer toxic or oxidative [10]. Using the reverse peptide, Aβ(40-1), which is nontoxic, others showed that a Tyr free radical could be formed [78]. That is, a central feature required in mechanisms that involve Cu(II) reduction as a cardinal paradigm occur only in a peptide that is nontoxic [78].

Oxidative modification of methionine 35 to methionine sulfoxide constitutes a major component of the various amyloid β-peptides isolated from AD brain [83–85], consistent with the role of methionine in the oxidative properties of Aβ(1-42). In vitro oxidation of methionine to methionine sulfoxide has been shown to abolish the oxidative stress and neurotoxic properties of Aβ(1-42) after a 24-h incubation with neurons. Mitochondrial dysfunction as measured by MTT reduction was also observed [73]. This finding was confirmed in a recent study [80]. However, after a 96-h treatment, the methionine sulfoxide of Aβ(1-42) reportedly resulted in neuronal death as observed by phase contrast microscopy. Aβ(1-42) containing methionine sulfoxide does not associate itself with the lipid bilayer due to the hydrophilic oxidized sulfur atom [80]. It is conceivable that Aβ(1-42) containing methionine sulfoxide may not form fibrils readily but does so after a long enough period. Thus, toxicity of Aβ(1-42) containing methionine sulfoxide may occur via a different mechanism than with native Aβ(1-42), that is, fibril formation conceivably could activate the receptor for advanced glycation end products (RAGE) leading to oxidative stress and neurotoxicity [86, 87].

5.4 Conclusions

Aβ(1-42) plays a critical role in the oxidative stress present in AD brain and, consequently, may play a central role in the pathogenesis of the disease. Aβ(1-42) induces protein oxidation and lipid peroxidation both in vitro and in vivo. Methionine 35 has been shown to play a vital role in the oxidative stress and neurotoxic properties of Aβ(1-42). Ongoing proteomic studies will lead to the identification of proteins that are specifically oxidatively modified by Aβ(1-42), providing insight into mechanisms of Aβ(1-42)-induced neurodegeneration and, consequently, a greater insight into the role that Aβ(1-42) plays in the pathogenesis of this dementing disorder.

Acknowledgments This work was supported by grants from NIH (AG-05119; AG-10836).

References

1. Katzman R and Saitoh T. Advances in Alzheimer's disease. FASEB J 1991; 5:278-86.
2. Selkoe DJ. Alzheimer's disease results from the cerebral accumulation and cytotoxicity of amyloid beta-protein. J Alzheimers Dis 2001; 3:75-80.
3. Butterfield DA and Lauderback CM. Lipid peroxidation and protein oxidation in Alzheimer's disease brain: potential causes and consequences involving amyloid beta-peptide-associated free radical oxidative stress. Free Radic Biol Med 2002; 32:1050-60.
4. Butterfield DA, Drake J, Pocernich C, et al. Evidence of oxidative damage in Alzheimer's disease brain: central role for amyloid beta-peptide. Trends Mol Med 2001; 7:548-54.
5. Hensley K, Hall N, Subramaniam R, et al. Brain regional correspondence between Alzheimer's disease histopathology and biomarkers of protein oxidation. J Neurochem 1995; 65:2146-56.
6. Markesbery MR. Oxidative stress hypothesis in Alzheimer's disease. Free Radic Biol Med 1997; 23:134-47.
7. Smith MA, Richey Harris PL, Sayre LM, et al. Widespread peroxynitrite-mediated damage in Alzheimer's disease. J Neurosci 1997; 17:2653-57.
8. Drake J, Link CD, Butterfield DA. Oxidative Stress precedes fibrillar deposition of Alzheimer's disease amyloid β-peptide (1-42) in a transgenic *Caenorhabditis elegans* model. Neurobiol Aging 2003; 24:415-20.
9. Subbarao KV, Richardson JS, Ang LC. Autopsy samples of Alzheimer's cortex show increased peroxidation in vitro. J Neurochem 1990; 55:342-45.
10. Yatin SM, Varadarajan S, Link CD, et al. In vitro and in vivo oxidative stress associated with Alzheimer's amyloid β-peptide (1-42). Neurobiol Aging 1999; 20:325-30.
11. Mohmmad-Abdul H, Wenk GL, Gramling M, et al. APP and PS-1 mutations induce brain oxidative stress independ of dietary cholesterol: implications for Alzheimer's disease. Neurosci Lett 2004; 368:148-50.
12. Butterfield DA. Amyloid beta-peptide (1-42)-induced oxidative stress and neurotoxicity: implications for neurodegeneration in Alzheimer's disease brain. A review. Free Radic Res 2002; 36:1307-13.

13. Butterfield DA. Amyloid beta-peptide [1-42]-associated free radical-induced oxidative stress and neurodegeneration in Alzheimer's disease brain: mechanisms and consequences. Curr Med Chem 2003; 10:2651-59.
14. Varadarajan S, Yatin S, Aksenova M, et al. Review: Alzheimer's amyloid β-peptide-associated free radical oxidative stress and neurotoxicity. J Struct Biol 2000; 130:184-08.
15. Lauderback CM, Hackett JM, Huang F, et al. The glial glutamate transporter, GLT-1, is oxidatively modified by 4-hydroxy-2-nonenal in the Alzheimer's disease brain: the role of Aβ(1-42). J Neurochem 2001; 78:413-16.
16. Yatin SM, Varadarajan S, Butterfield DA. Vitamin E prevents Alzheimer's amyloid β-peptide (1-42)-induced neuronal protein oxidation and reactive oxygen species production. J Alzheimers Dis 2000; 2: 123-31.
17. Castegna A, Lauderback CM, Mohmmad-Abdul H, et al. Modulation of phospholipid asymmetry in synaptosomal membranes by the lipid peroxidation products, 4-hydroxynonenal and acrolein: implications for Alzheimer's disease. Brain Res 2004; 1004:193-97.
18. Mohmmad Abdul H, Butterfield DA. Protection against amyloid beta-peptide (1-42)-induced loss of phospholipid asymmetry in synaptosomal membranes by tricyclodecan-9-xanthogenate (D609) and ferulic acid ethyl ester: implications for Alzheimer's disease. Biochim Biophys Acta 2005; 1741:140-148.
19. Lovell MA, Xie C, Markesbery WR. Acrolein is increased in Alzheimer's disease brain and is toxic to primary hippocampal cultures. Neurobiol Aging 2001; 22:187-94.
20. Markesbery WR and Lovell MA. Four-hydroxynonenal, a product of lipid peroxidation, is increased in the brain in Alzheimer's disease. Neurobiol Aging 1998; 19:33-36.
21. Subramaniam R, Roediger F, Jordan B, et al. The lipid peroxidation product, 4-hydroxy-2-trans-nonenal, alters the conformation of cortical synaptosomal membrane proteins. J Neurochem 1997; 69:1161-69.
22. LaFontaine MA, Mattson MP, Butterfield DA. Oxidative stress in synaptosomal proteins from mutant presenilin-1 knock-in mice: implications for familial Alzheimer's disease. Neurochem Res 2002; 27:417-21.
23. Mohmmad Abdul H, Sultana R, Keller JN, et al. Mutations in APP and PS1 genes increase the basal oxidative stress in murine neuronal cells and lead to increased sensitivity to oxidative stress mediated by Aβ(1-42), H_2O_2, and kainic acid: Implications for Alzheimer's disease. J Neurochem 2006; 96: 1322-1335.
24. Pocernich CB and Butterfield DA. Acrolien inhibits NADH-linked mitochondrial enzyme activity: implications for Alzheimer's disease. Neurotox Res 2003; 5:515-20.
25. Gabbita SP, Lovell MA, Markesbery WR. Increased nuclear DNA oxidation in the brain in Alzheimer's disease. J Neurochem 1998; 71:2034-40.
26. Drake J, Petroze R, Castegna A, et al. 4-Hydroxynonenal oxidatively modifies histones: implications for Alzheimer's disease. Neurosci Lett 2004; 356:155-58.
27. Mark RJ, Lovell MA, Markesberry WR, et al. A role for 4-hydroxynonenal, an aldehydic product of lipid peroxidation, in disruption of ion homeostasis and neuronal death induced by amyloid β-peptide. J Neurochem 1997; 68:255-64.
28. Sultana R, Butterfield DA. Oxidatively modified GST and MRP1 in Alzheimer's disease brain: implications for accumulation of reactive lipid peroxidation products. Neurochem Res 2004; 29:2215-2220.
29. Masliah E, Alford M, De Teresa R, et al. Deficient glutamate transport is associated with neurodegeneration in Alzheimer's disease. Ann Neurol 1995; 40:759-66.
30. Lovell MA, Xie C, Markesbery WR. Decreased glutathione transferase activity in brain and ventricular fluid in Alzheimer's disease. Neurology 1998; 51: 1562-1566.
31. Rosser MN, Svendsen C, Hunt SP, et al. The substantia innominata in Alzheimer's disease: a histochemical and biochemical study of cholinergic marker enzymers. Neurosci Lett 1982; 28:217-22.
32. Aksenov MY, Aksenova MV, Butterfield DA, et al. Oxidative modification of creatine kinase BB in Alzheimer's disease brain. J Neurochem 2000; 74:2520-27.
33. Butterfield DA and Stadtman ER. Protein oxidation processes in aging brain. Adv Cell Aging Gerontol 1997; 2:161-91.
34. Castegna A, Thongboonkerd V, Klein JB, et al. Proteomic identification of nitrated proteins in Alzheimer's disease brain. J Neurochem 2003; 85: 1394-01.
35. Good PF, Werner P, Hsu A, et al. Evidence for neuronal oxidative damage in Alzheimer's disease. Am J Pathol 1996; 149:21-27.
36. Sultana R, Poon HF, Cai J, et al. Identification of nitrated proteins in Alzheimer's disease using a redox proteomics approach. Neurobiol Dis 2006; 22:76-87.
37. Tohgi H, Abe T, Yamazaki K, et al. Alterations of 3-nityrosine concentration in the cerebrospinal fluid during aging and in patients with Alzheimer's disease. Neurosci Lett 1999; 269:52-54.

38. Sultana R, Newman S, Mohammad-Abdul H, et al. Protective effect of the xanthate, D609, on Alzheimer's amyloid β-peptide (1-42)-induced oxidative stress in primary neuronal cells. Free Radic Res 2004; 38:449-58.
39. Sultana R, Ravagna A, Mohmmad-Abdul H, et al. Ferulic acid ethyl ester protects neurons against amyloid β-peptide (1-42)-induced oxidative stress and neurotoxicity: relationship to antioxidant activity. J Neurochem 2005; 92:749-758.
40. Drake J, Sultana R, Aksenova M, et al. Elevation of mitochondrial glutathione by gamma-glutamylcysteine ethyl ester protects mitochondria against peroxynitrite-induced oxidative stress. J Neurosci Res 2003; 74:917-27.
41. Drake, J, Kanski, J, Varadarajan, S, et al. Elevation of brain glutathione by g-glutamylcysteine ethyl ester protects against peroxynitrite-induced oxidative stress. J Neurosci Res 2002; 68:776-84.
42. Koppal, T, Drake, J, Yatin, S, et al. Peroxynitrite-induced alterations in synaptosomal membrane proteins: insight into oxidative stress in Alzheimer's Disease. J Neurochem 1999; 72:310-17.
43. Castegna A, Aksenov M, Aksenova M, et al. Proteomic identification of oxidatively modified proteins in Alzheimer's disease brain part I: creatine kinase BB, glutamine synthetase, and ubiquitin carboxy-terminal hydrolase L-1. Free Radic Biol Med 2002; 33:562-71.
44. Yatin SM, Aksenov M, Butterfield DA. The antioxidant vitamin E modulated amyloid β-peptide-induced creatine kinase activity inhibition and increased protein oxidation: implications for the free radical hypothesis of Alzheimer's disease. Neurochem Res 1999; 24:427-35.
45. Butterfield DA. Proteomics: a new approach to investigate oxidative stress in Alzheimer's disease brain. Brain Res 2004; 1000:1-7.
46. Castegna A, Aksnov M, Thongboonkerd V, et al. Proteomics identification of oxidatively modified proteins in Alzheimer's disease brain part II: dihydropyrimidinase related protein II, a-enolase, and heat schock cognate 71. J Neurochem 2002; 82:1524-32.
47. Sultana R, Boyd-Kimball D, Poon HF, et al. Oxidative modification and down-regulation of Pin1 in Alzheimer's disease hippocampus: a redox proteomics analysis. Neurobiol Aging 2006; 27: 918-925.
48. Sultana R, Boyd-Kimball D, Poon HF, et al. Redox proteomics identification of oxidized proteins in Alzheimer's disease hippocampus and cerebellum: An approach to understand pathological and biochemical alterations in AD. Neurobiol Aging 2006; in press.
49. Sultana R, Perluigi M, Butterfield DA. Redox proteomics identification of oxidatively modified proteins in Alzheimer's disease brain and in vivo and in vitro models of AD centered around Aβ(1-42). J Chromatogr B Analyt Technol Biomed Life Sci 2006; 833:3-11.
50. Butterfield DA, Boyd-Kimball D, Castegna A. Proteomics in Alzheimer's disease: insights into mechanisms of neurodegeneration. J Neurochem 2003; 86:1313-1327.
51. Blass JP and Gibson GE. The role of oxidative abnormalities in the pathophysiology of Alzheimer's disease. Rev Neurol 1991; 147:513-525.
52. Scheltens P and Korf ESC. Contribution of neuroimaging in the diagnosis of Alzheimer's disease and other dementias. Curr Opin Neurol, 2000; 13:391-96.
53. Shringarpure R, Grune T, Davies KJ. Protein oxidation and 20S proteasome-dependent proteolysis in mammalian cells. Cell Mol Life Sci 2001; 58:1442-50.
54. Keller JN, Hanni KB, Markesbery WM. Impaired proteasome function in Alzheimer's disease. J Neurochem 2000; 75:436-39.
55. Lubec G, Nonaka M, Krapfenbauer K, et al. Expression of the dihydropyrimidinase related protein 2 (DRP-2) in Down syndrome and Alzheimer's disease brain is downregulated at the mRNA and dysregulated at the protein level. J Neural Transm Suppl 1999; 57:161-77.
56. Schonberger SJ, Edgar PF, Kydd R, et al. Proteomic analysis of the brain in Alzheimer's disease: Molecular phenotype of a complex disease process. Proteomics 2001; 1:1519-28.
57. Tsuji T, Shiozaki A, Kohno R, et al. Proteomic profiling and neurodegeneration in Alzheimer's disease. Neurochem Res 2002; 27:1245-53.
58. Coleman PD and Flood DG. Neuron numbers and dendritic extent in normal aging and Alzheimer's disease. Neurobiol Aging 1987; 8:521-45.
59. Ojika K, Tsugu Y, Mitake S, et al. NMDA receptor activation enhances the release of a cholinergic differentiation peptide (HCNP) from hippocampal neurons in vitro. Neuroscience 1998; 101:341-352.
60. Giacobini E. Cholinergic function and Alzheimer's disease. Int J Geriatr Psychiatry 2003; 8:S1-S5.
61. Boyd-Kimball D, Poon HF, Lynn BC, et al. Proteomic identification of proteins specifically oxidized by intracerebral injection of Aβ(1-42) into rat brain: implications for Alzheimer's disease. Neuroscience 2005; 132:313-324.
62. Boyd-Kimball D, Sultana R, Poon HF, et al. γ-Glutamylcysteine ethyl ester protection from Aβ (1-42)-mediated oxidative stress in neuronal cells: a proteomics approach. J Neurosci Res 2005; 79: 707-713.

63. Boyd-Kimball D, Castegna A, Sultana R, et al. Proteomic identification of proteins oxidized by Aβ(1-42) in synaptosomes: implications for Alzheimer's disease. Brain Res 1044:206-215.
64. Boyd-Kimball D, Poon HF, Lynn BC, et al. Proteomic identification of proteins specifically oxidized in *Caenorhabditis elegans* expressing human Aβ(1-42): implications for Alzheimer's disease. Neurobiol Aging 2006; in press.
65. Poon HF, Farr SA, Banks WA, et al. Proteomic identification of brain proteins in aged senescence accelerated mice that have decreased oxidative modification following administration of antisense oligonucleotide directed at the Aβ region of amyloid precursor protein. Mol Brain Res 2005; 138:8-16.
66. Walsh DM, Klyubin I, Fadeeva JV, et al. Naturally secreted oligomers of amyloid beta protein potently inhibit hippocampal long-term potentiation in vivo. Nature 2002; 416:535-39.
67. Klein WL, Stine WB Jr, Teplow DB. Small assemblies of unmodified amyloid beta-protein are the proximate neurotoxin in Alzheimer's disease. Neurobiol Aging 2004; 25:569-80.
68. Bitan G, Tarus B, Vollers SS, et al. A molecular switch in amyloid assembly: Met35 and amyloid beta-protein oligomerization. J Am Chem Soc 2003; 125:15359-65.
69. Butterfield DA, Boyd-Kimball D. The critical role of methionine 35 in Alzheimer's amyloid β-peptide (1-42)-induced oxidative stress and neurotoxicity. Biochim Biophys Acta 2005; 1703:149-156.
70. Kanski J, Aksenova M, Butterfield DA. The hydrophobic environment of Met35 of Alzheimer's Aβ(1-42) is important for the neurotoxic and oxidative properties of the peptide. Neurotox Res 2002; 4:219-223.
71. Curtin CC, Ali F, Volitakis I, et al. Alzheimer's disease amyloid-beta binds copper and zinc to generate an allosterically ordered membrane-penetrating structure containing superoxide dismutase-like subunits. J Biol Chem 2001; 276:20466-73.
72. Kanski J, Aksenova M, Butterfield DA. Substitution of isoleucine-31 by helical-breaking proline abolishes oxidative and neurotoxic properties of Alzheimer's amyloid beta-peptide. Free Radic Biol Med 2002; 32:1205-11.
73. Varadarajan S, Kanski J, Aksenova M, et al. Different mechanisms of oxidative stress and neurotoxicity for Alzheimer's Aβ(1-42) and Aβ(25-35). J Am Chem Soc 2001; 123:5625-31.
74. Schoneich C. Methionine oxidation by reactive oxygen species: reaction mechanisms and relevance to Alzheimer's disease. Biochim Biophys Acta 2005; 1703:111-119.
75. Butterfield DA and Bush AI. Alzheimer's amyloid β-peptide (1-42): involvement of methionine residue 35 in the oxidative stress and neurotoxicity properties of this peptide. Neurobiol Aging 2004; 25:563-68.
76. Pogocki D and Schöneich C. Redox properties of Met35 in neurotoxic b-amyloid peptide. A molecular modeling study. Chem Res Toxicol 2002; 15:408-18.
77. Schöneich C, Pogocki D, Hug GL, et al. Free radical reactions of methionine in peptides: mechanisms relevant to beta-amyloid oxidation and Alzheimer's disease. J Am Chem Soc 2003; 125:13700-13.
78. Kadlcik V, Sicard-Roselli C, Mattioli T, et al. One-electron oxidation of β-amyloid peptide: sequence modulation of reactivity. Free Radic Biol Med 2004; 37:881-91.
79. Huang X, Cuajungco MP, Atwood CS, et al. Cu(II) potentiation of alzheimer abeta neurotoxicity. Correlation with cell-free hydrogen peroxide production and metal reduction. J Biol Chem. 1999; 274:37111-16.
80. Barnham KJ, Ciccotosto GD, Tickler AK, et al. Neurotoxic, Redox-competent Alzheimer's β-amyloid is released from lipid membrane by methionine oxidation. J Biol Chem 2003; 278:42959-65.
81. Boyd-Kimball D, Abdul-Mohmmad H, Reed T, et al. Role of phenylalanine 20 in Alzheimer's Amyloid b-peptide (1-42)-induced oxidative stress and neurotoxicity. Chem Res Toxicol 2004; 17:1743-1749.
82. Boyd-Kimball D, Sultana R, Abdul-Mohammad H, et al. Rodent Aβ(1-42) exhibits oxidative stress properties similar to that of human Aβ(1-42): Implications for proposed mechanisms of toxicity. J Alzheimers Dis 2004; 6:515-525.
83. Dong J, Atwood CS, Anderson VE, et al. Metal binding and oxidation of Amyloid-β within isolated senile plaque cores: Raman microscopic evidence. Biochemistry 2003; 42:2768-73.
84. Kou YM, Kokjohn TA, Beach TG, et al. Comparative analysis of amyloid-β chemical structure and amyloid plaque morphology of transgenic mouse and Alzheimer's disease brains. J Biol Chem 2001; 276:12991-98.
85. Naslund J, Schierhorn A, Hellman U, et al. Relative abundance of Alzheimer Aβ amyloid peptide variants in Alzheimer's disease and normal aging. Proc Natl Acad Sci U S A 1994; 91:8378-82.
86. Yan SD, Zhu H, Zhu A, et al. Receptor-dependent cell stress and amyloid accumulation in systemic amyloidosis. Nat Med 2000; 6:633-34.
87. Hou L, Kang I, Marchant RE, et al. Methionine 35 oxidation reduces fibril assembly of the amyloid abeta-(1-42) peptide of Alzheimer's disease. J Biol Chem 2002; 277:40173-76.

6
Amyloid Toxicity, Synaptic Dysfunction, and the Biochemistry of Neurodegeneration in Alzheimer's Disease

Judy Ng, Marie-Isabel Aguilar, and David H. Small

6.1 Introduction

Despite considerable progress over the past few years in our understanding of β-amyloid protein (Aβ) production, aggregation, and degradation, little is known about the mechanism of Aβ-mediated neurotoxicity. Although numerous targets of Aβ's action have been reported [1], it has been difficult to determine which, if any, of these targets is important for disease causation. In this article, we review what is known about the cellular and biochemical mechanisms involved in Aβ neurotoxicity (Fig. 6.1).

6.2 Cellular Mechanisms of Neurotoxicity: Cell Loss versus Synaptic Dystrophy

Considerable attention has been paid to the mechanisms by which Aβ causes neuronal cell death. Studies have implicated a variety of mechanisms (e.g., generation of reactive oxygen species, caspase activation, disturbanced in calcium homeostasis) in Aβ-induced cell death [1]. However, although the number of neurons is lower in the AD brain compared with age-matched brains, there are good reasons to believe that cell loss does not play an important role in cognitive decline in AD. First, cell loss is only a minor neuropathologic feature of AD, and it is poorly correlated with cognitive decline [2]. Most of the brain atrophy can be accounted for by synaptic loss, rather than a decrease in the number of cell bodies [2]. Second, it may be argued on purely theoretical grounds that the pattern of retrograde amnesia that occurs in AD is unlikely to be caused by cell death. Computational studies involving attractor neural network models of memory suggest that synaptic dysfunction is more likely to be the mechanism that causes memory loss [1].

In contrast with cell death, neuritic dystrophy is an important diagnostic and pathologic feature of AD. Amyloid plaques are commonly surrounded by neurofibrillary tangle-bearing dystrophic neurites. Aberrant neuronal sprouting can be seen in areas of synaptic loss in the hippocampal formation and neocortex [3]. The dystrophic neurites are a characteristic of AD brains and are typically, but not exclusively, associated with Aβ deposition. Aβ has been reported to induce neurite dystrophy in culture [4] as well as in mutant mouse models [5]. For example, Tsai et al. [6] have recently demonstrated that microdeposits of Aβ amyloid can cause neuritic dystrophy and the breakage of neuronal branches in an APP transgenic mouse model of AD.

6.3 Aβ Aggregation: The Search for Neurotoxic Species

Aggregation of Aβ is a key step in the generation of neurotoxic Aβ species. Aβ neurotoxicity is increased when the peptide is incubated over many hours to days, a process known as aging [7]. Although there is a relationship between aggregation and toxicity, the major toxic form of Aβ in AD

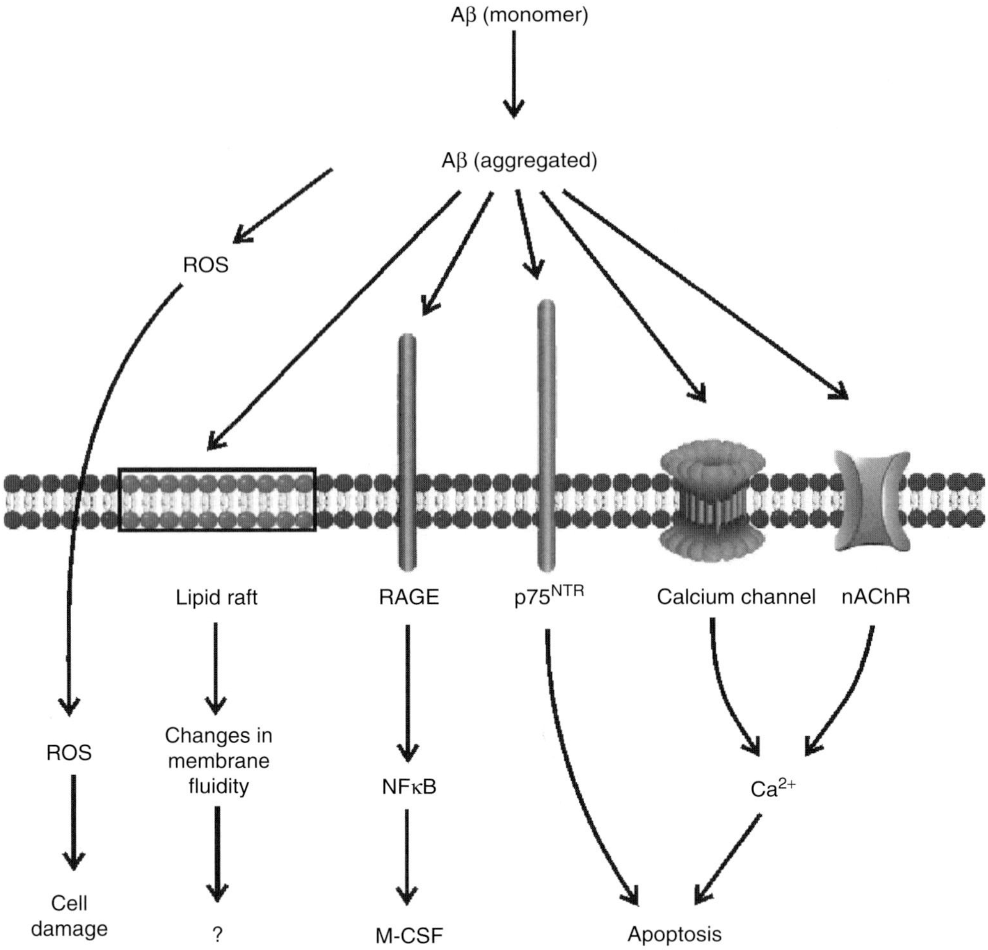

FIGURE 6.1. Possible mechanisms of Aβ-mediated neurotoxicity. A variety of different mechanisms have been proposed to explain the neurotoxic effects of Aβ. These mechanisms include the generation of ROS; binding to p75NTR, RAGE, or nAChRs. The interaction of Aβ with lipid rafts may disturb membrane fluidity and alter the function of membrane proteins such as calcium channels. It is still not clear which, if any, of these mechanisms may contribute to the synaptic dysfunction that is thought to underlie the cognitive decline in AD.

is not known. It has been demonstrated that aggregated Aβ in fibrillar form has neurotoxic properties in cell culture as well as in vivo. However, more recent findings suggest a toxic role of Aβ oligomeric species [8]. In vitro studies have shown that oligomeric Aβ, particularly diffusible low-molecular-weight species, are neurotoxic [9, 10]. This idea is reinforced by genetic studies, which demonstrate that familial AD mutations favor the production Aβ species that aggregate more readily [11].

Aβ aggregation is a complex process that is influenced by incubation time, concentration, temperature, pH, and ionic strength. Initially, monomeric Aβ probably develops an abnormal conformation, after which a variety of different aggregated structures, including oligomers, protofibrils, spheroids, and mature amyloid fibrils, can be produced. Protofibrils are thin 3- to 4-nm-diameter nonbranching linear aggregates [12], whereas fibrils are ~6 to 10 nm in diameter and are long and semiflexible [13]. Fibril formation proceeds with a lag time, which has been interpreted as a nucleation-dependent process, where oligomer formation takes place through the initial formation of nuclei or seeds [14, 15]. This idea is supported by studies where prepolymerized Aβ was added to monomeric protein, which led to the immediate onset of fibril formation [7, 14].

In the past, it was thought that only fibrillar Aβ was pathogenic. However, new evidence supports the hypothesis that prefibrillar structures may be even more important in AD. Brain cell damage and dementia do not correlate well with plaque location and quantity [16]. However, soluble Aβ oligomers are found in human AD cerebrospinal fluid, and the soluble Aβ content of human brain is better correlated with the severity of the disease than plaque density [17, 18]. Oxidative stress has been shown to precede fibrillar deposition of Aβ, suggesting that oxidative stress observed in the AD brain may be caused by nonfibrillar forms of Aβ [19]. It has even been suggested that plaques may not be toxic, and that instead, they may have a protective role in AD by decreasing the amount of the more toxic prefibrillar Aβ species [20].

6.4 Biochemical Effects of Aβ

The exact sequence of events whereby Aβ causes neurodegeneration in AD is not known. In vitro, Aβ can cause oxidative stress, mitochondrial dysfunction, disturbances in calcium homeostasis, and microglial activation [1]. However, the relative contribution of these biochemical changes to neurodegeneration in vivo *is unclear.*

6.5 Oxidative Stress and Mitochondrial Dysfunction

Aβ neurotoxicity is associated with oxidative stress and mitochondrial dysfunction [21]. Changes in mitochondrial enzymes have been described in the AD brain [22]. For example, cytochrome oxidase activity is decreased in AD [23], and defects in mitochondrial energy metabolism can lead to increased production of reactive oxygen species (ROS). Increased Aβ is associated with increased nitric oxide (NO) and reduced ATP levels [24]. NO can, in turn, interact with superoxide radicals to form peroxynitrite, which can damage cells by promoting membrane lipid peroxidation and apoptosis [25].

The interaction of metal ions with Aβ has been proposed to accelerate peptide aggregation and initiate hydrogen peroxide generation [26], although there is not yet strong evidence for metal-Aβ interactions in vivo. During the process of aggregation in vitro, Aβ can generate hydrogen peroxide and free radicals in the presence of Cu^{+} or Fe^{2+} [27]. The binding of Aβ to Zn^{2+} does not generate ROS, although Zn^{2+} competes with Cu^{+} or Fe^{2+} for binding to Aβ and therefore Zn^{2+} could inhibit the oxidizing properties of metal-bound Aβ [28]. The production of these ROS induces membrane lipid peroxidation, which can impair the function of membrane enzymes [29, 30], which in turn can cause an elevation in intracellular calcium [29]. The ability of antioxidants to prevent the loss of membrane enzyme function as well as to stabilize calcium homeostasis in vitro supports the role of membrane lipid peroxidation by Aβ [31, 32]. The major antioxidant glutathione (GSH) is greatly reduced in astrocytes and neurons exposed to Aβ [33, 34].

The role of oxidation in Aβ-induced neurodegeneration in vivo still remains very unclear. Notwithstanding the success of the in vitro experiments and evidence from epidemiological studies that antioxidants may be of value for the treatment of vascular dementia [35], antioxidants have yet to prove themselves in clinical trials for the treatment of AD [36]. There are many possible reasons for this failure. For example, the right drug may not yet have been found. However, it is also possible that the oxidative changes seen in vivo are the *consequence* of the neurodegeneration rather being than the underlying *cause*.

6.6 The Role of the Endoplasmic Reticulum

Some studies suggest that neuronal dysfunction in AD could arise from a defect in the endoplasmic reticulum (ER). As the ER is involved in protein folding and assembly, ER dysfunction could contribute to abnormal protein folding. It has been suggested that ER dysfunction could be due to a defect in the presenilins [37, 38]. Indeed, cells expressing mutant presenilins have an impaired ER response to stress [39]. However, presenilin mutations may also cause an increase in Aβ production [13, 38], which is known to be linked to AD pathogenesis. It is still unclear what role ER dysfunction plays in familial AD caused by presenilin mutations.

6.7 Aβ-Membrane Interactions

The binding of Aβ to a component of the plasma membrane may be the first event in Aβ-mediated neurotoxicity [1]. Aβ has been shown to interact either directly or indirectly with a number of different membrane components including lipids, carbohydrates, ion channels, and receptors. This section describes some of the interactions and their potential roles in neuronal dysfunction.

6.7.1 Interaction of Aβ with Membrane Lipids

Membrane lipids are localized in different domains: exofacial and cytofacial leaflets, cholesterol pools, annular lipids, and lipid rafts [40]. Aβ can interact strongly with the lipid bilayer [41, 42]. This binding causes an increase in Aβ fibrillogenesis and modifications of bilayer properties [42]. Aβ binds strongly to gangliosides and lipid rafts [43], which are also rich in cholesterol. Lipid rafts containing a ganglioside cluster serve as a conformational catalyst or chaperone, helping to seed Aβ oligomerization after binding [44, 45]. In mice, Aβ dimers appear in lipid rafts at 6 months of age and then continue to accumulate by 24–28 months of age [46].

Although it has been observed that Aβ binds preferentially to acidic lipids, it has also been suggested that charge-charge interactions are not required for Aβ-membrane interactions [47]. However, this idea is not supported by the results of Subasinghe et al. [42], which demonstrate that Aβ binds exclusively to lipid membranes through charge-charge interactions. Liposomes composed of phosphatidylserine and phosphatidylcholine induce rapid formation of Aβ aggregates [48].

The consequences of Aβ binding to membranes for cell function are unclear. Biological membranes are fluid in nature, and membrane fluidity is important for the proper functioning of integral membrane proteins and signal transduction pathways. Aβ may disturb the acyl chain layer of the membrane [49]. Aβ reportedly decreases membrane fluidity so the membrane has a more rigid structure, with the presence of gangliosides increasing this effect [50]. The addition of oligomeric Aβ to cultured neurons also causes the release of lipid particles such as cholesterol, phospholipids, and monosialogangliosides [51], although the significance of this effect for the pathogenesis of AD is unclear.

6.7.2 Effects of Aβ on Membrane Calcium Permeability

Insertion of Aβ into the lipid membrane may set off a series of independent events including disruption of Ca^{2+} homeostasis and free-radical formation, catalyzed by perturbation of the conformation of membrane proteins [52]. Aβ-mediated disruption of calcium homeostasis may in turn produce downstream effects [53]. Aβ may increase membrane permeability by interacting with membrane components to destabilize the structure of the membrane [54, 55], or it may be directly inserted into the membrane to form a pore [56, 57]. Aβ aggregation is associated with enhanced ion permeability [58]. Sustained increases in intracellular calcium may also enhance the production and release of Aβ [59, 60]. Aβ-induced destabilization of calcium can lead to caspase activation and apoptosis [61], however this effect may be caused by changes in the ER transport of calcium rather than from calcium transported across the plasma membrane. Reduction of calcium release from the ER may provide partial protection from Aβ toxicity by reducing stress signals in the ER and decreasing the increase in calcium triggered by Aβ [62].

6.7.3 Effect of Aβ on Membrane Receptors

Aβ may exert a toxic effect by binding to or altering the normal function of cell-surface receptors. A number of receptors have been found to interact directly or indirectly with Aβ. There receptors include the α7-nicotine acetylcholine receptor, the receptor for advanced glycation end products (RAGE), and the p75 neurotrophin receptor.

6.7.3.1 α7 Nicotinic Acetylcholine Receptor

The nicotinic acetylcholine receptor (nAChR) is a member of the pentameric ligand-gated ion channel family of receptors [63]. In the central nervous system, most nicotinic receptors are of the α4β2 or homomeric α7 subtype. α7 nAChR receptors are of particular interest for AD because of their high calcium permeability, which suggests an important role in neuronal plasticity and cognition [64]. α7 nAChRs are mainly located at nerve terminals and are believed to be involved in regulating the neurotransmitter release that mediates fast cholinergic neurotransmission [65, 66].

Several studies have shown that Aβ can bind to and influence the activity of α7 nAChRs [67–69]. α7 nAChRs are present in senile plaques and Aβ42 selectively and competitively binds α7 nAChRs with high affinity [67]. This binding may have functional consequences because Aβ40 and Aβ42 can impair cholinergic signaling and acetylcholine release [70]. Although Aβ can block α7 nAChRs on neurons in culture [68], other studies suggest that, under certain conditions, Aβ may activate α7 nAChRs.

The interaction of Aβ with α7 nAChRs may explain some of the biochemical changes that occur in the AD brain. For example, although acetylcholinesterase (AChE) is decreased in the brain of AD patients, AChE is increased around the amyloid plaques [71]. Fodero et al. [72] have demonstrated that this increase may be due to interactions between Aβ and the α7 nAChR. In primary cortical neurons, Aβ42 is more potent than Aβ40 in its ability to increase AChE [72]. Studies by Wang et al. [73] suggest that the binding of Aβ to α7 nAChRs may also influence phosphorylation pathways leading to increased tau phosphorylation.

6.7.3.2 p75 Neurotrophin Receptor

The p75 neurotrophin receptor ($p75^{NTR}$) is a member of the tumor necrosis factor receptor family that binds neurotrophins nonselectively and mediates neuronal apoptosis and survival [74]. $p75^{NTR}$ can bind Aβ and may thereby mediate some forms of Aβ toxicity [75–77]. However, notwithstanding these findings, levels of $p75^{NTR}$ have been found to correlate inversely with the degree of cognitive impairment in early AD, supporting the view that $p75^{NTR}$ may be protective for AD [78]. The idea that $p75^{NTR}$ is neuroprotective for AD is further supported by the observation that there are increased levels of $p75^{NTR}$ in the presence of extracellular Aβ deposits [79], that low concentrations of Aβ increase the level of $p75^{NTR}$ in primary cultures of neurons, and that this increase protects neurons from Aβ-induced toxicity [80].

6.7.3.3 RAGE

The receptor for advanced glycation and end products (RAGE) is a member of the immunoglobulin family of cell-surface molecules that exhibits a wide tissue distribution and interacts with a range of ligands. Aβ can bind to RAGE, and this binding may influence neuronal and microglial function [81]. Aβ is not the only protein that binds to RAGE, as the receptor interacts broadly with β-sheet fibrils [82]. The interaction of Aβ with RAGE expressed on endothelial cells, neurons, and microglia reportedly causes oxidative stress and activation of the transcription factor nuclear factor kappa B (NF-κB) [81], which in turn enhances expression of macrophage-colony stimulating factor (M-CSF) [83]. Aβ-mediated M-CSF expression has also been described in microglia, and anti-RAGE antibodies can block this effect. These findings suggests a feedback loop may exist, whereby Aβ-RAGE–mediated microglial activation enhances the expression of M-CSF and RAGE [84].

6.8 Conclusions

We still have a relatively poor understanding of the mechanism(s) by which Aβ causes neurotoxicity. There is increasing evidence to suggest that Aβ toxicity is caused by synaptic dysfunction rather than cell death. It is clear that aggregation of Aβ is a key step in the generation of neurotoxic species. However, whether the toxic species are fibrils, protofibrils, amyloidβ derived diffusible ligands (ADDLs), or some other aggregated form of Aβ remains to be established. It is also clear that Aβ can promote the formation of ROS as well as increase oxidation. The central question is whether these changes in oxidation are the underlying cause of synaptic dysfunction or simply the effect of some neurodegenerative mechanism.

References

1. Small, D.H., Mok, S.S., and Bornstein, J.C., Alzheimer's disease and Aβ toxicity: from top to bottom. Nat. Rev. Neurosci., 2001; 2:595-598.
2. Terry, R.D., Cell death or synaptic loss in Alzheimer's disease. J. Neuropathol. Exp. Neurol., 2000; 59:1118-1119.
3. Tolnay, M. and Probst, A., Review: tau protein pathology in Alzheimer's disease and related disorders. Neuropathol. Appl. Neurobiol., 1999; 25:171-187.
4. Postuma, R.B., He, W., Nunan, J., et al., Substrate-bound β-amyloid peptides inhibit cell adhesion and neurite outgrowth in primary neuronal cultures. J. Neurochem., 2000; 74:1122-1130.
5. Phinney, A.L., Deller, T., Stalder, M., et al., Cerebral amyloid induces aberrant axonal sprouting and ectopic terminal formation in amyloid precursor protein transgenic mice. J. Neurosci., 1999; 19:8552-8559.
6. Tsai, J., Grutzendler, J., Duff, K., et al., Fibrillar amyloid deposition leads to local synaptic abnormalities and breakage of neuronal branches. Nat. Neurosci., 2004; 7:1181-1183.
7. Jarret, J.T. and Lansbury, P.T.J., Seeding "one-dimensional crystallization" of amyloid: A pathogenic mechanism in Alzheimer's disease and scrapie. Cell, 1993; 73:1055-1058.
8. Hartley, D.M., Walsh, D.M., Ye, C.P., et al., Protofibrillar intermediates of amyloid β-protein induce acute electrophysiological changes and progressive neurotoxicity in cortical neurons. J. Neurosci., 1999; 19(20):8876-8884.
9. Lambert, M.P., Barlow, A.K., Chromy, B.A., et al., Diffusible, nonfibrillar ligands derived from Aβ1-42 are potent central nervous system neurotoxins. Proc. Natl. Acad. Sci. U. S. A., 1998; 95(11):6448-6453.
10. Gong, Y., Chang, L., Viola, K.L., et al., Alzheimer's disease-affected brain: presence of oligomeric A β ligands (ADDLs) suggests a molecular basis for reversible memory loss. Proc. Natl. Acad. Sci. U. S. A., 2003; 100(18):10417-10422.
11. Scheuner, D., Eckman, C., Jensen, M., et al., Secreted amyloid β-protein similar to that in the senile plaques of Alzheimer's disease in increased in vivo by the presenilin 1 and 2 and APP mutations linked to familial Alzheimer's disease. Nat. Med., 1996; 2:864-870.
12. Stine, W.B.J., Snyder, S.W., Ladror, U.S., et al., The nanometer-scale structure of amyloid-β visualised by atomic force microscopy. J. Protein Chem., 1996; 15: 193-203.
13. Fraser, P.E., Duffy, L.K., O'Malley, M.B., et al., Morphology and anitibody recognition of synthetic β-amyloid peptides. J. Neurosci. Res., 1991; 28: 475-485.
14. Harper, J.D., Wong, S.S., Leiber, C.M., et al., Observation of metastable Aβ amyloid protofibrils by atomic force microscopy. Chem. Biol., 1997; 4: 119-125.
15. Lansbury, P.T.J., The molecular mechanism of amyloid formation in Alzheimer's disease. Eur. J. Med. Chem., 1995; 30 (Suppl.): 621S-633S.
16. Terry, R.D., The neuropathology or Alzheimer's disease and the structural basis of its cognitive alterations. In: Terry, R.D., Katzman, R., Bick, K.L., Sisodia, S.S. (Eds.), Alzheimer's disease. Lippincott Williams & Wilkins, Philidelphia, 1999:187-206.
17. Kuo, Y.M., Emmerling, M.R., Vigo-Pelfrey, C., et al., Water-soluble Aβ (N-40, N-42) oligomers in normal and Alzheimer's disease brains. J. Biol. Chem., 1996; 271(8):4077-4081.
18. McLean, C.A., Cherny, R.A., Fraser, F.W., et al., Soluble pool of Aβ amyloid as a determinant of severity of neurodegeneration in Alzheimer's disease. Ann. Neurol., 1999; 46(6):860-866.
19. Drake, J., Link, C.D., and Butterfield, D.A., Oxidative stress precedes fibrillar deposition of Alzheimer's disease amyloid β-peptide (1-42) in a trangenic Caenorhabditis elegans model. Neurobiol. Aging, 2003; 24:415-420.
20. Obrenovich, M.E., Joseph, J.A., Atwood, C.S., et al., Amyloid-β: a (life) preserver for the brain. Neurobiol. Aging, 2002; 23(6):1097-1099.
21. Pike, C.J., Walencewicz, A.J., Glabe, C.G., et al., In vitro aging of β-amyloid protein causes peptide aggregation and neurotoxicity. Brain Res., 1991; 563:311-314.
22. Gibson, G.E., Sheu, K.F., and Blass, J.P., Abnormalities of mitochondrial enzymes in Alzheimer's disease. J. Neural Transm., 1998; 105:855-870.
23. Swerdlow, R.H. and Kish, S.J., Mitochondria in Alzheimer's disease. Int. Rev. Neurobiol., 2002; 53: 341-385.
24. Keil, U., Bonert, A., Marques, C.A., et al., Amyloid-β induced changes in nitric oxide production and mitochondrial activity lead to apoptosis. J. Biol. Chem., 2004; in press.
25. Butterfield, D.A., Castegna, A., Lauderback, C.M., et al., Evidence that amyloid β-peptide-induced lipid peroxidation and its sequelae in Alzheimer's disease brain contribute to neuronal death. Neurobiol. Aging, 2002; 23:655-664.
26. Barnham, K.J., Masters, C.L., and Bush, A.I., Neurodegeneration diseases and oxidative stress. Nat. Rev. Drug Discov., 2004; 3:205-214.
27. Hensley, K., Carney, J.M., Mattson, M.P., et al., A model for β-amyloid aggregation and neurotoxicity

base on free radical generation by the peptide: relevance to Alzheimer's disease. Proc. Natl. Acad. Sci. U. S. A., 1994; 91:3270-3274.
28. Cuajungco, M.P., Goldstein, L.E., Nunomura, A., et al., Evidence that the β-amyloid plaques of Alzheimer's disease represent the redox-silencing and entombment of Aβ by zinc. J. Biol. Chem., 2000; 275:19439-19442.
29. Mark, R.J., Hensley, K., Butterfield, D.A., et al., Amyloid β-peptide impairs ion-motive ATPase activities: evidence for a role in loss of neuronal Ca^{2+} homeostasis and cell death. J. Neurosci., 1995; 15: 6239-6249.
30. Mark, R.J., Pang, Z., Geddes, J.W., et al., Amyloid β-peptide impairs glucose uptake in hippocampal and cortical neurons: involvement of membrane lipid peroxidation. J. Neurosci., 1997; 17:1046-1054.
31. Goodman, Y. and Mattson, M.P., Secreted forms of β-amyloid precursor protein protect hippocampal neurons against amyloid β-peptide-induced oxidative injury. Exp. Neurol., 1994; 128:1-12.
32. Goodman, Y., Bruce, A.J., Cheng, B., et al., Estrogens attenuate and corticosterone exacerbates excitotoxicity, oxidative injury and amyloid β-peptide toxicity in hippocampal neurons. J. Neurochem., 1996; 66:1836-1844.
33. Abramov, A.Y., Canevari, L., and Duchen, M.R., Changes in intracellular calcium and glutathione in astrocytes as the primary mechanism of amyloid neurotoxicity. J. Neurosci., 2003; 23:5088-5095.
34. Keelan, J., Allen, N.J., Antcliffe, D., et al., Quantitative imaging of glutathione in hippocampal neurons and glia in culture using monochlorobimane. J. Neurosci. Res., 2001; 66:873-884.
35. Masaki, K.H., Losonczy, K.G., Izmirlian, G., et al., Association of vitamin E and C supplement use with cognitive function and dementia in elderly men. Neurology, 2000; 54:1265-1272.
36. Thal, L.J., Grundman, M., Berg, J., et al., Idebenone treatment fails to slow cognitive decline in Alzheimer's disease. Neurology, 2003; 61:1498-1502.
37. Katayama, T., Imaizumi, K., Sato, N., et al., Presenilin-1 mutations downregulate the signalling pathway of the unfolded-protein response. Nat. Cell Biol., 1999; 1(8):479-485.
38. Sato, N., Imaizumi, K., Manabe, T., et al., Increased production of β-amyloid and vulnerability to endoplasmic reticulum stress by an aberrant spliced form of presenilin 2. J. Biol. Chem., 2001; 276(3):2108-2114.
39. Guo, Q., Sopher, B.L., Furukawa, K., et al., Alzheimer's presenilin mutation sensitizes neural cells to apoptosis induced by trophic factor withdrawal and amyloid-peptide: involvement of calcium and oxyradicals. J. Neurosci., 1997; 17:4212-4222.
40. Tsui-Pierchala, B.A., Encinas, M., Milbrandt, J., et al., Lipid rafts in neuronal signaling and function. Trends Neurosci., 2002; 25:412-417.
41. Terzi, E., Holzemann, G., and Seelig, J., Interaction of Alzheimer β-amyloid peptide(1-40) with lipid membranes. Biochemistry (Moscow). 1997; 36: 14845-14852.
42. Subasinghe, S., Unabia, S., Barrow, C.J., et al., Cholesterol is necessary both for the toxic effect of Aβ peptides on vascular smooth muscle cells and for Aβ binding to vascular smooth muscle cell membranes. J. Neurochem., 2003; 84:471-479.
43. Kakio, A., Nishimoto, S., Kozutsumi, Y., et al., Formation of a membrane-active form of amyloid β-protein in raft-like model membranes. Biochem. Biophys. Res. Commun., 2003; 303:514-518.
44. Kakio, A., Nishimoto, I., Kozutsumi, Y., et al., Formation of a membrane-active form of amyloid β-protein in raft-like membranes. Biochem. Biophys. Res. Commun., 2003; 303:514-518.
45. Yip, C.M., Darabie, A.A., and McLaurin, J., Aβ42-peptide assembly on lipid bilayers. J. Mol. Biol., 2002; 318:97-107.
46. Kawarabayashi, T., Shoji, M., Younkin, L.H., et al., Dimeric amyloid β protein rapidly accumulates in lipid rafts followed by apolipoprotein E and phosphorylated tau accumulation in the Tg2576 mouse model of Alzheimer's disease. Neurobiol. Dis., 2004; 24:3801-3809.
47. Kremer, J.J., Sklansky, D.J., and Murphy, R.M., Profile of changes in lipid bilayer structure caused by β-amloid peptide. Biochemistry (Moscow). 2001; 40:8563-8571.
48. Zhao, H., Tuominen, E.K.J., and Kinnunen, P.K.J., Formation of amyloid fibres triggered by phosphotidylserine-containing membranes. Biochemistry (Moscow). 2004; 43:10302-10307.
49. Muller, W.E., Kirsch, C., and Eckert, G.P., Membrane-disordering effects of β-amyloid peptides. Biochem. Soc. Trans., 2001; 29:617-623.
50. Kremer, J.J., Pallitto, M.M., Sklansky, D.J., et al., Correlation of β-amyloid aggregate size and hydrophobicity with decreased bilayer fluidity of model membranes. Biochemistry (Moscow). 2000; 39:10309-10318.
51. Michikawa, M., Gong, J.S., Fan, Q.W., et al., A novel action of alzheimer's amyloid β-protein (Aβ): oligomeric Aβ promotes lipid release. J. Neurosci., 2001; 21:7226-7235.
52. Kanfer, J.N., Sorrentino, G., and Sitar, D.S., Amyloid β peptide membrane perturbation is the basis for its biological effects. Neurochem. Res., 1999; 24:1621-1630.
53. Mattson, M.P., Cheng, B., Davis, D., et al., β-Amyloid peptides destabilize calcium homeostasis

and render human cortical neurons vulnerable to excitotoxity. J. Neurosci., 1992; 12:376-389.
54. Muller, W.E., Koch, S., Eckert, A., et al., β-Amyloid peptide decreases membrane fluidity. Brain Res., 1995; 674:133-136.
55. Mason, R.P., Estermyer, J.D., Kelly, J.F., et al., Alzheimer's disease amyloid β peptide 25-35 in localized in the membrane hydrocarbon core: X-ray diffraction analysis. Biochem. Biophys. Res. Commun., 1996; 222:78-82.
56. Kawahara, M., Arispe, N., Kuroda, Y., et al., Alzheimer's disease amyloid β-protein forms Zn^{2+}-sensitive cation-selective channels across excited membrane patches from hypothalamic neurons. Biophys. J., 1997; 73:67-75.
57. Arispe, N., Rojas, E., and Pollard, H.B., Alzheimer's disease amyloid β-protein forms calcium channels in bilayer membranes: blockade by tromethamine and aluminium. Proc. Natl. Acad. Sci. U. S. A., 1993; 90: 567-571.
58. Hirakura, Y., Lin, M.C., and Kagan, B.L., Alzheimer amyloid Aβ 1-42 channels: effects of solvent, pH, and Congo red. J. Neurosci. Res., 1999; 57:458-466.
59. Querfurth, H.W. and Selkoe, D.J., Calcium ionophore increases amyloid β peptide production by cultured cells. Biochemistry (Moscow). 1994; 33: 4550-4561.
60. Pierrot, N., Ghisdal, P., Caumont, A., et al., Intraneuronal amyloid-β42 production triggered by sustained increase of cytosolic calcium concentration induces neuronal death. J. Neurochem., 2004; 88: 1140-1150.
61. Ferreiro, E., Oliveira, C.R., and Pereira, C., Involvement of endoplasmic reticulum Ca^{2+} release through ryanodine and inositol 1,4,5-triphosphate receptors in the neurotoxic effects induced by the amyloid-β peptide. J. Neurosci. Res., 2004; 76: 872-880.
62. Suen, K.C., Lin, K.F., Elyaman, W., et al., Reduction of calcium release from the endoplasmic reticulum could only provide partial neuroprotection again β-amyloid peptide toxicity. J. Neurochem., 2003; 87: 1413-1426.
63. Broide, R.S. and Leslie, F.M., The alpha7 nicotinic acetylcholine receptor in neuronal plasticity. Mol. Neurobiol., 1999; 20:1-16.
64. Small, D.H. and Fodero, L.R., Cholinergic regulation of synaptic plasticity as a therapeutic target in Alzheimer's disease. J. Alzheimers Dis., 2002; 4: 349-355.
65. Gray, R., Rajan, A.S., Radcliffe, K.A., et al., Hippocampal synaptic transmission enhanced by low concentrations of nicotine. Nature, 1996; 383: 713-716.
66. Chang, K.T. and Berg, D.K., Nicotinic acetylcholine receptors containing α7 subunits are required for reliable synaptic transmission in situ. J. Neurosci., 1999; 19:3701-3710.
67. Wang, H.Y., Lee, D.H., D'Andrea, M.R., et al., b-Amyloid (1-42) binds to alpha7 nicotinic acetylcholine receptor with high affinity. Implications for Alzheimer's disease pathology. J. Biol. Chem., 2000; 275:5626-5632.
68. Liu, Q., Kawai, H., and Berg, D.K., β-Amyloid peptide blocks the response of α7-containing nicotinic receptors on hippocampal neurons. Proc. Natl. Acad. Sci. U. S. A., 2001; 98:4734-4739.
69. Dineley, K.T., Bell, K.A., Bui, D., et al., β-Amyloid peptide activates α7 nicotinic acetylcholine receptors expressed in Xenopus oocytes. J. Biol. Chem., 2002; 277:25056-25061.
70. Kar, S., Issa, A.M., Seto, D., et al., Amyloid β-peptide inhibits high-affinity choline uptake and acetylcholine release in rat hippocampal slices. J. Neurochem., 1998; 70:2179-2187.
71. Atack, J.R., Perry, E.K., Bonham, J.R., et al., Molecular forms of acetylacholinesterase in senile dementia of Alzheimer type: selective loss of the intermediate (10S) form. Neurosci. Lett., 1983; 40: 199-204.
72. Fodero, L.R., Mok, S.S., Losic, D., et al., α7-Nicotinic acetylcholine receptors mediate an Aβ1-42-induced increase in the level of acetylcholinesterase in primary cortical neurones. J. Neurochem., 2004; 88:1186-1193.
73. Wang, H.Y., D'Andrea, M.R., and Nagele, R.G., Cerebellar diffuse amyloid plaques are derived from dendritic Aβ42 accumulations in Purkinje cells. Neurobiol. Aging, 2002; 23:213-223.
74. Barker, P.A., p75NTR is positively promiscuous: novel partners and new insights. Neuron, 2004; 42: 529-533.
75. Rabizadeh, A., Bitler, C.M., Butcher, L.L., et al., Expression of the low-affinity nerve growth factor receptor enhances β-amyloid peptide toxicity. Proc. Natl. Acad. Sci. U. S. A., 1994; 91:10703-10706.
76. Yaar, M., Zhai, S., Pilch, P., et al., Binding of b-amyloid to the p75 neurotrophin receptor induces apoptosis. A possible mechanism for Alzheimer's disease. J. Clin. Invest., 1997; 100:2333-2340.
77. Tsukamoto, E., Hashimoto, Y., Kanekura, K., et al., Characterization of the toxic mechanism triggered by Alzheimer's amyloid-β peptides bia p75 neurotrophin receptor in neuronal hybrid cells. J. Neurosci. Res., 2003; 73:627-636.
78. Mufson, E.J. and Kordower, J.H., Cortical neurons express nerve growth factor receptors in advanced

age and Alzheimer's disease. Proc. Natl. Acad. Sci. U. S. A., 1992; 89:569-573.
79. Jaffar, S., Counts, S.E., Ma, S.Y., et al., Neuropathology of mice carrying mutant APP_{swe} and/or PS_{1M146L} transgenes: alterations in the $p75^{NTR}$ choinergic basal forebrain septohippocampal pathway. Exp. Neurol., 2001; 170:277-243.
80. Zhang, Y., Hong, Y., Bounhar, Y., et al., p75 Neurotrophin receptor protects primary cultures of human neurons against extracellular amyloid β peptide cytotoxicity. J. Neurosci., 2003; 23:7385-7394.
81. Yan, S.D., Chen, X., Fu, J., et al., RAGE and amyloid-β peptide neurotoxicity in Alzheimer's disease. Nature, 1996; 382:685-691.
82. Yan, S.D., Zhu, H., Zhu, A., et al., Receptor-dependent cell stress and amyloid accumulation in systemic amyloidosis. Nat. Med., 2000; 6:643-651.
83. Du Yan, S., Zhu, H., Fu, J., et al., Amyloid-β peptide-receptor for advanced glycation endproduct interaction elicits neuronal expression of macrophage-colony stimulating factor: a proinflammatory pathway in Alzheimer's disease. Proc. Natl. Acad. Sci. U. S. A., 1997; 94:5296-5301.
84. Lue, L.F., Walker, D.G., Brachova, L., et al., Involvement of microglial receptor for advanced glycation endproducts (RAGE) in Alzheimer's disease: identification of a cellular activation mechanism. Exp. Neurol., 2001; 171:29-45.

7 Aβ Variants and Their Impact on Amyloid Formation and Alzheimer's Disease Progression

Laszlo Otvos, Jr.

7.1 Introduction

Alzheimer's disease (AD) is characterized pathologically by abnormal accumulation of amyloid plaques and neurofibrillary tangles in vulnerable brain regions [1]. Although the main proteinaceous component of the plaques is the amyloid β peptide (Aβ), the tangles are primarily made up from hyperphosphorylated versions of the microtubule-associated protein tau [2]. Emerging evidence for the overlap in the pathological and clinical features of patients with brain amyloidosis suggests that the plaques and tangles may be linked mechanistically [3]. Increased levels of Aβ peptides in brain can promote the formation of intracellular tau aggregates, although the mechanism for this process is still unclear. These results indicate that one form of amyloid can directly or indirectly impact the formation of another form of amyloid composed of different protein, likely contributing to the overlap in clinical and pathological features. Aβ is an approximately 4-kDa peptide with a strong potential to aggregate during electrophoresis [4] and when isolated from amyloid deposits or control brain tissue represents a family of numerous peptide species [5].

It is increasingly believed that Aβ amyloidogenesis and Alzheimer's disease are causally related, and this notion derives from both genetic and cellular observations. On one hand, all four genes definitively linked to inherited forms of the disease to date have been shown to increase the production and/or deposition of Aβ in the brain [6]. On the other hand, drugs known to reduce the prevalence of Alzheimer's disease in epidemiological studies also reduce Aβ levels in cultured cells [7]. In general, Aβ aggregates can directly and indirectly mediate neurotoxic effects, inflammatory responses, and abnormal tau phosphorylation, the hallmarks of Alzheimer's disease [8]. In spite of this correlation, no major differences in Aβ concentration between samples acquired from diseased or normal tissues could initially be identified, at least not from the cerebrospinal fluid [9]. The explanation may rest in the insensitivity of early Aβ analytical methodology [10] or more likely from the heterogeneity of the samples in Alzheimer's disease–affected or normal brains.

Aβ was originally isolated and sequenced as a 42 (43) residue-long peptide with no sequence homology to proteins available at that time [11]:

H-Asp1-Ala2-Glu3-Phe4-Arg5-His6-Asp7-Ser8-Gly9-Tyr10-Glu11-Val12-His13-His14-Gln15-Lys16-Leu17-Val18-Phe19-Phe20-Ala21-Glu22-Asp23-Val24-Gly25-Ser26-Asn27-Lys28-Gly29-Ala30-Ile31-Ile32-Gly33-Leu34-Met35-Val36-Gly37-Gly38-Val39-Val40-Ile41-Ala42-(Thr43)-OH

Ensuing biochemical characterization and comparison of soluble Aβ secreted by cells, soluble Aβ in the cerebrospinal fluid, and insoluble Aβ isolated from the brains of affected individuals has revealed that there are numerous Aβ species with extensive amino and carboxyl-terminal heterogeneity as well as featuring a series of mid-chain amino acid alterations [12]. As soon as the alterations were discovered, these genetic mutations or post-translational modifications, including oxidation by radicals, truncations, isomerization, and racemization, were speculated as modifiers of Aβ metabolism and/or enhancers of aggregation and hence as progression factors for familiar and sporadic cases of Alzheimer's disease. This article tries to unify the divergent views and provide a comprehensive account for the impact of Aβ variations in the development of amyloid diseases. Table 7.1 lists all known major Aβ sequence modifications and their relevance in molecular or clinical pathogenesis.

After a short analysis into the origin of modified Aβ forms in tissues and cultured cells, we will concentrate on the major properties of the amyloid protein, as regulated by the amino acid alterations. The two dominant attributes of Aβ, the golden standards to which every derivative is compared, are fibrillogenesis [13] and neurotoxicity [14], this latter frequently related to oxidative stress [15]. Fibril formation can be viewed directly as true aggregation [16] or indirectly as the ability of the peptide to assume β-pleated sheet conformation, the prerequisite for fibrillogenesis [17]. More precisely, the characteristic α-helix/random coil → β-pleated sheet conformational transition is considered an easily observable sign of increased ability to form aggregates [18]. Neurotoxicity can also be studied as direct killing of cells [19] or as an outcome of long-lived protein variants, unable to turn over within the life cycle of cells [20].

7.2 The Origin of Modified Aβ Forms

Mid-chain modifications, concentrated around residue Glu22, are clearly due to mutations in the precursor gene. Aβ is a normally secreted proteolytic product [21] of the amyloid precursor protein (APP), a 677–770 reside-long type 1 integral membrane protein [22]. A constitutive secretory metabolic pathway involves APP cleavage at Aβ position 16 by the α-secretase enzyme producing two halves of Aβ. When the γ-secretase further cleaves the product, a carboxy-terminal Aβ 17-40/42 fragment is formed, named p3 [23]. During an alternative proteolytic pathway, a third enzyme, the β-secretase, cleaves APP at the amino-terminus of Aβ [24] followed by γ-secretase action at the C-terminus producing the full-length amyloid peptide. C-terminal alterations are thought to originate from mutations in the APP gene. Processed from wild-type APP, the major 4-kDa Aβ species in both conditioned medium and human cerebrospinal fluid is Aβ 1-40 (>60–70%), although some Aβ 1-42 is also present (≈15%) along with minor amounts of other Aβ fragments [10]. However, when the APP gene includes mutations immediately downstream of the Aβ coding region, the production level of Aβ 1-42 significantly increases [25].

TABLE 7.1. Aβ variations known to affect Alzheimer's disease development.

	N-terminal truncations and isomerizations	
First residue in truncated Ab	**Species abbreviation in text**	**Presence in amyloid forms**
D-Asp1	rD-1	In plaques of controls with atherosclerosis
isoAsp1	iD-1	Increased amyloid in parenchyma
pGlu3	pGlu3-Nterm	Fifty percent in senile plaques
isoAsp7	iD-7	Increased amyloid in parenchyma
pGlu11	pGlu11-Nterm	Thirty percent in serum
Leu17	p3	Early deposits in Down syndrome
	Mid-chain genetic mutations	
Mutated residue	**Species abbreviation in text**	**Clinical phenotype**
Ala2 → Thr	Thr2	Stroke and myocardial infarction
Ala21→ Gly	Flemish type	Presenile dementia and cerebral hemorrhage
Glu22 → Gln	Dutch type	Cerebral hemorrhage
Glu22 → Gly	Arctic type	Early-onset Alzheimer's disease
Glu22 → Lys	Italian type	Presenile dementia and cerebral hemorrhage
Asp23 → Asn	Iowa type	Early-onset Alzheimer's disease
Ala42 → Val	Val42	Schrizophrenia
Ala42 → Thr	Thr42	Early-onset Alzheimer's disease
	C-terminal truncation	
Last residue in truncated Aβ	**Species abbreviation in text**	**Present**
Val40	1-40	When the precursor protein is not mutated downstream

In wild-type APP, the fourth residue after Aβ Ala42 is a valine; in familiar Alzheimer's disease in Anglo-Saxon, Italian, and Japanese kindreds, this Val is substituted with Ile, Phe, or Gly, respectively [26–28]. We compared Aβ production in human neuroblastoma (M17) cells transfected with constructs expressing wild-type APP or the APP717 mutants by either isolation of metabolically labeled Aβ from conditioned medium, digestion with cyanogen bromide, and analysis of the carboxyl-terminal peptides released, or by analysis of the amyloid peptide in conditioned medium with immunosorbent assays that discriminate Aβ 1-40 and 1-42. Both methods demonstrated that Aβ released from wild-type βAPP is primarily, but not exclusively, 40 residues long. The APP717 mutations consistently caused a 1.5- to 1.9-fold increase in the percentage of 42-residue Aβ generated. The pathological consequences of longer Aβ assembly will be discussed later.

In general, peptides are subjected to endopeptidase and exopeptidase cleavages with amino- and carboxy-peptidases being the major culprits for peptide degradation [29]. Carboxy-terminal truncations may theoretically occur from the cleaved Aβ 1-42 [43] peptides in tissues, but apparently genetic processing of APP is a more common explanation for explaining heterogeneity at the C-terminus [30]. Indeed, a novel expression system was developed, one that in the secretory pathway selectively generates Aβ 1-40 or Aβ 1-42 fused to the transmembrane BRI protein. Significantly, expression of Aβ 1-42 results in no increase in secreted Aβ 1-40, suggesting that the majority of Aβ 1-42 is not trimmed by carboxypeptidase to Aβ 1-40. Yet, as the identity and role of secretases responsible for APP processing in the human brain have yet to be clarified [31], the search for enzyme activities capable of cleaving native brain APP in human hippocampus is underway. A 40-kDa protein with proteolytic activity that degrades native brain APP in vitro was purified and characterized; molecular analysis identified it as a novel protease belonging to the carboxypeptidase B family [32]. PC12 cells overexpressing this protease generate a major 12-kDa Aβ-bearing peptide in cytosol, a peptide that has also been detected in a cell-free system using purified brain APP as substrate. Having said this, carboxypeptidase processing of longer Aβ variants enjoy much less attention than exopeptidase activity at the amino-terminus.

The amino acid sequence of wild-type Aβ starts with an N-terminal Asp residue, and a Glu residue is found two positions downstream; these amino acids are the main substrates of aminopeptidase A [33]. When the activity of aminopeptidases as a function of age or sex was studied, significant age-related increases were observed in glutamic aminopeptidase A activity in both human genders and in aspartic aminopeptidase A activity in females [34]. This may reflect the evolution of susceptible circulating substrates during development and aging. In support, when specific soluble and membrane-bound aspartyl-hydrolyzing activities were assayed in brain subcellular fractions from rat fetuses (19–20 days of gestation), and from 1- to 260-week-old rats, significant age-related changes were observed in all fractions for both enzymatic activities [35]. Taken together, it is well conceivable that the amino terminal Asp1 and Glu3 residues in Aβ undergo enzymatic degradation.

Alternatively, Asp is subject to a completely nonenzymatic processing pathway. It was hypothesized that Alzheimer's disease is initiated by a protein aging-related structural transformation in soluble Aβ [36]. According to this theory, spontaneous chemical modification of aspartyl residues in Aβ to transient succinimide induces a non-native conformation in a fraction of soluble Aβ, rendering it amyloidogenic and neurotoxic. As shown later, conformationally altered Aβ is characterized by increased stability in solution and the presence of a non-native β-turn that determines folding. Formation of the succinimide from Asp is a result of an intramolecular nucleophilic attack of the peptide amide-nitrogen on the side-chain carbonyl group of Asp (Fig. 7.1). Hydrolysis of succinimide leads to accumulation of stable isoaspartyl sites (isoAsp) in which a peptide bond is formed by the side-chain carboxyl group of Asp. A competing hydrolysis pathway leads to the production of peptides containing D-aspartic acid.

7.3 Different Aβ Variants in Space and Time

In order to identify the proteolytic enzymes responsible for the formation of the distinct Aβ forms and the organelles in which diverse forms of Aβ are generated and from which they are secreted, the Aβ

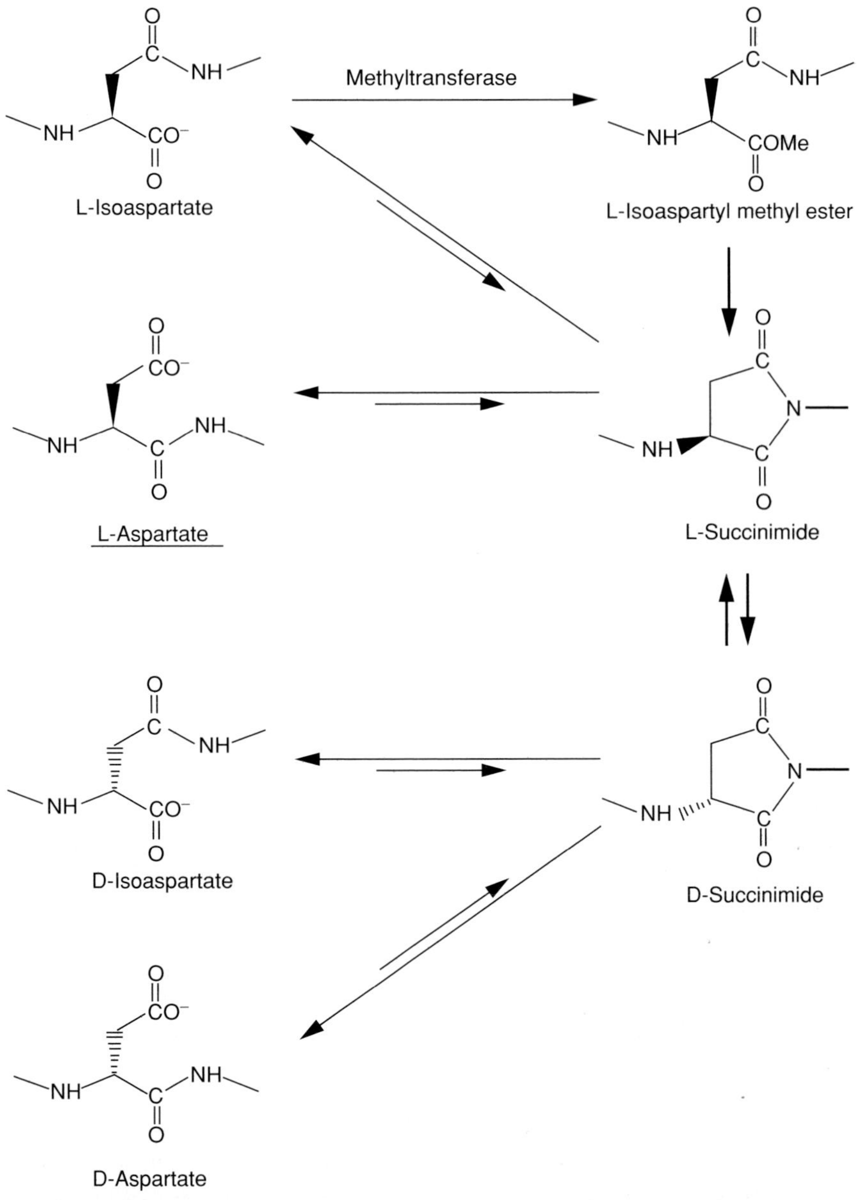

FIGURE 7.1. Formation of succinimide through spontaneous cyclization of aspartyl residues. Hydrolysis of the cyclic products leads to D-aspartate and L- and D-isoaspartates together with the unmodified L-asparate forms. Reprinted from Ref. 36, with permission of the Federation of American Societies for Experimental Biology.

compositions of subcellular compartments were investigated together with the compartments from which the Aβ variants were secreted [37]. It was found that Aβ 1-40 (or Aβ x-40) is generated exclusively within the trans-Golgi network and packaged into post–trans-Golgi network secretory vesicles; Aβ x-42 is made and retained within the endoplasmic reticulum in an insoluble state; all Aβ 42 forms are made in the trans-Golgi network and packaged into secretory vesicles; and finally the amyloid peptides formed consist of two pools (a soluble population extractable with detergents and a detergent-insoluble form). It was concluded that cell-free Aβ generation assays may distinguish between intracellular insoluble peptides and secreted soluble analogues.

To this extent, soluble Aβ and its variants, produced by mouse neuroblastoma cells, were selectively isolated by immunoprecipitation with anti-Aβ monoclonal antibodies, and the identities of these isolated amyloid peptides were determined by measuring their molecular masses using matrix-assisted laser desorption/ionization time-of-flight mass spectrometry. The relative signal intensities were used to estimate the concentrations of Aβ[10]. Although pharmacologically mass spectrometry without chromatographic quantitation steps is not fully defendable [38], this approach detected several novel Aβ variants and successfully quantified soluble Aβ in conditioned media of cultured mammalian cells. The identified 64 Aβ-related peptides (44 from human and 20 from murine amyloid sequences) included a cascade of N- and C-terminal truncations with little preference of a given structural motif. The human APP samples featured an increased abundance of peptides starting with Ala2 and Phe4 (in agreement with the hypothesized aminopeptidase A activity on Asp1 and Glu3) but without major statistical significance. At least, analysis of degradation products of synthetic human Aβ peptides revealed four primary cleavage sites (C-terminal to His13, Phe19, Lys28 and Gly33) with three different endopeptidase substrate specificities. These Aβ variants may contribute to the low levels of certain Aβ subpopulations normally observed in cell culture media of transfected cells.

Of course, these findings raise the question as to which residues promote aggregation and which endorse soluble Aβ derivatives. Because this review is concerned with natural Aβ variants, listing of all designer Aβ analogues falls outside the scope of this article. Yet, one study that claims to represent an unbiased search for sequence determinants of Aβ amyloidogenesis may fit the bill. This screen is based on the finding that fusions of the wild-type Aβ 1-42 sequence to green fluorescent protein form insoluble aggregates in which the green fluorescent protein is inactive. Cells expressing such fusions do not fluoresce as opposed to Aβ with reduced tendencies to aggregate, which can be constructed and screened from randomly mutated Aβ 1-42 green fluorescence protein libraries [39]. Not surprisingly, most of the observed solubility-enhancer residues are replacements of hydrophobic amino acids in the Leu17-Phe19, Ile31-Ile32, Leu34-Val36, and Val39-Ala42 fragments. The only notable finding is that some conservative amino acid changes (Val18 → Ala, Phe19 → Leu, and Ile32 → Val) also increase solubility, and these curiously fall into or proximal to the detected primary enzymatic cleavage sites of the previous paragraph.

7.4 Animal Models

A major obstacle to the pharmaceutical development of Aβ aggregation inhibitors is the lack of appropriate small animal models [40]. In most of the current mouse models of Alzheimer's disease, the animals contain amyloid plaques in their brain, but the amyloidosis is not accompanied by extensive tangle formation or massive neuronal loss. This is partially understandable if we compare the Aβ sequences in different animal species and their ability to form aggregates. When the Aβ sequences of human, dog, polar bear, rabbit, cow, sheep, pig, and guinea-pig are compared with the corresponding rodent sequences and a phylogenetic tree is generated, it is obvious that the Aβ amino acid sequence of human, dog, and polar bear and other mammals that may form amyloid plaques is conserved, and the mice and rats where amyloid has not been detected may be evolutionarily a distinct group [41, 42]. In addition, the predicted secondary structure of mouse and rat Aβ lacks the propensity to form a β-pleated sheet secondary structure.

Compared with human Aβ, the amino acid sequence of mouse Aβ differs at three positions: Arg5 is replaced with Gly, Tyr10 is replaced with Phe, and His13 is replaced with Arg [43], with the rat sequence being identical to that of mouse [44]. To study the preferred β-pleated sheet forming ability of the human peptide compared with the rodent analogue, we synthesized, purified, and characterized the two different Aβ sequences [45]. Circular dichroism (CD) and Fourier-transformed infrared spectroscopy were used with various membrane-mimicking solvents, different peptide concentrations, and variable pH to identify those environmental conditions that promoted β-pleated sheet formation of the human versus rodent amyloid peptides. We found that higher β-pleated sheet content was observed for the rodent sequence in acetonitrile/water mixtures. In contrast, more β-pleated sheets were detected for the human Aβ in trifluoroethanol/water mixtures at neutral pH. Remarkably, at relatively low peptide concentrations, only the human sequence assumed an extended secondary structure (Fig. 7.2). These data suggest that subtle inter-species amino-acid differences may account for the inability of the rodent peptide to form amyloid fibrils *in situ*, when only low amounts of soluble peptides are available for aggregation. However, if fibrils once formed, these N-terminal amino acid differences have virtually no effect on the morphology or organization of the fibrils [46]. It needs to be added that in the current article, altered peptide conformations are considered as factors that promote disease pathogenesis. However, the opposite can be equally true: differences in Aβ secondary structure may be a consequence of disease progression.

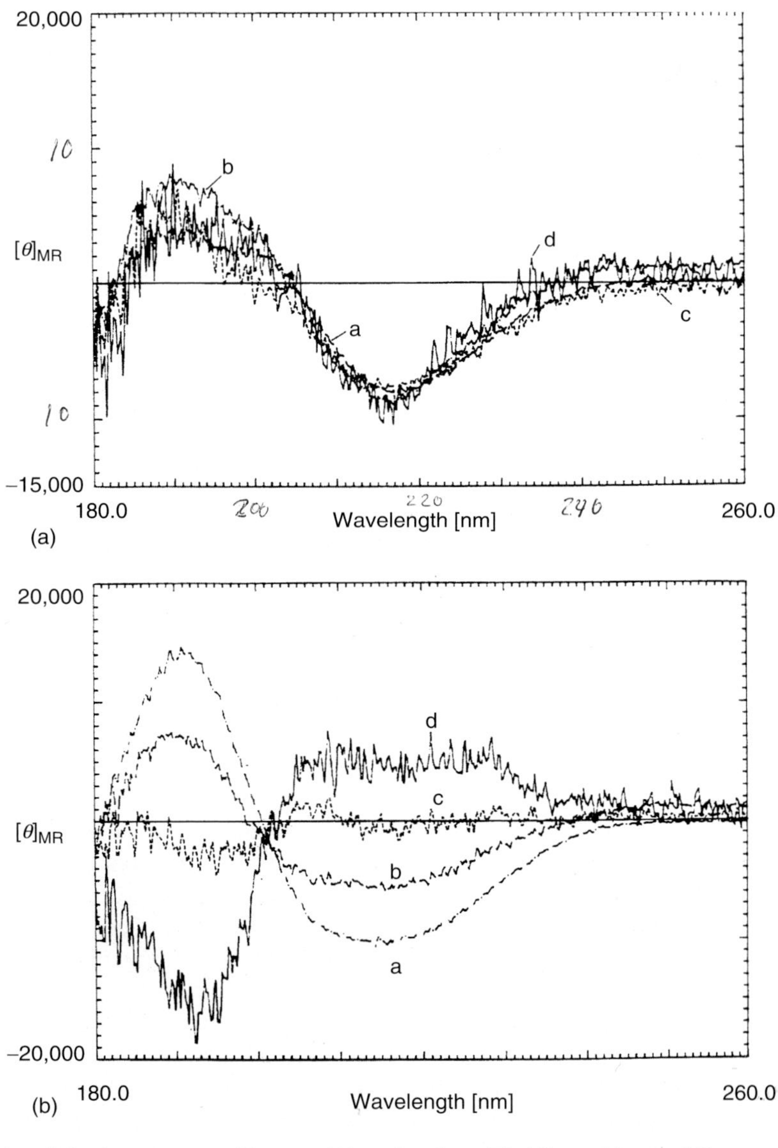

FIGURE 7.2. Circular dichroism spectra of human (A) and rodent (B) Aβ peptides at different concentrations. The rodent analogue forms β-pleated sheets at significantly higher concentration than the human version does: a, 0.5 mg/mL; b, 0.25 mg/mL; c, 0.125 mg/mL; d, 0.0625 mg/mL. Reprinted from Ref. 45, with permission of the Federation of European Biochemical Society.

Earlier we briefly mentioned that in human Alzheimer brain, the major C-terminal variant that forms amyloid fibers is Aβ 1-42. In contrast, the major fibrillar aggregates that present Congo red birefrigence in rat brain consist of the Aβ 1-40 peptide, whereas Aβ 1-42 aggregates as a nonfibrillar amorphous material [47]. Thus, instead of the lack of deposition process *per se*, factors might exist in the rat brain that inhibit the fibrillar assembly of the most pathogenic soluble Aβ 1-42 variant. In support of differences in fibril assembly rather than postsecretory processing, freshly solubilized human Aβ 1-40 or Aβ 1-42 were injected into rat brains, and it was shown that both peptides were equally processed at their amino-termini to yield variants starting at pGlu3 and at their C termini to yield variants ending at Val40 and at Val39 [47]. Contradictory to the previous argument, normal rat brain can produce enzymes that mediate the conversion of Aβ 1-40/1-42 into processed variants similar to those in Alzheimer's disease.

Obviously, the loss of the side-chain positive charge at position 5 in the native rodent Aβ analogue can influence metal-binding, a well-studied risk factor in Aβ aggregation [48] and fibril formation [49]. Indeed, Cu(II) (at concentrations lower than that associated with amyloid plaques) induces the generation of dityrosine cross-linked, sodium dodecyl sulfate–resistant oligomers of human, but not rat, Aβ peptides [50], and the alteration must involve Tyr10 (also missing in rodent Aβ) because no detectable peroxidative modifications are observed with Aβ 12-28 [51]. The coordination of metal ions for human and mouse N-terminal Aβ fragments starts from the N-terminal Asp residue, which stabilizes significantly the 1N complex as a result of chelation through the side-chain carboxylate group [52]. In a wide pH range of 4–10, the imidazole nitrogen of His6 is coordinated to form a macrochelate. Results show that, in the pH range 5–9, the human fragments form the complex with different coordination mode compared with that of the mouse fragments. The low pK(1)(amide) values (approximately 5) obtained for the mouse N-terminal Aβ fragments may suggest the coordination of the amide nitrogen of His6 while in case of the human fragments the coordination of the amide nitrogen of Ala2 is a more likely scenario. The Gly → Arg residue replacement in position 5 of the Aβ peptide sequence changes the coordination modes of a peptide to metal ion in the physiological pH range. The mouse fragments of Aβ are much more effective in Cu(II) binding than the human fragments.

Human and rat variants of Aβ 1-42 were compared to determine whether they produce the same amount of neuronal loss when combined with iron [53]. Coinjection of iron with either Aβ variant caused significantly more neuronal loss than the Aβ peptide alone, suggesting that iron may contribute to the toxicity associated with senile plaques. Rat Aβ 1-42 combined with iron was as toxic as iron alone, whereas iron combined with human Aβ 1-42 was significantly less toxic. This latter finding indicates that fibrillar human Aβ is able to reduce iron-induced neurotoxicity in vivo and raises the interesting possibility that senile plaques in Alzheimer's disease may represent a neuroprotective response to the presence of elevated metal ions.

When the human sequence is introduced into rodents, a thorough chemical and morphological comparison of the Aβ molecules and the amyloid plaques present in the brains of APP transgenic mice and human Alzheimer's disease patients show that despite an apparent overall structural resemblance to Alzheimer pathology, transgenic mice produce amyloid cores that are completely soluble in buffers containing sodium dodecyl sulfate, whereas human amyloid plaques are highly resistant to chemical and physical disruption [54]. It was suggested that Aβ chemical alterations account for the extreme stability of Alzheimer plaque core amyloid. Curiously, the corresponding lack of post-translational modifications such as N-terminal degradation, isomerization, racemization, pyroglutamyl formation, oxidation, and covalently linked dimers, all the alterations we review in this article, in transgenic mouse Aβ may provide an explanation for the differences in solubility between human and APP transgenic mouse plaques. It was hypothesized that either insufficient time is available for Aβ structural modifications to take place or the complex species-specific environment of the human disease is not precisely replicated in the transgenic mice. The appraisal of therapeutic agents or protocols in these animal models must be judged in the context of the lack of complete equivalence between the transgenic mouse plaques and human Alzheimer's disease lesions.

However, perhaps there is light at the end of the tunnel. In transgenic mice overexpressing the London mutant of human APP, N- and C-terminally modified Aβ peptides were detected, similar to the modified Aβ versions in humans [55]. The ratios of deposited Aβ 1-42/1-40 were of the order 2–3 for human and 8–9 for mouse peptides, indicating a preferential tendency for the deposition of the longer amyloid peptide. In protein extracts from soluble and insoluble brain fractions, the most prominent peptides were truncated either at the carboxyl- or the amino-termini yielding Aβ 1-38 and Aβ 11-42, respectively, and the latter was strongly enriched in the extracts of deposited peptides. These data indicate that plaques of APP-London transgenic mice consist of aggregates of multiple human and mouse Aβ variants, possibly indeed characteristic for those in the brains of Alzheimer's disease patients.

Most recently, a similar transgenic mouse model, named APP(SL)PS1KI, was presented [56]. This transgenic mouse model carries knocked-in mutations in the presenilin-1 gene and overexpresses mutated human APP. Just like in the human cases, Aβ (x-42) is the major form of Aβ species present in this model with progressive development of a complex pattern of N-truncated variants and dimers, similar to those observed in Alzheimer's disease brain. Significantly, an extensive neuronal loss (>50%) is present in the CA1/2 hippocampal pyramidal cell layer at 10 months of age together with strong reactive astrogliosis. Due to the appearance of the critical Aβ variations, APP(SL)PS1KI mice may provide a long-awaited tool to investigate therapeutic strategies designed to prevent neurodegeneration in Alzheimer's disease.

7.5 N-Terminal Truncations and Modifications

After so much about the modifications in general, let's look at the variant human Aβ peptides in detail. We start with N-terminal modifications, followed by mid-chain alterations; finally, a brief discussion of the differing fibrillogenesis by the C-terminal Aβ variants will be presented.

In a seminal report, Aβ peptides were isolated from the compact amyloid cores of neuritic plaques and separated from minor glycoprotein components by size-exclusion high-performance liquid chromatography [57]. Parenchymal Aβ was shown to have a maximal length of 42 residues, but shorter forms with "ragged" amino-termini were also present. Most of the heterogeneity was found in Aβ 1-5 and Aβ 6-16 fragments, each of which eluted as four peaks. Amino acid composition and sequence analyses, mass spectrometry, enzymatic methylation, and stereoisomer determinations revealed that these multiple peptide forms resulted from structural rearrangements of Asp1 and Asp7. The L-isoaspartyl form predominated at each of these positions, whereas the D-isoaspartyl, L-aspartyl, and D-aspartyl forms were present in lesser amounts. Aβ purified from the leptomeningeal microvasculature contained the same structural alterations as parenchymal Aβ, but at the C-terminus ended at Val40. It was suggested that the abundance of structurally altered aspartyl residues affect the conformation of the Aβ peptide within plaque cores and thus significantly impact normal catabolic processes designed to limit its deposition.

To this end, in a series of consecutive papers, we reported on the conformation-modifying effect of aspartic acid isomerization in general, and at the amino terminus of Aβ in particular. First we used circular dichroism and Fourier-transform infrared spectroscopy to characterize the conformational changes on human Aβ upon substitution of Asp1 and Asp7 to isoaspartic residues [58]. We found that the intermolecular β-pleated sheet content is markedly increased for the post-translationally modified peptide compared with that in the corresponding unmodified human or rodent Aβ sequences both in aqueous solutions in the pH 7–12 range and in membrane-mimicking solvents (such as aqueous octyl-β-D-glucoside or aqueous acetonitrile solutions). These findings underline the importance of the originally α-helical N-terminal regions of the unmodified Aβ peptides in defining its secondary structure and may offer an explanation for the selective aggregation and retention of the isomerized Aβ variants in Alzheimer's disease–affected brains. For identifying the general effect of isoaspartic acid–bond formation on peptide conformation, we selected five sets of synthetic model peptides, each representing one of the major secondary structures as the dominant spectroscopically determined conformation: a type I β-turn, a type II β-turn, short segments of α- or

3_{10}-helices, or extended β-strands. We found that both types of turn structures are stabilized by the aspartic acid–bond isomerization. The isomerization at a terminal position did not affect the helix propensity, but placing it in mid-chain broke the helix structure [59]. Interestingly, when Asp was already part of a β-pleated sheet, this structure was also destabilized.

The physical-chemical explanation for the conformational changes in Aβ upon isoAsp1 and isoAsp7 incorporation into the amino-terminal decapeptide fragment was provided based on molecular mechanics calculations [60]. The modeling showed that insertion of the extra $-CH_2-$ group into the decapeptide backbone results in the formation of stable reverse-turns and destabilizes the helical conformer that competes with the extended structure at the full-sized peptide level (Fig. 7.3). The molecular modeling also revealed a limited propensity of the Asp1, Asp7 diisomerized peptide

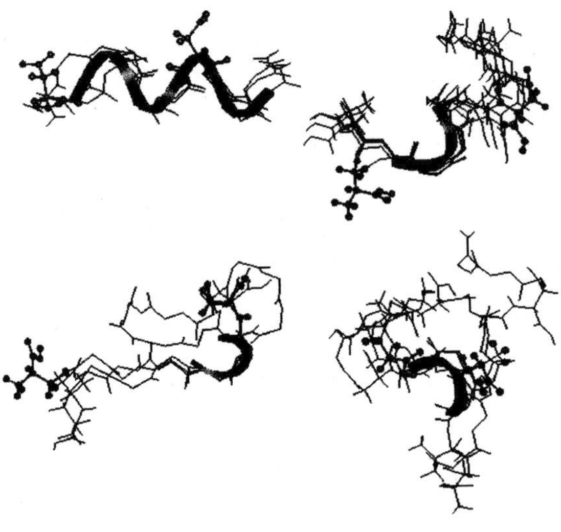

FIGURE 7.3. Low-energy conformers of wild-type Aβ 1-10 and Aβ 1-10 containing isoaspartyl residues in positions 1 and 7. The conformers for each subset are superimposed, and their peptide backbones are displayed as a line. For each conformer, the C^{α} trace of helical or β-turn regions are indicated by a ribbon and Asp and isoAsp residues in positions 1 and 7 by a ball and stick plot. Upper right: Type I β-turn with Glu in position i+1. Lower right: Type III β-turn with Phe at position i+1. Upper left: Aβ 1-10 with residues 3–9 and 5–9 positioned in a helix. Lower left: Type III β-turn with Arg in position i+1. Reprinted from Ref. 60, with permission from Blackwell Publishing.

to form extended structure directly. These basic findings were later confirmed by reports from other research groups. To test how changes in the aspartate forms influence peptide conformation, a series of designed peptides having the sequence VTVKVXAVKVTV, where X represents aspartic acid or its derivatives, were synthesized [61]. Studies using circular dichroism showed that neutralization of the aspartate residue through the formation of a methyl ester or an amide, or replacement of aspartate with glutamate led to an increased β-sheet content at neutral and basic pH. A higher content of β-sheet structure correlated with increased propensity for fibril formation and decreased solubility at neutral pH [61].

Anti-Aβ polyclonal antibody 2332 is more sensitive for the non-isomerized status of the decapeptide than that of the full-sized peptide [59]. Monoclonal antibody 6E10, raised against unmodified Aβ recognizes only the unmodified decapeptide or the peptide isomerized at the first aspartic acid in a conformation-dependent manner but does not recognize the mid-chain isomerized or diisomerized decapeptide in any circumstance. The diisomerized decapeptide was used as immunogen to generate polyclonal antibody 14943 that is not selective for the isomerized status of either the full-size peptide or the decapeptide but recognizes the isomerized peptides preferentially when the peptide antigen structures are conserved during the enzyme-linked immunoassay procedure [62]. Owing to the poor peak shape of the full-sized Aβ peptide during standard reversed-phase chromatography [63], serum stability studies that indicate extracellular stability can be more effectively performed on the decapeptide fragments. Remarkably, the diisomerized Aβ 1-10 peptide exhibits a significantly increased stability toward serum peptidases than the unmodified or monoisomerized peptides, suggesting a possible mechanism of the retention of the isomerized Aβ peptide in the affected brains.

More contemporary techniques are able to identify and quantitate the various Aβ forms with higher accuracy. Although the protein is not directly Alzheimer's disease related, serum amyloid α-1 can be detected in serum as full-length protein, as well as its well-characterized des-arginine and des-arginine/des-serine variants at the N-terminus by surface-enhanced laser desorption ionization mass spectroscopy [64]. The method is sensitive enough to detect a low-abundant variant with the first five N-terminal amino acids missing. Mass spectroscopy is reproducible, fast, and simple mode for the discovery and analysis of marker proteins of various diseases or for quality control of synthetic products.

This leads us to the quantification of the various Aβ forms in cells and tissues. We performed two-site enzyme-linked immunosorbent assay with antibodies specific for isomerized (i.e., Aβ with L-isoAsp at positions 1 and 7) and pGlu-modified (i.e., Aβ beginning with pyroglutamic acid at position 3) forms of Aβ to quantitate the levels of these different Aβ peptides in formic acid extracts of Alzheimer's disease frontal cortex [65]. The major species of Aβ in these samples were Aβ pGlu3-42 as well as Aβ x-42, whereas isomerized Aβ was a minor species. More specifically, across a panel of 14 samples, the μg/g wet tissue weight of the various Aβ species were as follows: Aβ 1-40 (1,7 di-isoAsp), 0.03; Aβ pGlu3-40, 0.14; Aβ 1-42 (1,7 di-isoAsp), 0.61; Aβ x-40 (where x is 1 or 2), 1.66; Aβ x-42, 3.14; and Aβ pGlu3-42, 3.18. As seen, the forms ending with Ala42 greatly exceeded those ending with Val40. This study was in line with an earlier report on cortical sections from 28 aged individuals with a wide range in senile plaque density. According to these results, the major Aβ molecular species deposited in the brain contain PGlu3 as the N-terminal amino acid residue [66]. The abundance of the pGlu N-terminal forms suggests that these Aβ variants can play important roles in the deposition of amyloid in Alzheimer's disease brains.

Of course, all quantitative data have to be viewed in light of the availability of the given Aβ analogue in the given sample. However, the hydrophobicity of the modified peptides is greatly different giving rise to potential inaccuracy in concentration-determination. After many years of trouble with reversed-phase chromatographic analysis of Aβ peptides, a new protocol was developed that uses high column temperature for optimal peak shape and separation [67]. Coupled with mass spectroscopy, the method is suitable for the quantification of Aβ isoforms in solution. Upon identical separation conditions, the recovery of the different Aβ species from the hydrophobic column were Aβ 1-40, 36%; Aβ pGlu11-40, 34%; Aβ pGlu3-40, 22%; and p3, 14%. It is obvious that the more hydrophobic the samples were, the lower recovery yield was obtained. If this experiment can be extrapolated to tissue samples, there is a good possibility that the total quantity of the less hydrophilic variants is regularly underestimated.

How would the increase the pGlu3 amino-terminal forms influence the two major properties of Aβ, aggregation and neurotoxicity? Using circular dichroism spectroscopy, it was determined that the pyroglutamic acid–containing peptides form β-sheet structure more readily than the corresponding full-length Aβ peptides, both in aqueous solutions and in 10% sodium dodecyl sulfate micelles [68]. CD spectra taken in aqueous trifluoroethanol solutions indicated that the relative β-sheet to α-helical stability is higher for the pGlu-containing peptides. The conformational differences were mirrored by alterations in the level of precipitated Aβ species and the kinetics of the sedimentation (Fig. 7.4). According this, pGlu3 and pGlu11-N-terminal Aβ 1-40 peptides have greater aggregation propensities

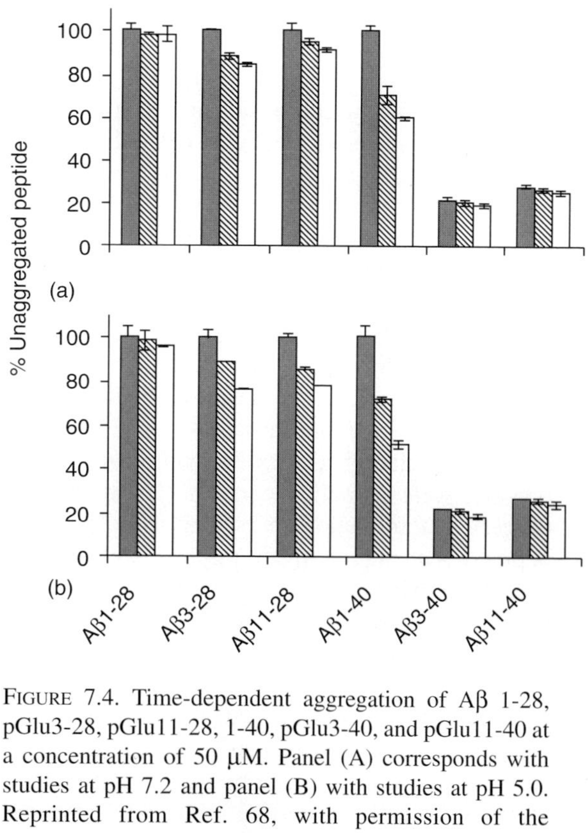

FIGURE 7.4. Time-dependent aggregation of Aβ 1-28, pGlu3-28, pGlu11-28, 1-40, pGlu3-40, and pGlu11-40 at a concentration of 50 μM. Panel (A) corresponds with studies at pH 7.2 and panel (B) with studies at pH 5.0. Reprinted from Ref. 68, with permission of the American Chemical Society, Copyright 1999.

than the corresponding nonmodified peptides, with about 4- to 5-fold reduction in the unaggregated form at various pH and after three different incubation periods. Comparison between peptides ending with Val40 or Lys28 (the carboxy-terminal end of the extracellular domain) indicated that the greater β-sheet forming and aggregation propensities of the pyroglutamyl peptides are not simply due to an increase in hydrophobicity [68]. As for the mechanistic explanation, it was suggested that the loss of N-terminal charges may facilitate β-sheet formation by decreasing the level of unfavorable interstrand charge repulsion, as long as the Aβ fibril is a hydrogen-bonded parallel β-sheet as previously suggested [69]. In addition, the loss of the Asp and Glu side-chain negative charges may destabilize helix formation by eliminating favorable charge dipole interactions [70].

In another study, the toxic properties, fibrillogenic capabilities, and in vitro degradation profile of Aβ 1-40, Aβ 1-42, Aβ pGlu3-40, and Aβ pGlu3-42 were compared [71]. The data show that the fiber morphology of the Aβ peptides is greatly influenced by the C-terminus while toxicity, interaction with cell membranes, and degradation are influenced by the N-terminus. Aβ pGlu3-40 induces significantly more cell loss than the other species both in neuronal and glial cell cultures. The numerical values are 23% decrease relative to controls at 0.1 μM, 31% loss at 1 μM, and 51% at 10 μM, well within the range of modified Aβ level in tissues (compare with the Aβ tissue concentrations above). Aggregated Aβ peptides starting with pyroglutamic acid in position 3 were heavily distributed on plasma membrane and within the cytoplasm of treated cells. The Aβ pGlu3-40/42 peptides showed a significant resistance to degradation by cultured astrocytes, while unmodified peptides were partially degraded. These findings suggest that formation of N-terminally modified peptides enhance both β-amyloid aggregation and neurotoxicity, likely worsening the onset and progression of Alzheimer's disease.

The question arises whether the isomerized/racemized forms are spatially and/or temporally separated from the unmodified Aβ isoform. Neuritic plaques in Alzheimer's disease brain typically immunostain with antibodies against nonisomerized Aβ and Aβ starting with pGlu3, but not Aβ starting with Leu17 (p3) or Asp1 racemized Aβ. Neuritic deposits in nondemented individuals with atherosclerotic and vascular hypertensive changes could be identified with all three Aβ isoforms [72]. The presence of Aβ with racemized Asp1 in neuritic plaques in nondemented individuals with atherosclerosis or hypertension, but not in Alzheimer's disease, suggests a different evolution of the plaques in the two conditions. In another antibody-based assay, the amino- and carboxyl-terminal properties of the various Aβ peptides deposited in diffuse plaques, one of the earliest forms of amyloid deposition, were examined [73]. It was concluded that the amino termini of the Aβ species that initially deposit in diffuse plaques begin with Asp1 with or without structural modifications (isomerization and racemization), as well as with pGlu3, and terminate preferentially at Aβ 1-42(43) rather than Aβ 40. This last paper well represents a research trend that looks at modifications in multiple positions along the Aβ sequence. In the end of this review, this approach will be scrutinized in detail. Finally, here is an interesting observation regarding the spatial relationship between a 100-kDa unidentified "AMY" protein and N-terminally modified Aβ peptides: AMY immunoreactive plaques colocalized with amyloid plaques labeled by antibodies to Aβ starting at position 3 with a pGlu, however AMY immunoreactive deposits colocalized to a lesser degree with amyloid plaques labeled by antibodies to other variants of the Aβ peptide [74] supporting the well-known finding that automatic water loss on natural and synthetic peptides with glutamine amino terminus leads to massive pGlu production.

Isomerized Aβ variants are not restricted to the amino-terminus of the peptide. A specific antibody recognizing isoAsp23 of Aβ suggests the isomerization of Aβ at Asp23 in vascular amyloid as well as in the core of senile plaques [75]. The widespread isomerization of aspartic acids in Alzheimer's disease is quite interesting, as biochemical analyses of neurofibrillary tangles also revealed L-isoaspartate at Asp193, Asn381, and Asp387 [76], indicating a modification, other than phosphorylation, that differentiates between normal tau and tau found in the paired helical filaments of Alzheimer's disease. Protein L-isoaspartyl methyltransferase is suggested to play a role in the repair of isomerized proteins containing L-isoAsp [77]. This enzyme is upregulated in neurodegenerative neurons and colocalizes in neurofibrillary tangles [75]. Taken together with the enhanced protein isomerization in Alzheimer's disease brains, it is implicated that upregulated isoaspartyl methyltransferase activity may associate with increased protein isomerization in Alzheimer's disease. It needs to be added that aspartic acid isomerization occurs during synthetic glycosylation reactions of tau fragments a well, suggesting a chemical rather than enzymatic modification in aged and post-translationally modified proteins [78]. Indeed, isomerization and racemization of aspartyl residues are often considered as products of spontaneous nonenzymatic reactions that give rise to many aspartyl forms, including L- and D-isoAsp and D-Asp [79].

7.6 Abundant Alterations at Mid-Chain Positions

The appearance of isoaspartate at position 23 takes us to Aβ modifications in mid-chain positions. Assays with the isoAsp23-specific antibody documented that Aβ isomerized at position 23 is deposited on plaques and vascular amyloids [80]. In vitro experiments showed that isomerization at position 23, but not position 7, enhanced aggregation. Furthermore, Aβ with the Dutch-type mid-chain mutation (Gln22), but not the Flemish-type mutation (Gly21), also showed greatly enhanced aggregation. These results suggest that mutations or modifications at unmodified Aβ positions Glu22 and Asp23 have a pathogenic role in amyloid deposition. The development and progression of sporadic Alzheimer's disease may be accelerated by spontaneous isomerization at position 23. However, the pathological consequences of the genetic mutation leading to the Flemish-type Aβ variant need alternative explanation as the Flemish mutation fails to show potent aggregation properties [80].

The previous study also showed that the aggregation rate of the Dutch-type mutation is more extensive than that of unmodified Aβ in the presence of Cu and Zn ions [80]. In support, in 8–28 residue Aβ fragments, the Dutch-type mutation accelerated fibril formation, this time around without metal ion addition [81]. The Gln22 Dutch, Asn23 Iowa, and Gln22, Asn23 Dutch/Iowa double mutant Aβ 1-40 peptides rapidly assembled in solution to form fibrils, whereas wild-type and Gly21 Flemish Aβ 1-40 peptides exhibited little fibril formation [82]. Similarly, the Dutch- and Iowa-type peptides, especially the double mutant form, were found to induce robust pathologic responses in cultured human cerebrovascular smooth muscle cells, including elevated levels of cell-associated APP, proteolytic breakdown of smooth muscle cell α-actin, and cell death. These data suggest that the different mid-chain mutations in Aβ may exert their pathogenic effects through different mechanisms. Whereas the Gly21 Flemish mutation appears to enhance Aβ production, the Gln22 Dutch and Asn23 Iowa mutations enhance fibrillogenesis and the pathogenicity of Aβ toward cultured cells. Very similar results with basically identical conclusions were reported based on an experiment in which the kinetics of aggregation was followed by reversed-phase high-performance liquid chromatography at 37°C at pH 7.4 [83].

Using size-exclusion chromatography and circular dichroism spectroscopy, kinetic and secondary structural characteristics were compared with other Aβ 1-40 peptides and the extracellular Aβ12-28 fragment, all having single amino acid substitutions in position 22 [84]. The Aβ 1-40 Gly22 protofibrils are a group of comparatively stabile β-sheet–containing oligomers with a heterogeneous size distribution, ranging from >100 kDa to >3000 kDa. Salt promotes protofibril formation. When all the Glu22 substitutions were compared, the rank order of protofibril formation of Aβ 1-40 and its variants was Val22 > Ala22 > Gly22 > Gln22 > Glu22 and correlated with the degree of hydrophobicity of the substituent in position 22. The conclusion was drawn that the physical properties of Aβ 1-40 Gly22 suggest an important role for the peptide in the neuropathogenesis in the Arctic form of Alzheimer's disease [84]. In support, a membrane-mimicking environment generated in the presence of detergents or a ganglioside is sufficient *per se* for amyloid fibril formation from soluble Aβ and hereditary variants of the Aβ peptide, including the Dutch, Flemish, and Arctic types. The peptides exhibit mutually different aggregation behavior in these environments [85]. Notably, the Arctic-type Aβ peptide, in contrast with the wild-type and other variant forms, shows a markedly rapid and higher level of amyloid fibril formation in the presence of sodium dodecyl sulfate or GM1 ganglioside. While in the presence of a zwitterionic detergent, unmodified Aβ forms 8- to 10-nm helical fibrils, and the Dutch- and Flemish-type variants grow rather thin 6- to 7-nm fibers. The Arctic-type Aβ peptide forms short and curved fibers with a diameter of 6–7 nm, and these can be defined as protofibrils (Fig. 7.5). These results underline the importance of favorable local environments for fibrillogenesis of the amyloid peptide.

This last report surveyed additional potential changes in the biochemical and biophysical properties of Aβ, brought upon mid-chain modifications [85]. In addition to the more extensively studied aggregation properties, the possible alterations included the formation of more toxic oligomeric and fibrillar Aβ species corresponding with the Dutch- and Arctic-type variants [86] or alteration in sensitivities to peptidase degradation [87]. The Dutch, Flemish, Italian, and Arctic mutations apparently make Aβ resistant to proteolysis by neprilysin, the peptidase with the most important role in catabolism of Aβ in the brain. Monomeric Aβ wild-type, Flemish, Italian (Lys22), and Iowa variants were readily degraded by a rat insulin-degrading enzyme, an important component of the Aβ clearance process [88], with similar efficiency. However, the proteolysis of Dutch- and Arctic-type Aβ variants was significantly less extensive as compared with the unmodified or the rest of the mutant peptides [89]. All of the Aβ variants were cleaved between Glu3-Phe4 and Phe4-Arg5 in addition to the previously described major endopeptidase sites around positions 13–15 and 18–21. Detergent-stable Aβ dimers were highly resistant to proteolysis regardless of the variant, suggesting that the insulin-degrading enzyme recognizes a conformation that is available for interaction only in monomeric Aβ.

What are the conformational differences between unmodified and Dutch-type Aβ peptides? We used Fourier-transform infrared and circular dichroism spectroscopies on synthetic peptides to demonstrate that the Glu22 → Gln mutation results in altered secondary structure in membrane mimicking solvents, characterized by a considerably higher β-structure content for the Dutch-type peptide [90]. Moreover, extreme high and low pH were less effective in eliminating the β-conformation for the Dutch-variant than for the normal human sequence (Fig. 7.6). The differences in the strength and stability of the aggregates are attributed to the

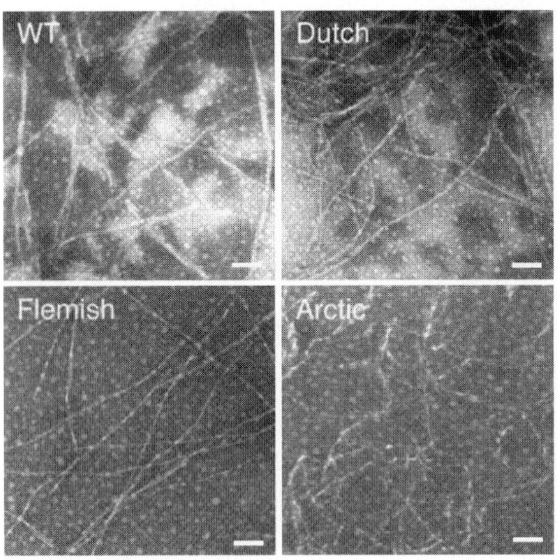

FIGURE 7.5. Electron micrographs of Aβ 1-40 solutions including wild-, Dutch-, Flemish-, and Arctic-type variants, incubated 24 h in the presence of 0.02% Zwittergent 3-14. Reprinted from Ref. 85, with permission of the International Society for Neurochemistry.

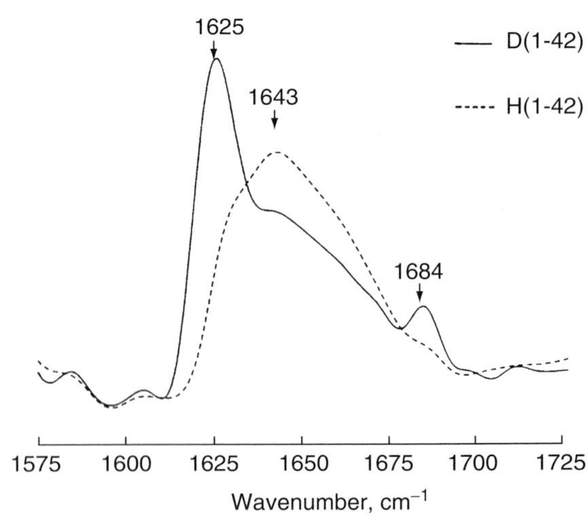

FIGURE 7.6. Infrared absorbance spectra of unmodified (broken line) and Dutch-type (solid line) Aβ 1-42 peptides in D_2O at pH 11. Reprinted from Ref. 91, with permission of the Society for Applied Spectroscopy.

presence of varying (small) proportions of the classical secondary structures [91]. Infrared spectra of material from autopsied human Alzheimer's disease brain show spectral features indicative of the formation of similar aggregates, which may be related to plaque formation. These results were later confirmed by additional spectroscopic, microscopic, and biochemical assays [92]. According to these, in the Dutch-type peptide the propensity of the Aβ N-terminal domain to adopt an α-helical structure is decreased, with a concomitant increase in amyloid formation. It was proposed that Aβ exists in an equilibrium between two species: one "able" and another "unable" to form amyloid, depending on the secondary structure adopted by the N-terminal domain. Thus, manipulation of the Aβ secondary structure with therapeutic compounds that promote the α-helical conformation may provide a tool to control the amyloid deposition observed in Alzheimer's disease patients.

In a more recent study, the cytotoxic properties of the Dutch- and Italian-type (Lys22) Aβ variants were compared with the unmodified peptide on cultured human cerebral endothelial cells after flow cytometry analysis [93]. Under the conditions tested, the Dutch-type Gln22-modified analogue exhibited the highest content of β-sheet conformation and the fastest aggregation/fibrillization properties. The Dutch variant also induced apoptosis of cerebral endothelial cells at a concentration of 25 μM, whereas the wild-type Aβ and the Italian mutant had no effect. The data suggest that different amino acids at position 22 confer distinct structural properties to the peptides that appear to influence the onset and aggressiveness of the disease rather than the phenotype.

7.7 C-Terminal Forms: Aβ 1-40 and Aβ 1-42

One of the studies concentrating on the amino-terminal modifications compared the fiber types as regulated by the length of the Aβ peptide [71]. Peptides ending with Ala42 grew to a mature fiber type regardless of the N-terminal residue, forming a dense meshwork of long fibrils by the end of the aggregation process. In contrast, Aβ variants ending with Val40 assembled more slowly to generate short, curly fibers.

To quantitate the various Aβ C-terminal forms present in the brains of patients with Alzheimer's disease, cerebral cortex was homogenized in 70% formic acid, and the supernatant was analyzed by sandwich enzyme-linked immunoabsorbent assays specific for various forms of Aβ [94]. In 9 of 27 brains examined, there was minimal congophilic angiopathy and virtually all Aβ (96%) ended at Ala42 (Thr43). The other 18 Alzheimer's disease brains contained increasing amounts of Aβ ending at Val40. From this set, 6 brains with substantial congophilic angiopathy were separately analyzed. In these brains, the amount of Aβ 1-42(43) was much the same as in brains with minimal congophilic angiopathy, but a large amount of Aβ 1-40 (76% of total Aβ) was also present. Immunocytochemical analysis with monoclonal antibodies selective for the various Aβ C-terminal forms confirmed that, in brains with minimal congophilic angiopathy, virtually all Aβ species ended at Ala42 (Thr43) and this Aβ variant was deposited in senile plaques of all types. In the remaining brains, Aβ 1-42(43) accumulated in a similar fashion in plaques, but, in addition, widely varying amounts of Aβ 1-40 were also deposited, primarily in blood vessel walls. The blood vessel also contained some Aβ 1-42(43) variants. These observations indicate that Aβ ending at Val42 (Thr43), which are a minor component of the Aβ in human cerebrospinal fluid and plasma, are critically important in Alzheimer's disease where they deposit selectively in plaques of all kinds.

A postmortem cross-sectional study comparing the deposition of Aβ variants in the prefrontal cortex of 79 nursing home residents having no, questionable, mild, moderate, or severe dementia revealed that all three Aβ forms, 1-40, 1-42, and 1-43 deposited in large quantities and the Aβ accumulation level could be correlated with the severity of the dementia [95]. The deposition of Aβ x-42 and Aβ x-43 occurred very early in the disease process before Alzheimer's disease could be actually diagnosed. Levels of accumulated Aβ x-43 appeared surprisingly high given the low amounts that are constitutively synthesized. These data indicate that Aβ x-42/43 are important species associated with early disease progression and suggest that the physiochemical properties of the Aβ species may be a major determinant in amyloid deposition. The results support an important role for Aβ in mediating initial pathogenic events in Alzheimer's disease dementia and reinforce that treatment strategies targeting the formation, accumulation, or cytotoxic effects of Aβ should be equally pursued.

Incubation of Aβ solutions at 37°C and pH 7.4 produces soluble oligomers in a concentration-dependent manner [96]. On one hand, fresh Aβ 1-42 solutions rapidly form soluble oligomers, whereas Aβ 1-40 solutions require prolonged incubation to produce oligomeric structures. On the other hand, fresh Aβ 1-42 solutions are more toxic to human neuroblastoma SH-SY5Y cells than Aβ 1-40 solutions, possibly mediated by soluble oligomers. Thus, differences in solution-phase toxicity between Aβ 1-42 and Aβ 1-40 could explain the association of the longer form with familial early-onset Alzheimer's disease.

Because Aβ 1-42/43 appear early in the deposition process, the question was asked whether the appearance of the other Aβ forms is dependent upon the longest form [97]. Aβ ending at residues Val40, Ala42, and Thr43 have been identified in neuritic deposits, while the peptide in vascular amyloid appears to terminate at residue Val39 or Val40. Kinetic studies of aggregation by three naturally occurring Aβ variants (1-39, 1-40, 1-42) and four model peptides (Aβ 26-39, Aβ 26-40, Aβ 26-42, and Aβ 26-43) demonstrate that amyloid formation, like crystallization, is a nucleation-dependent phenomenon [98]. The length of the C-terminus is a critical determinant of the rate of amyloid formation ("kinetic solubility") but has only a minor effect on the thermodynamic solubility. Amyloid formation by the kinetically soluble peptides (e.g., Aβ 1-39, 1-40, 26-39, or 26-40) can be nucleated, or "seeded," by peptides that include the critical C-terminal residues (Aβ 1-42, 26-42, 26-43, and 34-42). These results suggest that nucleation may be the rate-determining step of in vivo amyloidogenesis and confirm that Aβ 1-42/43, rather than Aβ 1-40, is the pathogenic protein(s) in Alzheimer's disease.

All we have left is a brief survey of the environment in which the various C-terminal Aβ variants form. We mentioned in the beginning of this review that the carboxy-terminus of Aβ is generally released from the precursor by γ-secretase. Whether the production of all Aβ peptide species requires the action of γ-secretase was investigated by a combination of surface-enhanced laser desorption/ionization time-of-flight mass spectrometry and a specific inhibitor of γ-secretase [99]. Using this approach, it was demonstrated that the production of all truncated Aβ peptides except those released by the action of the non-amyloidogenic α-secretase enzyme or potentially β-site APP cleaving enzyme 2 depends on γ-secretase activity. This indicates that none of these peptides are generated by a separate enzyme entity, and a specific inhibitor of the γ-secretase should have the potential to block the generation of all amyloidogenic variants. The majority of the early onset Alzheimer's disease cases is inherited as autosomal dominant disorders and cosegregate with mutations in the presenilin genes 1 and 2 [100, 101]. Mutations in presenilin (PS) 1 and 2 were found to be causative in ≈50% of pedigrees with early-onset familiar Alzheimer's disease [102]. It was shown that the ratio of Aβ 1-42(43) to Aβ 1-40 in conditioned media of N2a cell lines expressing three familiar Alzheimer's disease–linked PS-1 variants is uniformly elevated relative to cells expressing similar levels of wild-type PS1 [103]. Similarly, the Aβ 1-42 (43)/Aβ1-40 ratio is elevated in the brains of young transgenic animals coexpressing a chimeric amyloid precursor protein and a PS-1 variant compared with brains of transgenic mice expressing APP alone or transgenic mice coexpressing wild-type human PS-1 and APP. These studies provide compelling support for the view that one mechanism by which these mutant PS-1 cause Alzheimer's disease is by increasing the extracellular concentration of Aβ peptides terminating at 42(43), species that foster Aβ deposition.

7.8 Multiple Mutations May Point to a Unified Picture

As all the studies cited above indicate, single Aβ alterations affect various properties of the wild-type peptides without a clear view of the pathological consequences of the modifications. We suggested that some Aβ species feature multiple amino acid residue changes, and the coexistence of these alterations may better define the role of certain changes in the deposition or neurotoxic processes. The first, and quite obvious, double modification represents the appearance of cyclized

Asp residues (succinimidyl) both at the amino-terminus and in the middle of the Aβ chain, at positions 7 and 23. A potential consequence of succinimide formation is a significant increase in the water accessibility to the backbone and α-carbon atoms of the succinimidyl-modified Asp7 and Asp23 residues [104]. If cell toxicity of Aβ is mediated by soluble forms [105], this would explain the increased neurotoxicity of the multiply modified peptide. It was also suggested that spontaneous Asp → Suc transformation might lead to an increase of the racemization rates due to the higher accessibility of water at these sites [104]. Moreover, adjacent residues may influence the selectivity of the racemization to given Asp residues, and these residues may indirectly control the water accessibility at the modification sites.

Increased solubility influences amyloidogenic properties of the Flemish Aβ variant [106]. Comparative biophysical and neurotoxicity studies on wild-type and Flemish (Gly21) Aβ 1-40, Aβ 5-40, and Aβ 11-40 revealed that the Flemish amino acid substitution increases the solubility of each form of peptide, decreases the rate of formation of thioflavin-T-positive assemblies, and increases the sodium dodecyl sulfate stability of peptide oligomers. Although the kinetics of peptide assembly are altered by the Ala21 → Gly substitution, all three Flemish variants form fibrils, as do the wild-type peptides. The N-terminally truncated peptides were chosen on the basis of earlier cell culture studies, which detected increased amounts of N-terminally truncated peptides secreted by cells transfected with the Flemish APP [107]. Importantly, toxicity studies using cultured primary rat cortical cells showed that the Flemish assemblies were as potent a neurotoxin as were the wild-type assemblies regardless of peptide length. These results are consistent with a pathogenetic process in which conformational changes in Aβ induced by the Gly21 form would facilitate peptide adherence to the vascular endothelium, creating nidi for amyloid growth. Increased peptide solubility and assembly stability would favor formation of larger deposits and inhibit their elimination [108]. In addition, increased concentrations of neurotoxic assemblies would accelerate neuronal injury and death.

The effects of amino-terminal truncations on the Dutch-(Gln22) and Flemish-type Aβ peptides were also compared with more conclusive data on the toxicity induced by the various N-terminal forms [109]. At a concentration of 5 μM, the aggregation of the Aβ peptides followed the order Aβ 1-42 unmodified > Aβ 12-42 normal mid-section >Aβ 12-42 Flemish type> Aβ 12-42 Dutch type. The lower level of aggregation of the shorter peptides, especially for the Dutch variant, could be due to the formation of smaller Aβ fibrils, and this is in accordance with previous studies that observed shorter and stubbier fibrils for the Dutch version [110]. Apoptosis was induced in neuronal cells by the truncated Aβ wild-type and Flemish peptides at concentrations as low as 1–5 μM, as evidenced by propidium iodide staining, DNA laddering, and caspase-3 activity measurements. Even when longer incubation times and higher peptide concentrations were applied, the N-truncated Dutch-type peptide did not induce apoptosis. Apoptosis induced by the full-length Aβ 1-42 peptide was weaker than that induced by its N-truncated variant. These data suggest that N-truncation enhanced the cytotoxic effects of unmodified Aβ and Flemish-type peptides, which may play a role in the accelerated progression of dementia. When the effects of the modifications at different parts of the Aβ peptide are compared, it can be concluded that while loss of charge at Glu22 (for either Gln or Ala) enhances the pathogenic effects on cerebrovascular smooth muscle cells, the N-terminal residues in the wild-type variant confer a neuroprotective effect, partially in agreement with earlier findings [111].

This latter study leads us to double modifications at the two termini. Aβ variants starting with Asp1, Phe4, Ser8, Val12, and Leu17 and ending with Val40 or Ala42 were synthesized and their aggregation and neurotoxic properties were compared [111]. The N-terminally truncated peptides exhibited enhanced peptide aggregation relative to full-length species, as quantitatively assessed by sedimentation analyses. The sedimentation levels were greater for peptides terminating at residue 42 than for those terminating at residue 40. The increased aggregation properties of the N-terminal short and C-terminal long peptides were accompanied by increased β-pleated sheet conformation, fibrillar morphology under transmission electron microscopy, and toxicity in cultures of rat hippocampal neurons. Indeed, decreased level of in vitro solubility of N-terminally truncated Aβ peptides were noted earlier [112], but the negative relationship between peptide solubility and toxicity reported here is in contrast with the positive relationship of these properties as discussed at the beginning of this section. It has to be noted that assessing the solubility and hydrophobic properties of different Aβ variants is not easy. In 8 M urea, the otherwise α-helical or β-pleated sheet Aβ peptide becomes 100% random coil and remains monomeric [113]. However, during electrophoresis in this medium, the peptide and its truncated variants do not obey the law of mass/mobility relationship that most proteins—including Aβ peptides—follow in conventional sodium dodecyl sulfate gel electrophoresis. Rather, the smaller carboxy-terminally truncated Aβ 1-38 or 1-40 peptides migrate slower than the larger Aβ 1-42 full-length peptide, while the amino terminally truncated Aβ 13-42 peptide does migrate faster than the full-length Aβ variant. Thus, despite their small size (2–4 kDa) and minor differences between their lengths, the Aβ peptides display a wide separation in this low-porosity (12% acrylamide) gel. It was found that this unusual electrophoretic mobility in 8 M urea is due to the fact that the quantity of labeled detergent bound to the Aβ peptides, instead of being proportional to the total number of amino acids, is rather proportional to the sum of the hydrophobicity consensus indices of the constituent amino acids. In turn, this underlines the importance of the total number and each individual charged residue in the sequence in defining the three-dimensional shape and physical relationship with the immediate environment.

Photo-induced cross-linking was used to evaluate systematically the oligomerization of 34 physiologically relevant Aβ variants, including those containing familial Alzheimer's disease–linked amino acid substitutions, naturally occurring N-terminal truncations, and modifications altering the charge, the hydrophobicity, or the conformation of the peptide [114]. The most important structural feature controlling early oligomerization was the length of the C-terminus. Specifically, the side-chain of Ile41 in Aβ 1-42 was found to be important both for effective formation of paranuclei and for self-association of paranuclei into larger oligomers. The side-chain of Ala42, and the C-terminal carboxyl group, affected paranucleus self-association. Aβ 1-40 oligomerization was particularly sensitive to substitutions of Glu22 or Asp23 and to truncation of the N-terminus but not to substitutions of Phe19 or Ala21. Aβ 1-42 oligomerization, in contrast, was largely unaffected by substitutions at positions 22 or 23 or by N-terminal truncations but was affected significantly by substitutions of Phe19 or Ala21. These results reveal how specific regions and residues control Aβ oligomerization and show that these controlling elements differ between diverse Aβ C-terminal forms.

Both mid-chain and C-terminal Aβ modifications were made in synthetic peptides to explain the increase of cerebral amyloid angiopathy in familiar Alzheimer's disease [115]. All Aβ 1-40 mutants at positions 22 and 23, including those corresponding with the Dutch (Gln22), Arctic (Gly22), Italian (Lys22), and Iowa (Asn23) types, showed stronger neurotoxicity than wild-type Aβ 1-40. Similar tendency was observed for Aβ 1-42 mutants at positions 22 and 23 whose toxic effects were 50–200 times stronger than that of the corresponding Aβ 1-40 variants, suggesting that these Aβ 1-42 species are the ones that are mainly involved in the pathogenesis of cerebral amyloid angiopathy. While the aggregation of Arctic- and Iowa-type Aβ 1-42 was similar to that of wild-type Aβ 1-42, Gln22- and Lys22-containing Aβ peptides aggregated extensively, supporting the clinical evidence that Dutch and Italian patients are diagnosed as hereditary cerebral hemorrhage with amyloidosis. In contrast, the Flemish Gly21 mutation needs alternative explanation with the exception of altered physicochemical properties. Although attenuated total reflection–Fourier transform infrared spectroscopy spectra suggested that the β-pleated sheet content correlated with Aβ aggregation, the enhanced β-turn around positions 22 and 23 in the mutated versions also enhanced the aggregative ability [115].

A noteworthy feature of the last report is the exceptional purity of the synthetic Aβ peptides, supported by mass spectroscopy data. It had previously been reported that Gln22 Aβ 1-40 rather than Gln22 Aβ 1-42 plays a significant role in Dutch-type cerebral amyloid angiopathy because the Dutch-type Aβ 1-42 did not show any cytotoxic effects [116]. However, the newer report clearly demonstrates the most potent cytotoxicity of Gln22

Aβ 1-42 among all the Aβ 1-42 variants. In addition, in the newer paper, wild-type Aβ 1-42 aggregated far more rapidly than wild-type Aβ 1-40, differing from earlier data published by other groups [47, 117]. Potentially novel and reliable synthetic methods of pure Aβ 1-42 peptides [118, 119] allowed more reliable measurements. If this is indeed true, the varying purity levels of synthetic Aβ peptide preparations might be one of the major reasons of the discrepancies in the biological data.

7.9 Conclusions

It is an undeniable fact that different Aβ variants populate the tissues in different amyloid diseases and the N-terminal, mid-chain, or C-terminal modifications are likely to contribute to the development of a given clinical phenotype. Due to the lack of naturally occurring material in quantities large enough for detailed biochemical, biophysical, and biological analysis, synthetic peptides corresponding with the isolated Aβ forms are prepared, and the potential role of the modifications in the pathogenesis of the disease, mostly Alzheimer's disease, is investigated on these synthetic products. In general, Aβ mutations enhance both typical properties of the amyloid peptide: fibrillogenesis and neurotoxicity. The first is quite understandable because deletion of the amino-terminal hydrophilic residues, addition of two carboxy-terminal hydrophobic residues, or elimination of charged side-chains in mid-chain positions all likely contribute to the reduction of the α-helical conformer and to an increased β-pleated sheet formation as well as aggregation. Less clear is the effect of the changes on cell toxicity, especially as contrasting views are present on the requirement for neurotoxic properties. Peptide solubility is certainly one factor, and while most modifications are expected to decrease aqueous solubility, N-terminal cyclization of aspartyl residues actually increases it. Moreover, toxic properties associated with interactions with the cell membrane or other hydrophobic cell-originated components may play a role in the ability of the modified Aβ variants to disrupt cellular functions.

The modified Aβ forms are partly due to post-translational processing of the unmodified peptide; however, the mutations themselves may lend to decreased sensitivity to further proteolytic degradation hence delayed turnover. One aspect is certain: The Aβ peptide is a very difficult compound to prepare and purify, and the purity of the synthetic products (and we are usually dealing with single amino acid mutations) can significantly influence the results of comparative biological assays. Aβ peptides are notorious for irregular behavior during chromatography or other separation techniques, and single amino acid modifications, often of charged residues as they are present in the Dutch-, Italian-, Arctic-, or Iowa-type Aβ variants, may dramatically change the physical behavior of the peptide and this reflects in controversial biochemical data.

The development of reliable and reproducible synthetic, separation, and analytical Aβ protocols as well as the refinement of characteristic assays for fibrillogenesis and cell toxicity will allow the views on the effects of the various Aβ forms to unify and provide clues for molecular or cellular therapeutic interventions to eliminate the pathogenic Aβ species.

References

1. Klucken, J., McLean, P.J., Gomez-Tortosa, E., et al. Neuritic alterations and neural system dysfunction in Alzheimer's disease and dementia with Lewy bodies. Neurochem. Res. 2003; 28:1683-1691.
2. Ghiso, J., and Frangione, B. Amyloidosis and Alzheimer's disease. Adv. Drug Deliv. Rev. 2002; 54:1539-1551.
3. Giasson, B.I., Lee, V.M.Y., and Trojanowski, J.Q. Interactions of amyloidogenic proteins. Neuromol. Med. 2003; 4:49-58.
4. Knauer, M.F., Soreghan, B., Burdick, D., et al. Intracellular accumulation and resistance to degradation of the Alzheimer amyloid A4/β protein. Proc. Natl. Acad. Sci. U. S. A. 1992; 89:7437-7441.
5. Tabaton, M., Nunzi, M.G., Xue, R., et al. Soluble amyloid β-protein is a marker of Alzheimer amyloid in brain but not in cerebrospinal fluid. Biochem. Biophys. Res. Commun. 1994; 200:1598-1603.
6. Selkoe, D.J. Alzheimer's disease: genes, proteins, and therapy. Physiol. Rev. 2001; 81:741-766.
7. Weggen, S., Eriksen, J.L., Das, P., et al. A subset of NSAIDs lower amyloidogenic Aβ42 independently of cyclooxygenase activity. Nature 2001; 414:212-216.
8. Yankner, B.A. Mechanisms of neuronal degeneration in Alzheimer's disease. Neuron 1996; 16:921-932.

9. Southwick, P.C., Yamagata, S.K., Echols, C.L., et al. Assessment of amyloid β-protein in cerebrospinal fluid as an aid in the diagnosis of Alzheimer's disease. J. Neurochem. 1996; 66:259-265.
10. Wang, R., Sweeney, D., Gandy, S.E., and Sisodia, S.S. The profile of soluble amyloid β protein in cultured cell media. J. Biol. Chem. 1996; 271:31894-31902.
11. Glenner, G.G., and Wong, C.W. Alzheimer's disease: initial report of the purification and characterization of a novel cerebrovascular amyloid protein. Biochem. Biophys. Res. Commun. 1984; 120:885-890.
12. Golde, T.E., Eckman, C.B., and Younkin, S.G. Biochemical detection of Aβ isoforms: implications for pathogenesis, diagnosis, and treatment of Alzheimer's disease. Biochim. Biophys. Acta 2000; 1502:172-187.
13. Gorman, P.M., and Chakrabartty, A. Alzheimer β-amyloid peptides: structures of amyloid fibrils and alternate aggregation products. Biopolymers 2001; 60:381-394.
14. Paradisi, S., Sacchetti, B., Balduzzi, M., et al. Astrocyte modulation of in vitro β-amyloid neurotoxicity. Glia 2004; 46:252-260.
15. Butterfield, D.A. Amyloid β-peptide (1-42)-induced oxidative stress and neurotoxicity: implications for neurodegeneration in Alzheimer's disease brain. A review. Free Radic. Res. 2002; 36:1307-1313.
16. Harper, J.D., and Lansbury, P.T. Jr. Models of amyloid seeding in Alzheimer's disease and scrapie: mechanistic truths and physiological consequences of the time-dependent solubility of amyloid proteins. Annu. Rev. Biochem. 1997; 66:385-407.
17. Barrow, C.J., and Zagorski, M.G. Solution structures of β peptide and its constituent fragments: relation to amyloid deposition. Science 1991; 253:179-182.
18. Watson, A.A., Fairlie, D.P., and Craik, D.J. Solution structure of methionine-oxidized amyloid β-peptide (1-40). Does oxidation affect conformational switching? Biochemistry 1998; 37:12700-12706.
19. Meda, L., Cassatella, M.A., Szendrei, G.I., et al. Activation of microglial cells by β-amyloid protein and interferon-γ. Nature 1995; 374:647-650.
20. Fonseca, M.I., Head, E., Velazquez, P., et al. The presence of isoaspartic acid in β-amyloid plaques indicates plaque age. Exp. Neurol. 1999; 157:277-288.
21. Shoji, M., Golde, T.E., Ghiso, J., et al. Production of the Alzheimer amyloid β-protein by normal proteolytic processing. Science 1992; 258:126-129.
22. Goldgaber, D., Lerman, M.I., McBride, O.W., et al. Characterization and chromosomal localization of a cDNA encoding brain amyloid of Alzheimer's disease. Science 1987; 235:877-880.
23. Busciglio, J., Gabuzda, D.H., Matsudaira, P., and Yankner, B.A. Generation of β-amyloid in the secretory pathway in neuronal and nonneuronal cells. Proc. Natl. Acad. Sci. U. S. A. 1993; 90:2092-2096.
24. Seubert, P., Oltersdorf, T., Lee, M.G., et al. Secretion of β-amyloid precursor protein cleaved at the amino terminus of the β-amyloid peptide. Nature 1993; 361:260-263.
25. Suzuki, N., Cheung, T.T., Cai, X.D., et al. An increased percentage of long amyloid β-protein secreted by familial amyloid β-protein precursor (βAPP717) mutants. Science 1994; 264:1336-1340.
26. Goate, A., Chartier-Harlin, M.C., Mullan, M., et al. Segregation of a missense mutation in the amyloid precursor protein gene with familial Alzheimer's disease. Nature 1991; 349:704-706.
27. Murrell, J. Farlow, M., Ghetti, B., and Benson, M.D. A mutation in the amyloid precursor protein associated with hereditary Alzheimer's disease. Science 1991; 254:97-99.
28. Chartier-Harlin, M.C., Crawford, F., Houlden, H., et al. Early-onset Alzheimer's disease caused by mutations at codon 717 of the β-amyloid precursor protein gene. Nature 1991; 353:844-846.
29. Medeiros, M.S., and Turner, A.J. Post-secretory processing of regulatory peptides: the pancreatic polypeptide family as a model example. Biochimie 1994; 76:283-287.
30. Lewis, P.A., Piper, S., Baker, M., et al. Expression of BRI-amyloid β peptide fusion proteins: a novel method for specific high-level expression of amyloid β peptides. Biochim. Biophys. Acta 2001; 1537:58-62.
31. Hooper, N.M., and Turner, A.J. Protein processing mechanisms: from angiotensin-converting enzyme to Alzheimer's disease. Biochem. Soc. Trans. 2000; 28: 441-446.
32. Matsumoto, A., Itoh, K., and Matsumoto, R. A novel carboxypeptidase B that processes native β-amyloid precursor protein is present in human hippocampus. Eur. J. Neurosci. 1990; 12:227-238.
33. Wu, Q., Li, L., Cooper, M.D., et al. Aminopeptidase A activity of the murine B-lymphocyte differentiation antigen BP-1/6C3. Proc. Natl. Acad. Sci. U. S. A. 1991; 88:676-680.
34. Martinez, J.M., Prieto, I., Ramirez, M.J., et al. Sex differences and age-related changes in human serum aminopeptidase A activity. Clin. Chim. Acta 1998; 274:53-61.
35. Arechaga, G., Sanchez, B., Alba, F., et al. Developmental changes of soluble and membrane-bound aspartate aminopeptidase activities in rat brain. Rev. Esp. Fisiol. 1996; 52:149-154.

36. Orpiszewski, J., Schormann, N., Kluve-Beckerman, B., et al. Protein aging hypothesis of Alzheimer's disease. FASEB J. 2000; 14:1255-1263.
37. Greenfield, J.P., Tsai, J., Gouras, G.K., et al. Endoplasmic reticulum and trans-Golgi network generate distinct populations of Alzheimer β-amyloid peptides. Proc. Natl. Acad. Sci. U. S. A. 1999; 96:742-747.
38. Prokai, L., Zharikova, A.D., Janaky, T., and Prokai-Tatrai, K. Exploratory pharmacokinetics and brain distribution study of a neuropeptide FF antagonist by liquid chromatography/atmospheric pressure ionization tandem mass spectrometry. Rapid Commun. Mass Spectrom. 2000; 14:2412-2418.
39. Wurth, C., Guimard, N.K., and Hecht, M.H. Mutations that reduce aggregation of the Alzheimer's Aβ42 peptide: an unbiased search for the sequence determinants of Aβ amyloidogenesis. J. Mol. Biol. 2002; 319:1279-1290.
40. Janus, C., Chishti, M.A., and Westaway, D. Transgenic mouse models of Alzheimer's disease. Biochim. Biophys. Acta 2000; 1502:63-75.
41. Selkoe, D.J. Biochemistry of altered brain proteins in Alzheimer's disease. Annu. Rev. Neurosci. 1989; 12: 463-490.
42. Johnstone, E.M., Chaney, M.O., Norris, F.H., et al. Conservation of the sequence of the Alzheimer's disease amyloid peptide in dog, polar bear and five other mammals by cross-species polymerase chain reaction analysis. Brain Res. Mol. Brain Res. 1991; 10: 299-305.
43. Yamada, T., Sasaki, H., Furuya, H., et al. Complementary DNA for the mouse homolog of the human amyloid β protein precursor. Biochem. Biophys. Res. Commun. 1987; 149:665-671.
44. Shivers, B.D., Hilbich, C., Multhaup, G., et al. Alzheimer's disease amyloidogenic glycoprotein: expression pattern in rat brain suggests a role in cell contact. EMBO J. 1988; 7:1365-1370.
45. Otvos, L. Jr., Szendrei, G.I., Lee, V.M., and Mantsch, H.H. Human and rodent Alzheimer β-amyloid peptides acquire distinct conformations in membrane-mimicking solvents. Eur. J. Biochem. 1993; 211:249-257.
46. Fraser, P.E., Nguyen, J.T., Inouye, H., et al. Fibril formation by primate, rodent, and Dutch-hemorrhagic analogues of Alzheimer amyloid β-protein. Biochemistry 1992; 31:10716-10723.
47. Shin, R.W., Ogino, K., Kondo, A., et al. Amyloid β-protein (Aβ) 1-40 but not Aβ1-42 contributes to the experimental formation of Alzheimer's disease amyloid fibrils in rat brain. J. Neurosci. 1997; 17:8187-8193.
48. Huang, X., Cuajungco, M.P., Atwood, C.S., et al. Alzheimer's disease, β-amyloid protein and zinc. J. Nutr. 2000; 130:1488S-1492S.
49. Harris, J.R. In vitro fibrillogenesis of the amyloid β 1-42 peptide: cholesterol potentiation and aspirin inhibition. Micron 2002; 33:609-626.
50. Atwood, C.S., Perry, G., Zeng, H., et al. Copper mediates dityrosine cross-linking of Alzheimer's amyloid-β. Biochemistry 1998; 43:560-568.
51. Galeazzi, L., Ronchi, P., Franceschi, C., and Giunta, S. In vitro peroxidase oxidation induces stable dimers of β-amyloid (1-42) through dityrosine bridge formation. Amyloid 1999; 6:7-13.
52. Kowalik-Jankowska, T., Ruta-Dolejsz, M., Wisniewska, K., and Lankiewicz, L. Cu(II) interaction with N-terminal fragments of human and mouse β-amyloid peptide. J. Inorg. Biochem. 2001; 86:535-545.
53. Bishop, G.M., and Robinson, S.R. Human Aβ1-42 reduces iron-induced toxicity in rat cerebral cortex. J. Neurosci. Res. 2003; 73:316-323.
54. Kuo, Y.M., Kokjohn, T.A., Beach, T.G., et al. Comparative analysis of amyloid-β chemical structure and amyloid plaque morphology of transgenic mouse and Alzheimer's disease brains. J. Biol. Chem. 2001; 276:12991-12998.
55. Pype, S., Moechars, D., Dillen, L., and Mercken, M. Characterization of amyloid β peptides from brain extracts of transgenic mice overexpressing the London mutant of human amyloid precursor protein. J. Neurochem. 2003; 84:602-609.
56. Casas, C., Sergeant, N., Itier, J.M., et al. Massive CA1/2 neuronal loss with intraneuronal and N-terminal truncated Aβ42 accumulation in a novel Alzheimer transgenic model. Am. J. Pathol. 2004; 165:1289-1300.
57. Roher, A.E., Lowenson, J.D., Clarke, S., et al. Structural alterations in the peptide backbone of β-amyloid core protein may account for its deposition and stability in Alzheimer's disease. J. Biol. Chem. 1993; 268:3072-3083.
58. Fabian, H., Szendrei, G.I., Mantsch, H.H., et al. Synthetic post-translationally modified human Aβ peptide exhibits a markedly increased tendency to form β-pleated sheets in vitro. Eur. J. Biochem. 1994; 221:959-964.
59. Szendrei, G.I., Fabian, H., Mantsch, H.H., et al. Aspartate-bond isomerization affects the major conformations of synthetic peptides. Eur. J. Biochem. 1994; 226:917-924.
60. Szendrei, G.I., Prammer, K.V., Vasko, M., et al. The effects of aspartic acid-bond isomerization on in vitro properties of the amyloid β-peptide as modeled with N-terminal decapeptide fragments. Int. J. Pept. Protein Res. 1997; 47:289-296.
61. Orpiszewski, J., and Benson, M.D. Induction of -sheet structure in amyloidogenic peptides by

neutralization of aspartate: a model for β-amyloid nucleation. J. Mol. Biol.1999; 289:413-428.
62. Lang, E., Szendrei, G.I., Lee, V.M., and Otvos, L. Jr. Spectroscopic evidence that monoclonal antibodies recognize the dominant conformation of medium-sized synthetic peptides. J. Immunol. Methods 1994; 170:103-115.
63. Barrow, C.J., Yasuda, A., Kenny, P.T., and Zagorski, M.G. Solution conformations and aggregational properties of synthetic amyloid β-peptides of Alzheimer's disease. Analysis of circular dichroism spectra. J. Mol. Biol. 1992; 225:1075-1093.
64. Tolson, J., Bogumil, R., Brunst, E., et al. Serum protein profiling by SELDI mass spectrometry: detection of multiple variants of serum amyloid α in renal cancer patients. Lab. Invest. 2004; 84:845-856.
65. Hosoda, R., Saido, T.C., Otvos, L. Jr., et al. Quantification of modified amyloid β peptides in Alzheimer's disease and Down syndrome brains. J. Neuropathol. Exp. Neurol. 1998; 57:1089-1095.
66. Saido, T.C., Iwatsubo, T., Mann, D.M., et al.. Dominant and differential deposition of distinct β-amyloid peptide species, Aβ N3(pE), in senile plaques. Neuron 1995; 14:457-466.
67. Thompson, A.J., Lim, T.K., and Barrow, C.J. On-line high-performance liquid chromatography/mass spectrometric investigation of amyloid-β peptide variants found in Alzheimer's disease. Rapid Commun. Mass. Spectrom. 1999; 13:2348-2351.
68. He, W., and Barrow, C.J. The Aβ 3-pyroglutamyl and 11-pyroglutamyl peptides found in senile plaque have greater β-sheet forming and aggregation propensities in vitro than full-length Aβ. Biochemistry 1999; 38:10871-10877.
69. Benzinger, T.L., Gregory, D.M., Burkoth, T.S., et al. Propagating structure of Alzheimer's β-amyloid (10-35) is parallel β-sheet with residues in exact register. Proc. Natl. Acad. Sci. U. S. A. 1998; 95:13407-13412.
70. Houston, M.E. Jr., Campbell, A.P., Lix, B., et al. Lactam bridge stabilization of α-helices: the role of hydrophobicity in controlling dimeric versus monomeric α-helices. Biochemistry 1996; 35: 10041-10050.
71. Russo, C., Violani, E., Salis, S., et al. Pyroglutamate-modified amyloid β-peptides -AβN3(pE) -strongly affect cultured neuron and astrocyte survival. J. Neurochem. 2002; 82:1480-1489.
72. Tekirian, T.L., Saido, T.C., Markesbery, W.R., et al. N-terminal heterogeneity of parenchymal and cerebrovascular Aβ deposits. J. Neuropathol. Exp. Neurol. 1998; 57:76-94.
73. Iwatsubo, T., Saido, T.C., Mann, D.M., et al. Full-length amyloid-β (1-42(43)) and amino-terminally modified and truncated amyloid-β 42(43) deposit in diffuse plaques. Am. J. Pathol. 1996; 149:1823-1830.
74. Schmidt, M.L., Saido, T.C., Lee, V.M., and Trojanowski, J.Q. Spatial relationship of AMY protein deposits and different species of Aβ peptides in amyloid plaques of the Alzheimer's disease brain. J. Neuropathol. Exp. Neurol. 1999; 58:1227-1233.
75. Shimizu, T., Watanabe, A., Ogawara, M., et al. Isoaspartate formation and neurodegeneration in Alzheimer's disease. Arch. Biochem. Biophys. 2000; 381:225-234.
76. Watanabe, A., Takio, K., and Ihara, Y. Deamidation and isoaspartate formation in smeared τ in paired helical filaments. Unusual properties of the microtubule-binding domain of τ. J. Biol. Chem. 1999; 274:7368-7378.
77. Clarke, S. Protein carboxyl methyltransferases: two distinct classes of enzymes. Annu. Rev. Biochem. 1985; 54:479-506.
78. Hoffmann, R., Craik, D.J., Bokonyi, K., et al. High level of aspartic acid-bond isomerization during the synthesis of an N-linked tau glycopeptide. J. Pept. Sci. 1999; 5:442-456.
79. Geiger, T., and Clarke, S. Deamidation, isomerization, and racemization at asparaginyl and aspartyl residues in peptides. Succinimide-linked reactions that contribute to protein degradation. J. Biol. Chem. 1987; 262:785-794.
80. Shimizu, T., Fukuda, H., Murayama, S., et al. Isoaspartate formation at position 23 of amyloid β peptide enhanced fibril formation and deposited onto senile plaques and vascular amyloids in Alzheimer's disease. J. Neurosci. Res. 2002; 70:451-461.
81. Wisniewski, T., Ghiso, J., and Frangione, B. Peptides homologous to the amyloid protein of Alzheimer's disease containing a glutamine for glutamic acid substitution have accelerated amyloid fibril formation. Biochem. Biophys. Res. Commun. 1991; 179:1247-1254.
82. Van Nostrand, W.E., Melchor, J.P., Cho, H.S., et al. Pathogenic effects of D23N Iowa mutant amyloid β-protein. J. Biol. Chem. 2001; 276:32860-32866.
83. Clements, A., Walsh, D.M., Williams, C.H., and Allsop, D. Effects of the mutations Glu22 to Gln and Ala21 to Gly on the aggregation of a synthetic fragment of the Alzheimer's amyloid β/A4 peptide. Neurosci. Lett. 1993; 161:17-20.
84. Paivio, A., Jarvet, J., Graslund, A., et al. Unique physicochemical profile of β-amyloid peptide variant Aβ1-40E22G protofibrils: conceivable neuropathogen in arctic mutant carriers. J. Mol. Biol. 2004; 339:145-159.
85. Yamamoto, N., Hasegawa, K., Matsuzaki, K., et al. Environment- and mutation-dependent aggregation

behavior of Alzheimer amyloid β-protein. J. Neurochem. 2004; 90:62-69.
86. Dahlgren, K.N., Manelli, A.M., Stine, W.B., et al. Oligomeric and fibrillar species of amyloid-β peptides differentially affect neuronal viability. J. Biol. Chem. 2002; 277:32046-32053.
87. Tsubuki, S., Takaki, Y., and Saido, T.C. Dutch, Flemish, Italian, and Arctic mutations of APP and resistance of Aβ to physiologically relevant proteolytic degradation. Lancet 2003; 361:1957-1958.
88. Eckman, E.A., Reed, D.K., and Eckman, C.B. Degradation of the Alzheimer's amyloid β peptide by endothelin-converting enzyme. J. Biol. Chem. 2001; 276:24540-24548.
89. Morelli, L., Llovera, R., Gonzalez, S.A., et al. Differential degradation of amyloid β genetic variants associated with hereditary dementia or stroke by insulin-degrading enzyme. J. Biol. Chem. 2003; 278: 23221-23226.
90. Fabian, H., Szendrei, G.I., Mantsch, H.H., and Otvos, L., Jr. Comparative analysis of human and Dutch-type Alzheimer β-amyloid peptides by infrared spectroscopy and circular dichroism. Biochem. Biophys. Res. Commun. 1993; 191:232-239.
91. Fabian, H., Choo, L-P., Szendrei, G.I., et al. Infrared spectroscopic characterization of Alzheimer's plaques. Applied Spectrosc. 1993; 47:1513-1518.
92. Soto, C., Castano, E.M., Frangione, B., and Inestrosa, N.C. The α-helical to β-strand transition in the amino-terminal fragment of the amyloid β-peptide modulates amyloid formation. J. Biol. Chem. 1995; 270:3063-3067.
93. Miravalle, L., Tokuda, T., Chiarle, R., et al. Substitutions at codon 22 of Alzheimer's Aβ peptide induce diverse conformational changes and apoptotic effects in human cerebral endothelial cells. J. Biol. Chem. 2000; 275:27110-27116.
94. Gravina, S.A., Ho, L., Eckman, C.B., et al. Amyloid β protein (Aβ) in Alzheimer's disease brain. Biochemical and immunocytochemical analysis with antibodies specific for forms ending at Aβ 40 or Aβ 42(43). J. Biol. Chem. 1995; 270: 7013-7016.
95. Parvathy, S., Davies, P., Haroutunian, V., et al. Correlation between Aβx-40-, Aβx-42-, and Aβx-43-containing amyloid plaques and cognitive decline. Arch. Neurol. 2001; 58:2025-2032.
96. El-Agnaf, O.M., Mahil, D.S., Patel, B.P., and Austen, B.M. Oligomerization and toxicity of β-amyloid-42 implicated in Alzheimer's disease. Biochem. Biophys. Res. Commun. 2000; 273:1003-1007.
97. Jarrett, J.T., Berger, E.P., and Lansbury, P.T., Jr. The C-terminus of the β protein is critical in amyloidogenesis. Ann. N. Y. Acad. Sci. 1993; 695:144-148.
98. Jarrett, J.T., Berger, E.P, and Lansbury, P.T., Jr. The carboxy terminus of the β amyloid protein is critical for the seeding of amyloid formation: implications for the pathogenesis of Alzheimer's disease. Biochemistry 1993; 32:4693-4697.
99. Beher, D., Wrigley, J.D., Owens, A.P., and Shearman, M.S. Generation of C-terminally truncated amyloid-β peptides is dependent on γ-secretase activity. J. Neurochem. 2002; 82:563-575.
100. St George-Hyslop, P., Haines, J., Rogaev, E., et al. Genetic evidence for a novel familial Alzheimer's disease locus on chromosome 14. Nat. Genet. 1992; 2:330-334.
101. Levy-Lahad, E., Wasco, W., Poorkaj, P., et al. Candidate gene for the chromosome 1 familial Alzheimer's disease locus. Science 1995; 269:973-977.
102. Schellenberg, G.D. Genetic dissection of Alzheimer's disease, a heterogeneous disorder. Proc. Natl. Acad. Sci. U. S. A. 1995; 92:8552-8559.
103. Borchelt, D.R., Thinakaran, G., Eckman, C.B., et al. Familial Alzheimer's disease-linked presenilin 1 variants elevate Aβ1-42/1-40 ratio in vitro and in vivo. Neuron 1996; 17:1005-1013.
104. Lins, R.D., Soares, T.A., Ferreira, R., and Longo, R.L. Solvent accessibility to aspartyl and succinimidyl residues at positions 7 and 23 in the amyloid β 1-28 peptide. Z. Naturforsch. 1999; 54: 264-270.
105. Gentile, M.T., Vecchione, C., Maffei, A., et al. Mechanisms of soluble β-amyloid impairment of endothelial function. J. Biol. Chem. 2004; 279: 48135-48142.
106. Walsh, D.M., Hartley, D.M., Condron, M.M., et al. In vitro studies of amyloid β-protein fibril assembly and toxicity provide clues to the aetiology of Flemish variant (Ala692 → Gly) Alzheimer's disease. Biochem. J. 2001; 355:869-877.
107. Haass, C., Hung, A.Y., Selkoe, D.J., and Teplow, D.B. Mutations associated with a locus for familial Alzheimer's disease result in alternative processing of amyloid β-protein precursor. J. Biol. Chem. 1994; 269:17741-17748.
108. Clements, A., Allsop, D., Walsh, D.M., and Williams, C.H. Aggregation and metal-binding properties of mutant forms of the amyloid Aβ peptide of Alzheimer's disease. J. Neurochem. 1996; 66:740-747.
109. Demeester, N., Mertens, C., Caster, H., et al. Comparison of the aggregation properties, secondary structure and apoptotic effects of wild-type, Flemish and Dutch N-terminally truncated amyloid β peptides. Eur. J. Neurosci. 2001; 13:2015-2024.
110. Sian, A.K., Frears, E.R., El-Agnaf, O.M., et al. Oligomerization of β-amyloid of the Alzheimer's

and the Dutch-cerebral-haemorrhage types. Biochem. J. 2000; 349:299-308.

111. Pike, C.J., Overman, M.J., and Cotman, C.W. Amino-terminal deletions enhance aggregation of β-amyloid peptides in vitro. J. Biol. Chem. 1995; 270:23895-23898.
112. Hilbich, C., Kisters-Woike, B., Reed, J., et al. Aggregation and secondary structure of synthetic amyloid β A4 peptides of Alzheimer's disease. J. Mol. Biol. 1991; 218:149-163.
113. Kawooya, J.K., Emmons, T.L., Gonzalez-DeWhitt, P.A., et al. Electrophoretic mobility of Alzheimer's amyloid-β peptides in urea-sodium dodecyl sulfate-polyacrylamide gel electrophoresis. Anal. Biochem. 2003; 323:103-113.
114. Bitan, G., Vollers, S.S., and Teplow, D.B. Elucidation of primary structure elements controlling early amyloid β-protein oligomerization. J. Biol. Chem. 2003; 278:34882-34889.
115. Murakami, K., Irie, K., Morimoto, A., et al. Neurotoxicity and physicochemical properties of Aβ mutant peptides from cerebral amyloid angiopathy: implication for the pathogenesis of cerebral amyloid angiopathy and Alzheimer's disease. J. Biol. Chem. 2003; 278:46179-46187.
116. Davis, J., and van Nostrand, W.E. Enhanced pathologic properties of Dutch-type mutant amyloid β-protein. Proc. Natl. Acad. Sci. U. S. A. 1996; 93: 2996-3000.
117. Kirkitadze, M.D., Condron, M.M., and Teplow, D.B. Identification and characterization of key kinetic intermediates in amyloid β-protein fibrillogenesis. J. Mol. Biol. 2001; 312:1103-1119.
118. Fukuda, H., Shimizu, T., Nakajima, M., et al. Synthesis, aggregation, and neurotoxicity of the Alzheimer's Aβ1-42 amyloid peptide and its isoaspartyl isomers. Bioorg. Med. Chem. Lett. 1999; 9: 953-956.
119. Tickler, A.K., Barrow, C.J., and Wade, J.D. Improved preparation of amyloid-β peptides using DBU as Nα-Fmoc deprotection reagent. J. Pept. Sci. 2001; 7: 488-494.

8 Copper Coordination by β-Amyloid and the Neuropathology of Alzheimer's Disease

Cyril C. Curtain and Kevin J. Barnham

8.1 Introduction

It is nearly two decades since high concentrations of the redox active transition metal ions Cu^{2+} and Fe^{3+} found in β-amyloid plaques were first proposed to play an important role in the pathology of Alzheimer's disease (AD) (see review by Bush [1]). Over this time, a new field of metallo-neurobiology relating to AD and other neurodegenerative diseases has arisen with approximately 250 original papers and more than 1000 references in secondary publications to date. At first, many neuroscientists failed to recognize the importance of this growing literature. However, a recent pilot Phase II clinical trial of a blood-brain barrier permeable metal protein attenuating compound (MPAC), clioquinol, in patients with moderately severe AD has shown promising results [2]. In a randomized sample of 36 subjects, the effect of treatment was significant in the more severely affected group, where those treated with clioquinol showed minimal deterioration in their cognitive scores (Alzheimer's disease Assessment Scale ≥25) compared with substantial worsening of the scores for the placebo group. Although subjected to the usual cautions applied to small-scale trials, this is an encouraging result that renders even more urgent the full elucidation of the possible role of transition metals, particularly Cu and Zn, in AD. It must be stressed that, although there is much experimental evidence on various aspects of the interaction between Cu, Zn, and the constituent of the amyloid plaques, the β-amyloid peptide (Aβ), the structural biology and elucidation of the neuropathological significance of metal binding are very much works in progress.

The naturally occurring Aβ1–42, 1–41 and 1–39 peptides (sequence of Aβ1–42 given in Fig. 8.1) represent part of the putative trans-membrane domain of the amyloid precursor protein, liberated from the membrane by proteolytic (secretase) action. Although its sequence is generally highly conserved, the rat sequence has Arg5, Tyr10, and His13 of human Aβ replaced by Gly5, Phe10, and Arg13 (see highlighted residues in Fig. 8.1). Because the murine species do not develop amyloid plaques in the brain with aging, it was recognized that these substitutions could be an important pointer to mechanisms of plaque formation in human beings. The coordination of transition metals by Aβ has been linked variously to their role in promoting peptide aggregation to form amyloid plaques, in the production of cytotoxic reactive oxygen species (ROS), and in promoting potentially cytotoxic interactions with cell membranes.

8.2 Cu^{2+} and Zn^{2+} Induced Aggregation of Aβ

Transition metal ion homeostasis is severely dysregulated in the AD brain [3, 4] and the role of these metals has been the subject of continuing study [5–11]. The transition metal ions Cu^{2+}, Fe^{3+}, and Zn^{2+} have been reported to occur at high concentrations in the neocortical parenchyma of healthy brain (total dry weight concentrations of 70, 340, and 350 μM, respectively). These concentrations may seem high but are not surprising when one considers the intense bioenergetics of the brain and the fact that the transition metal ions are an

Human DAEFRHDSGYEVHHQKLVFFAEDVGSNKGAIIGLMVGGVVIA

Rat DAEFGHDSGFEVRHQKLVFFAEDVGSNKGAIIGLMVGGVVIA

FIGURE 8.1. Sequence of human Aβ compared with that of the rat.

essential part of the redox systems involved. Their levels are far higher in the neuropil of the AD-affected brain, where they reach 0.4 and 1.0 mM for Cu and Fe/Zn, respectively in the amyloid plaque deposits [12]. It is of interest that these have been termed "trace metals," an evident misnomer because their concentrations in the gray matter are of the same order of magnitude as Mg (0.1–0.5 mM).

Miller et al [13] have imaged the *in situ* secondary structure of the amyloid plaques in AD brain tissue. Using synchrotron Fourier transform infrared micro-spectroscopy and a synchrotron x-ray fluorescence microprobe on the same sample, they showed a strong spatial correlation between elevated β-sheet content in Aβ plaques and accumulated Cu^{2+} and Zn^{2+}, emphasizing an association of metal ions with amyloid formation in AD. There was also a strong spatial correlation between the two ions. Higher Zn^{2+} concentrations have also been seen histologically in plaque deposits [14], and the importance of Zn^{2+} in plaque formation has been emphasized by the finding that age- and female sex-related plaque formation in APP2576 transgenic mice was greatly reduced upon the genetic ablation of the zinc transporter 3 protein, which is required for zinc transport into synaptic vesicles [15].

Bush et al. [16] found that Aβ coordinated Cu^{2+}, Zn^{2+}, and Fe^{3+} with high affinity [17, 18], which would explain the presence of these metals in amyloid plaques. This study also showed stabilization of an apparent Aβ1–40 dimer by Cu^{2+} on gel chromatography suggesting an interaction between Cu^{2+} and Aβ1–40. Clements et al. [19] observed displacement of $^{65m}Zn^{2+}$ from Aβ when co-incubated with excess Cu^{2+}, while Yang et al. [20] found that Cu^{2+} and Zn^{2+} shared a common binding site. Atwood et al. [21] found that Cu^{2+} was bound to soluble Aβ via histidine residues and that the precipitation of soluble Aβ by Cu^{2+} was reversibly modulated by pH with mildly acidic conditions (pH 6.6) greatly promoting Cu^{2+}-mediated precipitation, whereas raising the pH dissolved precipitated Aβ:Cu^{2+} complexes. Cherny et al. [22] observed that Zn^{2+} induced aggregation of soluble Aβ at pH 7.4 in vitro, which was totally reversible with chelation. They also found that marked Cu^{2+}-induced aggregation of Aβ1–40 occurred as the solution pH was lowered from 7.4 to 6.8 and that the reaction was completely reversible with either chelation or raising the pH. Aβ1–40 was reported to bind three to four Cu^{2+} ions when precipitated at pH 7.0. Rapid, pH-sensitive aggregation occurred at low nanomolar concentrations of both A β1–40 and Aβ1–42 with submicromolar concentrations of Cu^{2+}. Unlike Aβ1–40, Aβ1–42 was precipitated by submicromolar Cu^{2+} concentrations at pH 7.4. Rat Aβ1–40 and histidine-modified human Aβ1–40 were not aggregated by Zn^{2+}, Cu^{2+}, or Fe^{3+}, indicating that histidine residues are essential for metal-mediated Aβ assembly. Cherny et al. [23] also showed that Cu^{2+}- and Zn^{2+}-selective chelators enhanced the dissolution of amyloid deposits in postmortem brain specimens from AD subjects and from amyloid precursor protein overexpressing transgenic mice, confirming the part played by these metal ions in cerebral amyloid assembly. In particular, Zn^{2+} efficiently induces aggregation of synthetic Aβ under conditions similar to the physiological ones in the normal brain, that is, at nanomolar and submicromolar concentrations of Aβ and free Zn^{2+}, respectively [15–17].

Recently, it has been demonstrated that Aβ will not precipitate when trace metal ions are rigorously excluded [24]. On the other hand, the very strong precipitating effect of Zn^{2+} implies that there are some factors protecting Aβ from Zn^{2+}-induced aggregation in the normal brain. Certain metal ions such as Mg^{2+} and Ca^{2+}, which do not exhibit a precipitating effect, have been hypothesized to have this protective effect [25]. However, the inhibition of Zn^{2+}-induced Aβ aggregation by these metal ions has not yet been verified. The effect of Cu^{2+} on the aggregation of Aβ is ambiguous compared with Zn^{2+}. Cu^{2+} has been shown to be a strong inducer of

Aβ aggregation under certain conditions [24]. In contrast with the Zn^{2+}-induced Aβ aggregation that occurs over a wide pH range (5.5–7.5), the Cu^{2+}-induced aggregation occurs primarily at mildly acidic pH [21].

Atwood et al. [21] determined a half-maximal binding of Cu^{2+} for Aβ in the micromolar range (4.0 μM for Aβ1–40 and 0.3 μM for Aβ1–42) by indirect spectrophotometric analysis. However, this analysis of binding affinities was limited by the sensitivity of the spectrophotometric technique and the lack of competitive binding factors in the incubation that would emulate the physiological situation more closely. Garzon-Rodriguez et al. [26] used a more sensitive fluorescence technique and a single tryptophan (F4W) mutant of Aβ1-40 to show that the relative affinities were Fe < Cu > Zn. Syme et al. [27] used the competitive effects of glycine and L-histidine to measure Cu^{2+} affinity for Aβ by fluorescence spectroscopy. Adding Cu^{2+} to Aβ1–28 caused marked quenching of the tyrosine fluorescence signal at 307 nm. Added glycine competes with Aβ for the Cu^{2+}, and the tyrosine fluorescence signal reappears at a sufficiently high glycine levels. Cu^{2+} coordinates to glycine via the amino and carboxylate groups with an apparent pH-adjusted K_a of 1.8×10^6 M^{-1}, and two glycine residues will bind to a single Cu^{2+} ion [27]. It took more than 100 mol equivalents of glycine to cause the tyrosine fluorescence signal to completely return to its maximal strength. Half of the maximal quenching is achieved at approximately 18 ± 2 eq. of glycine. Finally, Huang et al. [24] had shown that binding of Cu^{2+} to Aβ1–42 promoted precipitation with so high an affinity that it was hard to avoid aggregation unless buffers were most rigorously treated with chelating agents. Even then, it is difficult to remove the last traces of metal ion, which may account for many of the inconsistencies reported in the Aβ metal binding literature. Extremely small changes in free or exchangeable Cu^{2+} concentration are also likely to have a significant effect on Aβ solubility in vivo.

8.3 Aβ Structures

The structure of the metal binding site of Aβ must be considered in the context of the structure of the whole molecule. Because it has been widely held that Aβ exerts its neurotoxic action via interactions with neuronal membranes additionally to or in concert with its redox activity, there have been many studies on its structure in a variety of membrane mimetic systems. A major obstacle to the determination of definitive structures is the difficulty of obtaining reproducibly a random-structured starting material or, alternatively, of mimicking its transmembrane conformation immediately after secretase cleavage. Furthermore, because aqueous solutions of Aβ accumulate significant amounts of aggregates within a few hours, NMR studies can be difficult. Nevertheless, early NMR studies of human Aβ1-40 showed a random coil structure in aqueous solution (pH 4) at micromolar concentration [28]. The secondary structure of Aβ40 peptide in 40% TFE buffered at pH 2.8 with 50 mM potassium phosphate was also studied by NMR. Under these conditions, there was aggregation only after a week and the NMR spectra were well resolved. Solution structures of Aβ1–40 in perdeuterated sodium dodecyl sulfate (SDS-d_{25}) micelles obtained by Coles et al. [29] showed two α-helical segments. The helical arrangement of residues 15–25 and 29–37 was confirmed by intense NOE connectivity (3–4 residues) while medium-range NOE for residues 25–29 were either weak or not observed. The "break" between the two helices was suggested by D_2O exchange experiments, where protons on residues 25–29 were shown to exchange rapidly and, from quantitative structural and dihedral angle restraint calculation prediction, a kink was seen at residues 26–27 acting as a "hinge" for the two helices.

Shao et al. [30] showed two α-helical regions between Tyr10-Val24 and Lys28-Val36 for both Aβ1–40 and Aβ1–42 in SDS-d_{25} at pH 7.2. The data were supported by structural calculations indicating α-helices between residues 10–24 and 28–42 with the region Gly25-Asn27 as a connecting loop. Similar downfield shifts of Aβ1–40 and Aβ1–42 at Val39-Val40 and Val40-Ile41, respectively, suggested a structural preference for the peptides at their C-terminus. This may be related to conformational averaging between a micelle bound α-helical structure and β-sheet when the peptides leave the micelle surface.

Most NMR studies in solution were done in either trifluoroethanol (TFE) [31] or SDS-d_{25}/D_2O to mimic a membrane environment, although an early study by Sorimachi and Craik [32] showed some α-helical structure in dimethyl sulfoxide (DMSO). The α-helical conformation found by NMR was further supported by far ultraviolet circular dichrosim (CD) spectroscopy which showed that Aβ1–28 in the presence of charged membrane-like surfaces, especially negatively charged SDS, preferred a helical structure. Other membrane-like species, zwitterionic dodecylphosphocholine (DPC) and dodecyltrimethylammonium chloride (DTAC), with heterogenous amphiphilic environments similar to biological systems have been used. Fletcher and Keire [33] used solution NMR and CD to study the conformation of Aβ12–28 in dodecylphosphocholine (DPC) and SDS micelles as a function of pH and lipid type. Interaction with micelles was weak but changed the conformation when compared with aqueous buffer alone. However, the peptide interacted strongly with anionic SDS micelles, where it was mostly bound, was α-helical from Lys16 to Val24, and aggregated slowly. The pH-dependent conformational changes of the peptide in solution occurred in the pH range at which the side-chain groups of Asp22, Glu23, His13, and HisI4 are deprotonated (pKs ~ 4 and 6.5). The authors concluded that the interaction of Aβ12–28 with SDS micelles altered the pH-dependent conformational transitions of the peptide whereas the weak interaction with DPC micelles caused little change.

These conformational changes indicate a relationship between peptide structure and electrostatic interactions involving protonation and deprotonation of the micelle lipid head groups at different pH. In experiments using Aβ1–40 with the imidazole side chains of the histidine residues 6, 13, 14 methylated, Tickler et al. [34] found that the peptide-lipid interaction was modulated by the histidine residues and, therefore, would be pH sensitive. Aβ1–28 appears to associate with the surface of the membrane based on an irregular pattern in the amide chemical shift temperature coefficient dependence, suggesting that the amide backbone is situated at the water and micelle interface. Narrower NMR line widths indicated conformational mobility at the micelle surface and the concentration of Aβ1–28 not affecting CD and NMR data suggested that the α-helical structure is more likely to be stabilized by rapid exchange [33].

Jin et al. [35] used NMR spectroscopy to determine the solution structure of rat Aβ1–28 (see Fig. 8.1) and its binding constant for Zn^{2+}. They found that the three-dimensional solution structure of rat Aβ1–28 was more stable than that of human Aβ1–28 in DMSO-d_6 and that a helical region from Gln15 to Val24 existed in the rat Aβ1–28. The affinity of Zn^{2+} for rat Aβ1–28 was lower than that for human Aβ1–28, and Arg13, His6, and His14 residues provide the primary binding sites for Zn^{2+}. They also found that Zn^{2+} binding to rat Aβ1–28 caused the peptide to change to a more stable conformation.

Gröbner et al. [36] have outlined a method for structure determination of Aβ in membrane systems. First, they used CD and ^{31}P magic angle spinning (MAS) NMR spectroscopies to characterise the peptide in a dimyristoyl phosphatidyl choline/dimyristoyl phosphatidyl glycerol vesicle system. Their most notable finding was that they could get Aβ1–40 to give an α-helical structure if the peptide were dialyzed from TFE solution into the vesicles. That is, it was given no opportunity to form β-structure inducing fibrils by contact with water. Second, they used rotational resonance ^{13}C CP MAS NMR recoupling techniques to show that the membrane-penetrant part of the peptide was α-helical before major aggregation had occurred. To gain further insights, these authors concluded, future MAS studies would have to be made on multiple uniformly labeled peptides. Further advances in spectral resolution and sensitivity are vital, as is development of labeling methodologies. The development of pulse sequences and appropriate algorithms to extract multiple distance and torsion angle constraints from these systems would also be needed. Thus, the determination of the structure of Aβ by NMR in a membrane environment is still incomplete.

8.4 The Structure of Aβ in Fibrils

Conventionally, the supramolecular structure of β-sheet entities such as amyloid plaques can be considered to be either parallel or antiparallel. Which mode is likely to be important for determining the residues involved in the metal-bridged cross-links that occur in amyloid plaques and for the subsequent redox chemistry. ^{13m}C multiple quantum SS-NMR has been used to probe the structure of the full-length Aβ peptide [37]. Internuclear distances of approximately 4.8 Å would be observed for ^{13m}C-labeled residues if the β-sheets form an in-register parallel structure. An antiparallel structure, on the other hand, would have nearest neighbor residues exhibiting far larger distances than 4.8 Å. Using these NMR techniques, Aβ1–40 was shown to form a parallel β-structure [35]. This finding is similar to that of Benziger et al. [38] for Aβ10–35. Comparison with their data shows evidence that Aβ10–35 fibrils have parallel β-sheet organization beyond dimers. However, SS-NMR studies on Aβ34–42 fibrils suggested an antiparallel β-structure, which was also observed for Aβ16–22 capped at both ends [39]. Lansbury et al. [40] characterized fibrils made from the C terminal fragment Aβ34–42. They found the alignment of Aβ34–42 fibrils to be antiparallel and two residues out of register using rotational resonance experiments on doubly ^{13}C-labeled samples. Therefore, SS-NMR studies have presented evidence for both parallel and antiparallel alignments of Aβ fragments, depending on the peptide sequence studied and the methodology employed.

In a different approach, Egnaczyk et al. [41] used photo cross-linking. They synthesized a photoreactive Aβ1–40 ligand by substituting L-*p*-benzoylphenylalanine (Bpa) for phenylalanine at position 4. This peptide was incorporated into synthetic amyloid fibrils and exposed to near-UV radiation. Analysis of the fibrils showed a Bpa4-Met35 intermolecular cross-link, which was consistent with an antiparallel alignment of Aβ peptides within amyloid fibrils. Together, the above results show that fibrils can adopt different supramolecular structures depending on the peptide length and properties of the residues present. The differences are of considerable significance. For example, the photo cross-linking data show that the Met35 could be very close to the metal binding site, thus favoring redox reactions with the Met as an electron donor. On the other hand, it is quite conceivable that parallel alignment would greatly favor metal-peptide cross-linking. It is possible that physiologically both kinds of alignment could occur, the proportions being affected by different environments, such as extracellular or membrane associated, the presence/absence of metal ions or differing ratios of Zn to Cu.

8.5 The Metal-Binding Sites and the Structure of Aβ

The randomness of the Aβ peptide in aqueous solution makes it difficult to determine the nature of the metal-binding sites. The problem has been approached using various spectroscopic techniques, such as Raman, CD, and magnetic resonance. Miura et al. [42] used Raman spectroscopy to study the binding modes of Zn^{2+} and Cu^{2+} to Aβ in solution and insoluble aggregates. They found two different modes of metal-Aβ binding, one characterized by metal binding to the imidazole N_τ atom of histidine, producing insoluble aggregates, the other involving metal binding to the N_π, but not the N_τ, atom of histidine as well as to main-chain amide nitrogens, giving soluble complexes. Zn^{2+} binds to Aβ only via the N_τ regardless of pH, while the Cu^{2+} binding mode is pH dependent. At mildly acidic pH, Cu^{2+} binds to Aβ in the former mode, whereas the latter mode is predominant at neutral pH. Miura et al. [42] proposed that the transition from one binding mode to the other explained the strong pH dependence of Cu^{2+}-induced Aβ aggregation. Dong et al. [43] also employed Raman microscopy to study the metal-binding sites in amyloid plaque cores, using the spectra-structure correlations for Aβ–transition metal binding. They observed that Zn^{2+} was coordinated to the histidine N_τ and the Cu^{2+} to the N_π, confirming that the metal binding mode was the same in both the synthetic peptide and its aggregates and the naturally occurring plaques.

Huang et al. [44] used multifrequency EPR (L-band, X- and Q-band) to show that copper coordinates tightly to Aβ1–40 and that an approximately equimolar mixture of peptide and $CuCl_2$ produced a single Cu^{2+}-peptide complex. Computer simulation of the L-band spectrum with an axially symmetrical spin Hamiltonian and the *g* and *A* matrices ($g_\parallel$, 2.295; $g_\perp$, 2.073; $A_\parallel$, 163.60; $A_\perp$, 10.0×10^{-4} cm^{-1}) suggested a tetragonally distorted geometry, which is commonly found in type 2 copper proteins. Expansion of the $M_I = -1/2$ resonance revealed nitrogen ligand hyperfine coupling. Computer simulation of these resonances indicated the presence of at least three nitrogen atoms. This and the magnitude of the $g_\parallel$ and $A_\parallel$ values, together with Peisach and Blumberg [45] plots, are consistent with a fourth equatorial ligand binding to copper via an oxygen rather than a sulfur donor atom. Thus, the coordination sphere for the copper-peptide complex was considered to be 3N1O. These authors also used EPR spectroscopy to measure residual Cu^{2+} remaining after incubating stoichiometric ratios of $CuCl_2$ with Aβ1–40. There was a 76% loss of the Cu^{2+} signal, compatible with peptide-mediated reduction of Cu^{2+} to diamagnetic Cu^{+}, which is undetectable by EPR, agreeing with the corresponding concentration of Cu^{+} measured by bioassay. There was no evidence of free, uncoordinated Cu^{2+} remaining after addition of the peptide, because unbound Cu^{2+} itself gives a different multiple resonance signal.

Using a combination of NMR and EPR spectroscopies, Curtain et al. [46] proposed a structure for the high-affinity site and drew some conclusions about the interaction of the peptide with lipids and its modification by Cu^{2+}, Zn^{2+}, and pH. NMR studies on Aβ1–28 and Aβ1–40/2 indicated that both peptides were undergoing significant conformational exchange in aqueous solution. NMR and EPR spectra were also recorded for Aβ1–28 where the Nε2 nitrogens of the imidazole ring of the His residues 6, 13, and 14 were methylated (Me-Aβ1–28). The NMR spectra of Me-Aβ1–28 were virtually identical to Aβ1–28, the only significant differences being three strong singlets in the ^1H spectrum at 3.80, 3.82, and 3.83 ppm from the methyl groups attached to the His imidazole rings. A precipitate formed when Zn^{2+} was added to the solutions of Aβ1–28 or Aβ in PBS. NMR spectra of the supernatant of Aβ1–28 treated with Zn^{2+} showed that peaks assigned to C2H and C4H of His6, His13, and His14 of Aβ1–28 had broadened significantly. However, there was little or no change in the rest of the spectrum compared with Aβ1–28 prior to the addition of Zn^{2+}. This broadening of the NMR histidine residue peaks is the result of the interaction of these residues with Zn^{2+}.

The histidyl side chain is a well-established ligand of zinc in proteins and peptides [47], so this result suggested that three of the ligands bound to Zn^{2+} were most likely to be the imidazole rings of the histidine residues [48]. Indeed, His13 had been established by Liu et al. [49] as a crucial residue in the Zn^{2+}-mediated aggregation of Aβ. The broadening of these peaks is the result of chemical exchange between free and metal-bound states or among different metal-bound states. The extent of broadening of the peaks indicated intermediate exchange, which on the NMR timescale suggests that the metal-binding affinity is in the micromolar range, in agreement with the low-affinity site described by Bush et al. [16]. The absence of any change in the rest of the spectrum suggested that the metal-bound form of the peptide was monomeric and that there was little or no significant amount of soluble oligomer in solution, because higher order aggregates would have resulted in significantly broadened resonances.

When Cu^{2+} or Fe^{3+} was titrated into an aqueous solution of Aβ1–28, similar changes were observed in the ^1H spectrum, with the peaks assigned to the C2H and C4H of His6, His13, and His14 disappearing from the spectrum (Fig. 8.2). A slight broadening of all peaks in the spectrum (associated with the paramagnetism of Cu^{2+} and Fe^{3+}) was also observed, but there were no other major changes after the addition of the metal ions. Metal-induced precipitation blocked attempts to saturate the metal-binding site. The precipitate made the collection of NMR spectra difficult, and few conclusions could be drawn from spectra of peptide remaining in solution. When Cu^{2+} was added to an aqueous solution of Me-Aβ1–28, the changes observed in the spectrum were identical to those

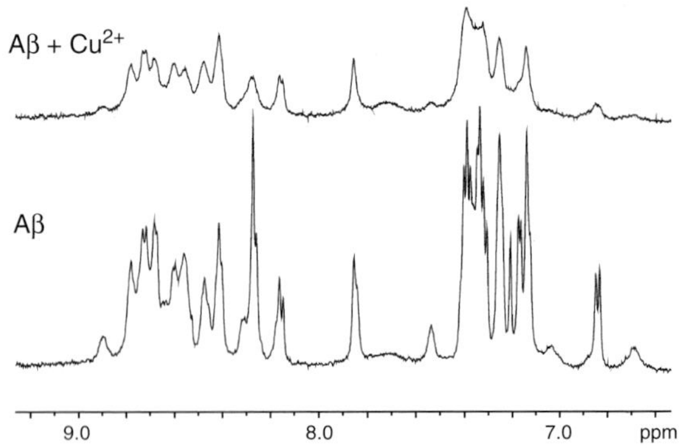

FIGURE 8.2. Amide and aromatic region of the 600 MHz ^1H NMR spectra of Aβ in aqueous PBS solution and following the addition of Cu^{2+}. Peaks caused by the C2H and C4H of histidines 6, 13, and 14 have been broadened beyond detection because of coordination to the copper. There is a generalized broadening of the rest of the spectrum due to the paramagnetism of the added Cu^{2+}. After Curtain et al. [46].

observed for Cu^{2+} added to Aβ1–28, but there was no visible precipitate. In aqueous solution and lipid environments, coordination of metal ions to Aβ is the same, with His6, His13, and His14 all involved.

The X-band EPR spectrum of Cu^{2+} bound to the peptides had the unsplit intense $g_{\perp}$ resonance characteristic of an axially symmetric square planar 3N1O or 4N coordination, $g_{\parallel} = 2.28$ and $g_{\perp} = 2.03$, $A_{\parallel} = 173.8$ gauss. Similar parameters were found for Cu^{2+} coordination by Aβ1–16, Aβ1–40, and Aβ1–42, indicating that the site was not affected by the size of the C-terminal regions of the peptides. A notable finding with peptides of all lengths was that increasing the Cu^{2+} above ~0.3 mol/mol peptide caused line broadening in the Cu^{2+} EPR spectra, over a pH range of 5.5 to 7.5, suggesting the presence of dipolar or exchange effects (Fig. 8.3). These would be observed if two or more Cu ions were within approximately 6 Å of each other. These effects could be explained if at Cu^{2+}/peptide molar ratios >0.3, Aβ coordinated a second Cu^{2+} atom cooperatively. They were abolished if the histidine residues were methylated at either Nδ1 or Nε2, suggesting that bridging histidine residues were being formed (Fig. 8.4) [32, 46].

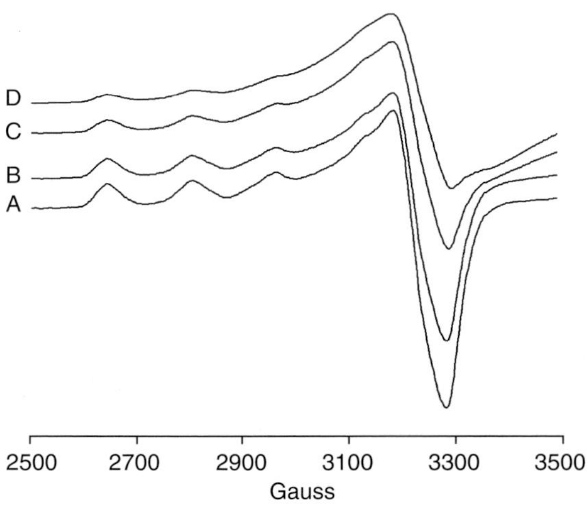

FIGURE 8.3. EPR spectra (9.7 GHz) of Aβ1–28 to which had been added respectively: A, 0.2/1 M/M; B, 0.4/1 M/M; C, 0.6/1 M/M; D, 0.8/1 M/M Cu^{2+}/peptide. All spectra recorded at 130 K in pH 7.4 phosphate-buffered saline. Spectra C and D show significant broadening of the $g_{\perp}$ line. All lengths of Aβ studied give identical spectra (Curtain et al. [46, 79]).

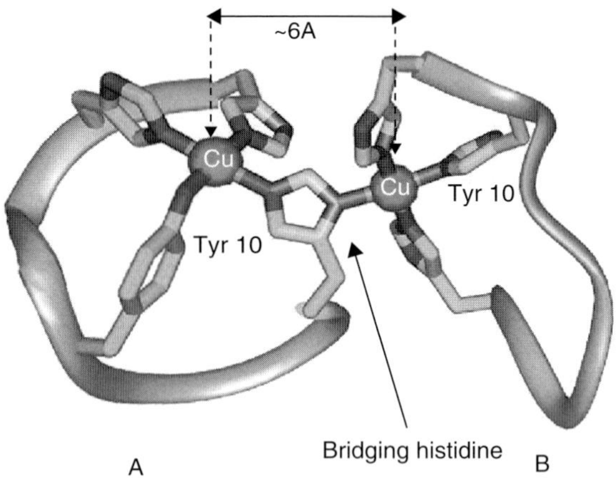

FIGURE 8.4. Model showing how two Aβ strands (A and B) could be linked by two copper atoms through a bridging histidine. The 6 Å distance between the copper atoms is within the range at which we would expect to see dipolar broadening of Cu^{2+} EPR spectra of the type seen in Figure 8.2.

One consequence of coordination by a metal ion to the Nδ1 of a histidine residue is a reduction in the pK_a of Nε2 NH, making this nitrogen more suitable for metal binding [48], resulting in a histidine residue that can bridge metal ions; a good example being His63 at the active site of superoxide dismutase [50]. Similar bridging histidine residues have been proposed in the octarepeat region of the prion protein [51], which has been shown to possess significant SOD activity in the presence of Cu^{2+} [52]. The line-broadening effects observed in the EPR spectra at Cu^{2+}/Aβ molar fractions up to 1.0 by Curtain et al. [46] were not observed by Syme et al. [27], Huang et al. [44] or Antzutkin [52]. It is relevant that Huang et al. [54] along with Narayanan and Reif [55] have shown that NaCl has a marked effect on metal-induced aggregation of Aβ. Huang et al. [44] and Curtain et al. [46] obtained their spectra from samples in phosphate-buffered saline at pH 7.4, Antzutkin [53] adjusted the pH of his sample to pH 7.4 and dialyzed against distilled water, while Syme et al. [27] used ethyl morpholine buffers.

Similar line-broadening phenomena to that observed by Curtain et al. [46] have been observed in the EPR spectra of imidazole-bridged copper complexes designed as SOD mimetics [56]. The bridging histidine may be responsible for the reversible metal-induced aggregation that is observed when Aβ is metallated with Cu^{2+} and Zn^{2+}. The bridging histidine residues may also explain the multiple metal-binding sites observed for each peptide and the high degree of cooperativity evident for subsequent metal binding. With three histidines bound to the metal center, a large scope exists for metal-mediated cross-linking of the peptides leading to aggregation, which will be reversible when the metal is removed by chelation. It should be noted here that the bridging histidine hypothesis of peptide association would favor a parallel over an antiparallel β-sheet structure for the fibrils and plaques. It is quite possible that metal-induced precipitation of Aβ is quite different from that induced by prolonged incubation of monomeric peptide in the putative absence of metal. For example, Miura et al. [42] strongly suggested that the metal-induced aggregation of Aβ was promoted by cross-linking of the peptides through metal-His[N_τ] bonds, most likely through His[N_τ]-metal-His[N_τ] bridges at three histidine residues.

Observations that rat Aβ, which differs from human Aβ by three substitutions (Fig. 8.1) [57], does not reduce Cu^{2+} and Fe^{3+}, is not readily precipitated by Zn^{2+} or Cu^{2+}, does not produce ROS as strongly as the human sequence, and does not produce plaques highlight the importance of the three histidines. Rat Aβ forms a metal complex via two histidine residues and two oxygen ligands rather than three histidine residues and one oxygen ligand, compared with human Aβ where the side chain of His13 of human Aβ is ligated to the metal ion. This was borne out by the EPR spectrum, which was typical of a square planar 2N2O Cu^{2+} coordination [44].

Syme et al. [27] and Antzutkin [53] both used X-band EPR to study the interaction of Aβ with Cu^{2+} in solution, confirming the axially symmetric binding site. Syme et al. [27] obtained EPR spectra at pH 7.4 and higher that showed heterogeneity attributed to a second high-affinity binding site. This site became much more prominent when the pH was raised to 10.0. The heterogeneity at pH 7.4 was not observed by Huang et al. [44], Curtain et al. [46], or by Antzutkin [53] and warrants further investigation. It is possible that the second binding site is a buffer ion effect. In order to define the binding site, Syme et al. [27] also prepared mutants of Aβ1–28 in which each of the histidine residues had been replaced by alanine or in which the N-terminus was acetylated, and their data suggested that the N-terminus and His13 and His14 are crucial for Cu^{2+} binding and that H6 also played a part. On this basis, they proposed a square planar model with the Cu^{2+} coordinated to His13, His14, His6, and the amino N of the N-terminus. Although a 4N model may be fitted to Syme et al.'s [27] X-band spectra, it is not compatible with the conclusions derived by Huang et al. [44] from L-band spectra and their superhyperfine structure that point to a 3N1O coordination.

Karr et al. [58] found that Aβ peptides lacking one to three N-terminal amino acids but containing His6, His13, and His14 and Tyr10 did not coordinate Cu^{2+} in the same environment as the native peptide, suggesting that these N-terminal residues are significant for Cu^{2+} binding. They also confirmed that the coordination is identical with any length of peptide (Aβ1–16, Aβ1–28, Aβ1–40, Aβ1–42) that contained the first 16 amino acids. These authors also showed [59] that the coordination of Cu^{2+} did not change during organization of monomeric Aβ into fibrils and that neither soluble nor fibrillar forms of Aβ1–40 contained antiferromagnetically exchange-coupled binuclear Cu^{2+} sites in which two ions were bridged by an intervening ligand. The latter conclusion was based on a temperature-dependence study of the EPR spectra for Cu^{2+} bound to soluble or fibrillar Aβ showing that the Cu^{2+} center displayed normal Curie behavior, indicating that the site was mononuclear.

Further advances in understanding the N coordination of Cu^{2+} will require more sophisticated EPR techniques than have been used so far, supported by input from other methods such as XAFS. Equally, there remains uncertainty as to the nature of the potential O ligand. Proton NMR data obtained by Syme et al. [27] agreed with the findings of Huang et al. [44] and Curtain et al. [46] that histidine residues are involved in Cu^{2+} coordination, but they found that Tyr10 was not involved. Further, Karr et al. [58] found that the coordination of Cu^{2+} in the Y10F mutant of Aβ remained 3N1O with EPR spectra identical to the wild-type spectra. Isotopic labeling experiments showed that water was not the O-atom donor to Cu^{2+} in Aβ fibrils or in the Y10F mutant. However, the Raman data of

Miura et al. [42] suggest that the ligand was the O of the tyrosine hydroxyl. They were able to assign the 1504 cm^{-1} band in the Raman spectra of insoluble Cu^{2+}-Aβ1–16 aggregates to Cu^{2+}-bound tyrosinate, and the high intensity of the 1604 cm^{-1} band was attributed to a contribution from the Y8a band of tyrosinate. Unlike Zn^{2+}, Cu^{2+} binds to tyrosine in the insoluble aggregates of Aβ1–16. When the deprotonated phenolic oxygen of tyrosinate is bound to a transition metal ion such as Cu^{2+} and Fe^{3+}, the Y19a band shifts to about 1500 cm^{-1} and gains intensity through resonance with a ð (phenolate) f *d* (metal) charge-transfer transition in the visible. Such charge transfer does not occur for Zn^{2+}, the *d* orbitals of which are fully occupied. It should be noted that in these experiments, Miura et al. [42] used phosphate-buffered saline, which might have had the effect of encouraging peptide association [54, 55].

In conclusion, although there is general agreement as to the nature of the monomeric binding site insofar as it is type two Cu^{2+} with a 3N1O coordination, varying buffer conditions, peptide concentration, and conformation make it difficult to compare one set of published data with another. There is a similarity here with the studies on the alignment of the peptide in fibrils. In considering the issue of monomeric versus dimeric Cu^{2+}, it is important to remember that Aβ may form oligomers and multimers in a variety of ways, some more relevant to its neurotoxicity than others [60–64].

8.6 Aβ Redox Activity and the Role of Metal Coordination

Oxidative stress markers characterize the neuropathology both of Alzheimer's disease and of amyloid-bearing transgenic mice. The neurotoxicity of Aβ has been linked to hydrogen peroxide generation in cell cultures by a mechanism that is still being fully described but is likely to be dependent on Aβ coordinating redox active metal ions. Huang et al. [65] showed that human Aβ directly produces hydrogen peroxide (H_2O_2) by a mechanism that involves the reduction of metal ions, Fe^{3+} or Cu^{2+}. They used spectrophotometry to show that the Aβ peptide reduced Fe^{3+} and Cu^{2+} to Fe^{2+} and Cu^{+} and that molecular oxygen is then trapped by Aβ and reduced to H_2O_2 in a reaction that is driven by sub-stoichiometric amounts of Fe^{2+} or Cu^{+}. In the presence of Cu^{2+} or Fe^{3+}, Aβ produced a positive thiobarbituric-reactive substance, compatible with the generation of the hydroxyl radical [OH^*]. Tabner et al. [66] used the 5,5-dimethyl-1-pyrroline *N*-oxide (DMPO) spin-trap to identify the radical produced by Aβ in the presence of Fe^{2+} and concluded that it was OH^*. However, they also found OH^* was produced in the presence of Fe^{2+} by Aβ25–35, which does not contain a strong metal binding site. Because Fe^{2+} with trace amounts of Cu^{2+} as low as 0.01 mol%, corresponding with the amount of adventitious Cu found in the average peptide preparation, will produce an OH^* adduct with the DMPO spin trap [Curtain et al., unpublished], Tabner et al.'s [66] results should be treated with caution even though they appear to confirm the findings of Huang et al. [65].

In the course of metal-catalyzed redox activity, Aβ may undergo under a number of changes. Atwood et al. [67] found that Cu^{2+} induced the formation of SDS-resistant oligomers of Aβ that gave a fluorescence signal characteristic of the cross-linking of the peptide's Tyr10. This finding was confirmed by directly identifying the dityrosine by electrospray ionization mass spectrometry and by the use of a specific dityrosine antibody. The addition of H_2O_2 strongly promoted Cu^{2+}-induced dityrosine cross-linking of Aβ1–28, Aβ1–40, and Aβ1–42, and it was suggested that the oxidative coupling was initiated by interaction of H_2O_2 with a Cu^{2+} tyrosinate. The dityrosine modification is significant because it is highly resistant to proteolysis and would be important in increasing the structural strength of the plaques. Schoneich and Williams [68], however, were unable to find any evidence of tyrosine oxidation. They used ascorbate/Cu^{2+}-induced oxidation and electrospray ionization-time-of-flight MS/MS analysis to study the oxidation products of Aβ1–16, Aβ1–28, and Aβ1–40. Initial oxidation targets were His13 and His14, which were converted to 2-oxo-His, while His6 and Tyr10 were unchanged, although His6 was oxidized after longer oxidation times. The formation of 2-oxo-His suggests that a transient 2C centered His radical might have been formed. Such radicals have been described in a number of biological redox systems [69, 70], although not so far in any of neuropathological significance.

Schoneich and Williams [68] explained the insensitivity of His6 to initial oxidation by suggesting that histidine bridging of two Cu^{2+}-Aβ molecules lowered the electron density on His6, comparable with similar results on a Cu^{2+}- and Zn^{2+}-bridging His61 residue of bovine Cu,Zn superoxide dismutase.

Barnham et al. [71] used density functional theory calculations to elucidate the chemical mechanisms underlying the catalytic production of H_2O_2 by Aβ/Cu and the production of dityrosine. Here, Tyr10 was identified as the critical residue. This finding accords with the growing awareness that the O_2 activation ability of many cupro-enzymes is also coupled to the redox properties of tyrosine and the relative stability of tyrosyl radicals. The latter play important catalytic roles in photosystem II, ribonucleotide reductase, COX-2, DNA photolyase, galactose oxidase, and cytochrome-*c* oxidase [72].

With ascorbate as the electron donor, the first step in the catalytic production of H_2O_2 is the reduction of Cu^{2+} to Cu^+. Barnham et al. [71] proposed that the transfer could take place via a proton-coupled electron transfer (PCET) mechanism. Reactions involving PCET are being increasingly implicated in a range of biological systems, including charge transport in DNA and enzymatic oxygen production [73]. In this system, the electron transfer involves both p- and d-orbitals on the ascorbate, Tyr10, and the copper ion, while proton transfer involves p-orbitals on the O_2-atom of ascorbate, and the side-chain oxygen of Tyr10 (Figs. 8.5A and 8.5B). The significant change in electron spin on the copper ion going from the ground state to the transition state suggests that the proton and the electron are transferred within different molecular orbitals, as is predicted to be necessary for PCET to occur [73]. The activation energy for this one electron reduction step was computed to be only 0.9 kcal/mol.

Barnham et al. [71] tested the Cu/tyrosinate hypothesis using an Aβ1–42 peptide with Tyr10 substituted with alanine (Y10A). Both peptides gave rise to similar ^{65m}Cu EPR spectra with the strong single $g_\perp$ resonance characteristic of an axially symmetric square planar complex, although there was a significant increase in the $g_\parallel$ value of Y10A. The increase was probably due to some distortion of the coordination sphere because the

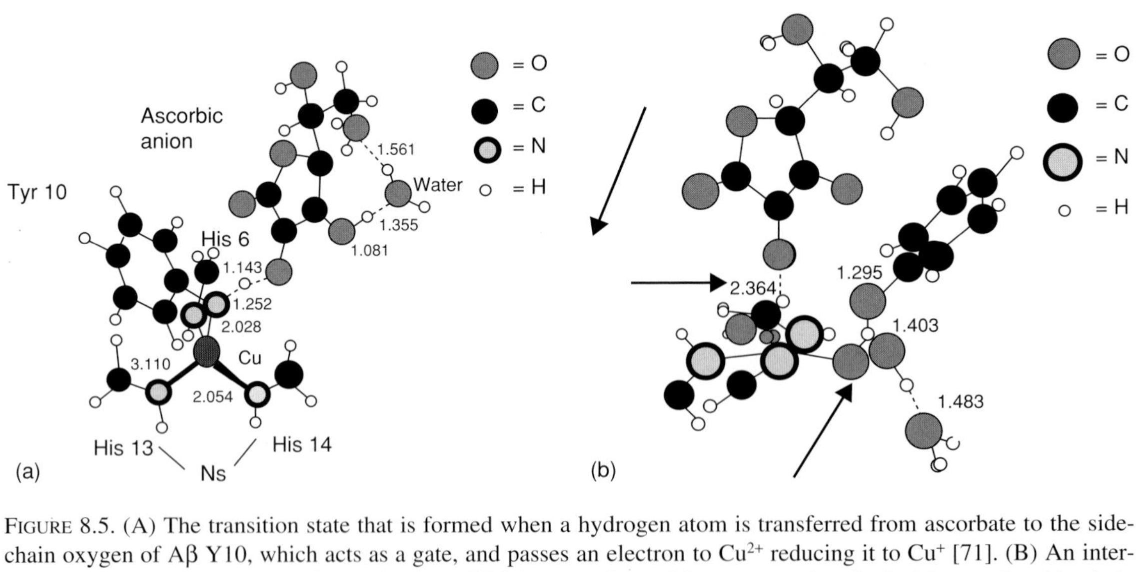

FIGURE 8.5. (A) The transition state that is formed when a hydrogen atom is transferred from ascorbate to the side-chain oxygen of Aβ Y10, which acts as a gate, and passes an electron to Cu^{2+} reducing it to Cu^+ [71]. (B) An intermediate formed along the reaction path where Y10 has transformed into a tyrosyl radical giving up its side-chain hydroxyl hydrogen atom to $O_2{}^{\bullet-}{}_\perp$ via hydrogen atom transfer. Simultaneously, H_3O^+ has donated its proton to $O2^{\bullet-}$ via proton transfer, whereupon H_2O_2 has formed. Formed tyrosyl radical and water molecule are hydrogen bonded to H_2O_2. Ascorbyl radical anion coordinates via its O1-oxygen anion in an apical position to Cu^{2+}. Figures based on data of Barnham et al. [71].

oxygen ligand, which was possibly from Tyr10, was now derived from another oxygen donor (e.g., phosphate, or carboxylate from the peptide). While wild-type Aβ1–42 rapidly reduces Cu^{2+} to Cu^{+} in aqueous solution, with near-complete reduction taking 80 min, the mutation of Tyr10 to alanine markedly decreased the ability of Aβ to reduce Cu^{2+}. Further, spin trapping studies also confirmed the DFT observation that Tyr10 acts as a gate that facilitates the electron transfer needed to reduce Cu^{2+} to Cu^{+}. When the spin trap 2-methyl-2-nitrosopropane (2MNP) [74] was added to the reaction mixture *w.t.* Aβ1– 42/Cu^{2+}/ascorbate, a broad line triplet characteristic of a trapped carbon-centered radical bound to a peptide appeared in the EPR spectra. However, if Y10A peptide were substituted for the *w.t.*, formation of this triplet was inhibited (Fig. 8.6). Although this is not conclusive evidence that the radical is on Tyr10, the possibility that a His radical was trapped by the 2MNP can be discounted because the A_N value (15.5) of the spectrum in Figure 8.6 is closer to those found for Tyr adducts [75] than for C-centered His, which furthermore show marked superhyperfine structure [69]. It is likely that which transient radical is trapped in a given Cu:Aβ redox system will depend on a number of experimental variables only some of which may be biologically relevant.

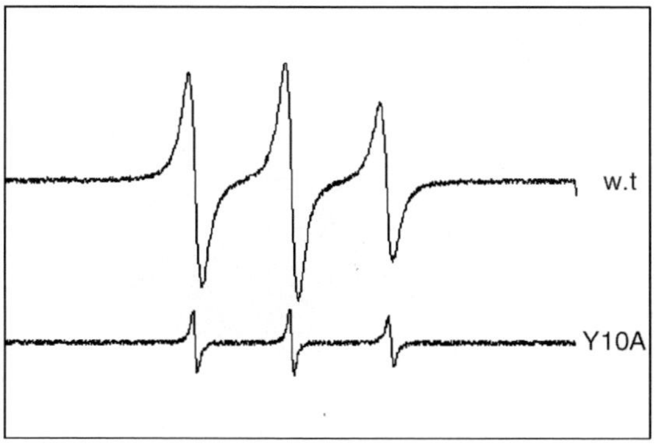

FIGURE 8.6. X-band EPR spectra of adducts formed after the addition of 100 mM spin trap 2-methyl, 2-nitrosopropane to respectively wild-type Aβ (50 mM incubated at 20°C for 30 min with 25 mM Cu^{2+} in pH 7.4 PBS) and the Y10A mutant of Aβ at the same Cu/peptide ratio and pH. End-to-end width of spectrum, 100 gauss. Figure based on data of Barnham et al. [71].

8.7 The Effect of Metal Binding on the Interaction of Aβ with Membranes

An alternative explanation of Aβ neurotoxicity, not necessarily excluding the production of ROS, is based on the peptide's interaction with membranes and/or membrane proteins. Numerous reports have described the effects of Aβ on membranes and lipid systems and their possible roles in its neurotoxicity. The NMR studies cited earlier in this chapter showed considerable variation in peptide conformation in different membrane-mimetic systems. There is much experimental evidence from CD and Fourier transform infrared spectroscopies that the Aβ peptides can be membrane associated in the β-configuration [75], although there are reports of membrane-associated α-helices being found in the presence of gangliosides [76], cholesterol [77], and Cu^{2+} or Zn^{2+} [46]. This variability under different conditions can be understood because most of the amyloidogenic peptides have been identified, along with viral fusion peptides, as being exceptionally pleiomorphic in structure [78]. This identification was based on the high prevalence of alanine and glycine residues within a hydrophobic sequence.

As the cell membrane is a mosaic of lipids and protein segments, it is possible that the peptides will exhibit different structures with different properties in different parts of the mosaic. The pleiomorphism is highly relevant to the cytotoxicity of the peptide, because factors influencing it could act as switches to determine whether the peptide is a β-sheet with the potential to form amyloid or be membrane surface seeking, or a membrane-penetrant α-helix.

Curtain et al. [46, 79] used a combination of EPR and CD spectroscopies to study the effect of metal ions, pH, and cholesterol on the interaction of Aβ with bilayer membranes. EPR spectroscopy, using spin-labeled lipid chains or protein segments, has been used extensively to study translational and rotational dynamics in biological membranes. Lipids at the hydrophobic interface between lipid and transmembrane protein segments and peptides in their monomeric and oligomeric states have their rotational motion restricted [80]. This population of lipids can be resolved in the EPR spectrum as a motionally restricted component distinct from the

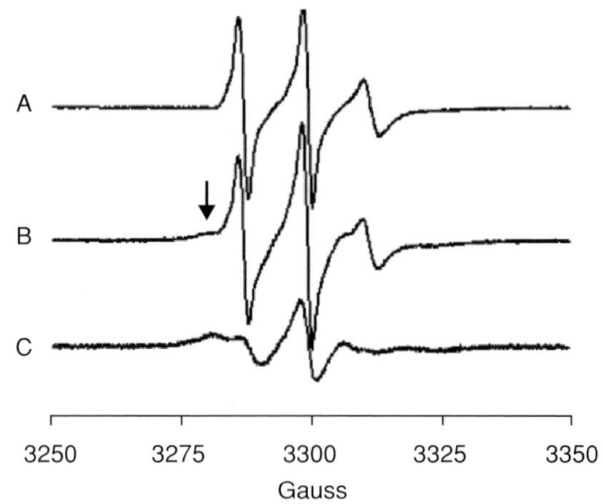

FIGURE 8.7. A: X-band EPR spectrum recorded at 305 K of the negatively charged spin probe 1-palmitoyl-2-(16-doxyl stearoyl) phosphatidyl serine in negatively charged LUV made from 50% palmitoyl oleoyl phophatidyl serine and 50% palmitoyl oleoyl phophatidyl choline (probe/lipid 1/300). B: X-band spectra of system A at the same temperature after the addition of Cu^{2+}Aβ1–42 (peptide/lipid 1/50), showing a shoulder (marked with arrow) to the left of the low field line. This is typical of peptide penetration into the bilayer core [73, 75]. C: The difference spectrum × 5 obtained when spectrum A is subtracted from spectrum B. This spectrum represents the motionally restricted lipid in the boundary. Original data given in Curtain et al. [46, 79].

fluid bilayer lipids (Fig. 8.7), which can be quantified to give both the stoichiometry and selectivity of the first shell of lipids interacting directly with membrane-penetrant peptides. The stoichiometric data can give an estimate of the number of subunits in a membrane-penetrant oligomeric structure. Using this approach, it was shown that Aβ1–40 and Aβ1–42 bound to Cu^{2+} or Zn^{2+} penetrated bilayers of negatively charged, but not zwitterionic lipid, giving rise to such a partly immobilized component in the spectrum (see Fig. 8.7 and its caption) [46, 79].

When the peptide:lipid ratio was increased, the relationship between the mole fraction of peptide and proportion of slow component was linear. Even at a fraction of 15%, all of the peptide was associated with the lipid, suggesting that the structure penetrating the membrane lipid was well defined, although at such a high peptide:lipid ratio further study would be needed to confirm whether the lipid still retained a lamellar structure. Formation of non-lamellar structures in regions of the membrane associated with Aβ could well be the cause of the peptide's cytotoxicity. From the spin-label data, the first shell lipid:peptide was approximately 4:1. This stoichiometry can be satisfied by 6 helices arranged in a pore surrounded by 24 boundary lipids. This hypothetical structure gains credibility from atomic force microscopy studies of Aβ1–42 reconstituted in a planar lipid bilayer that showed multimeric channel-like structures, many resembling hexamers, similar to that modeled in Figure 8.8 [81]. It was found [46] that in the presence of Zn^{2+}, Aβ1–40 and Aβ1–42 both inserted into the bilayer over the pH range 5.5–7.5, as did Aβ1–42 in the presence of Cu^{2+}. However, Aβ40 only penetrated the lipid bilayer in the presence of Cu^{2+} at pH 5.5–6.5; at higher pH, there was a change in the Cu^{2+} coordination sphere that inhibited membrane insertion. The addition of cholesterol up to 0.2 mole fraction of the total lipid inhibited insertion of both peptides under all con-

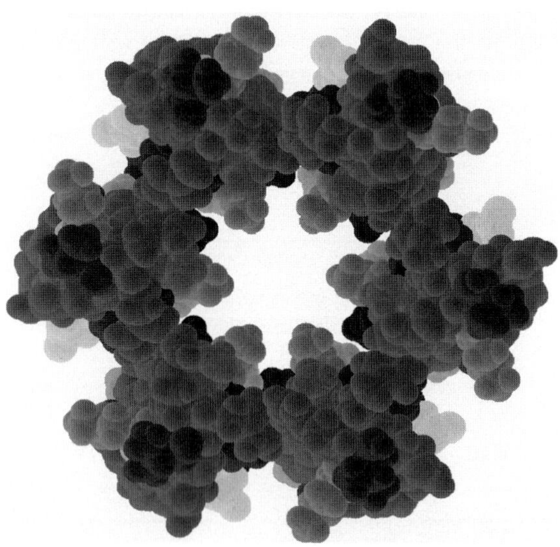

FIGURE 8.8. Animation of hexameric pore formed by Aβ1–40 helices calculated from annular lipid stoichiometry as determined from the EPR data shown in Figure 8.5. Polar residues are shown as dark and nonpolar as light. View from N-terminus. Peptide coordinates (in SDS) obtained from Barrow and Zagorski [31]. Model prepared using Sculpt® by aligning hydrophobic contacts between helices and orienting nonpolar residues in sequence 21–40 to annular lipid.

ditions investigated. CD spectroscopy revealed that the Aβ peptides had a high α-helix content when membrane penetrant, but were predominantly β-strand when not. Simulation of the spectra and calculation of the on-off rates suggested that the peptide was most likely penetrating as an α-helix [82].

In membrane-mimetic environments, coordination of the metal ion is the same as in aqueous solution, with the three-histidine residues, at sequence positions 6, 13, and 14, all involved in the coordination, along with an oxygen ligand. As had been observed at Cu^{2+}/peptide molar ratios >0.3 in aqueous solution, line broadening was detectable in the EPR spectra, indicating that the peptide was coordinating a second Cu^{2+} atom in a highly cooperative manner at a site 6 Å from the initial binding site. So, there appear to be two switches, metal ions (Zn^{2+} and Cu^{2+}) and negatively charged lipids, needed to change the conformation of the peptide from β-strand nonpenetrant to α-helix penetrant. The closest parallel to this behavior is that observed with the B18 fusogenic sequence of the fertilization protein bindin [83] that, like Aβ, possesses three histidine residues strategically placed to coordinate metals. In the absence of Zn^{2+}, this peptide forms nonfusing β-sheet amyloid fibrils. In the presence of Zn^{2+}, an α-helical conformation is imposed on its backbone and it forms fusogenic oligomers.

8.8 The Relevance of Membrane Binding to Aβ Cytoxicity: The Role of Methionine 35

In vitro, the methionine at position 35 can act as an electron donor, and its conversion to the sulfoxide form has been the subject of several studies, given that the Met(O)Aβ peptide has been isolated from AD amyloid brain deposits [84, 85]. Furthermore, the Raman spectroscopic study by Dong et al. [43] of senile plaque cores isolated from diseased brains has shown that much of the Aβ in these deposits contained methionine sulfoxide with copper and zinc coordinated to the histidine residues.

Although there are several potential electron donors such as GSH and ascorbic acid, in vivo it is likely that Met35 occupies a privileged position being part of the Aβ sequence. When it is missing as in Aβ1–28, the addition of exogenous methionine permits redox action to proceed, but with slower kinetics [46]. When Met35 is sequestered within a lipid environment, there is also no metal reduction. Its oxidation also alters the physical properties of the peptide. Met(O)Aβ is more soluble in aqueous solution, and there is a disruption of the local helical structure when the peptide is dissolved in SDS micelles [86].

The formation of trimers and tetramers by Met(O)Aβ is significantly attenuated and fibril formation is inhibited [87, 88]. Barnham et al. [89] showed by solid-state NMR that when Aβ coordinates and reduces Cu^{2+} to Cu^{+}, the Met35 is oxidized. Although the Cu^{2+} coordination of the oxidized peptide is identical to nonoxidized Aβ and it will produce H_2O_2, it cannot penetrate lipid bilayers either in the presence or absence of Cu^{2+} or Zn^{2+}. On the other hand, Met(O)Aβ is toxic to neuronal cell cultures, a toxicity that is rescued by catalase and the MPAC clioquinol. These results suggest that fibril formation and membrane penetration by Aβ could be epiphenomena, and that the main requirement for cytotoxicity is redox competence. In this connection, it is important to note that the oxidized M35 has the potential for further reduction to the sulfone [90] and could thus still act as a Cu^{2+} reductant, acting in vivo in concert with agents such as ascorbic acid and GSH.

It might be legitimately asked whether Met35 could act as a Cu^{2+} ligand. After all, there are many instances of copper proteins where the ion is coordinated to a thioether, giving in most cases a type 1 binding site [91]. Such coordination involving two nitrogens and an oxygen in addition to the sulfur is generally distorted tetrahedral rather than square planar and would favor Cu^{+} over Cu^{2+}. Because the former is EPR silent, the possibility of this coordination might have been overlooked. However, in their Raman spectroscopic studies, Miura et al. [42] were unable to detect any Cu-S bonds.

Ciccotosto et al. [92] further probed the role of Met35 by preparing Aβ1–42 in which it was replaced with valine (AβM35V). The neurotoxic activity on primary mouse neuronal cortical cells of this peptide was enhanced, and this diminished cell viability occurred at a much faster rate compared with Aβ1–42. When cortical cells were treated with the peptides for only a short 1-h duration so as to minimize the incidence of cell death, and the amount of peptide bound to cortical cell extracts was quantitated by Western blotting, it was found that twice as much AβM35V compared with wild-type Aβ peptide bound to the cells after a 1-h cell exposure. It was suggested that the increased toxicity was related to the increased binding.

AβM35V bound Cu^{2+} with the same coordination sphere as *w.t.* Aβ and produced similar amounts of H_2O_2 as Aβ1–42 in vitro. The neurotoxic activity was rescued by catalase. The redox activity of the mutated peptide was followed by measuring the decline in time of the strength of the Cu^{2+}-AβM35V EPR signal, which showed that the reduction of Cu^{2+} to the EPR silent Cu^{+} was much slower compared with Aβ1-42, confirming that the M35 residue in Aβ42 plays an important part in the redox behavior of this peptide in solution. Like Cu^{2+}-Aβ1–42, Cu^{2+}-AβM35V inserted into a spin-labeled lipid bilayer gave a partially immobilized component in the EPR spectrum. This component had a narrower linewidth than that found for the similar component obtained with *w.t.* Cu^{2+}-Aβ1–42, suggesting that the valine substitution made the mutant peptide less rigid in the bilayer region and possibly easier to insert, thus explaining the increased cell membrane binding. The on- and off-rate constants estimated from the simulation experiments showed that AβM35V had a higher affinity for the lipid bilayer as compared with Aβ42. CD analysis showed that AβM35V had a higher proportion of β-sheet structure and random coil than Aβ1–42, which would also suggest a more flexible structure in the bilayer [80, 82]. In summary, these and the results described above tell us that the wild-type Aβ, its oxidized form, Met(O)Aβ, and the mutant peptide, AβM35V, induce cell death via similar pathways that are metal-dependent and can generate H_2O_2 in the absence of a methionine residue. Fibril formation as a toxic species is not responsible for cell death. Membrane association *per se* may play a part in localizing the peptide, perhaps in domains particularly susceptible to oxidative damage. It follows, therefore, that elucidating the metal ion binding site of Aβ may provide a promising new therapeutic target for AD.

References

1. Bush AI. Copper, zinc, and the metallobiology of Alzheimer's disease. Alzheimer Dis Assoc Disord 2003;17:147-50.
2. Ritchie CW, Bush, AI, Mackinnon, A et al. Metal-protein attenuation with iodochlorhydroxyquin (clioquinol) targeting Abeta amyloid deposition and toxicity in Alzheimer's disease: a pilot phase 2 clinical trial. Arch Neurol 2003;60:1685-91.
3. Hershey CO, Hershey LA, Varnes A et al. Cerebrospinal fluid trace element content in dementia: clinical, radiologic and pathologic correlations. Neurology 1983;33:1350-53.
4. Ehmann WD, Markesbery WR, Alauddin M, et al. Brain trace elements in Alzheimer's disease. Neurotoxicology 1986;7:195-206.
5. Thompson CM, Markesbery WR, Alaudin M et al. Regional brain trace-element studies in Alzheimer's disease. Neurotoxicology 1988;9:1-8.
6. Basun H, Forssell LG, Wetterberg L, et al. Metals and trace elements in plasma and cerebrospinal fluid in normal aging and Alzheimer's disease. J Neural Transm Park Dis Dement Sect 1991;3:231-58.
7. Samudralwar DL, Diprete CC, Ni BF, et al. Elemental imbalances in the olfactory pathway in Alzheimer's disease. J Neurol Sci 1995;130:139-45.
8. Deibel MA, Ehmann WD, Markesbery WR. Copper, iron, and zinc imbalances in severely degenerated brain regions in Alzheimer's disease: possible relation to oxidative stress. J Neurol Sci 1996;143:137-42.
9. Cornett CR, Markesbery WR, Ehmann WD. Imbalances of trace elements related to oxidative damage in Alzheimer's disease brain. Neurotoxicology 1998;19:339-45.
10. González C, Martin T, Cacho J, et al. Serum zinc, copper, insulin and lipids in Alzheimer's disease epsilon 4 apolipoprotein E allele carriers. Eur J Clin Invest 1999;29:637-42.
11. Atwood, CS, Huang, X, Moir, RD, et al. Role of free radicals and metal ions in the pathogenesis of Alzheimer's disease. Met Ions Biol Syst 1999;36:309-64.
12. Lovell MA, Robertson JD, Teesdale WJ, et al. Copper, iron and zinc in Alzheimer's disease senile plaques. J Neurol Sci 1998;158:47-52.
13. Miller LM, Wang Q, Telivala TP, et al. Synchrotron-based infrared and X-ray imaging shows focalized accumulation of Cu and Zn co-localized with beta-amyloid deposits in Alzheimer's disease. J Struct Biol 2006; 155:30-37.
14. Suh SW, Jensen KB, Jensen MS, et al. Histological evidence implicating zinc in Alzheimer's disease. Brain Res 2000;852:274-78.

15. Lee JY, Cole TB, Palmiter RD, et al. Contribution by synaptic zinc to the gender disparate plaque formation in human Swedish mutant APP transgenic mice. Proc Natl Acad Sci USA 2002;99:7705-10
16. Bush AI, Pettingell WH Jr, Paradis MD, et al. Modulation of Abeta adhesiveness and secretase site cleavage by zinc. J Biol Chem 1994;269:12152-58.
17. Bush AI, Pettingell WH, Multhaup G, et al. Rapid induction of Alzheimer A beta amyloid formation by zinc. Science 1994;265:1464-67.
18. Bush AI, Moir RD, Rosenkranz KM, et al. Zinc and Alzheimer's disease -response. Science 1995; 268:1921-23.
19. Clements A, Allsop D, Walsh DM, et al. Aggregation and metal-binding properties of mutant forms of the amyloid Aβ peptide of Alzheimer's disease. J Neurochem 1996;66:740-47.
20. Yang DS, McLaurin J, Qin K, et al. Examining the zinc binding site of the amyloid-beta peptide. Eur J Biochem 2000;267:6692-8.
21. Atwood CS, Moir RD, Huang X, et al. Dramatic aggregation of Alzheimer A-beta by Cu(II) is induced by conditions representing physiological acidosis J Biol Chem 1998;273:12817-26.
22. Cherny RA, Atwood CS, Xilinas ME, et al. Treatment with a copper-zinc chelator markedly and rapidly inhibits beta-amyloid accumulation in Alzheimer's disease transgenic mice. Neuron 2001;30:665-76..
23. Cherny RA, Legg JT, McLean CA, et al. Aqueous dissolution of Alzheimer's disease A-beta amyloid deposits by biometal depletion. J Biol Chem 1999;274:23223-28
24. Huang X, Atwood CS, Moir RD, et al. Trace metal contamination initiates the apparent auto-aggregation, amyloidosis, and oligomerization of Alzheimer's A-beta peptides. J Biol Inorg Chem 2004;9:954-60.
25. Basun H, Forssell LG, Wetterberg L, Winblad B. Metals and trace elements in plasma and cerebrospinal fluid in normal aging and Alzheimer's disease. J Neural Transm Park Dis Dement Sect 1991;3:231-58.
26. Garzon-Rodriguez W, Yatsimirsky AK, Glabe CG. Binding of Zn(II), Cu(II), and Fe(II) ions to Alzheimer's A beta peptide studied by fluorescence. Bioorg Med Chem Lett 1999;9:2243-8.
27. Syme CD, Nadal RC, Rigby, SEJ, et al. Copper binding to the amyloid-beta (Abeta) peptide associated with Alzheimer's disease: folding, coordination geometry, pH dependence, stoichiometry and affinity of Abeta-(1-28) :insights from a range of complementary spectroscopic techniques. J Biol Chem 2004;279:18169-77.
28. Zagorski MG, Barrow CJ. NMR studies of amyloid beta-peptides: proton assignments, secondary structure and mechanism of an alpha-helix-beta-sheet conversion for a homologous, 28-residue, N-terminal fragment. Biochemistry 1992;31:5621-31.
29. Coles M, Bicknell W, Watson AA, et al. Solution structure of amyloid beta-peptide (1-40) in a water-micelle environment. Is the membrane-spanning domain where we think it is? Biochemistry 1998;37:11064-77.
30. Shao H, Jao S.-C, Ma K, et al. Solution structures of micelle-bound amyloid beta-(1-40) and beta-(1-42) peptides of Alzheimer's disease. J Mol Biol 1999;285:755-73
31. Barrow CJ, Zagorski MG. Solution structures of beta peptide and its constituent fragments: relation to amyloid deposition. Science 1991;253: 179-82
32. Sorimachi K, Craik DJ. Structure determination of extracellular fragments of amyloid proteins involved in Alzheimer's disease and Dutch-type hereditary cerebral haemorrhage with amyloidosis. Eur J Biochem 1994;219:237-51
33. Fletcher TG, Keire DA. The interaction of beta-amyloid protein fragment (12-28) with lipid environments. Protein Sci 1997;6: 666-75
34. Tickler AK, Smith, DG, Ciccotosto, GD, et al. Methylation of imidazole side chains of the Alzheimer's disease amyloid beta peptide results in abolition of SOD-like structures and inhibition of neurotoxicity. J Biol Chem 2005;280:13355-63
35. Jin H, Yong Y, Jun L, et al. The solution structure of rat Abeta-(1-28) and its interaction with zinc ion: insights into the scarcity of amyloid deposition in aged rat brain. J Biol Inorg Chem 2004;9:627-35.
36. Gröbner G, Glaubitz C, Williamson PTF, et al. Structural insight into the interaction of amyloid-beta peptide with biological membranes by solid state NMR. Focus Struct Biol 2001;1:203-14.
37. Antzutkin ON, Balbach JJ, Leapman RD, et al. Multiple quantum solid-state NMR indicates a parallel, not antiparallel, organization of β-sheets in Alzheimer's beta-amyloid fibrils. Proc Natl Acad Sci USA 2000;97:13045-50
38. Benzinger T, Gregory DM, Burkoth TS, et al. Propagating structure of Alzheimer's beta-amyloid 10-35 is parallel beta-sheet with residues in exact order. Proc Natl Acad Sci USA 1998;95:13407-12.
39. Balbach JJ, Ishii Y, Antzutkin ON, et al. Amyloid fibril formation by $Abeta_{16-22}$, a seven residue fragment of the Alzheimer's beta-amyloid peptide, and structural characterization by solid state NMR. Biochemistry 2000;39:13748-59.
40. Lansbury PT, Costa PR, Griffiths JM, et al. Structural model for the beta-amyloid fibril based on inter-

strand alignment of an antiparallel-sheet comprising a C-terminal peptide. Nat Struct Biol 1995;2:990-98.
41. Egnaczyk GF Greis KD, Stimson ER, et al. Photoaffinity cross-linking of Alzheimer's disease amyloid fibrils reveals interstrand contact regions between assembled beta-amyloid peptide subunits. Biochemistry 2001;40:11706-14
42. Miura T, Suzuki K, Kohata N, et al. Metal binding modes of Alzheimer's amyloid beta-peptide in insoluble aggregates and soluble complexes. Biochemistry 2000;39:7024-31.
43. Dong J, Atwood CS, Anderson VE, et al. Metal binding and oxidation of amyloid-beta within isolated senile plaque cores: Raman microscopic evidence. Biochemistry 2003;42:2768-73.
44. Huang X, Cuajungco MP, Atwood CS, et al. Cu(II) potentiation of Alzheimer abeta neurotoxicity. Correlation with cell-free hydrogen peroxide production and metal reduction. J Biol Chem 1999;274:37111-16.
45. Peisach J, Blumberg WE. Structural implications derived from the analysis of electron paramagnetic resonance spectra of natural and artificial copper proteins. Arch Biochem Biophys 1974;165:691-708.
46. Curtain CC, Ali F, Volitakis I, Cherny RA, et al. Alzheimer's disease amyloid-beta binds copper and zinc to generate an allosterically ordered membrane-penetrating structure containing superoxide dismutase-like subunits. J Biol Chem 2001;276:20466-73.
47. Alberts IL, Nadassy K, Wodak SJ. Analysis of zinc binding sites in protein crystal structures. Protein Sci 1998;7:1700-16
48. Sundberg RJ, Martin RB. Interactions of histidine and other imidazole derivatives with transition metal ions in chemical and biological systems. Chem Rev 1974;74:471-517.
49. Liu S-T, Howlett G, Barrow CJ. Histidine-13 is a crucial residue in the zinc ion-induced aggregation of the Aβ peptide of Alzheimer's disease. Biochemistry 1999;38:9373-78.
50. Parge HE, Hallewell RA, Tainer JA. Atomic structures of wild-type and thermostable mutant recombinant human Cu, Zn superoxide dismutase. Proc Natl Acad Sci USA 1992;89:6109-13.
51. Viles JH, Cohen, FE, Prusiner SB, et al. Copper binding to the prion protein: Structural implications of four identical cooperative binding sites. Proc Natl Acad Sci USA 1999;96:2042-47.
52. Brown DR, Wong BS, Hafiz F, et al. Normal prion protein has an activity like that of superoxide dismutase. Biochem J 1999;344:Pt 1:1-5.
53. Antzutkin ON. Amyloidosis of Alzheimer's A peptides: solid-state nuclear magnetic resonance, electron paramagnetic resonance, transmission electron microscopy, scanning transmission electron microscopy and atomic force microscopy studies. Magn Reson Chem 2004;42:231-46.
54. Huang X, Atwood CS, Moir RD, et al. Zinc-induced Alzheimer's Abeta1-40 aggregation is mediated by conformational factors. J Biol Chem 1997;272: 26464-70.
55. Narayanan S, Reif B. Characterization of chemical exchange between soluble and aggregated states of beta-amyloid by solution-state NMR upon variation of salt conditions. Biochemistry 2005;44:1444-52.
56. Ohtsu H, Shimazaki Y, Odani A, et al. Synthesis and characterization of imidazolate-bridged dinuclear complexes as active site models of Cu, Zn-SOD. J Am Chem Soc 2000;122:5733-41.
57. Shivers BD, Hilbich C, Multhaup G, et al. Alzheimers-disease amyloidogenic glycoprotein expression pattern in rat-brain suggests a role in cell contact. EMBO J 1988;7:1365-70.
58. Karr JW, Akintoye H, Kaupp LJ, Szalai VA. N-Terminal deletions modify the Cu2+ binding site in amyloid-beta. Biochemistry 2005 12;44:5478-87
59. Karr JW, Kaupp LJ, Szalai VA. Amyloid-beta binds Cu2+ in a mononuclear metal ion binding site J Am Chem Soc 2004 20;126:13534-8
60. Roher AE, Chaney MO, Kuo YM, et al. Morphology and toxicity of Abeta-(1-42) dimer derived from neuritic and vascular amyloid deposits of Alzheimer's disease. J Biol Chem 1996;271:20631-5.
61. Walsh DM, Klyubin I, Fadeeva JV, et al. Naturally secreted oligomers of amyloid beta protein potently inhibit hippocampal long-term potentiation in vivo. Nature 2002 ;416:535-9.
62. Roher AE, Chaney MO, Kuo YM, et al. Morphology and toxicity of Abeta-(1-42) dimer derived from neuritic and vascular amyloid deposits of Alzheimer's disease. J Biol Chem 1996;271:20631-5.
63. Cleary JP, Walsh DM, Hofmeister JJ, et al. Natural oligomers of the amyloid-beta protein specifically disrupt cognitive function. Nat Neurosci 2005;8:79-84
64. Karr JW, Akintoye H, Kaupp LJ, Szalai VA. Copper is implicated in the in vitro formation and toxicity of Alzheimer's disease amyloid plaques containing the beta-amyloid (A-beta) peptide. Proc Natl Acad Sci USA 2003;100:11934-40
65. Huang X, Atwood CS, Hartshorn MA, et al. The A beta peptide of Alzheimer's disease directly produces hydrogen peroxide through metal ion reduction. Biochemistry 1999;38:7609-16
66. Tabner BJ, Turnbull S, El-Agnaf OM, et al. Formation of hydrogen peroxide and hydroxyl radicals from A(beta) and alpha-synuclein as a possible mechanism of cell death in Alzheimer's disease and

Parkinson's disease. Free Radic Biol Med 2002;32:1076-83.
67. Atwood CS, Perry G, Zeng H, et al. Copper mediates dityrosine cross-linking of Alzheimer's amyloid-beta. Biochemistry 2004;43:560-68.
68. Schoneich C, Williams TD. Cu(II)-catalyzed oxidation of beta-amyloid peptide targets His13 and His14 over His6: Detection of 2-Oxo-histidine by HPLC-MS/MS. Chem Res Toxicol 2002;15:717-22.
69. Gunther MR, Peters, JA, Sivaneri MK. Histidinyl radical formation in the self-peroxidation reaction of bovine copper-zinc superoxide dismutase. J Biol Chem 2002;277:9160–66
70. Alvarez B, Demicheli V, Durán R, Trujillo M, et al. Inactivation of human Cu,Zn superoxide dismutase by peroxynitrite and formation of histidinyl radical Free Radic Biol Med 2004;37: 813–22.
71. Barnham KJ, Haeffner F, Ciccotosto GD, et al. Tyrosine gated electron transfer is key to the toxic mechanism of Alzheimer's disease β-amyloid. FASEB J 2004;18:1427-9.
72. Whittaker, JW. Free radical catalysis by galactose oxidase. Chem Rev 2003;103:2347-63.
73. Cukier RI, Nocera DG. Proton-coupled electron transfer. Annu Rev Phys Chem 1998;49:337-69.
74. Davies MJ, Hawkins CL. EPR spin trapping of protein radicals. Free Radic Biol Med 2004;36:1072-86.
75. Choo-Smith LP, Surewicz WK. The interaction between Alzheimer amyloid beta(1-40) peptide and ganglioside GM1-containing membranes. FEBS Lett 1997;402:95-98.
76. McLaurin Jo-A, Franklin T, Fraser PE, et al. Structural transitions associated with the interaction of Alzheimer β–amyloid peptides with gangliosides. J Biol Chem 1998;273:4506-15.
77. Ji S-R, Wu Y, Sui S-F. Cholesterol is an important factor affecting the membrane insertion of beta-amyloid peptide (A beta 1-40), which may potentially inhibit the fibril formation. J Biol Chem 2002;277:6273-79.
78. Del Angel VD, Dupuis F, Mornon J-P, et al. Viral fusion peptides and identification of membrane-interacting segments. Biochim Biophys Res Commun 2002;293:1153-60.
79. Curtain CC, Ali FE, Smith DG, et al. Metal ions, pH, and cholesterol regulate the interactions of Alzheimer's disease amyloid-β peptide with membrane lipid. J Biol Chem 2003;278:2977-82.
80. Marsh D, Horváth LI Structure, dynamics and composition of the lipid-protein interface. Perspectives from spin-labelling. Biochim Biophys Acta 1998;1376:267-96.
81. Lin H, Bhatia R, Lal R. Amyloid beta protein forms ion channels: implications for Alzheimer's disease pathophysiology. FASEB J 2001;15: 2433-44.
82. Horváth LI, Brophy PJ, Marsh D. Exchange rates at the lipid-protein interface of myelin proteolipid protein studied by spin-label electron spin resonance. Biochemistry 1988;27:46-52.
83. Ulrich AS, Tichelaar W, Förster G, et al. Ultrastructural characterization of peptide-induced membrane fusion and peptide self-assembly in the lipid bilayer. Biophys J 1999;77:829-41
84. Kuo YM, Kokjohn TA, Beach TG, et al. Comparative analysis of amyloid-beta chemical structure and amyloid plaque morphology of transgenic mouse and Alzheimer's disease brains. J Biol Chem 2001;276: 12991-98.
85. Naslund J, Schierhorn A, Hellman U, et al. Relative abundance of Alzheimer A beta amyloid peptide variants in Alzheimer's disease and normal aging. Proc Natl Acad Sci USA 1994;91:8378-82.
86. Watson AA, Fairlie DP, Craik DJ. Solution structure of methionine oxidized amyloid beta-peptide (1-40). Does oxidation affect conformational switching? Biochemistry 1998;37:12700-06.
87. Palmblad M, Westlind-Danielsson A, Bergquist J. Oxidation of methionine 35 attenuates formation of amyloid beta-peptide 1-40 oligomers. J Biol Chem 2002;277:19506-10.
88. Hou L, Kang I, Marchant RE, et al. Methionine 35 oxidation reduces fibril assembly of the amyloid A-beta-(1-42) peptide of Alzheimer's disease. J Biol Chem 2002;277:40173-76.
89. Barnham KJ, Ciccotosto GD, Tickler AK, et al. Neurotoxic, redox-competent Alzheimer's beta-amyloid is released from lipid membrane by methionine oxidation. J Biol Chem 2003;278:42959-65.
90. Ali FE, Separovic F, Barrow CJ, et al. Methionine regulates copper/hydrogen peroxide oxidation products of Abeta. J Pept Sci 2005;11:353-60.
91. Boas JF. Electron paramagnetic resonance of copper proteins. In: Lontie R, editor. Copper Proteins and Copper Enzymes. Boca Raton, FL: CRC Press, 1984: 5-62.
92. Ciccotosto GD, Tew D, Curtain CC, et al. Enhanced toxicity and cellular binding of a modified amyloid beta peptide with a methionine to valine substitution. J Biol Chem 2004;279:42528-34.

9
Cholesterol and Alzheimer's Disease

Joanna M. Cordy and Benjamin Wolozin

9.1 Introduction

Recent studies indicate that cholesterol plays an important part in the regulation of amyloid-β peptide (Aβ) production, with high cholesterol levels being linked to increased Aβ generation and deposition. The mechanisms underlying the role(s) of cholesterol are not fully understood at present, but from the evidence currently available, it appears that there are many different ways in which abnormalities in cholesterol metabolism can affect the development of Alzheimer's disease (AD). Polymorphisms in genes involved in cholesterol catabolism and transport have been associated with an increased level of Aβ and are therefore potential risk factors for the disease. The best known of these genes is the apolipoprotein E gene (apoE), which encodes a protein involved in cholesterol transport. The existence of a particular allele of apoE, ε4, is the major genetic risk factor known for late-onset AD. Other genes implicated include cholesterol 24-hydroxylase (Cyp46), the LDL receptor related protein (LRP), the cholesterol transporters ABCA1 and ABCA2, acyl-CoA:cholesterol acetyl transferase (ACAT), and the LDL receptor (LDLR).

In addition to this genetic evidence, epidemiological and biochemical findings also demonstrate relationships between cholesterol and AD and/or Aβ. The prevalence of AD has been shown to be reduced among people taking 3-hydroxy-3-methylglutaryl (HMG)-CoA reductase inhibitors, such as lovastatin, which inhibit *de novo* cholesterol synthesis, while levels of serum low-density lipoprotein (LDL) and total cholesterol have been reported to correlate with Aβ levels in the AD brain. These studies are supported by work on transgenic mice overexpressing the amyloid precursor protein (APP), demonstrating that increased dietary cholesterol results in higher levels of Aβ, and also by experiments showing that cholesterol loading or depletion of cells in culture leads to an increase or decrease, respectively, in Aβ production.

In this chapter, all the evidence described above will be discussed in more detail to provide a picture of our current understanding of the ways in which cholesterol may affect the production of Aβ and the development of AD.

9.2 Cholesterol Metabolism

9.2.1 Synthesis

Cholesterol performs many important functions within cells, particularly as a structural component of cell membranes and as a precursor for the generation of steroid hormones and bile salts. It is vital, however, that a balance is maintained between cholesterol synthesis, uptake, and catabolism, as an excess of cholesterol is a major risk factor for the development of atherosclerosis.

Within the body, cholesterol is only synthesized in the liver and brain and is the product of a complex multi-enzyme pathway. This pathway begins with the condensation of acetyl-CoA with acetoacetyl-CoA to form HMG-CoA. This is then converted to mevalonate by HMG-CoA reductase, in the rate-limiting step of the process [1]. A cascade of other reactions then occurs to produce cholesterol (Fig. 9.1), and this pathway generates

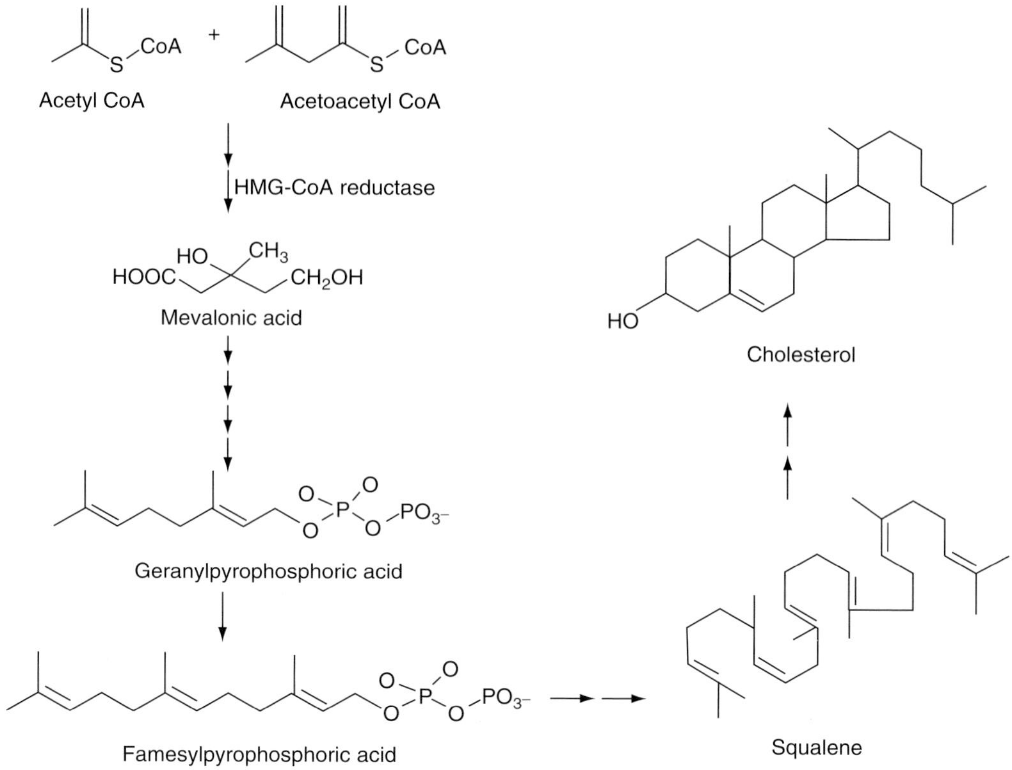

FIGURE 9.1. The biosynthesis of cholesterol. The synthesis of cholesterol begins with the condensation of acetyl-CoA with acetoacetyl-CoA, to form HMG-CoA, which is then converted to mevalonate. A cascade of other reactions occurs to produce cholesterol and many biologically important intermediate molecules.

many intermediate molecules that have important biological functions. For example, dolichol, which is involved in synthesis of the oligosaccharide chains of glycoproteins, and ubiquinone, a component of the electron transport chain, are both synthesized from farnesyl pyrophosphate, a cholesterol intermediate.

After synthesis in the endoplasmic reticulum (ER), cholesterol builds up in membranes through the Golgi apparatus to the plasma membrane, which has the highest cholesterol content. Within these membranes, the distribution of cholesterol is not uniform, but instead it clusters in regions known as lipid rafts, which are also enriched in glycosphingolipids and particular proteins [2–4]. These domains will be discussed in more detail below.

In addition to the *de novo* synthesis of cholesterol by the brain and liver, dietary cholesterol can also be absorbed from the gut. The identity of the transporter(s) involved in this process is elusive, but one protein recently shown to have a critical role is the Niemann-Pick C1 like 1 (NPC1L1) protein [5]. This protein shows ~50% homology to NPC1, the protein that is defective in the cholesterol storage disease Niemann-Pick type C [6].

9.2.2 Transport and Uptake

Cholesterol is insoluble in the blood and therefore must be transported to and from cells by carriers known as lipoproteins. Absorbed dietary cholesterol in the intestine is assembled into chylomicrons, which then enter the bloodstream, while cholesterol from the liver is released in very-low-density lipoproteins (VLDL). These particles contain triacylglycerols, phospholipids, and proteins known as apolipoproteins in addition to having cholesterol. VLDL, LDL, and other lipoproteins contain varying ratios of protein to lipid and also different species of apolipoproteins. ApoB, which is present in VLDL and LDL, is the most important lipoprotein in the periphery and is responsible for binding to the LDL receptor. ApoD, E, and J are also important, although animals with defects in the apoD gene show normal cholesterol levels, while cholesterol uptake is impaired if apoB or E are knocked out [7–9].

After their synthesis in the liver or intestine, both VLDL and chylomicrons are converted, through the loss of triacylglycerol, to LDL, which is the primary carrier of plasma cholesterol to extrahepatic tissues. LDL is then taken up into cells via interaction with the LDL receptor, and cholesterol is released into the cells after degradation of the LDL particle by lysosomal enzymes.

9.2.3 Storage and Catabolism

Cholesterol within the cell can either be stored as free cholesterol (FC) in the membrane or it can be converted to cholesteryl esters (CEs) and stored in cytoplasmic droplets. An equilibrium exists between these two pools of cholesterol controlled by acyl-CoA:cholesterol acyltransferase (ACAT), which catalyzes the formation of CEs from FC. ACAT is activated by a rise in FC levels, and conversely, low FC levels promote the hydrolysis of CEs back to FC.

An alternative route of elimination of FC from cells is oxidation. In the periphery, the majority of cholesterol is oxidized at the 7α position (Fig. 9.2) and is then glycosylated and secreted as bile acids. Oxidation can also occur at the 24 or 27 positions by the mitochondrial enzymes cholesterol 24 or 27 hydroxylase (Cyp46 and Cyp27, respectively). This generates oxysterols, which diffuse from cells into the extracellular fluids and vasculature. Oxysterols play an important role in cholesterol biology by acting as transcriptional regulators. They bind to and activate the liver X receptor (LXR), which then can dimerize with the retinoic acid receptor or retinoic X receptor to stimulate transcription of genes important in cholesterol metabolism. Genes regulated by LXR include apoE [10] and the ABCA1 transporter [11].

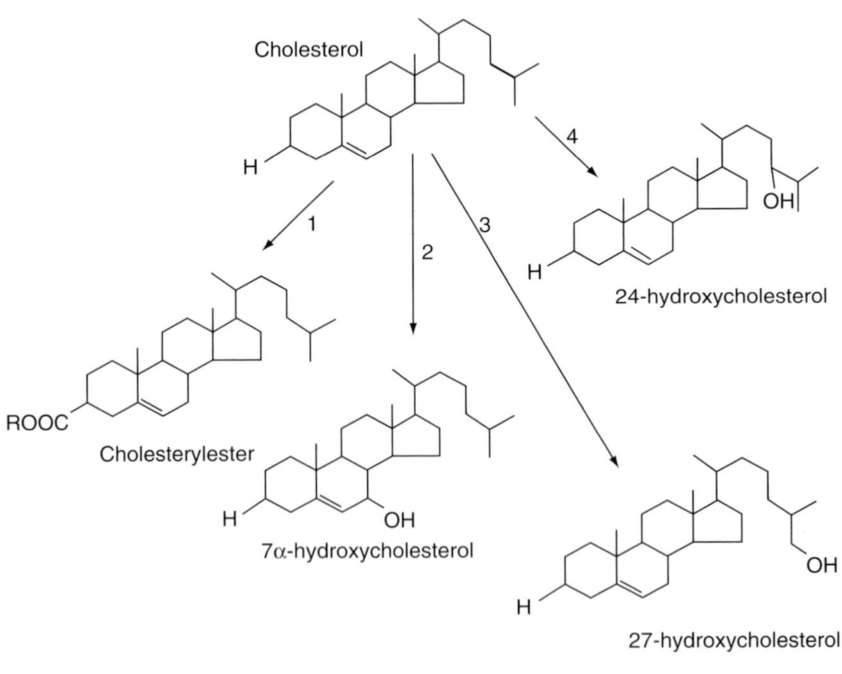

FIGURE 9.2. Cholesterol catabolism. Cholesterol can be converted into cholesteryl esters by the action of ACAT (1), or alternatively it can be converted into oxysterols by oxidation at the 7α position by cholesterol 7α hydroxylase (2), or the 24 or 27 position by Cyp 46 (4) or Cyp 27 (3), respectively.

9.2.4 Cholesterol Metabolism in the Brain

The brain contains approximately 20% of the total cholesterol in the body, despite only accounting for 2% of body mass. The majority of this cholesterol is found in myelin membranes, with some also present in neurons and glial cells. Compared with the periphery, the turnover of cholesterol in the human brain is very slow, with a half-life of almost a year, as opposed to a matter of hours in plasma, and this is largely due to the stability of the myelin sheaths. Most brain cholesterol is synthesized *in situ*, and production of cholesterol in the brain is largely independent of plasma cholesterol levels. The extent of regulatory separation between brain and periphery, though, might differ depending on the species or conditions. Mice fed a high-lipid diet exhibit increased cholesterol levels in the CNS as well as in plasma [12]. However, changes in dietary cholesterol do not appear to affect apoE levels [13].

Cholesterol metabolism in the brain differs from that in the periphery. Cholesterol is mainly generated in glia and then transported to neurons. After synthesis and secretion from glia via the ABCA1 transporter, cholesterol is packaged into lipoprotein particles resembling HDL. These HDL particles differ from those in the periphery in that they contain apoE but no apoB, as occurs in the periphery. HDL is taken up into neurons through recognition of ApoE by a variety of lipoprotein receptors including the LDL receptor (LDLR), the LDL receptor related protein (LRP), the apoE receptor, as well as other lipoprotein receptors. Elimination of cholesterol from the brain occurs mainly via oxidation at the 24 and 27 positions to produce a class of compounds termed oxysterols, rather than being oxidized at the 7α position by Cyp7a to produce bile acids, as occurs in the periphery. The two oxysterols 24(S) hydroxycholesterol and 27 hydroxycholesterol are produced by enzymes Cyp46 and Cyp27, respectively. As mentioned above, 24(S) hydroxycholesterol is predominantly made in the brain, and within the brain, predominantly made by neurons. In contrast, 27 hydroxycholesterol is produced by many cells including neurons and oligodendrocytes [14]. Oxysterols are far more soluble than cholesterol and diffuse across the blood-brain barrier (BBB) where they enter the peripheral circulation for excretion. Although the enzymes that represent the first step in bile acid production, Cyp7a, is present in the brain, bile acids are not a major mechanism of cholesterol catabolism in the CNS [15].

9.3 The Genetics of AD and Cholesterol Metabolism

9.3.1 ApoE

Three genes associated with early-onset AD have been identified to date. These are the APP gene on chromosome 21 [16–18] and the genes encoding presenilin 1 and 2 on chromosomes 14 and 1, respectively [19–21]. The only gene, however, that has been unequivocally linked to late-onset AD is the ApoE gene [22]. This gene, found on chromosome 19, has three common variants, ε2, ε3, and ε4, and it is the presence of the ε4 allele (apoE4) that is the most potent known risk factor for late-onset AD, after age. The lifetime risk of AD for an individual without the ε4 allele is approx. 9%, whereas the presence of at least one ε4 allele is believed to increase the risk to approximately 29% [23] and also to lower the average age of onset of the disease [22, 24]. Conversely, the presence of the ε2 allele delays the onset of the disease and is thought to have a protective effect [24].

The strongest hypothesis explaining how apoE impacts on AD derives from the effects of apoE on Aβ deposition and clearance. ApoE is believed to act as a chaperone protein and accelerate the formation of Aβ fibrils [25], with the apoE4 isoform being most efficient at promoting fibrillogenesis in vitro (Fig. 9.3) [26]. Results obtained from studies with transgenic mice also support these data, showing that mice expressing apoE4 and APP have accelerated Aβ deposition compared with mice expressing other apoE isoforms or no apoE [27, 28]. More recently, experimental studies demonstrate that blocking the interaction of Aβ and apoE using a synthetic peptide not only reduces Aβ fibril formation in vitro but also reduces Aβ load and plaque formation in a mouse model of AD [29]. These studies provide experimental evidence that the ability of apoE4 to accelerate Aa aggregation and deposition represents an important mechanism by which apo E4 accelerates the progression of

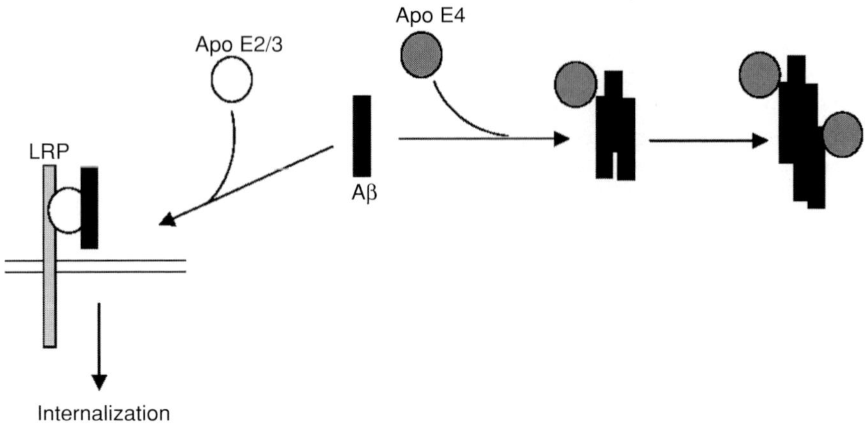

FIGURE 9.3. Possible roles of apoE isoforms in amyloid metabolism. The apoE4 isoform accelerates the aggregation and deposition of Aβ fibrils, whereas the apo E2 and E3 isoforms promote clearance of Aβ via LRP.

AD. ApoE is also involved in Aβ clearance, in an isoform specific manner, with apo E2 and E3, but not E4 being important for the removal of Aβ from the extracellular space (Fig. 9.3) [30].

The importance of apoE in cholesterol metabolism, though, remains a striking phenomenon that raises the possibility that the presence of different apoE isoforms may alter cholesterol homeostasis in the brain and thereby influence the progression of AD. ApoE genotype is known to correlate with plasma cholesterol levels, with apoE4 being associated with the highest LDL cholesterol levels [31], a believed risk factor for AD [32, 33]. However, whether the association between apoE4 and AD derives from its effect on cholesterol metabolism remains a source of debate. Some studies suggest that the effects of apolipoprotein E4 on AD are independent of cholesterol while others show a relationship between cholesterol, apoE, and AD [32, 34–36]. In the periphery, apoE4 appears to associate predominantly with VLDL particles, which contain a high percentage of cholesterol, whereas apoE2 prefers to associate with the less cholesterol-rich high-density lipoprotein particles [37–39]. It is not known whether different apoE isoforms associate with different lipid particles in the brain, but the occurrence of a similar effect could alter cholesterol metabolism and help to explain the increased risk of AD associated with apoE4.

9.3.2 Other Genes Linked to Late-Onset AD and Cholesterol Metabolism

9.3.2.1 Cyp46

Cholesterol 24-hydroxylase, encoded by the Cyp46 gene on chromosome 14, is expressed almost exclusively in the brain, with only very low levels of mRNA found in other tissues such as liver and testis [40]. The enzyme is a member of the cytochrome P450 family and is responsible for the catabolism of nearly all CNS cholesterol to 24S-hydroxycholesterol. Knockout of the gene in mice results in a decrease of more than 98% in the level of 24S-hydroxycholesterol in the brain, however total brain cholesterol remains unchanged, perhaps because there is a compensatory downregulation of *de novo* cholesterol synthesis by approximately 40% [41]. Not surprisingly, knockout of Cyp46 produces no appreciable differences in the levels of peripheral cholesterol and lipoproteins in these mice.

In AD, and in mild cognitive impairment, the levels of 24S-hydroxycholesterol in cerebral spinal fluid are elevated [42], however other studies suggest that plasma levels are decreased or unchanged [43–45]. The reason for the discrepancy might lie in the dependence of plasma 24(S) hydroxycholesterol levels on a variety of factors including disease state, cerebral injury, brain size, cerebro-vascular blood flow, and so forth. The integration of all of these factors might produce effects that counteract each other and limit the linkage between serum 24(S) hydroxycholesterol and Alzheimer's disease. Our own studies demonstrate that Cyp46 is selectively expressed around neuritic plaques, perhaps reflecting the need of neurons to remove excess cholesterol from degenerating neuritis [14]. Recently, a number of studies have investigated the link between polymorphisms in the Cyp46 gene and late-onset AD, with varied results. Two different intronic polymorphisms with potential association with AD, Aβ levels and/or phosphorylated tau have been identified [46, 47], and these results have since been corroborated in other populations [48, 49]. The genotyping results are ambiguous, though, because other studies have failed to detect any associations between Cyp46 polymorphisms and AD [50, 51]. These contradictory findings leave the role of *Cyp46* in AD development controversial.

9.3.2.2 ABCA1

The adenosine triphosphate–binding cassette transporter ABCA1 functions to secrete cholesterol from the cell and is an important regulator of cholesterol metabolism. The gene encoding this protein, on chromosome 9, is another gene with a potential link to AD. In the periphery, ABCA1 transports free cholesterol out of cells, and lack of this protein results in reduced plasma HDL levels and an increased risk of cardiovascular disease [52–54]. Overexpression of the transporter in mice leads to opposite effects [55, 56]. In the brain, ABCA1 is also important in cholesterol trafficking, and it has been shown that its expression in cerebral endothelial cells can be stimulated by 24S-hydroxycholesterol, suggesting a role in the removal of excess brain cholesterol [57].

An increasing number of studies suggest that ABC proteins are important to the pathophysiology of AD. A polymorphism in the ABCA1 gene, already known to be linked to a modified risk of coronary heart disease [58, 59], has recently been shown to delay onset of AD by 1.7 years [60], and a larger study has provided further evidence that variants of ABCA1 alter the risk of developing AD [61]. ABCA1 has also been shown to directly alter production of Aβ. Transfectng ABCA1 or inducting ABCA1 via LXR reduces Aβ generation, presumably by lowering cholesterol levels [62, 63]. Recently, a second ABC transporter that is expressed in the brain has been cloned. ABCA2 is expressed in the endolysosomal compartment, primarily in oligodendrocytes, but also in the cortex [64]. When expressed in cell culture ABCA2 strongly regulates formation of cholesterol esters and expression of other proteins implicated in cholesterol metabolism, such as the LDLR. A polymorphism in ABCA2 strongly increases the risk of AD, with a LOD score of 3.5 [65]. The association of two different ABC transporters with AD, combined with the direct evidence that these proteins modulate Aβ metabolism, suggests that these proteins could be particularly relevant to AD.

9.3.2.3 ACAT

Proteins like ABCA2 and Cyp46/LXR modulate many other proteins important to cholesterol catabolism or transport. One of these proteins is acyl-Coenzyme A:cholesterol acyl transferase (ACAT), which is a protein that converts cholesterol to cholesterol esters, which are highly insoluble and are thought to be used for storage. ACAT could be particularly important for AD because pharmacological inhibitors of ACAT are available, and these inhibitors have recently been shown to reduce Aβ production and decrease amyloid load in a transgenic mouse model of AD [66]. Because related compounds have also been investigated in human clinical trials and found to be safe, these compounds hold great promise for therapy of AD.

9.3.3 LRP and LDLR

LRP is a member of the LDL receptor family and, in brain, is expressed predominantly on neurons and reactive astrocytes [67, 68]. The main ligand for LRP in the brain is apoE, although it can also bind a number of different proteins, including LDLR, urokinase-type plasminogen activator, and lactoferrin [69]. The fact that LRP is an important neuronal receptor for apoE, which has long been implicated in AD, suggests that this protein may also be important in the disease. In addition, LRP and many of its ligands are found in senile plaques [70], suggesting that the function of LRP could be impaired in AD, resulting in this buildup. Another interesting link between LRP and AD is that it can bind APP and regulate its internalization and processing [71, 72], thereby potentially affecting production of Aβ, as well as its clearance via apoE.

More evidence for a role for LRP in AD comes from genetic association studies. A polymorphism in exon 3 of the gene has been identified, which is linked to reduced AD susceptibility and decreased amyloid burden [73]. This has since been corroborated by other studies [74–77]. In addition, another polymorphism in the LRP gene has also been identified and linked to AD [78] providing further genetic evidence for a connection between LRP and AD. A meta-analysis of LRP polymorphisms has recently been done at the Alzgene website (http://www.alzforum.org/res/com/gen/alzgene/default.asp), which suggests a slight increased risk of AD associated with the C allele of the rs1799986 polymorphism. However, the main message provided by the meta-analysis is that the effect of this polymorphism, if real, is much, much smaller than the effect of apoE4.

9.3.4 α2M

One of the ligands for LRP is α-2-macroglobulin (α2M), a protein capable of binding Aβ with high specificity [79, 80] and preventing its fibrillization. α2M is found in neuritic plaques in AD brain [81, 82] and it may play a role in Aβ clearance via LRP, as it is known to be able to bind other ligands and target them for internalization and degradation [83]. The gene encoding α2M has also been identified as a potential risk factor for AD in some studies, but the overwhelming majority of studies have failed to observe a linkage [84–86].

9.4 Cholesterol and APP Processing

9.4.1 In Vitro Studies

A large number of experiments performed on cells in culture demonstrate that cellular processing of APP and production of Aβ can be modulated by cholesterol metabolism (Table 9.1). Klein and colleagues were the first investigators to examine this issue. They added cholesterol complexed with methyl-β-cyclodextrin to the cell line HEK and demonstrated that the cholesterol decreased APP secretion [87]. Next, Simons et al. [88] used a combination of an HMG-CoA reductase inhibitor and methyl-β cyclodextrin to deplete cholesterol levels in hippocampal neurons by 70%. This caused a dramatic decrease in production of Aβ. Later studies using similar treatments confirmed these results [89, 90]. The system appears to be reciprocal with respect to cholesterol levels because adding exogenous cholesterol to cells in culture upregulates Aβ production [89]. The mechanism underlying the regulation appears to depend in part on activity of β-secretase, because cholesterol depletion reduces CTFβ [88, 90]. Regulation of APP processing by cholesterol is not limited to β-secretase activity; it appears to occur on multiple levels. For instance, α-secretase activity is also controlled by cholesterol, with low cholesterol levels stimulating production of sAPPα [91]. The third enzyme involved in APP processing, γ-secretase, could also be affected by cholesterol, as recent work has shown that disruptions in cholesterol trafficking cause a redistribution of the presenilins and an associated increase in Aβ generation [92, 93]. However, γ-secretase activity appears to be the least affected by cholesterol of all the enzymes regulating APP processing.

Cholesterol metabolism can also modulate APP processing through trafficking. There are many different pools of cholesterol, cholesteryl esters (CEs), or free cholesterol (FC) present in cells. In addition, APP processing also occurs in many different compartments. Modulation of particular enzymes in particular compartments or modulation of the distribution of APP among different vesicles can alter generation of Aβ and APPs. For instance, the enzyme responsible for controlling the interconversion of these cholesterol pools is the ER-resident enzyme ACAT, and it has been shown that the activity of this enzyme can regulate Aβ generation, suggesting that it may be the distribution of intracellular cholesterol that is important rather than the total amount [94]. This investigation by Puglielli and co-workers [94] showed that the level of Aβ was most closely correlated with cholesteryl ester levels, although they could not rule out the possibility that it may be the ratio of FC to CEs that is most important. It is likely that other types of cholesterol-related modulation also act by changing he vesicular distribution of components that affect APP processing.

TABLE 9.1. Summary of the effects of cholesterol modulation on amyloid precursor protein (APP) processing and amyloid-β peptide (Aβ) production.

In vitro / in vivo	Modulation of cholesterol		Effects	Reference
In vitro	↑ Cholesterol	Exogenous cholesterol added	↓ sAPPα production	87
		Exogenous cholesterol added	↑ Aβ production	89
	↓ Cholesterol	Cholesterol depleted using statin	↓ Aβ production	88
		Cholesterol depleted using statin	↓ β-secretase cleavage products	89
		Cholesterol depleted using statin	↓ Aβ production	90
		Cholesterol depleted using statin or methyl-β-cyclodextrin	↑ sAPPα production ↓ Aβ production	91
In vivo	↑ Cholesterol	Primates fed high-fat diet	↑ Aβ deposition	108
		APP Tg mice fed high-fat diet	↑ Aβ deposition Learning impairments	109
		APP Tg mice fed high-fat diet	↑ Aβ and CTFβ production ↓ sAPPα production	12
		APP Tg mice fed high-fat diet	↑ Aβ deposition	110
		APP Tg mice fed high-cholesterol diet	↓ Aβ and sAPPβ production ↓ sAPPα production	**112**
		APP Tg mice fed high-cholesterol diet	↓ Aβ deposition ↓ sAPPα production ↑ AICD	**113**
	↓ Cholesterol	Guinea pigs treated with simvastatin	↓ Aβ deposition	90
		APP Tg mice treated with cholesterol-lowering drug	↓ Aβ and CTFβ productionn ↑ sAPPα production	111
		APP Tg mice treated with lovastatin	↑ Aβ deposition in female mice No change in male mice	138

The *in vitro* studies suggest that increasing cholesterol levels results in an upregulation of amyloidogenic APP processing, whereas lowering cholesterol levels has the opposite effect. The majority of results from in vivo studies show the same pattern, however there are some reports (highlighted) that contradict this trend.

9.4.2 APP Processing and Lipid Rafts

A key to understanding how cholesterol might modulate APP processing lies in the concept of lipid rafts. Lipid rafts are small domains within cell membranes consisting of sphingolipids in the outer leaflet of the bilayer and phospholipids with saturated fatty acid chains in the inner leaflet, tightly packed together with cholesterol (Fig. 9.4). The surrounding bilayer is less tightly packed due to the unsaturated nature of the phospholipid hydrocarbon chains, with the result that the rafts form ordered, although still fluid, platforms within this liquid-disordered phase (for reviews, see Refs. 2–4, 95). As well as containing particular classes of lipids, rafts can bind certain proteins. Different proteins are found to be associated with raft domains to varying extents, for example proteins with a glycosylphosphatidyl inositol (GPI) membrane anchor and doubly acylated proteins such as Src family tyrosine kinases tend to reside in rafts constitutively [96], whereas many proteins are able to move in and out of rafts depending on ligand-binding, oligomerization, or palmitoylation [97, 98]. Because of this, the movement of proteins in and out of rafts, and their associations within these domains, can be tightly controlled.

Lipid rafts have been hypothesized to be involved in APP processing and could therefore help to explain how the connection between cholesterol and AD occurs [99]. Several proteins relevant to Aβ production have been shown to be present in raft domains including a small proportion of APP [100–103], the β-secretase BACE (β-site APP cleaving enzyme) [104, 105], the presenilins [101, 103, 106], and Aβ itself [101]. These results, which were obtained from several different cell-lines and from samples of human, mouse, and rat brain, prompted the hypothesis that amyloidogenic processing of APP may take place

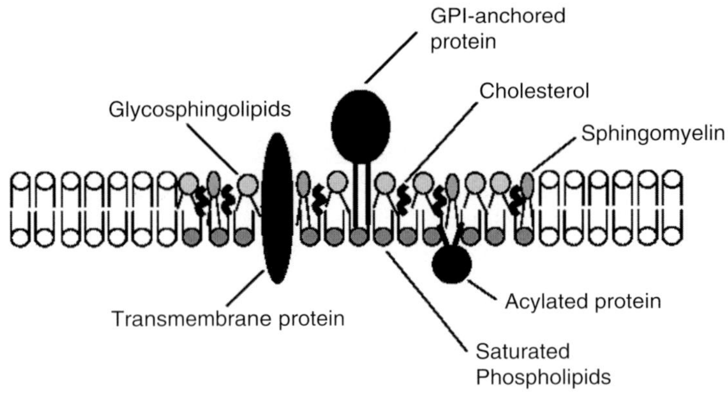

FIGURE 9.4. Schematic diagram of a lipid raft domain. The lipid raft is rich in cholesterol, sphingolipids, and sphingomyelin. Lipid-modified proteins such as acylated or GPI-anchored proteins tend to cluster in these regions, along with some transmembrane proteins.

within lipid rafts. The putative α-secretase ADAM10, however, is predominantly soluble after detergent extraction [91], leading to a model being proposed in which amyloidogenic and non-amyloidogenic processing of APP occur in separate membrane compartments [99]. The existence of two pools of APP within the cell membrane, one raft-localized and one present in phospholipid domains [100, 101], fits in with this theory by allowing APP access to both α-secretase and β- and γ-secretases. According to this model of APP cleavage, a high concentration of membrane cholesterol would therefore favor Aβ production, whereas a reduced cholesterol level would favor the non-amyloidogenic α-secretase pathway.

The studies described above, demonstrating that depletion of cellular cholesterol levels results in inhibition of Aβ production [88–90], support this hypothesis, as cholesterol removal disrupts lipid raft domains. Further evidence that amyloidogenic APP processing, particularly by BACE, occurs in lipid rafts comes from recent work showing that antibody cross-linking of APP and BACE causes them to copatch with known raft marker proteins, and that this dramatically increases production of Aβ [107]. In addition, the direct dependence of BACE activity on lipid rafts has been demonstrated by targeting BACE exclusively to these domains using a GPI-anchor [104]. The production of Aβ and sAPPβ was increased significantly by targeting BACE to lipid rafts, confirming that this environment is favorable for the amyloidogenic processing of APP [104].

9.4.3 In Vivo Studies

A number of studies suggest that cholesterol also modulates APP processing in vivo (Table 9.1), but when interpreting the studies, one must consider the added complexity of the in vivo situation. When analyzing in vivo and human data, one must distinguish between plasma cholesterol and cerebral cholesterol because the amount of cross-talk between the two pools of cholesterol and the mechanism of cross-talk is unclear. One must also distinguish between the type of animal being investigated because lipid metabolism differs among species such as mice, guinea-pigs, and humans. For instance, mice generally have high levels of LDL while humans tend to have higher levels of HDL.

Despite these differences, several groups have shown that changes in cholesterol metabolism induced by pharmacological means (e.g., statins) or by feeding alter cholesterol metabolism. This has been shown in primates [108] and transgenic mouse models of AD [12, 109, 110]. For example, Refolo et al. [12] showed that both β-cleaved C-terminal APP fragments (CTFβ) and Aβ were increased in the CNS of mice fed a high-cholesterol diet, whereas the production of α-cleaved soluble APP (sAPPα) was decreased, suggesting that cholesterol was regulating APP processing. Other in vivo studies have demonstrated that treatment of guinea-pigs or transgenic mice with cholesterol-lowering drugs resulted in lowered levels of Aβ

[90, 111] and also increased sAPPα and decreased CTFβ production [111]. Each of these studies presents cogent examples of the impact of cholesterol metabolism on APP processing in vivo.

Although the results from these in vivo studies indicate that hypercholesterolemia leads to an increase in the amyloidogenic processing of APP, whereas reduced cholesterol level has the opposite effect, some studies have observed contradictory evidence. Howland et al. [112] examined the effect of a high-cholesterol diet on a different transgenic mouse model of AD and found that levels of sAPPα, sAPPβ, and Aβ were all reduced. More recently, another study has shown a similar effect [113]; the reasons for these apparent discrepancies are not clear. Possible differences that could contribute to these conflicting results could lie in the transgenes present in the mouse models, the genetic backgrounds of the mouse models, variability in the ages, or differences in the sex of the animals studied. Interestingly, the study by George and colleagues [113] demonstrated that production of the APP intracellular domain (AICD) is increased in mice fed a high-cholesterol diet. This fragment appears to act as a transcriptional activator [114, 115] and can induce apoptosis in neurons [116], leading to the possibility that cholesterol could affect AD progression via the regulation of AICD production [113].

9.4.4 Aβ Aggregation and Toxicity

Cholesterol also appears to be important for the aggregation and toxicity of Aβ. Aggregated or fibrillar Aβ is widely believed to be more toxic to neurons than the monomeric peptide [117], and there is evidence to suggest that polymerization of Aβ is seeded by a species of the peptide that is tightly bound to GM1 ganglioside (GM1-Aβ) [118]. GM1-Aβ has been shown to accelerate amyloid fibril formation in vitro [119, 120], and the formation of this species appears to be sensitive to the lipid environment, with cholesterol being an important factor [121]. Kakio et al. [122] demonstrated that Aβ bound preferentially to clusters of GM1 molecules and that these clusters formed in cholesterol-rich environments such as lipid rafts, and this is supported by a study reporting that depletion of cellular cholesterol can protect cells from the toxic effects of Aβ [123]. More recently, Subasinghe and colleagues [124] have shown that binding of Aβ to membrane lipids is important for toxicity of the peptide and that both membrane-binding and toxicity were reduced by the removal of cholesterol.

9.5 Epidemiological and Clinical Evidence

9.5.1 Cholesterol Levels and AD

Despite the strong genetic and biochemical evidence that points to a strong connection between cholesterol and AD, epidemiological evidence linking plasma levels of cholesterol and lipoproteins with the development of AD is conflicting. Some studies have demonstrated a link between cholesterol level, particularly in mid-life, and AD. For example, Pappolla and colleagues [125] found that there was a strong correlation between total cholesterol level and amyloid deposition in subjects aged between 40 and 55 years, but this correlation became weaker as the age of the subjects increased. In another study, Finnish men who had displayed a high serum cholesterol level at age 40–59 were found to be three times more likely to have developed AD 30 years later [35]. Kivipelto et al. [126, 127] also demonstrated a correlation between mid-life cholesterol level and the risk of developing AD later in life. These results, and the fact that in the study by Notkola et al. [35] the cholesterol level of men who developed AD decreased before the disease manifested itself, suggest that hypercholesterolemia in mid-life could be a risk factor for AD, while cholesterol level in later life shows less correlation with the disease. Kuo et al. [33], however, examined serum levels of LDL and HDL cholesterol at postmortem and found significantly higher LDL cholesterol and lower HDL cholesterol in AD patients than in control subjects.

In contrast with these studies, which have found correlations between cholesterol levels and AD, other investigations have failed to find such a connection. Tan et al. [128] looked at total serum cholesterol levels from participants in the Framingham study and found no association between average cholesterol level over a 30-year period and development of AD 10–20 years later. Another study investigating a wide variety of serum markers in neurodegenerative diseases also found no correlation between serum cholesterol and AD [129], although, interestingly, the levels of precursors to cholesterol synthesis appeared to be significantly different in AD patients compared with controls.

9.5.2 Use of Statins

An alternate approach to addressing the issue of cholesterol and AD is to shift the question from whether abnormal cholesterol metabolism increases the risk of AD to the question of whether modulating cholesterol metabolism can alter the incidence or progression of AD. HMG-CoA reductase inhibitors, known collectively as statins, were developed in the 1970s and have been widely used since the late 1980s to lower cholesterol levels in patients at risk of coronary heart disease. Examples of statins that are currently available include lovastatin (Mevacor, currently off patent), pravastatin (Pravacor), simvastatin (Zocor), rosuvastatin (Crestor), and atorvastatin (Lipitor). In 2000, two retrospective studies suggested that the prevalence of AD was reduced by approximately 70% among patients taking statins compared with control subjects [130, 131]. Similar studies have since corroborated these findings in different groups of patients [132, 133].

More variable results have been obtained by prospective studies examining the use of statins as potential therapeutic agents in AD. Simons et al. [134] observed a decrease in the CSF $A\beta_{40}$ levels of patients suffering from mild AD after treatment with simvastatin for 26 weeks, but this was not seen in patients with a more severe form of the disease. Cognitive decline appeared to be slowed in both groups compared with subjects receiving a placebo. Another small study of AD patients found that CSF levels of sAPPα and sAPPβ were decreased after a 12-week treatment with simvastatin, but $A\beta_{42}$ levels were unaltered [135]. Two larger studies, looking primarily at the cardiovascular benefits of longer term (3- to 5-year) statin treatment, found that cognitive decline was not prevented by statins [136, 137], however, a recent pilot study of the effects of atorvastatin, reported at the American Heart Association's Scientific Sessions 2004, has shown that it appears to slow mental decline and improve cognitive symptoms in AD patients (www.americanheart.org). These studies have used a variety of statins with differing lipophilicities, suggesting that the variable results cannot be explained by the ability of the drug to cross the blood-brain barrier (BBB). The reason for the mixed results obtained is unknown but have to do with the severity of AD or the cholesterol level in the patients examined or the methods used to test for cognitive function. Other clinical trials of statins in AD, such as the Cholesterol Lowering Agent to Slow Progression (CLASP) of AD Study, sponsored by the NIA, are currently in progress, so these should provide more information about the possible therapeutic benefits of these drugs.

9.6 Future Directions

Despite the current interest in determining the association between cholesterol and AD, there are still many crucial questions that need to be addressed before a complete picture of this complex relationship emerges. The effects of statins on Aβ production appear to be clear in cell culture, but the effects in vivo and the role of cholesterol in the pathogenesis of AD are by no means clear-cut, and if these drugs are to be used in the treatment of AD, many issues still need to be resolved. One important factor that has recently come to light is a possible gender-related difference in response to statin treatment. When male and female APP transgenic mice were treated with lovastatin, both groups showed the expected reduction in cholesterol levels, but female mice showed an increase in both Aβ production and plaque load [138]. No changes were seen in the male mice. These results suggest that it will be important to reexamine the results from other studies and trials involving statins, to take into account gender differences. Another issue that is currently being investigated is whether the neuroprotective effects of statins are due less to their role as inhibitors of cholesterol synthesis and more to other effects such as their anti-inflammatory properties [139, 140].

The fact that ageing leads to alterations in the lipid and cholesterol distribution within membranes could affect the number and stability of lipid rafts. Currently, however, no data exist regarding changes in raft number, size, or composition during aging or AD progression. If this issue could be addressed, the results would be valuable in assessing exactly how lipid rafts are involved in

APP processing. Unfortunately, native rafts are very difficult to study, as detergent isolation can cause individual rafts to coalesce [141] providing an inaccurate picture of the actual organization of rafts within the membrane. The development of new technologies to study lipid rafts may be required before this question can be answered satisfactorily.

Despite all of these questions, there continues to be a great deal of promise for cholesterol modulation in therapy of AD. Whether statins modulate Aβ in vivo remains a question, but increasing data suggest that statins have potent anti-inflammatory properties, which could be valuable in treating AD [142]. Other means of modulating cholesterol metabolism also appear to be promising. For instance, ACAT inhibitors appear to be very effective in reducing Aβ and plaque load in vivo. Other matters that require further investigation include the relationship between plasma and brain cholesterol. A better understanding of brain cholesterol metabolism is required to clarify how modulating plasma cholesterol using diet or drugs could affect Aβ production or deposition in the brain. In addition, the contribution of different forms of cholesterol, free cholesterol, or cholesteryl esters, to the overall effect of cholesterol in AD needs to be examined further.

9.7 Conclusions

Over the past few years, an increasing amount of evidence has accumulated suggesting that cholesterol metabolism is strongly connected to the development of Alzheimer's disease. This evidence includes studies showing linkages between genes involved in cholesterol metabolism, such as apoE and cyp46, and AD and epidemiological evidence that drugs aimed at lowering cholesterol levels may be useful for treating AD. Additionally, there are a large number of biochemical studies indicating that cholesterol is involved in APP processing, possibly by providing a favorable membrane environment in which the amyloidogenic secretase enzymes can act, and also in Aβ aggregation and toxicity. This evidence has led to the possibility that drugs affecting cholesterol metabolism, such as statins and ACAT inhibitors, or the modulation of cholesterol levels by dietary control, may be beneficial in the treatment of AD.

Despite this growing amount of evidence, we do not currently have a clear picture of the relationships between cholesterol and AD, and more work is needed to confirm the importance of cholesterol in the progression of the disease and to elucidate the molecular basis of the relationship. The advances in our knowledge that will surely come over the next few years may lead to the development of new strategies for both prevention and treatment of Alzheimer's disease.

References

1. Bloch K. The biological synthesis of cholesterol. Science 1965;150(692):19-28.
2. Hooper NM. Detergent-insoluble glycosphingolipid/cholesterol-rich membrane domains, lipid rafts and caveolae (review). Mol Membr Biol 1999;16(2):145-56.
3. Simons K, Ikonen E. Functional rafts in cell membranes. Nature 1997;387(6633):569-72.
4. Simons K, Ikonen E. How cells handle cholesterol. Science 2000;290(5497):1721-6.
5. Altmann SW, Davis HR Jr, Zhu LJ, et al. Niemann-Pick C1 Like 1 protein is critical for intestinal cholesterol absorption. Science 2004;303(5661):1201-4.
6. Davies JP, Levy B, Ioannou YA. Evidence for a Niemann-pick C (NPC) gene family: identification and characterization of NPC1L1. Genomics 2000;65 (2):137-45.
7. Homanics GE, Smith TJ, Zhang SH, et al. Targeted modification of the apolipoprotein B gene results in hypobetalipoproteinemia and developmental abnormalities in mice. Proc Natl Acad Sci U S A 1993; 90(6):2389-93.
8. Ishibashi S, Herz J, Maeda N, et al. The two-receptor model of lipoprotein clearance: tests of the hypothesis in "knockout" mice lacking the low density lipoprotein receptor, apolipoprotein E, or both proteins. Proc Natl Acad Sci U S A 1994;91(10): 4431-5.
9. Srivastava RA, Toth L, Srivastava N, et al. Regulation of the apolipoprotein B in heterozygous hypobetalipoproteinemic knock-out mice expressing truncated apoB, B81. Low production and enhanced clearance of apoB cause low levels of apoB. Mol Cell Biochem 1999;202(1-2):37-46.
10. Laffitte BA, Repa JJ, Joseph SB, et al. LXRs control lipid-inducible expression of the apolipoprotein E gene in macrophages and adipocytes. Proc Natl Acad Sci U S A 2001;98(2):507-12.
11. Venkateswaran A, Laffitte BA, Joseph SB, et al. Control of cellular cholesterol efflux by the nuclear

oxysterol receptor LXR alpha. Proc Natl Acad Sci U S A 2000;97(22):12097-102.
12. Refolo LM, Pappolla MA, Malester B, et al. Hypercholesterolemia accelerates the Alzheimer's amyloid pathology in a transgenic mouse model. Neurobiol Dis 2000;7(4):321-31.
13. Lafarga M, Crespo P, Berciano MT, et al. Apolipoprotein E expression in the cerebellum of normal and hypercholesterolemic rabbits. Brain Res Mol Brain Res 1994;21(1-2):115-23.
14. Brown J, Theisler C, Silberman S, et al. Differential expression of cholesterol hydroxylases in Alzheimer's disease. J Biol Chem 2004;279:34674-81.
15. Lathe R. Steroid and sterol 7-hydroxylation: ancient pathways. Steroids 2002;67(12):967-77.
16. Chartier-Harlin M, Crawford F, Houlden H, et al. Early-onset Alzheimer's disease caused by mutations at codon 717 of the a-amyloid precursor protein gene. Nature 1991;353:844-6.
17. Goate A, Chartier-Harlin MC, Mullan M, et al. Segregation of a missence mutation in the amyloid precursor protein gene with familial Alzheimer's disease. Nature 1991;349:704-6.
18. Murrell J, Farlow M, Ghetti B, Benson M. A mutation in the amyloid precursor protein associated with hereditary Alzheimer's disease. Science 1991;254:97-9.
19. Levy-Lahad E, Wasco W, Poorkaj P, et al. Candidate gene for the chromosome 1 familial Alzheimer's disease locus. Science 1995;269(5226):973-7.
20. Rogaev EI, Sherrington R, Rogaeva EA, et al. Familial Alzheimer's disease in kindreds with missense mutations in a gene on chromosome 1 related to the Alzheimer's disease type 3 gene. Nature 1995;376(6543):775-8.
21. Sherrington R, Rogaev E, Liang Y, et al. Cloning of a gene bearing missense mutations in early-onset familial Alzheimer's disease. Nature 1995;375:754-60.
22. Corder E, Saunders A, Strittmatter W, et al. Gene dose of apolipoprotein E type 4 allele and the risk of Alzheimer's disease in late onset families. Science 1993;261:921-3.
23. Seshadri S, Drachman DA, Lippa CF. Apolipoprotein E epsilon 4 allele and the lifetime risk of Alzheimer's disease. What physicians know, and what they should know. Arch Neurol 1995;52(11):1074-9.
24. Strittmatter WJ, Roses AD. Apolipoprotein E and Alzheimer's disease. Proc Natl Acad Sci U S A 1995;92(11):4725-7.
25. Tomiyama T, Corder EH, Mori H. Molecular pathogenesis of apolipoprotein E-mediated amyloidosis in late-onset Alzheimer's disease. Cell Mol Life Sci 1999;56(3-4):268-79.
26. Wisniewski T, Castano EM, Golabek A, et al. Acceleration of Alzheimer's fibril formation by apolipoprotein E in vitro. Am J Pathol 1994;145(5):1030-5.
27. Carter DB, Dunn E, McKinley DD, et al. Human apolipoprotein E4 accelerates beta-amyloid deposition in APPsw transgenic mouse brain. Ann Neurol 2001;50(4):468-75.
28. Holtzman DM, Bales KR, Tenkova T, et al. Apolipoprotein E isoform-dependent amyloid deposition and neuritic degeneration in a mouse model of Alzheimer's disease. Proc Natl Acad Sci U S A 2000;97(6):2892-7.
29. Sadowski M, Pankiewicz J, Scholtzova H, et al. A synthetic peptide blocking the apolipoprotein E/beta-amyloid binding mitigates beta-amyloid toxicity and fibril formation in vitro and reduces beta-amyloid plaques in transgenic mice. Am J Pathol 2004;165(3):937-48.
30. Yang DS, Small DH, Seydel U, et al. Apolipoprotein E promotes the binding and uptake of beta-amyloid into Chinese hamster ovary cells in an isoform-specific manner. Neuroscience 1999;90(4):1217-26.
31. Menzel HJ, Kladetzky RG, Assmann G. Apolipoprotein E polymorphism and coronary artery disease. Arteriosclerosis 1983;3(4):310-5.
32. Jarvik GP, Wijsman EM, Kukull WA, et al. Interactions of apolipoprotein E genotype, total cholesterol level, age, and sex in prediction of Alzheimer's disease: a case-control study. Neurology 1995;45(6):1092-6.
33. Kuo YM, Emmerling MR, Bisgaier CL, et al. Elevated low-density lipoprotein in Alzheimer's disease correlates with brain abeta 1-42 levels. Biochem Biophys Res Commun 1998;252(3):711-5.
34. Evans RM, Hui S, Perkins A, et al. Cholesterol and APOE genotype interact to influence Alzheimer's disease progression. Neurology 2004;62(10):1869-71.
35. Notkola IL, Sulkava R, Pekkanen J, et al. Serum total cholesterol, apolipoprotein E epsilon 4 allele, and Alzheimer's disease. Neuroepidemiology 1998;17(1):14-20.
36. Prince M, Lovestone S, Cervilla J, et al. The association between APOE and dementia does not seem to be mediated by vascular factors. Neurology 2000;54(2):397-402.
37. Gregg RE, Zech LA, Schaefer EJ, et al. Abnormal in vivo metabolism of apolipoprotein E4 in humans. J Clin Invest 1986;78(3):815-21.
38. Steinmetz A, Jakobs C, Motzny S, Kaffarnik H. Differential distribution of apolipoprotein E isoforms in human plasma lipoproteins. Arteriosclerosis 1989;9(3):405-11.
39. Weisgraber KH. Apolipoprotein E distribution among human plasma lipoproteins: role of the cysteine-arginine interchange at residue 112. J Lipid Res 1990;31(8):1503-11.

40. Lund EG, Guileyardo JM, Russell DW. cDNA cloning of cholesterol 24-hydroxylase, a mediator of cholesterol homeostasis in the brain. Proc Natl Acad Sci U S A 1999;96(13):7238-43.
41. Lund EG, Xie C, Kotti T, et al. Knockout of the cholesterol 24-hydroxylase gene in mice reveals a brain-specific mechanism of cholesterol turnover. J Biol Chem 2003;278(25):22980-8.
42. Papassotiropoulos A, Lutjohann D, Bagli M, et al. 24S-hydroxycholesterol in cerebrospinal fluid is elevated in early stages of dementia. J Psychiatr Res 2002;36(1):27-32.
43. Bretillon L, Lutjohann D, Stahle L, et al. Plasma levels of 24S-hydroxycholesterol reflect the balance between cerebral production and hepatic metabolism and are inversely related to body surface. J Lipid Res 2000;41(5):840-5.
44. Heverin M, Bogdanovic N, Lutjohann D, et al. Changes in the levels of cerebral and extracerebral sterols in the brain of patients with Alzheimer's disease. J Lipid Res 2004;45(1):186-93.
45. Kolsch H, Heun R, Kerksiek A, et al. Altered levels of plasma 24S- and 27-hydroxycholesterol in demented patients. Neurosci Lett 2004;368(3):303-8.
46. Kolsch H, Lutjohann D, Ludwig M, et al. Polymorphism in the cholesterol 24S-hydroxylase gene is associated with Alzheimer's disease. Mol Psychiatry 2002;7(8):899-902.
47. Papassotiropoulos A, Streffer JR, Tsolaki M, et al. Increased brain beta-amyloid load, phosphorylated tau, and risk of Alzheimer's disease associated with an intronic CYP46 polymorphism. Arch Neurol 2003;60(1):29-35.
48. Borroni B, Archetti S, Agosti C, et al. Intronic CYP46 polymorphism along with ApoE genotype in sporadic Alzheimer's Disease: from risk factors to disease modulators. Neurobiol Aging 2004;25(6):747-51.
49. Johansson A, Katzov H, Zetterberg H, et al. Variants of CYP46A1 may interact with age and APOE to influence CSF Abeta42 levels in Alzheimer's disease. Hum Genet 2004;114(6):581-7.
50. Desai P, DeKosky ST, Kamboh MI. Genetic variation in the cholesterol 24-hydroxylase (CYP46) gene and the risk of Alzheimer's disease. Neurosci Lett 2002;328(1):9-12.
51. Kabbara A, Payet N, Cottel D, et al. Exclusion of CYP46 and APOM as candidate genes for Alzheimer's disease in a French population. Neurosci Lett 2004;363(2):139-43.
52. Bodzioch M, Orso E, Klucken J, et al. The gene encoding ATP-binding cassette transporter 1 is mutated in Tangier disease. Nat Genet 1999;22(4):347-51.
53. Brooks-Wilson A, Marcil M, Clee SM, et al. Mutations in ABC1 in Tangier disease and familial high-density lipoprotein deficiency. Nat Genet 1999;22(4):336-45.
54. Rust S, Rosier M, Funke H, et al. Tangier disease is caused by mutations in the gene encoding ATP-binding cassette transporter 1. Nat Genet 1999;22(4):352-5.
55. Singaraja RR, Bocher V, James ER, et al. Human ABCA1 BAC transgenic mice show increased high density lipoprotein cholesterol and ApoAI-dependent efflux stimulated by an internal promoter containing liver X receptor response elements in intron 1. J Biol Chem 2001;276(36):33969-79.
56. Singaraja RR, Fievet C, Castro G, et al. Increased ABCA1 activity protects against atherosclerosis. J Clin Invest 2002;110(1):35-42.
57. Panzenboeck U, Balazs Z, Sovic A, et al. ABCA1 and scavenger receptor class B, type I, are modulators of reverse sterol transport at an in vitro blood-brain barrier constituted of porcine brain capillary endothelial cells. J Biol Chem 2002;277(45):42781-9.
58. Clee SM, Kastelein JJ, van Dam M, et al. Age and residual cholesterol efflux affect HDL cholesterol levels and coronary artery disease in ABCA1 heterozygotes. J Clin Invest 2000;106(10):1263-70.
59. Clee SM, Zwinderman AH, Engert JC, et al. Common genetic variation in ABCA1 is associated with altered lipoprotein levels and a modified risk for coronary artery disease. Circulation 2001;103(9):1198-205.
60. Wollmer MA, Streffer JR, Lutjohann D, et al. ABCA1 modulates CSF cholesterol levels and influences the age at onset of Alzheimer's disease. Neurobiol Aging 2003;24(3):421-6.
61. Katzov H, Chalmers K, Palmgren J, et al. Genetic variants of ABCA1 modify Alzheimer's disease risk and quantitative traits related to beta-amyloid metabolism. Hum Mutat 2004;23(4):358-67.
62. Koldamova RP, Lefterov IM, Ikonomovic MD, et al. 22R-hydroxycholesterol and 9-cis-retinoic acid induce ATP-binding cassette transporter A1 expression and cholesterol efflux in brain cells and decrease amyloid beta secretion. J Biol Chem 2003;278(15):13244-56.
63. Sun Y, Yao J, Kim TW, Tall AR. Expression of liver X receptor target genes decreases cellular amyloid beta peptide secretion. J Biol Chem 2003;278(30):27688-94.
64. Chen ZJ, Vulevic B, Ile KE, et al. Association of ABCA2 expression with determinants of Alzheimer's disease. FASEB J 2004;18(10):1129-31.
65. Mace S, Cousin E, Ricard S, et al. ABCA2 is a strong genetic risk factor for early-onset Alzheimer's disease. Neurobiol Dis 2005;18(1):119-25.
66. Hutter-Paier B, Huttunen HJ, Puglielli L, et al. The ACAT inhibitor CP-113,818 markedly reduces amyloid pathology in a mouse model of Alzheimer's disease. Neuron 2004;44(2):227-38.

67. Bu G, Maksymovitch EA, Nerbonne JM, Schwartz AL. Expression and function of the low density lipoprotein receptor-related protein (LRP) in mammalian central neurons. J Biol Chem 1994;269 (28):18521-8.
68. Ishiguro M, Imai Y, Kohsaka S. Expression and distribution of low density lipoprotein receptor-related protein mRNA in the rat central nervous system. Brain Res Mol Brain Res 1995;33(1):37-46.
69. Strickland DK, Kounnas MZ, Argraves WS. LDL receptor-related protein: a multiligand receptor for lipoprotein and proteinase catabolism. FASEB J 1995;9(10):890-8.
70. Rebeck GW, Harr SD, Strickland DK, Hyman BT. Multiple, diverse senile plaque-associated proteins are ligands of an apolipoprotein E receptor, the alpha 2-macroglobulin receptor/low-density-lipoprotein receptor-related protein. Ann Neurol 1995;37(2):211-7.
71. Kounnas MZ, Moir RD, Rebeck GW, et al. LDL receptor-related protein, a multifunctional ApoE receptor, binds secreted beta-amyloid precursor protein and mediates its degradation. Cell 1995;82(2):331-40.
72. Ulery PG, Beers J, Mikhailenko I, et al. Modulation of beta-amyloid precursor protein processing by the low density lipoprotein receptor-related protein (LRP). Evidence that LRP contributes to the pathogenesis of Alzheimer's disease. J Biol Chem 2000;275(10):7410-5.
73. Kang DE, Saitoh T, Chen X, et al. Genetic association of the low-density lipoprotein receptor-related protein gene (LRP), an apolipoprotein E receptor, with late-onset Alzheimer's disease. Neurology 1997;49(1):56-61.
74. Baum L, Chen L, Ng HK, et al. Low density lipoprotein receptor related protein gene exon 3 polymorphism association with Alzheimer's disease in Chinese. Neurosci Lett 1998;247(1):33-6.
75. Hollenbach E, Ackermann S, Hyman BT, Rebeck GW. Confirmation of an association between a polymorphism in exon 3 of the low-density lipoprotein receptor-related protein gene and Alzheimer's disease. Neurology 1998;50(6):1905-7.
76. Kolsch H, Ptok U, Mohamed I, et al. Association of the C766T polymorphism of the low-density lipoprotein receptor-related protein gene with Alzheimer's disease. Am J Med Genet 2003;121B(1):128-30.
77. Wavrant-DeVrieze F, Perez-Tur J, Lambert JC, et al. Association between the low density lipoprotein receptor-related protein (LRP) and Alzheimer's disease. Neurosci Lett 1997;227(1):68-70.
78. Van Leuven F, Thiry E, Stas L, Nelissen B. Analysis of the human LRPAP1 gene coding for the lipoprotein receptor-associated protein: identification of 22 polymorphisms and one mutation. Genomics 1998;52(2):145-51.
79. Du Y, Ni B, Glinn M, et al. alpha2-Macroglobulin as a beta-amyloid peptide-binding plasma protein. J Neurochem 1997;69(1):299-305.
80. Hughes SR, Khorkova O, Goyal S, et al. Alpha2-macroglobulin associates with beta-amyloid peptide and prevents fibril formation. Proc Natl Acad Sci U S A 1998;95(6):3275-80.
81. Bauer J, Strauss S, Schreiter-Gasser U, et al. Interleukin-6 and alpha-2-macroglobulin indicate an acute-phase state in Alzheimer's disease cortices. FEBS Lett 1991;285(1):111-4.
82. Van Gool D, De Strooper B, Van Leuven F, et al. alpha 2-Macroglobulin expression in neuritic-type plaques in patients with Alzheimer's disease. Neurobiol Aging 1993;14(3):233-7.
83. Borth W. Alpha 2-macroglobulin, a multifunctional binding protein with targeting characteristics. FASEB J 1992;6(15):3345-53.
84. Blacker D, Wilcox M, Laird N, et al. Alpha-2 macroglobulin is genetically associated with Alzheimer's disease. Nat Gen 1998;19:357-60.
85. Myllykangas L, Polvikoski T, Sulkava R, et al. Genetic association of alpha2-macroglobulin with Alzheimer's disease in a Finnish elderly population. Ann Neurol 1999;46(3):382-90.
86. Rogaeva EA, Premkumar S, Grubber J, et al. An alpha-2-macroglobulin insertion-deletion polymorphism in Alzheimer's disease. Nat Genet 1999;22(1):19-22.
87. Bodovitz S, Klein WL. Cholesterol modulates alpha-secretase cleavage of amyloid precursor protein. J Biol Chem 1996;271(8):4436-40.
88. Simons M, Keller P, De Strooper B, et al. Cholesterol depletion inhibits the generation of β-amyloid in hippocampal neurons. Proc Natl Acad Sci U S A 1998;95:6460-4.
89. Frears ER, Stephens DJ, Walters CE, et al. The role of cholesterol in the biosynthesis of beta-amyloid. Neuroreport 1999;10(8):1699-705.
90. Fassbender K, Simons M, Bergmann C, et al. Simvastatin strongly reduces Alzheimer's disease Aβ42 and Aβ40 levels in vitro and in vivo. Proc Natl Acad Sci U S A 2001;98:5856-61.
91. Kojro E, Gimpl G, Lammich S, et al. Low cholesterol stimulates the nonamyloidogenic pathway by its effect on the alpha-secretase ADAM 10. Proc Natl Acad Sci U S A 2001;98(10):5815-20.
92. Burns M, Gaynor K, Olm V, et al. Presenilin redistribution associated with aberrant cholesterol transport enhances beta-amyloid production in vivo. J Neurosci 2003;23(13):5645-9.

93. Runz H, Rietdorf J, Tomic I, et al. Inhibition of intracellular cholesterol transport alters presenilin localization and amyloid precursor protein processing in neuronal cells. J Neurosci 2002;22(5):1679-89.
94. Puglielli L, Konopka G, Pack-Chung E, et al. Acyl-Coenzyme A: Cholesterol Acyltransferase (ACAT) modulates the generation of the amyloid b-peptide. Nat Cell Bio 2001;3:905-12.
95. Brown DA, London E. Functions of lipid rafts in biological membranes. Annu Rev Cell Dev Biol 1998;14:111-36.
96. Simons K, Toomre D. Lipid rafts and signal transduction. Nat Rev Mol Cell Biol 2000;1(1):31-9.
97. Harder T, Scheiffele P, Verkade P, Simons K. Lipid domain structure of the plasma membrane revealed by patching of membrane components. J Cell Biol 1998;141(4):929-42.
98. Zacharias DA, Violin JD, Newton AC, Tsien RY. Partitioning of lipid-modified monomeric GFPs into membrane microdomains of live cells. Science 2002;296(5569):913-6.
99. Wolozin B. A fluid connection: cholesterol and Abeta. Proc Natl Acad Sci U S A 2001;98(10):5371-3.
100. Bouillot C, Prochiantz A, Rougon G, Allinquant B. Axonal amyloid precursor protein expressed by neurons in vitro is present in a membrane fraction with caveolae-like properties. J Biol Chem 1996;271(13):7640-4.
101. Lee SJ, Liyanage U, Bickel PE, et al. A detergent-insoluble membrane compartment contains A beta in vivo. Nat Med 1998;4(6):730-4.
102. Parkin ET, Turner AJ, Hooper NM. Amyloid precursor protein, although partially detergent-insoluble in mouse cerebral cortex, behaves as an atypical lipid raft protein. Biochem J 1999;344(Pt 1):23-30.
103. Wahrle S, Das P, Nyborg AC, et al. Cholesterol-dependent gamma-secretase activity in buoyant cholesterol-rich membrane microdomains. Neurobiol Dis 2002;9(1):11-23.
104. Cordy JM, Hussain I, Dingwall C, et al. Exclusively targeting beta-secretase to lipid rafts by GPI-anchor addition up-regulates beta-site processing of the amyloid precursor protein. Proc Natl Acad Sci U S A 2003;100(20):11735-40.
105. Riddell DR, Christie G, Hussain I, Dingwall C. Compartmentalization of beta-secretase (Asp2) into low-buoyant density, noncaveolar lipid rafts. Curr Biol 2001;11(16):1288-93.
106. Parkin ET, Hussain I, Karran EH, et al. Characterization of detergent-insoluble complexes containing the familial Alzheimer's disease-associated presenilins. J Neurochem 1999;72 (4):1534-43.
107. Ehehalt R, Keller P, Haass C, Thiele C, Simons K. Amyloidogenic processing of the Alzheimer's beta-amyloid precursor protein depends on lipid rafts. J Cell Biol 2003;160(1):113-23.
108. Schmechel D, Sullivan P, Mace B, et al. High saturated fat diets are associated with abeta deposition in primates. Neurobiol Aging 2002;23:S323.
109. Li L, Cao D, Garber DW, et al. Association of aortic atherosclerosis with cerebral beta-amyloidosis and learning deficits in a mouse model of Alzheimer's disease. Am J Pathol 2003;163(6):2155-64.
110. Shie FS, Jin LW, Cook DG, et al. Diet-induced hypercholesterolemia enhances brain A beta accumulation in transgenic mice. Neuroreport 2002;13(4):455-9.
111. Refolo LM, Pappolla MA, LaFrancois J, et al. A cholesterol-lowering drug reduces beta-amyloid pathology in a transgenic mouse model of Alzheimer's disease. Neurobiol Dis 2001;8(5):890-9.
112. Howland DS, Trusko SP, Savage MJ, et al. Modulation of secreted beta-amyloid precursor protein and amyloid beta-peptide in brain by cholesterol. J Biol Chem 1998;273(26):16576-82.
113. George AJ, Holsinger RM, McLean CA, et al. APP intracellular domain is increased and soluble Abeta is reduced with diet-induced hypercholesterolemia in a transgenic mouse model of Alzheimer's disease. Neurobiol Dis 2004;16(1):124-32.
114. Cao X, Sudhof TC. A transcriptionally [correction of transcriptively] active complex of APP with Fe65 and histone acetyltransferase Tip60. Science 2001;293(5527):115-20.
115. Gao Y, Pimplikar SW. The gamma -secretase-cleaved C-terminal fragment of amyloid precursor protein mediates signaling to the nucleus. Proc Natl Acad Sci U S A 2001;98(26):14979-84.
116. Lu DC, Rabizadeh S, Chandra S, et al. A second cytotoxic proteolytic peptide derived from amyloid beta-protein precursor. Nat Med 2000;6(4):397-404.
117. Selkoe DJ. Translating cell biology into therapeutic advances in Alzheimer's disease. Nature 1999;399(6738 Suppl):A23-31.
118. Yanagisawa K, Odaka A, Suzuki N, Ihara Y. GM1 ganglioside-bound amyloid beta-protein (A beta): a possible form of preamyloid in Alzheimer's disease. Nat Med 1995;1(10):1062-6.
119. Choo-Smith LP, Garzon-Rodriguez W, Glabe CG, Surewicz WK. Acceleration of amyloid fibril formation by specific binding of Abeta-(1-40) peptide to ganglioside-containing membrane vesicles. J Biol Chem 1997;272(37):22987-90.
120. Kakio A, Nishimoto S, Yanagisawa K, et al. Interactions of amyloid beta-protein with various gangliosides in raft-like membranes: importance of GM1 ganglioside-bound form as an endogenous

seed for Alzheimer's amyloid. Biochemistry 2002;41(23):7385-90.
121. Mizuno T, Nakata M, Naiki H, et al. Cholesterol-dependent generation of a seeding amyloid beta-protein in cell culture. J Biol Chem 1999;274(21):15110-4.
122. Kakio A, Nishimoto SI, Yanagisawa K, et al. Cholesterol-dependent formation of GM1 ganglioside-bound amyloid beta-protein, an endogenous seed for Alzheimer's amyloid. J Biol Chem 2001;276(27):24985-90.
123. Wang SS, Rymer DL, Good TA. Reduction in cholesterol and sialic acid content protects cells from the toxic effects of beta-amyloid peptides. J Biol Chem 2001;276(45):42027-34.
124. Subasinghe S, Unabia S, Barrow CJ, et al. Cholesterol is necessary both for the toxic effect of Abeta peptides on vascular smooth muscle cells and for Abeta binding to vascular smooth muscle cell membranes. J Neurochem 2003;84(3):471-9.
125. Pappolla MA, Bryant-Thomas TK, Herbert D, et al. Mild hypercholesterolemia is an early risk factor for the development of Alzheimer's amyloid pathology. Neurology 2003;61(2):199-205.
126. Kivipelto M, Helkala EL, Laakso MP, et al. Midlife vascular risk factors and Alzheimer's disease in later life: longitudinal, population based study. Br Med J 2001;322(7300):1447-51.
127. Kivipelto M, Helkala EL, Laakso MP, et al. Apolipoprotein E epsilon4 allele, elevated midlife total cholesterol level, and high midlife systolic blood pressure are independent risk factors for late-life Alzheimer's disease. Ann Intern Med 2002;137(3):149-55.
128. Tan ZS, Seshadri S, Beiser A, et al. Plasma total cholesterol level as a risk factor for Alzheimer's disease: the Framingham Study. Arch Intern Med 2003;163(9):1053-7.
129. Teunissen CE, De Vente J, von Bergmann K, et al. Serum cholesterol, precursors and metabolites and cognitive performance in an aging population. Neurobiol Aging 2003;24(1):147-55.
130. Jick H, Zornberg GL, Jick SS, et al. Statins and the risk of dementia. Lancet 2000;356(9242):1627-31.
131. Wolozin B, Kellman W, Ruosseau P, et al. Decreased prevalence of Alzheimer's disease associated with 3-hydroxy-3-methyglutaryl coenzyme A reductase inhibitors. Arch Neurol 2000;57(10):1439-43.
132. Rockwood K, Kirkland S, Hogan DB, et al. Use of lipid-lowering agents, indication bias, and the risk of dementia in community-dwelling elderly people. Arch Neurol 2002;59(2):223-7.
133. Yaffe K, Barrett-Connor E, Lin F, Grady D. Serum lipoprotein levels, statin use, and cognitive function in older women. Arch Neurol 2002;59(3):378-84.
134. Simons M, Schwarzler F, Lutjohann D, et al. Treatment with simvastatin in normocholesterolemic patients with Alzheimer's disease: a 26-week randomized, placebo-controlled, double-blind trial. Ann Neurol 2002;52(3):346-50.
135. Sjogren M, Gustafsson K, Syversen S, et al. Treatment with simvastatin in patients with Alzheimer's disease lowers both alpha- and beta-cleaved amyloid precursor protein. Dement Geriatr Cogn Disord 2003;16(1):25-30.
136. Heart Protection Study Collaborative Group. MRC/BHF Heart Protection Study of cholesterol lowering with simvastatin in 20,536 high-risk individuals: a randomised placebo-controlled trial. Lancet 2002;360(9326):7-22.
137. Shepherd J, Blauw GJ, Murphy MB, et al. Pravastatin in elderly individuals at risk of vascular disease (PROSPER): a randomised controlled trial. Lancet 2002;360(9346):1623-30.
138. Park IH, Hwang EM, Hong HS, et al. Lovastatin enhances Abeta production and senile plaque deposition in female Tg2576 mice. Neurobiol Aging 2003;24(5):637-43.
139. Grip O, Janciauskiene S, Lindgren S. Pravastatin down-regulates inflammatory mediators in human monocytes in vitro. Eur J Pharmacol 2000;410(1):83-92.
140. Ortego M, Bustos C, Hernandez-Presa MA, et al. Atorvastatin reduces NF-kappaB activation and chemokine expression in vascular smooth muscle cells and mononuclear cells. Atherosclerosis 1999;147(2):253-61.
141. Mayor S, Maxfield FR. Insolubility and redistribution of GPI-anchored proteins at the cell surface after detergent treatment. Mol Biol Cell 1995;6(7):929-44.
142. Cordle A, Landreth G. 3-hydroxy-3-methylglutaryl-coenzyme A reductase inhibitors attenuate beta-amyloid-induced microglial inflammatory responses. J Neurosci 2005;25(2):299-307.

10 Amyloid β-Peptide and Central Cholinergic Neurons: Involvement in Normal Brain Function and Alzheimer's Disease Pathology

Satyabrata Kar, Z. Wei, David MacTavish, Doreen Kabogo, Mee-Sook Song, and Jack H. Jhamandas

10.1 Introduction

Alzheimer's disease (AD), the most common form of dementia affecting individuals over 65 years of age, is a progressive neurodegenerative disorder. It is characterized by a global deterioration of intellectual function that includes an amnesic type of memory impairment, deterioration of language, and visuospatial deficits. Motor and sensory abnormalities are uncommon until the late phases of the disease, and basic activities of daily living are gradually impaired as the disease enters advanced phases. Psychosis and agitation also develop during middle or later phases of the disease. The average course of AD from the onset of clinical symptoms to death is approximately a decade, but the rate of progression is variable [1, 2]. Epidemiological data have shown that AD afflicts about 8–10% of the population over 65 years of age, and its prevalence doubles every 5 years thereafter [3].

Although our understanding of the pathophysiology of AD still remains fragmentary, it is widely accepted that both genetic and environmental factors can contribute to the development of the disease. In the majority of cases, AD appears to occur as sporadic disease after the age of 65 years, but in a small proportion of cases the disease is inherited as an autosomal dominant trait and appears as an early-onset form prior to 65 years of age. To date, mutations within three genes—the amyloid precursor protein (APP) gene on chromosome 21, the presenilin 1 (PS1) gene on chromosome 14, and the presenilin 2 (PS2) gene on chromosome 1—have been identified as the cause of early-onset familial AD [4–6]. Although these findings are of importance in elucidating the biological pathogenesis of AD, it is vital to recognize that mutations in these three genes may only account for 30–50% of all autosomal dominant early-onset cases. The inheritance of late-onset AD is more complex than that of the early-onset form. Various factors, including concomitant pathology and limited sample sizes, make it difficult to identify genetic causes of late-onset disease by conventional linkage analysis. However, association studies have identified candidate genes that significantly increase the risk for late-onset disease. The ε4 allele of the apolipoprotein E (APOE) gene, on chromosome 19, is one such risk factor. Possessing a single copy of the allele may increase the chance of developing AD two- to fivefold, whereas having two ε4 alleles raises this probability to more than fivefold [5–8]. Despite these advances in understanding the genetics of AD, the vast majority of cases has not yet been associated with any of the four genes implicated to date, thus suggesting that additional causative mutations and genetic risk factors remain to be identified [4–6, 9]. Other factors that may play an important role in the pathogenesis of AD include age, head injury, and oxidative stress [10].

10.2 Neuropathological Features of AD

The neuropathological changes of AD are characterized by the presence of intracellular neurofibrillary tangles, extracellular parenchymal and cerebrovascular amyloid deposits, and loss of neurons and synaptic integrity in specific brain areas. These features are also seen in Down syndrome (DS) brains (<40 years of age) and, to a limited extent, in the normal aging brain [9–11].

10.2.1 Neurofibrillary Tangles and Neuritic Plaques

Neurofibrillary tangles in the AD brain are particularly abundant in the entorhinal cortex, hippocampus, amygdala, association cortices of the frontal, temporal, and parietal lobes, and certain subcortical nuclei. This abnormal pathology, which is evident in neuronal cell bodies, neuropil threads, and dystrophic neuritis, is composed of hyperphosphorylated form of microtubule-associated protein tau. Accumulation of phospho-tau reduces the ability of tau to stabilize microtubules, leading to disruption of neuronal transport and eventually to the death of affected neurons [12–15]. The extent of neurofibrillary pathology, and particularly the number of cortical neurofibrillary tangles, correlates positively with the severity of dementia. However, tangles are also found in a variety of other neurodegenerative diseases without any evidence of amyloid deposits [9, 12, 13, 16]. Neuritic plaques, on the other hand, are multicellular lesions containing a compact deposit of amyloid peptides in a milieu of reactive astrocytes, activated microglia, and dystrophic neurites. The major amyloid peptides that are found in the plaques are β-amyloid$_{1\text{-}42}$ (Aβ$_{1\text{-}42}$) and Aβ$_{1\text{-}40}$, peptides that are generated by proteolytic cleavage of APP. The time required to develop a neuritic plaque is not known, but these lesions are believed to evolve gradually over a period of time from "diffuse plaques" containing only Aβ$_{1\text{-}42}$ [9, 17–19]. The diffuse plaques are found in large numbers in areas that are not typically affected in AD pathology (e.g., cerebellum, striatum, and thalamus), whereas neuritic plaques are usually seen in areas affected by neurodegeneration such as entorhinal cortex, hippocampus, and association cortices [9, 17]. Neuritic plaque number does not itself correlate with the severity of dementia, although a clinical correlation between elevated levels of the total Aβ peptide in the brain and cognitive decline has been reported [20]. Recent investigations in animal models and human brain samples have placed a special emphasis on measurement of soluble Aβ species [9, 21, 22].

Diverse lines of evidence suggest that accumulation of Aβ peptide in the brain may, over time, initiate and/or contribute to AD pathogenesis. These include the association of some AD cases with inherited APP mutations [4, 9, 11]; the elevation of Aβ peptides and the appearance of amyloid plaques in advance of other pathology in AD and DS brains [23]; the inheritance of APOE e4 allele(s) leads to enhanced Aβ deposition in the brain [5, 6, 9]; the increased production of Aβ$_{1\text{-}42}$ in vivo and in vitro by pathogenic mutations in PS1 and PS2 [9]; and the in vitro neurotoxic potential of fibrillar Aβ peptides [9, 24, 25]. Recent studies of APP transgenic mice [26–29] and of intrathecally administered Aβ in nontransgenic adult animals [30–33] reinforce the notion that overexpression of Aβ peptide, or injection of aggregated Aβ, induces subcellular alterations or neuronal loss in selected brain regions. It has been suggested that overexpression or injection of Aβ peptide may potentiate the formation of neurofibrillary tangles in tau transgenic mice [34, 35], a relationship first inferred from consideration of familial AD kindreds. Although these results implicate a role for Aβ peptides in the neurodegenerative process, both the role of Aβ in the normal brain and the mechanisms by which it causes neuronal loss and tau abnormalities in AD remain poorly understood.

10.2.2 Loss of Basal Forebrain Cholinergic Neurons

Selective synapse loss along with neuronal dysfunction and death are part of the elemental lesions associated with AD pathology. Evidence suggests that degenerating neurons and synapses are predominantly located in neuroanatomic regions that either project to or from the brain areas displaying highest density of plaques and tangles. Regions that are severely affected in AD brains include the hippocampus, entorhinal cortex, amygdala, neocortex, some subcortical areas such as basal forebrain cholinergic neurons, serotonergic neurons of the dorsal raphe, and noradrenergic neurons of the locus coeruleus [36–38]. Biochemical investigations of biopsy and autopsy tissues indicate that various neurotransmitters/modulators, including acetylcholine (ACh), serotonin, glutamate, noradrenaline, and somatostatin, are differentially altered in AD brains [11, 36, 39]. One of the most consistently reproduced finding is a profound reduction in the activity of the ACh synthesizing enzyme choline acetyltransferase (ChAT) in the neocortex that correlates positively with the severity of dementia [36, 38, 40]. Reduced choline uptake, ACh release, and loss of cholinergic neurons from the basal forebrain region further indicate a selective presynaptic cholinergic deficit in the hippocampus and neocortex of AD brains [39, 41]. Some of the earlier studies have also reported that depletion of cholinergic markers in the cortical regions of the AD brain may occur early in the course of the disease, perhaps as initiating events. In contrast, the cholinergic markers of the striatum (originating from striatal interneurons) and of the thalamus (originating from the brain stem) are either spared or affected only in late stages of the disease [36, 38, 39]. Together with pharmacological evidence of cholinergic involvement in the affected cognitive processes, these findings led to the development of a "cholinergic hypothesis" of AD. This hypothesis posits the degeneration of the cholinergic neurons in the basal forebrain and the loss of cholinergic transmission in the cerebral cortex and other areas as the principal cause of cognitive dysfunction in AD patients [38, 39, 41–43]. The hypothesis is supported, in part, by evidence that drugs that potentiate central cholinergic function (such as donepezil, rivastigmine, and galantamine) have some value in symptomatic treatment during early stages of the disease [38, 44]. However, some of the recent reports, all based on elderly subjects, have challenged the assumption that the cholinergic depletion is an early event in AD pathology [45]. Two of these studies report that mild AD is not associated with a loss of cortical ChAT activity [46, 47], whereas the third report suggests that the neurons containing ChAT and vesicular ACh transporter protein may not be decreased in early AD [48]. Collectively, these studies have not only raised doubts over the validity of the cholinergic hypothesis as it applies to early AD but also raise the possibility that the modest efficacy of cholinesterase inhibitor drugs in mild-to-moderate AD may involve mechanisms other than simple upregulation of a central cholinergic deficit [49, 50]. While these studies have created a number of new questions related to the role of the cholinergic system in the prodromal stage of AD, further investigations using in vivo imaging techniques or biochemical analysis of autopsy tissue using complementary approaches are needed to evaluate other components of cholinergic function (e.g., high-affinity choline transporter and nicotinic receptors) during aging and the progression of AD.

The loss of basal forebrain cholinergic neurons has prompted extensive study of ACh receptors in AD brains [36, 38, 39, 41, 50, 51]. ACh exerts effects on the central nervous system by interacting with G-protein–coupled muscarinic and ligand-gated cation channel nicotinic receptors. Five distinct muscarinic receptor subtypes, m_1–m_5, have been cloned and shown to correspond with five pharmacologically defined M1–M5 muscarinic receptors. It is generally believed that M2 receptors, most of which are located on presynaptic cholinergic terminals, are reduced in AD brains [38, 51]. The density of postsynaptic M1 receptors remains unaltered, but there is some evidence for disruption of the coupling between the receptors, their G-proteins, and second messengers [50–52]. The profiles of M3 and M4 receptors in the AD brain remain equivocal [53, 54]. For the nicotinic receptor family, 11 genes encoding 8 α (α_2–α_9) and three β receptor subunits (β_2–β_4) have been identified [38, 55]. High-affinity central nervous system binding sites of the agonist nicotine are mostly composed of $\alpha_4\beta_2$ subunits, whereas homomers of the α_7 receptor subunit contribute to the high-affinity binding of the antagonist α-bungarotoxin (α-BgTx) [55, 56]. Epibatidine, a potent nicotine agonist, binds with high-affinity to a subtype of nicotinic receptor containing the α_3 subunit [55]. Nicotinic receptors are predominantly located on cholinergic terminals. High-affinity nicotinic binding sites are markedly reduced in the hippocampus and cortex of the postmortem AD brains, and these observations have been confirmed in vivo by positron emission tomography [39, 57]. There is also evidence of a significant decrease in α_7 protein expression and α-BgTx binding sites in the hippocampus of AD brains [58]. However, a recent immunocytochemical study demonstrated an increase in the proportion of astrocytes expressing α_7 immunoreactivity in the hippocampus and entorhinal cortex of the AD brain relative to the age-matched controls [59]. Notwithstanding these data, no muscarinic or nicotinic receptor–based therapeutic approaches have provided convincing evidence of an adequate level of efficacy and reliability in AD balanced with an acceptable burden of side effects. Whether alterations in cholinergic receptors play a pathogenic role in dysregulating APP processing or promoting tau phosphorylation associated with AD pathology remains an area of intense investigation.

10.3 Cholinergic System and APP Processing

10.3.1 APP Processing

Aβ peptides, the principal component of amyloid deposits, are a group of hydrophobic peptides of 39–43 amino acid residues. These peptides are derived by proteolytic cleavage of APP—a type 1 integral membrane protein with a long N-terminal extracellular region, a single membrane-spanning domain, and a short C-terminal cytoplasmic tail [9, 11, 19, 60]. Multiple isoforms are produced from a single APP gene by alternative mRNA splicing and encode proteins ranging from 365 to 770 amino acids. In the nervous system, APP_{695} isoform is expressed predominantly in neurons, whereas APP_{770} and APP_{751} isoforms are found in neuronal as well as non-neuronal cells [9, 18, 19]. Mature APP is proteolytically processed by mutually exclusive α-secretase or β-secretase pathways. The α-secretase activity cleaves the Aβ domain within Lys^{16} and Leu^{17} residues, thus precluding the formation of full-length Aβ peptide. This pathway yields a soluble N-terminal APPα and a 10-kDa C-terminal APP fragment that can be further processed by γ-secretase to generate $A\beta_{17\text{-}40}$ or $A\beta_{17\text{-}42}$, also known as the P3 peptides. Three members of the disintegrin metalloproteases family that can act as potential candidates for α-secretase are tumor necrosis factor alpha converting enzyme (TACE or ADAM-17), ADAM-10, and MDC-9 [9, 18]. The β-secretase pathway, which results in the formation of intact Aβ peptide, is carried out by the sequential actions of two distinct proteases namely, β-secretase and γ-secretase. The β-secretase cleavage is mediated by a novel aspartyl protease referred to as the β-site APP cleaving enzyme (BACE), which generates a truncated soluble APPβ and a membrane-bound Aβ-containing C-terminal fragment. Further proteolysis of the C-terminal fragment by γ-secretase yields the full-length $A\beta_{1\text{-}40}$ or $A\beta_{1\text{-}42}$ peptide and a recently described C-terminal fragment termed γ-CTF [9, 18, 19, 61]. γ-Secretase activity resides in a multimeric protein complex that contains PS, considered as a putative aspartyl protease [62] along with four components (nicastrin, PEN-2, APH-1, and CD147) that are required for substrate recognition, complex assembly, and targeting the complex to its site of action [63, 64].

Assimilated evidence suggests that the majority of $A\beta_{1\text{-}40/1\text{-}42}$ is generated in the endosomal recycling pathway, whereas only a minority of $A\beta_{1\text{-}40/1\text{-}42}$ is produced in the secretory pathway, within the endoplasmic reticulum and Golgi apparatus [9, 18, 19]. Once generated, Aβ peptide, depending on the concentrations, can exist in multiple forms, including monomers, dimers, higher oligomers and polymers; the latter includes the fibrils that accumulate in amyloid deposits [9]. At present, the mechanisms by which APP processing is regulated under normal or pathological conditions remain unclear. However, several lines of experimental data have clearly shown that the discrete APP processing pathways can be influenced by a variety of factors, including the stimulation of receptors for ACh, serotonin, glutamate, estrogen, neuropeptides, and growth factors [65, 66]. The influence of cholinergic stimulation on amyloid formation is of particular interest in view of the preferential vulnerability of the cholinergic basal forebrain in AD and the possibility that maintenance of this cholinergic tone might slow amyloid deposition in cholinergic terminal fields.

10.3.2 Cholinergic Regulations of APP Processing

Over the years, a clear connection has been established between the cholinergic system and APP metabolism. Nitsch and colleagues first demonstrated cholinergic regulation of APP processing in human embryonic kidney (HEK) 293 cell lines that were stably transfected with human muscarinic m_1, m_2, m_3, and m_4 receptors [67]. Carbachol, a nonselective muscarinic receptor agonist, significantly increased the release of soluble APPα in cells expressing m_1 and m_3, but not in cells expressing m_2 or m_4 receptor subtypes. This response was both atropine-sensitive and blocked by staurosporine, indicating the mediation of intracellular protein kinases in receptor-controlled APPα secretion [67]. Activation of muscarinic m_1 receptor–transfected cells not only enhanced soluble APPα secretion but also reduced the secretion of Aβ peptide, thus suggesting that cholinergic agents may activate the non-amyloidogenic α-secretase pathway with the potential to prevent amyloid formation. Similarly, muscarinic m_1 and m_3 receptor agonists stimulated soluble APPα release from rat cortical slices [68] as well as brain cultured neurons [69]. Both m_1 and m_3 receptors activate signaling cascades involving phosphatidylinositol hydrolysis/protein kinase C (PKC) as well as mitogen activated protein (MAP) kinase pathways [70]. Treating cells with phorbol esters mimicked the effect of agonist administration on soluble APPα secretion, and this effect was blocked by PKC inhibitors [65, 71]. There is also evidence from cultured SH-SY5Y cells that carbachol-mediated soluble APPα secretion could be mediated, at least in part, by a MAP kinase–dependent pathway [69]. The mechanism whereby PKC- or MAP kinase–dependent pathways increase soluble APPα secretion is still unknown but may involve additional kinase steps and the eventual activation of the proteases that mediate APP cleavage [65, 66, 69, 71]. Moreover, a variety of other neurotransmitter/hormone receptors that activate PKC- or MAP kinase–dependent signaling pathways, including the vasopressin, bradykinin, estrogen, serotonin, and metabotropic glutamate receptors, share this capacity to stimulate soluble APP secretion and inhibit Aβ formation [65, 69, 71, 72].

In addition to the muscarinic receptor, some studies have examined the influence of the nicotinic receptor on APP processing. Treatment of PC12 cells with nicotine increases the release of soluble APPα without affecting Aβ secretion or expression of APP mRNA [73]. The relative increase in soluble APPα was attenuated by the α_7 nicotinic receptor antagonist methyllycaconitine and also by EGTA, a Ca^{2+} chelator. The nicotine antagonist chlorisondamine blocked in vivo elevation of total soluble APP induced by exposure to a high dose (8 mg $kg^{-1}day^{-1}$) of nicotine [74]. A nicotine-induced increase in Ca^{2+} influx was found to correspond with the increase in soluble APP secretion, suggesting that Ca^{2+} influx through nicotinic receptors may be involved in enhanced secretion. This result is in agreement with the findings from several studies showing that increased cytoplasmic Ca^{2+} levels can stimulate soluble APP secretion [66, 71, 75].

A number of studies have investigated whether acetylcholinesterase (AChE) inhibitors, which improve central cholinergic neurotransmission, can influence APP processing with the potential to modulate the biochemical pathways involved in the AD pathogenesis. The effects of various AChE inhibitors on soluble APPα levels differ between cell types and depend upon the specific drug, duration of treatment and the dose tested. For example, metrifonate did not alter soluble APP or Aβ levels in human SK-N-SH neuroblastoma cells [76], whereas acute treatment of the inhibitor could increase the secretion of soluble APPα in SH-SY5Y neuroblastoma cells, presumably by increasing the availability of ACh and thereby stimulating muscarinic receptors [69, 77]. Donepezil, a reversible AChE inhibitor, was found to increase the secretion of soluble APPα in a neuroblastoma cell line and platelets from AD patients by altering the activity/trafficking of α-secretase enzyme [78, 79]. Physostigmine elevated soluble APPα secretion in rat cortical slices [80] but decreased soluble APP secretion without altering Aβ levels in SK-N-SH neuroblastoma cells [76]. Tacrine, a potent cholinesterase inhibitor, was found to attenuate secretion of soluble APPα in glial, fibroblast, and PC12 cells. The addition of tacrine to neuroblastoma cell lines resulted in reduction of the levels of total Aβ, $A\beta_{1\text{-}40/1\text{-}42}$ along with soluble APPα [81]. Other AChE inhibitors such as phenserine, cymserine, and tolserine decreased soluble APPα levels, whereas 3,4-diaminopyridine failed to affect soluble APPα levels in SK-N-SH neuroblastoma cells [76]. The differential effects of the AChE inhibitors on APP processing appear to be unrelated to their selectivity for the cholinesterase enzymes but may depend upon other mechanisms, such as their influence on APP synthesis, expression, turnover, trafficking, or the regulation of APP processing enzymes [69, 71, 76, 82].

10.4 Regulation of Cholinergic System by Aβ Peptides

10.4.1 Effects of Aβ on ACH Synthesis and Release

Several studies over the past decade have clearly shown that nM concentrations of Aβ peptides, under acute as well as chronic conditions, can negatively regulate various steps of ACh synthesis and release, without apparent neurotoxicity. The high potency and reversible nature of this effect, together with the fact that pM to nM concentrations of Aβ peptides are found constitutively in normal brain cells, suggest that Aβ-related peptides may act as a modulator of cholinergic function under normal conditions (Table 10.1; Fig. 10.1) [41, 71, 83–86]. A 1-h exposure to pM to nM concentrations of Aβ can inhibit K^+- or veratridine-evoked endogenous ACh release from rat hippocampal and cortical slices. This effect is tetrodotoxin-insensitive, suggesting that Aβ peptide may act at the level or in close proximity to the cholinergic terminals [87, 88]. Structure activity studies reveal that inhibitory effects of Aβ-related peptides on ACh release from rat hippocampal slices reside within the sequence $A\beta_{25-28}$ (GSNK; the C-terminal domain of the nontoxic $A\beta_{1-28}$ fragment). In contrast with the effects on hippocampal and cortical slices, striatal ACh release is relatively insensitive to Aβ peptides [87]. This regional selectivity indicates that factors other than transmitter phenotype, such as the distance over which cholinergic axons project to their terminal fields and regional variation in the expression of Aβ binding sites, may contribute to the differences in cellular responsiveness to Aβ-related peptides. However, the sensitivity to Aβ of cholinergic neurons in cortex, hippocampus, and striatum matches the pattern of regional vulnerability in AD.

The inhibitory effects of Aβ on ACh release have been confirmed in rat and guinea-pig cortical synaptosomes [89], rat retinal neurons [90], and in cholinergic synaptosomes from the electric organ of the electric ray *Narke japonica* [91]. These effects may be affected by age-related cognitive deficits. Higher levels of $A\beta_{1-40}$ were observed in the aged rat hippocampus than were found in young adult rats, and the cholinergic neurons of aged cognitively impaired rats may be more sensitive to Aβ-mediated inhibition of hippocampal ACh release than either cognitively unimpaired aged or young adult rats [92]. This is supported in part by recent data showing that administration of antibody to Aβ can increase ACh levels in the hippocampus of 12-month SAMP8 mice that exhibit age-related increases in Aβ levels and deficits in learning and memory [93]. Lee et al. reported that inhibition of ACh release by $A\beta_{25-35}$ could be reversed by ginkgolide B and certain ginseng saponins at concentrations that did not by themselves alter ACh release [94, 95]. This effect was tetrodotoxin-insensitive, suggesting a direct interaction of ginseng at the level of the cholinergic synapse.

At present, the cellular mechanisms by which Aβ-related peptides, under acute conditions, can attenuate ACh release from selected brain regions remain unclear. Given the nature and potency of the effects, several steps that are critical for ACh synthesis and release—ranging from precursor recruitment to vesicular fusion—could be impaired by Aβ peptides (Table 10.1; Fig. 10.1). Turnover of ACh in the cholinergic terminals is regulated so that increased transmitter release is associated with increased synthesis. When brain slices are exposed to submaximal concentrations of depolarizing agents such as K^+ or veratridine, ongoing synthesis of ACh keeps pace with release from the terminals [96]. ACh synthesis under these conditions depends on the high-affinity uptake of choline from extracellular sources to intracellular acetyl CoA and ChAT. The availability of choline is a rate-limiting determinant of ACh biosynthesis, whereas ChAT activity is not [96]. Under acute treatment conditions, pM to nM concentrations of $A\beta_{1-40/1-42}$ do not affect ChAT activity in tissue homogenates or in slice preparations from hippocampus, cortex, or striatum [88]. Additionally, it is also reported that soluble $A\beta_{25-35}$ did not affect ChAT activity, under acute conditions, in the adult or aged rat brain [97]. The phosphorylation of the ChAT enzyme in IMR32 neuroblastoma cells expressing human ChAT is known to be regulated by $A\beta_{1-42}$, but its significance to ACh synthesis and/or release remains unclear [98].

TABLE 10.1. Effects of Aβ-related peptides on cholinergic neurons.

Peptide fragment	Effect on	Concentration	Model	Refs
	ACh synthesis and release			
$A\beta_{1\text{-}42,\ 1\text{-}40,\ 1\text{-}28,\ 25\text{-}35}$	Decrease in choline uptake	pM to μM	Cortical and hippocampal synaptosomes	88, 99
$A\beta_{1\text{-}42}$	Decrease in PDH activity	nM	Primary septal cultures	102
$A\beta_{1\text{-}42,\ 1\text{-}40,\ 1\text{-}28,\ 25\text{-}35}$	Decrease in ChAT activity	nM to μM	SN56 cell line and primary septal cultures	100, 126
$A\beta_{1\text{-}42,\ 1\text{-}28,\ 25\text{-}35,\ 25\text{-}28}$	Decrease in ACh content	pM to nM	SN56 cell line and primary septal cultures	100–102
$A\beta_{1\text{-}42,\ 1\text{-}40,\ 1\text{-}28,\ 25\text{-}35}$	Decrease in ACh release	pM to μM	Cortical and hippocampal slices, cortical and electric organ synaptosomes, retinal neurons	87–95
	Neuronal excitability			
$A\beta_{1\text{-}42,\ 25\text{-}35}$	Decrease in whole-cell currents and increase in excitability	nM to μM	Dissociated cells from diagonal band of Broca	84
	ACh receptors			
$A\beta_{1\text{-}40,\ 25\text{-}35}$	Disrupt M1-like receptor signaling	nM to μM	Primary cortical cultures	120
$A\beta_{1\text{-}42}$	Interacts with nicotinic receptor	pM to nM	AD hippocampus, transfected cells, rat and guinea-pig hippocampus	108,109
$A\beta_{1\text{-}40,\ 1\text{-}42,\ 12\text{-}28}$	Inhibits nicotinic receptor currents	nM to μM	Rat hippocampal slices and cultured neurons, transfected cells, and *Xenopus* oocytes	110–114
$A\beta_{1\text{-}40,\ 1\text{-}42,\ 25\text{-}35}$	Stimulates nicotinic receptor currents	pM to μM	Dissociated cells from diagonal band of Broca and *Xenopus* oocytes	115, 116
	Neuronal vulnerability			
$A\beta_{1\text{-}42,\ 1\text{-}40,\ 25\text{-}35}$	Induce tau phosphorylation	μM	SN56 cell line and primary septal cultures	125, 126
$A\beta_{1\text{-}42,\ 1\text{-}40,\ 25\text{-}35}$	Induce toxicity	μM	SN56 cell line, RN46A cell line, and primary rat septal cultures	124–128

Aβ, β-amyloid peptide; ACh, acetylcholine; AD, Alzheimer's disease; ChAT, choline acetyltransferase; PDH, pyruvate dehydrogenase.

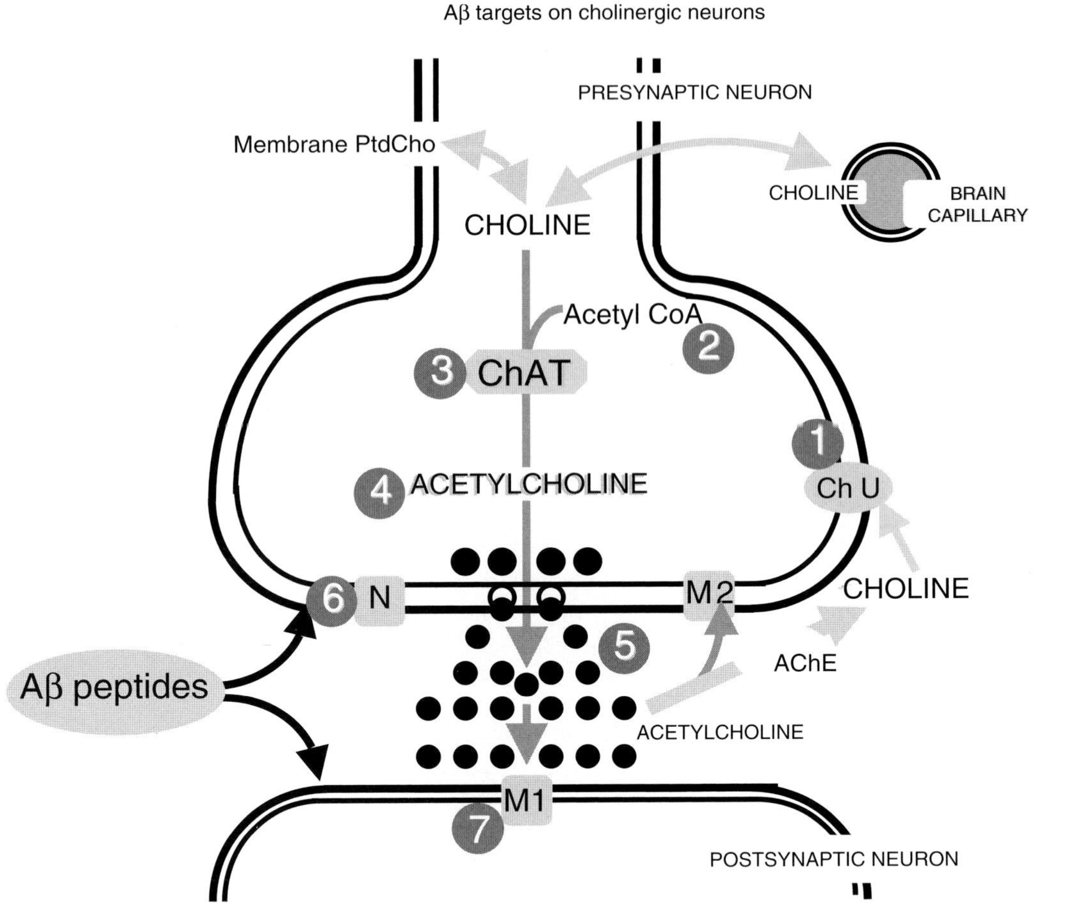

FIGURE 10.1. Targets of β-amyloid (Aβ) peptide on central cholinergic neurons. 1, Aβ reduces high-affinity uptake of choline; 2, Aβ reduces activity of pyruvate dehydrogenase (PDH), an enzyme that generates acetyl-CoA from pyruvate; 3, chronic exposure to Aβ reduces activity of the enzyme choline acetyltransferase (ChAT); 4, Aβ reduces acetylcholine (ACh) content; 5, Aβ reduces ACh release from presynaptic terminals; 6, Aβ interacts directly with nicotinic receptor; 7, Aβ impairs muscarinic M1-like signaling. AChE, acetylcholine-sterase; Ch U, site of choline uptake; M2, presynaptic muscarinic M2 receptor; N, presynaptic nicotinic receptor. Modified from Kar et al. [94].

In contrast with ChAT activity, high-affinity [^{3}H]choline uptake is found to be decreased after 20 minutes of preincubation with Aβ. This effect is particularly marked in tissues from the hippocampus and cortex, mirroring the effect of Aβ on ACh release in these regions [88]. Acute incubation of hippocampal synaptosomes with low nM $A\beta_{1-40}$ attenuates depolarization-induced high-affinity choline uptake as well as [^{3}H]hemicholinium-3 ([^{3}H]HC-3) binding [99]. Further analysis of these data indicates that changes in the transport are due to an alteration of V_{max}, whereas the changes in specific binding possibly involve alterations of both B_{max} and K_D. Micromolar concentrations of $A\beta_{1-40}$ decrease high-affinity choline uptake and the [^{3}H]HC-3 binding under basal conditions in a time-dependent manner [99]. These results indicate that Aβ can affect acute ACh release, at least in part, by regulating high-affinity choline uptake, but not the activity of the ChAT enzyme. The possible involvement of Aβ in the intracellular transport of newly synthesized ACh molecules and the fusion of ACh-containing vesicles with the presynaptic membrane remain to be investigated.

In addition to the acute effects, a 2-day exposure to pM to nM concentrations of $A\beta_{1-42}$, $A\beta_{1-28}$, $A\beta_{25-35}$, and to a lesser extent $A\beta_{25-28}$ was found to decrease intracellular ACh concentrations in the cholinergic hybrid SN56 cell line without causing toxicity (Table 10.1; Fig. 10.1). The decrease in ACh could be attributed to reduced biosynthesis, as it was accompanied by a reduction in ChAT activity. Interestingly, the observed decrease could be prevented by a cotreatment with *trans*-retinoic acid, a compound that increases ChAT mRNA expression in SN56 cells, or by coadministration of tyrosine kinase inhibitors [41, 100, 101]. However, inhibition of DNA synthesis or treatment with antioxidants did not alter ACh concentrations, thus suggesting that neither gene transcription nor free-radical production is involved in mediating the long-term effect of Aβ on the cholinergic SN56 cell line [101]. In keeping with these results, treatment of rat primary septal neurons with nM concentrations of $A\beta_{1\text{-}42}$ was found to decrease ACh production and reduce activity of the acetyl-CoA biosynthesizing enzyme pyruvate dehydrogenase (PDH) without affecting ChAT activity or neuronal survival. The decreased PDH activity possibly results from Aβ activation of the glycogen synthase kinase-3β (GSK-3β), which can phosphorylate and inactivate PDH [102]. Collectively these results suggest that chronic exposure to Aβ peptide may impair ACh synthesis/levels by reducing the availability of acetyl CoA and/or activity of the ChAT enzyme.

10.4.2 Effects of Aβ on Whole-Cell Currents in Cholineric Neurons

Apart from interacting with cholinergic terminals in the hippocampal and cortical regions, Aβ peptide can also act at the level of cell body of cholinergic neurons within the basal forebrain to increase neuronal excitability [84]. Application of 1 μM $A\beta_{1\text{-}42/25\text{-}35}$ to acutely dissociated rat neurons from the diagonal band of Broca decreased whole-cell voltage-sensitive currents in cholinergic neurons that were identified by single cell RT-PCR [84]. This reduction was observed for a suite of K^+ currents, including the Ca^{2+}-activated K^+ currents (BK or Ic), the delayed rectifier current (I_K), and transient outward current (I_A), but not for calcium or sodium currents. The responses were blocked by tyrosine kinase inhibitors, suggesting that Aβ induces phosphorylation-dependent cascades to alter these currents [84]. These results indicate that Aβ peptides acutely modulating K^+ currents at the level of the cell body can increase excitability of the basal forebrain cholinergic neurons. More recently, it has been demonstrated that the effects of Aβ peptide on whole-cell currents are similar to those evoked by human amylin, a 37-amino-acid pancreatic peptide that is deposited in the islet cells of patients with non-insulin-dependent diabetes mellitus. Aβ evoked responses can be occluded by human amylin and can be blocked by AC187—a specific amylin receptor antagonist. These data raise the intriguing possibility that the effects of Aβ on basal forebrain cholinergic neurons may be expressed through the amylin receptor [103].

10.4.3 Effects of Aβ on Cholinergic Receptors

Over the years, a variety of receptors (e.g., receptors for advanced glycation end products [RAGE], class A scavenger receptor [SR], the 75-kDa neurotrophin receptor [$p75^{NTR}$], amylin receptor, and serpin-enzyme complex receptors) have been shown to interact with Aβ in vitro [103–107]. These interactions have attracted attention both for the insights they may provide into the mechanism of Aβ action and also as potential targets for drug design. A number of recent studies suggest that $A\beta_{1\text{-}42}$ can interact with the nicotinic ACh receptors to mediate its acute as well as chronic effects. The first reported observation of an interaction between Aβ and α_7/α-BgTx nicotinic receptors showed that these proteins co-immunoprecipitated in samples from postmortem AD hippocampus, and α_7/α-BgTx nicotinic receptor antagonists compete for $A\beta_{1\text{-}42}$ binding to heterologously expressed α_7/α-BgTx nicotinic receptors [108]. A subsequent study indicated that $A\beta_{1\text{-}42}$ can bind with high affinity (Ki ~ 4–5 pM) to α_7/α-BgTx nicotinic receptors and with lower affinity (Ki ~ 20–30 nM) to $\alpha_4\beta_2$/cytisine nicotinic (but not muscarinic) receptors in the rat and guinea-pig hippocampus and cerebral cortex [109]. This is supported by the observation that nanomolar Aβ peptide was found to inhibit nicotine-evoked currents via the α_7/α-BgTx receptor and/or the non-α_7 nicotinic receptor in both rat hippocampal slices and cultured neurons, human SH-EP1 cells expressing $\alpha_4\beta_2$ nicotinic receptor subunits, and in *Xenopus* oocytes containing heterologously expressed rat or human α_7 nicotinic receptor subunits [110–114]. However, there is also evidence that Aβ peptide can directly activate acutely dissociated rat basal forebrain neurons via non-α_7 nicotinic receptors and in the case of *Xenopus* oocytes expressing α_7 nicotinic receptor subunit through the α_7/α-BgTx receptors [115, 116]. In addition, it has been reported that α_7/α-BgTx receptors can facilitate internalization of Aβ_{1-42} in transfected human SK-N-MC neuroblastoma cells [117] and can mediate Aβ-induced tau phosphorylation in cultured SK-N-MC cells and hippocampal synaptosomes [118]. The effects of Aβ on the nicotinic receptor are consistent with receptor involvement in Aβ-mediated inhibition of ACh release. In support of this notion, the inhibitory effects of Aβ_{1-40} on cortical ACh release were found to be restored by addition of α_7 agonist, such as nicotine and epibatidine, but not by $\alpha_4\beta_2$ nicotinic receptor agonist cytosine [119]. However, further studies are needed not only to define the precise role of the α_7 nicotinic receptor in regulating the inhibitory effects of Aβ peptides on ACh release but also to establish its significance in relation to AD pathology.

In addition to interacting with nicotinic ACh receptors, solubilized Aβ peptide has been shown to disrupt transduction of the muscarinic M1-like receptor signal [120]. A 4-h exposure to nM-μM Aβ_{1-40} reduced carbachol-induced GTPase activity in rat cortical cultured neurons without affecting muscarinic receptor ligand binding parameters. At higher concentrations, similar treatment with Aβ attenuated muscarinic M1 receptor signaling by decreasing intracellular Ca^{2+} and the accumulation of Ins(1)P, Ins(1,4)P_2, Ins(1,4,5)P_3, and Ins(1,3,4,5) P_4 [120]. Exposure of rat cortical cultured neurons to nM Aβ_{1-42}/Aβ_{25-35} inhibits carbachol-, but not glutamate-, induced increases in intracellular Ca^{2+} and Ins(1,4,5)P_3 indicating that selective disruption of the muscarinic M1-like signaling pathway is another means by which Aβ can affect the function of cholinoceptive neurons [121].

10.4.4 Effects of Aβ on Cholinergic Neuron Survival

A number of in vitro studies have shown that chronic exposure to Aβ peptides can induce toxicity in a variety of cell lines, as well as in primary rat and human cultured neurons. The toxicity of the peptide is considered to be related to its ability to form insoluble aggregates [24, 25]. However, recent evidence suggests that the most detrimental forms of Aβ peptides are the soluble oligomers and that the insoluble amorphous or fibrillar deposits represent a less harmful form of the peptide [9, 122]. Some neuronal phenotypes, such as GABAergic and serotonergic neurons, appear resistant to Aβ toxicity, and various cell lines differ in their degree of sensitivity [123, 124]. Differentiated SN56 cholinergic cell lines are a susceptible line for toxicity studies, and when exposed to Aβ_{1-40}, these cells exhibit retraction of neurites, cell shrinkage, and death [125]. When treated with ciliary neurotrophic factor, the RN46A cell line develops a cholinergic phenotype and is highly sensitive to Aβ peptides. In contrast, stimulation of RN46A differentiation with brain-derived neurotrophic factor yields an Aβ-insensitive cell population with a serotonergic transmitter phenotype.[124] Prolonged exposure of rat primary septal cultured neurons to μM Aβ peptides induces both cell death and a concomitant decrease in ChAT activity [126–128]. Collectively, these results suggest that cells expressing cholinergic transmitter phenotype are vulnerable to the toxic effects of Aβ peptide.

The mechanisms by which Aβ induces cholinergic cell death remains unclear but may involve alteration in intracellular calcium and/or the production of toxic and inflammatory mediators such as nitric oxide, cytokines, and reactive oxygen intermediates [129–131]. Studies on a variety of cell lines and primary cultured neurons suggest that Aβ toxicity might be mediated either by interaction with a hydroxysteroid dehydrogenase enzyme or by plasma membrane RAGE, SR, $p75^{NTR}$, amylin, or α_7 nicotinic receptors [105–109, 127]. A role for the death domain of $p75^{NTR}$ in Aβ-induced cell death was observed in neuroblastoma (SK-N-BE) cells expressing full-length or truncated forms of $p75^{NTR}$, but recent evidence from primary human cultured neurons suggest that overexpression of $p75^{NTR}$ can provide protection against Aβ-mediated toxicity by activating a phosphatidylinositide 3-kinase–dependent but Akt-independent pathway [132, 133]. Studies of transfected neuroblastoma (SK-N-MC) cells indicate that expression of α_7 nicotinic receptor may also have a critical role in the degeneration by facilitating internalization and accumulation of Aβ_{1-42} into neurons [117]. Given the marked expression of $p75^{NTR}$ and of the α_7 nicotinic receptor in the cholinergic basal forebrain, their role in cholinergic cell death bears further investigation. More recently, it has been demonstrated that the amylin receptor antagonist AC-187 can attenuate Aβ-induced toxicity in rat primary septal cultured neurons by inhibiting a caspase-dependent pathway thus suggesting a possible role for this receptor in mediating the toxic effects of Aβ [127].

Tau phosphorylation has long been considered to contribute to neuronal vulnerability by destabilizing microtubules and impaired axonal transport [125, 134–136]. Aggregated Aβ induces the phosphorylation of tau protein in SN56 cholinergic cell lines [125]. Studies with rat septal cultured neurons have indicated that aggregated Aβ increases levels of both total tau as well as phosphorylated tau [126]. Phosphorylated tau immunoreactivity could be detected primarily in the distal axons of untreated cells, whereas staining was evident in axons, soma, and dendrites of neurons exposed to Aβ [126]. Hyperphosphorylated tau protein can lead to the neuronal death via disruption of the cytoskeletal network [13–15]; it is likely that the increase in tau phosphorylation plays some role in Aβ-induced death of the cholinergic neurons. However, the mechanisms by which Aβ might induce the phosphorylation of the tau protein remain unclear. Reactive oxygen species and the lipid peroxidation product 4-hydroxynonenal may be involved in Aβ-neurotoxicity and cross-linking of tau proteins [137]. Additionally, Aβ might also affect tau phosphorylation by directly increasing relevant kinase activity or by decreasing phosphatase activity [125, 134, 138–140]. Activation of GSK-3β [136, 139, 141] and MAP kinase [138] induces tau protein phosphorylation and cell death in a variety of cultured neuron paradigms, and prolonged exposure of rat septal cultured neurons to μM Aβ peptide has been shown to induce tau phosphorylation by activating MAP kinase and GSK-3β [126]. Various kinases phosphorylate tau at discrete sites, and it is likely that the phosphorylation of tau protein in cholinergic neurons is regulated by multiple kinases, including MAP kinase and GSK-3β. Thus, it is important to explore both the biochemical potential of additional tau kinases, such as cyclin-dependent kinase 5, PKC, and calcium-calmodulin kinase to phosphorylate tau [13–16], and the particular cellular expression of these kinases by cholinergic neurons.

Tau phosphorylation can be regulated by cholinergic agonists, and control of tau hyperphosphorylation by muscarinic receptor activation may provide a side benefit of cholinomimetic therapeutics. Muscarinic agonists, carbachol and AF 102B, attenuate tau phosphorylation in cultured PC12 cells stably transfected with muscarinic m_1 receptors [142]. On the other hand, activation of the nicotinic receptor by nicotine and epibatidine increased the levels of phosphorylated as well as non-phosphorylated tau in SH-SY5Y human neuroblastoma cells [143]. The mechanisms by which muscarinic m_1 or nicotinic receptor activation modify tau phosphorylation remain unclear, but recent data suggest that stimulation of α_7/α-BgTx nicotinic receptors by $A\beta_{1\text{-}42}$ can induce tau phosphorylation in human neuroblastoma cells and hippocampal synaptosomes *via* extracellular receptor kinases (ERKs) and c-Jun N-terminal kinase (JNK-1) [118]. These activities may likely involve alteration of other protein kinase/protein phosphatase systems [71].

10.4.5 Effects of In Vivo Administration of Aβ on Cholinergic Neurons

Attempts have been made to measure the impact of intracerebroventricular or local administration of Aβ on cholinergic system under in vivo conditions. Several studies have reported that Aβ peptides can induce cholinergic hypofunction when administered to the brain [31, 41, 83, 144, 145]. Injection of $A\beta_{25\text{-}35/1\text{-}40}$ into the rat medial septum causes a reduction in ACh release from the hippocampus in the absence of toxicity [146]. Using a similar approach, Harkany et al [31]. demonstrated that $A\beta_{1\text{-}42}$ is toxic to cholinergic neurons, as indicated by reduction in ChAT-immunoreactive cell bodies in the basal forebrain and fibers in the cerebral cortex. This effect was partly antagonized by the *N*-methyl-D-aspartate (NMDA) receptor antagonist MK-801, thus suggesting a possible involvement of an excitotoxic pathway in mediating the effects of Aβ peptide [31]. More recently, it has been shown that aging and high-cholesterol diet can enhance in vivo toxicity of Aβ peptide on cholinergic neurons [145]. Other studies have reported that infusion of Aβ into the lateral ventricles of adult rats impairs performance on learning and memory tasks in a manner similar to the effect of cholinergic inhibition [30, 32, 83, 144]. Local injection of preaggregated $A\beta_{1\text{-}42}$ into the nucleus basalis magnocellularis (NBM) produces congophilic deposits and a strong inflammatory response, characterized by activation of astrocytes and microglia and by induction of microglial p38MAP kinase activity [147]. These changes were accompanied by a decrease in the number of cholinergic neurons around the congophilic amyloid deposit and hypofunction of the cortical cholinergic system [147]. Clearly, the influence of these astrocytic and microglial responses must be considered in assessing in vivo effects of Aβ peptides on cholinergic function.

10.4.6 Cholinergic System in Transgenic Mice Overexpressing Aβ Peptide

Over the past few years, the central cholinergic system has been examined extensively in a variety of mutant APP, PS1, or APP/PS1 transgenic mouse lines, all of which exhibit elevated Aβ levels [148–163].

In mice expressing the $hAPP_{V642I}$ London mutant transgene, a selective decrease was found in the size of medial septal cholinergic neurons, but not in NBM cholinergic neurons. At 17–22 months of age, this line exhibits both reorganization of AChE-positive fibers in the hippocampus and dystrophic AChE-positive fibers around amyloid plaques in the cortex [149]. Cerebral amyloidosis was found to cause a significant cholinergic fiber loss and severe disruption of neocortical cholinergic fiber networks in aged APP23 mice expressing $hAPP_{KM670/671NL}$ Swedish mutant transgene [148]. Although the cholinergic neurons of the medial septum and vertical limb of the diagonal band of Broca were smaller in APP23 transgenic mice than in non-transgenic controls, the number and volume of ChAT-positive neurons in the NBM complex were not affected. Hippocampal cholinergic fiber density in APP23 mice has yet to be reported [148]. Homozygous PDAPP mice expressing the $hAPP_{V717F}$ mutant transgene showed an age-dependent decrease in hippocampal and cortical cholinergic fiber density without any evident loss of basal forebrain cholinergic neurons compared with the non-transgenic controls. The degeneration of cholinergic nerve terminals in these transgenic mice was found to occur prior to the deposition of Aβ-containing neuritic plaques [159].

In another study, $hAPP_{KM670/671NL}$ mutant mice demonstrated an upregulation in the density of cholinergic synapses in the frontal cortex, parietal cortex, and the hippocampus, whereas $PS1_{M146L}$ transgenic mice showed no changes in either the size or density of cholinergic synapses. When crossed to yield $hAPP_{KM670/671NL}/PS_{1M146L}$ double transgenic mice, extensive amyloid plaques were found to be associated with decreased density and size of cholinergic synapses in the frontal cortex and hippocampus [150]. A significant inverse relationship was noted between the presynaptic cholinergic bouton density and size of Aβ-containing neuritic plaques located in the frontal cortex of the $hAPP_{KM670/671NL}/PS_{1M146L}$ double transgenic mice [160]. In one study, a selective increase in immunostaining for $p75^{NTR}$ (a marker of basal fore brain cholinergic neurons) was evident in the medial septum of 12-month-old $hAPP_{KM670/671NL}$ or $PS1_{M146L}$ single transgenic mice but not in $hAPP_{KM670/671NL}/PS_{1M146L}$ double transgenic mice. Staining of $p75^{NTR}$-immunoreactive fibers in hippocampus was more robust in single transgenic mice, relative to non-transgenic controls, while double transgenic mice displayed less intense $p75^{NTR}$ fiber staining [151]. Whether the increased immunostaining in singly transgenic mice indicates a trophic effect on the cholinergic neurons as a consequence of either $hAPP_{KM670/671NL}$ or $PS1_{M146L}$ gene overexpression remains to be investigated. However, a separate study revealed no differences between $hAPP_{KM670/671NL}$ mice and non-transgenic controls in ChAT activity, AChE activity, vesicular ACh transporter binding, or high-affinity choline uptake sites in cortex, hippocampus, striatum, or cerebellum at multiple times up to 23 months of age [152]. Interestingly, a recent study showed that extracellular hippocampal ACh levels, but not stimulated ACh release, were slightly but significantly reduced (~26% decrease) in knock-in mice carrying $hAPP_{KM670/671NL}/PS_{1M146L}$ transgenes compared with mice overexpressing $hAPP_{KM670/671NL}/PS_{wild-type}$ transgenes, thus suggesting expression of mutant APP/PS1 genes may induce subtle alteration in cholinergic transmission [164].

Densities of M1/[^{3}H]pirenzepine, M2/[^{3}H]AF-DX 384, or α_7 nicotinic/[^{125}I]α-BgTx receptor binding sites in all brain regions of mutant $PS1_{L286V}$ transgenic and wild-type PS1 transgenic mice are comparable with those found in non-transgenic controls [153]. In $hAPP_{KM670/671NL}$ mutant mice, a decrease in M1/[^{3}H]pirenzepine and $\alpha_4\beta_2$ nicotinic/[^{3}H]cytisine, but not M2/[^{3}H]AF-DX 384, receptor binding was evident in the hippocampus and cortex compared with non-transgenic controls [157]. However, in other studies, elevated hippocampal α_7 nicotinic receptor levels have been reported in hAPP$_{K670N/M671L}$ single and two lines (i.e., hAPP$_{K670N/M671L}$/PS1$_{A246E}$ and APP$_{KM670/671NL+V717F}$/PS1$_{M146L+L286V}$) of double transgenic mice [154, 156]. In triple transgenic mice harboring hAPP$_{KM670/671NL}$/PS1$_{M146V}$/Tau$_{P301L}$ transgenes, an age-dependent reduction of α_7/α-BgTx nicotinic receptor binding sites was observed in the hippocampus and cortical regions compared with non-transgenic mice. Additionally, chronic nicotine intake was found to exacerbate tau pathology in these transgenic mice, suggesting an in vivo role for the nicotinic receptor in the phosphorylation of tau protein [163]. Apart from receptor binding site, high-affinity [^{3}H]HC binding (i.e., choline uptake sites) was found to be reduced in cortical regions of 5- and 17-month-old hAPP$_{KM670/671NL}$ mutant mice, whereas [^{3}H]vesamicol binding (i.e., vesicular Ach transporter sites) was increased in 17-month-old but not in 5-month-old transgenic mice compared with littermate non-transgenic controls [162]. However, the significance of the changes in these presynaptic cholinergic markers and their association with the amyloid pathology remains unclear. In sum, increased expression of Aβ peptides produces a range of effects on cholinergic systems of mutant APP, PS1, or APP/PS1 transgenic mice. Establishing which of these effects are robustly related to the type of pathogenic mutation, the level of transgene expression, or to the intensity of amyloid deposits remains to be defined in future studies.

10.5 Significance of Amyloid Interactions with Cholinergic Neurons

Earlier results have shown that Aβ-related peptides are produced constitutively by brain cells and are found in the pM to nM range in the cerebrospinal fluid of normal individuals [9, 165–167]. These concentrations of Aβ can have a neuromodulatory role in the regulation of normal cholinergic functions, possibly through their negative effects on ACh biosynthesis and release. Conversely, there is evidence that ACh can regulate APP synthesis and processing. For example, lesions of the basal forebrain cholinergic neurons or transient inhibition of cortical ACh release could elevate local APP synthesis [65, 168–170], whereas agonist-induced activation of muscarinic m_1 and m_3 receptor subtypes increases the secretion of soluble APP derivatives and reduces the production of amyloidogenic Aβ peptides [65–71, 171]. These results suggest a reciprocal mechanism whereby normal cholinergic innervation participates in the nonamyloidogenic maturation of APP via the α-secretase pathway, while the amyloidogenic Aβ-related peptides depress the activity of cholinergic neurons. A shift in the balance between these activities may possibly be a key factor in the targeting of cholinergic neurons in AD. Insults that reduce cholinergic transmission, increase Aβ generation, or reduce Aβ clearance may enhance vulnerability of neurons to direct toxicity of Aβ peptide [9, 24, 25] or to choline limitation [83, 86, 88, 99, 172, 173]. Because cholinergic neurons utilize choline from membrane phosphatidylcholine to synthesize ACh, it is likely that Aβ-induced alteration in intracellular choline levels might lead to an autocannibalistic process in which membrane turnover is disrupted to sustain neurotransmission [173]. Given the evidence that Aβ deposits precede any other lesions in AD brains [23], it is possible that amyloid-induced tau phosphorylation may also play a critical role in neuronal loss. This is supported by some in vivo studies in which intrathecal administration, or transgene-delivered expression of Aβ peptides was shown to induce a loss of neurons, or a change in presynaptic cholinergic markers, within selected brain regions [30–33, 148–150, 159]. The selective interactions of Aβ with basal forebrain cholinergic neurons provide candidate mechanisms that may contribute, at least in part, to the vulnerability of these neurons and their projections in AD. It remains to be determined whether changes in cholinergic transmission alter APP processing pathways so as to further AD pathology. If so, appropriate cholinomimetic therapeutics might be expected both to provide symptomatic benefit and to abrogate AD pathogenesis.

Acknowledgments The authors gratefully acknowledge support of the Canadian Institutes of Health Research and the many contributions of Drs. D. Westaway, H.T. Mount, and R. Quirion to this research program. J.H.J. is a recipient of Canada Research Chair (CRC) in Alzheimer's Research, and S.K. is a recipient of CRC in Neurodegenerative Diseases and a Senior Scholar award from the Alberta Heritage Foundation for Medical Research.

References

1. Whitehouse PJ. Genesis of Alzheimer's disease. Neurology 1997;48(Suppl 7):S2-7.
2. Katzman R. The prevalence and malignancy of Alzheimer's disease. Arch Neurol 1976;33:217-8.
3. Cummings JL. Alzheimer's disease. N Engl J Med 2004;351:56-67.
4. Holmes C. Genotype and phenotype in Alzheimer's disease. Br J Psychiatry 2002;180:131-4.
5. Bertram L, Tanzi RE. The current status of Alzheimer's disease genetics: what do we tell patients? Pharmacol Res 2004;50:385-396.
6. St George-Hyslop PH, Petit A. Molecular biology and genetics of Alzheimer's disease. C R Biologies 2004;328:119-130.
7. Strittmatter WJ, Saunders AM, Schmeckel D, et al. Apolipoprotein E: High-avidity binding to β-amyloid and increased frequency of type 4 allele in late-onset familial Alzheimer's disease. Proc Natl Acad Sci USA 1993;90:1977-81.
8. Poirier J, Davingnon J, Bouthillier D, et al. Apolipoprotein E polymorphism and Alzheimer's disease. Lancet 1993;342:697-9.
9. Selkoe DJ. Alzheimer's disease: genes, proteins and therapy. Physiol Rev 2001;81:741-66.
10. Muller-Spahn F, Hock C. Risk factors and differential diagnosis of Alzheimer's disease. Eur Arch Psychiatry Clin Neurosci 1999;249(Suppl 3):III/37-III/42.
11. Price DL, Sisodia SS. Mutant genes in familial Alzheimer's disease and transgenic models. Annu Rev Neurosci 1998;21:479-505.
12. Lee VM. Disruption of the cytoskeleton in Alzheimer's disease. Curr Opin Neurobiol 1995;5:663-8.
13. Iqbal K, Alonso Adel C, Chen S, et al. Tau pathology in Alzheimer's disease and other tauopathies. Biochim Biophys Acta 2005;1739(2-3):198-210.
14. Brion JP, Anderton BH, Authelet M, et al. Neurofibrillary tangles and tau phosphorylation. Biochem Soc Symp 2001;67:81-8.
15. Billingsley ML, Kincaid RL. Regulated phosphorylation and dephosphorylation of tau protein: effects on microtubule interaction, intracellular trafficking and neurodegeneration. Biochem J 1997;323:577-91.
16. Bierer LM, Hof PR, Purohit DP, et al. Neocortical neurofibrillary tangles correlate with dementia severity in Alzheimer's disease. Arch Neurol 1995;52:81-8.
17. Dickson DW. The pathogenesis of senile plaques. J Neuropathol Exp Neurol 1997;56:321-39.
18. Clippingdale AB, Wade JD, Barrow CJ. The amyloid-β peptide and its role in Alzheimer's disease. J Peptide Sci 2001;7:227-49.
19. Wisniewski T, Ghiso J, Frangione B. Biology of Aβ amyloid in Alzheimer's disease. Neurobiol Dis 1997;4:313-28.
20. Naslund J, Haroutunian V, Mohs R, et al. Correlation between elevated levels of amyloid beta-peptide in the brain and cognitive decline. JAMA 2000;283: 1571-7.
21. McLean CA, Cherny RA, Fraser FW, et al. Soluble pool of Abeta amyloid as a determinant of severity of neurodegeneration in Alzheimer's disease. Ann Neurol 1999;46:860-6.
22. Lue LF, Kuo YM, Roher AE, et al. Soluble amyloid beta peptide concentration as a predictor of synaptic change in Alzheimer's disease. Am J Pathol 1999; 155:853-62.
23. Tanzi RE. Neuropathology in the Down's syndrome brain. Nat Med 1996;2:31-2.
24. Pike CJ, Burdick D, Walencewicz AJ, et al. Neurodegeneration induced by β-amyloid peptides in vitro: the role of peptide assembly state. J Neurosci 1993;13:1676-87.
25. Yankner BA. Mechanisms of neuronal degeneration in Alzheimer's disease. Neuron 1996;16:921-32.
26. Games D, Adams D, Alessandrini R, et al. Alzheimer's type neuropathology in transgenic mice overexpressing V717F β-amyloid precursor protein. Nature 1995;373:523-7.
27. Hsiao K, Chapman P, Nilsen S, et al. Correlative memory deficits, Aβ elevation, and amyloid plaques in transgenic mice. Science 1996;274:99-102.
28. Calhoun M, Wiederhold K, Abramowski D, et al. Neuron loss in APP transgenic mice. Nature 1998;395:755-6.
29. Bondolfi L, Calhoun M, Ermini F, et al. Amyloid-associated neuron loss and gliogenesis in the neocortex of amyloid precursor protein transgenic mice. J Neurosci 2002;22:515-22.
30. Giovannelli L, Casamenti F, Scali C, et al. Differential effects of amyloid peptides β-(1-40) and β-(25-35) injections into rat nucleus basalis. Neuroscience 1995;66:781-92.
31. Harkany T, Abraham I, Timmerman W, et al. β-amyloid neurotoxicity is mediated by a glutamate-triggered excitotoxic cascade in rat nucleus basalis. Eur J Neurosci 2000;12:2735-45.

32. Itoh A, Nitta A, Nadai M, et al. Dysfunction of cholinergic and dopaminergic neuronal systems in β-amyloid protein-infused rats. J Neurochem 1996;66: 1113-7.
33. Geula C, Wu CK, Saroff D, et al. Aging renders the brain vulnerable to amyloid β-protein neurotoxicity. Nat Med 1998;4:827-31.
34. Gotz J, Chen F, van Dorpe J, et al. Formation of neurofibrillary tangles in P301l tau transgenic mice induced by Abeta 42 fibrils. Science 2001;293:1491-5.
35. Lewis J, Dickson DW, Lin WL, et al. Enhanced neurofibrillary degeneration in transgenic mice expressing mutant tau and APP. Science 2001;293:1487-91.
36. Geula C, Mesulam MM. Cholinergic system and related neuropathological predilection patterns in Alzheimer's disease. In: Terry RD, Katzman R, Bick KL, editors. Alzheimer's Disease. New York: Raven Press Ltd.; 1994:263-91.
37. DeKosky ST, Scheff SW, Styren SD. Structural correlates of cognition in dementia: quantification and assessment of synapse change. Neurodegeneration 1996;5:417-21.
38. Lander CJ, Lee JM. Pharmacological drug treatment of Alzheimer's disease: the cholinergic hypothesis revisited. J Neuropathol Exp Neurol 1998;57:719-31.
39. Francis PT, Palmer AM, Snape M, et al. The cholinergic hypothesis of Alzheimer's disease: a review of progress. J Neurol Neurosurg Psychiatry 1999;66: 137-47.
40. Davies P, Maloney AJF. Selective loss of central cholinergic neurons in Alzheimer's disease. Lancet 1976;2:1403.
41. Blusztajn JK, Berse B. The cholinergic neuronal phenotype in Alzheimer's disease. Metab Brain Dis 2000;15:45-64.
42. Perry EK, Tomlinson BE, Blessed G, et al. Correlation of cholinergic abnormalities with senile plaques and mental test scores in senile dementia. Br Med J 1978;2:1457-9.
43. Bartus RT, Dean RLIII, Beer B, et al. The cholinergic hypothesis of geriatric memory dysfunction. Science 1982;217:408-17.
44. Trinh NH, Hoblyn J, Mohanty S, et al. Efficacy of cholinesterase inhibitors in the treatment of neuropsychiatric symptoms and functional impairment in Alzheimer's disease: a meta-analysis. JAMA 2003;289:210-6.
45. Morris JC. Challenging assumptions about Alzheimer's disease; mild cognitive impairment and the cholinergic hypothesis. Ann Neurol 2002;51: 143-4.
46. Davis KL, Mohs RC, Marin D, et al. Cholinergic markers in elderly patients with early signs of Alzheimer's disease. JAMA 1999;281:1401-6.
47. DeKosky ST, Ikonomovic MD, Styren SD, et al. Upregulation of choline acetyltransferase activity in hippocampus and frontal cortex of elderly subjects with mild cognitive impairment. Ann Neurol 2002;51:145-55.
48. Gilmor ML, Erickson JD, Varoqui H, et al. Preservation of nucleus basalis neurons containing choline acetyltransferase and the vesicular acetylcholine transporter in the elderly with mild cognitive impairment and early Alzheimer's disease. J Comp Neurol 1999;411:693-704.
49. Terry AV, Buccafusco JJ. The cholinergic hypothesis of age and Alzheimer's disease-related cognitive deficits: recent challenges and their implications for novel drug development. J Pharmacol Exp Ther 2003;306:821-7.
50. Mesulam M. The cholinergic lesions of Alzheimer's disease:pivotal factor or slide show? Learn Mem 2004;11:43-9.
51. Nordberg A, Alafuzoff I, Winbald B. Nicotinic and muscarinic receptor subtypes in the human brain: changes with aging and dementia. J Neurosci Res 1992;31:103-11.
52. Warpman U, Alafuzoff I, Nordberg A. Coupling of muscarinic receptors to GTP proteins in postmortem human brain – alterations in Alzheimer's disease. Neurosci Lett 1993;150:39-43.
53. Rodriguez-Puertas R, Pascual J, Vilaro T, et al. Autoradiographic distribution of M1, M2, M3, and M4 muscarinic receptor subtypes in Alzheimer's disease. Synapse 1997;26:341-50.
54. Mulugeta E, Karlsson E, Islam A, et al. Loss of muscarinic M4 receptors in hippocampus of Alzheimer's patients. Brain Res 2003;960:259-62.
55. Colquhoun LM, Patrick JW. Pharmacology of neuronal nicotinic acetylcholine receptor subtypes. Adv Pharmacol 1997;39:191-20.
56. Drisdel RC, Green WN. Neuronal α-bungarotoxin receptors are α_7 subunit homomers. J Neurosci 2000; 20:133-9.
57. Nordberg A, Lundqvist H, Hartvig P, et al. Kinetic analysis of regional (S)(-)^{11}C-nicotine binding in normal and Alzheimer's brains—in vivo assessment using positron emission tomography. Alzheimer's Dis Assoc Disord 1995;9:21-7.
58. Court J, Martin-Ruiz C, Piggott M, et al. Nicotinic receptor abnormalities in Alzheimer's disease. Biol Psychiatry 2001;49:175-84.
59. Teaktong T, Graham A, Court J, et al. Alzheimer's disease is associated with a selective increase in α_7 nicotinic acetylcholine receptor immunoreactivity in astrocytes. Glia 2003;41:207-11.
60. Kang J, Lemaire GH, Unterbeck A, et al. The precursor of Alzheimer's disease amyloid A4 protein resembles a cell surface receptor. Nature 1987;325:733-6.

61. Vassar R, Bennett BD, Babu-Khan S, et al. β-secretase cleavage of Alzheimer's amyloid precursor protein by the transmembrane aspartic protease BACE. Science 1999;286:735-41.
62. Kimberly WT, LaVoie MJ, Ostaszewski BL, et al. Gamma-secretase is a membrane protein complex comprised of presenilin, nicastrin, Aph-1, and Pen-2. Proc Natl Acad Sci USA 2003;100:6382-7.
63. Haass C, Steiner H. Alzheimer's disease γ-secretase: a complex story of GxGD-type presenilin proteases. Trends Cell Biol 2002;12:556-62.
64. Zhou S, Zhou H, Walian PJ, et al. CD147 is a regulatory subunit of the gamma-secretase complex in Alzheimer's disease amyloid beta-peptide production. Proc Natl Acad Sci USA 2005;102:7499-504.
65. Roberson MR, Harrell LE. Cholinergic and amyloid precursor protein metabolism. Brain Res Rev 1997;25:50-69.
66. Mills J, Reiner PB. Regulation of amyloid precursor protein cleavage. J Neurochem 1999;72:443-60.
67. Nitsch RM, Slack BE, Wurtman RJ, et al. Release of Alzheimer's amyloid precursor derivatives stimulated by activation of muscarinic cholinergic receptor. Science 1992;258:304-7.
68. Pittel Z, Heldman E, Barg J, et al. Muscarinic control of amyloid precursor protein secretion in rat cerebral cortex and cerebellum. Brain Res 1996;742:299-304.
69. Racchi M, Mazzucchelli M, Porrello E, et al. Acetylcholinesterase inhibitors:novel activities of old molecules. Pharmacol Res 2004;50:441-51.
70. Guo FF, Kumahara E, Saffen D. A CalDAG-GEFI/Rap1/B-Raf cassette couples M(1) muscarinic acetylcholine receptors to the activation of ERK1/2. J Biol Chem 2001;276:25568-81.
71. Hellstrom-Lindahl E. Modulation of β-amyloid precursor protein processing and tau phosphorylation by acetylcholine receptors. Eur J Pharmacol 2000;393: 255-63.
72. Manthey D, Heck S, Engert S, et al. Estrogen induces a rapid secretion of amyloid beta precursor protein via the mitogen-activated protein kinase pathway. Eur J Biochem 2001;268:4285-91.
73. Kim SH, Kim YK, Jeong SJ, et al. Enhanced release of secreted form of Alzheimer's amyloid precursor protein from PC12 cells by nicotine. Mol Pharmacol 1997;52:430-6.
74. Lahiri DK, Utsuki T, Chen D, et al. Nicotine reduces the secretion of Alzheimer's β-amyloid precursor protein containing β-amyloid peptide in the rat without altering synaptic proteins. Ann N Y Acad Sci 2002;965:364-372.
75. Buxbaum JD, Ruefli AA, Parker CA, et al. Calcium regulates processing of the Alzheimer's amyloid protein precursor in a protein kinase C-independent manner. Proc Natl Acad Sci USA 1994;91:4489-93.
76. Lahiri DK, Farlow MR, Hintz N, et al. Cholinesterase inhibitors, β-amyloid precursor protein and amyloid β-peptides in Alzheimer's disease. Acta Neurol Scand Suppl 2000;176:60-7.
77. Racchi M, Govoni S. The pharmacology of amyloid precursor protein processing. Exp Gerontology 2003;38:145-57.
78. Zimmermann M, Gardoni F, Marcello E, et al., Acetylcholinesterase inhibitors increase ADAM10 activity by promoting its trafficking in neuroblastoma cell lines. J Neurochem 2004;90:1489-99.
79. Zimmermann M, Borroni B, Cattabeni F, et al. Cholinesterase inhibitors influence APP metabolism in Alzheimer's disease patients. Neurobiol Dis 2005;19:237-42.
80. Mori F, Lai CC, Fusi F, et al. Cholinesterase inhibitors increase secretion of APPs in rat brain cortex. NeuroReport 1995;6:633-6.
81. Lahiri DK, Farlow MR, Sambamurti K. The secretion of amyloid beta-peptides is inhibited in the tacrine-treated human neuroblastoma cells. Mol Brain Res 1998;62:131-40.
82. Shaw KT, Utsuki T, Rogers J, et al. Phenserine regulates translation of beta-amyloid precursor protein mRNA by a putative interleukin-1 responsive element, a target for drug development. Proc Natl Acad Sci USA 2001;98:7605-10.
83. Auld DS, Kornecook TJ, Bastianetto S, et al. Alzheimer's disease and the basal forebrain cholinergic system: relations to beta-amyloid peptides, cognition, and treatment strategies. Prog Neurobiol 2002;68:209-45
84. Jhamandas JH, Cho C, Jassar B, et al. Cellular mechanisms for amyloid β-protein activation of rat basal forebrain neurons. J Neurophysiol 2001;86: 1312-20.
85. Dolezal V, Kasparova J. β-amyloid and cholinergic neurons. Neurochem Res 2003;28:499-506.
86. Kar S, Slowikowski SP, Westaway, Mount HT. Interactions between b-amyloid and central cholinergic neurons: implications for Alzheimer's disease. J Psychiatry Neurosci 2004;29:427-41.
87. Kar S, Seto D, Gaudreau P, et al. β-amyloid-related peptides inhibit potassium-evoked acetylcholine release from rat hippocampal formation. J Neurosci 1996;16:1034-40.
88. Kar S, Issa AM, Seto D, et al. Amyloid β-peptide inhibits high-affinity choline uptake and acetylcholine release in rat hippocampal slices. J Neurochem 1998;70:2179-87.

89. Wang HY, Wild KD, Shank RP, et al. Galanin inhibits acetylcholine release from rat cerebral cortex *via* a pertussis toxin-sensitive G_i protein. Neuropeptides 1999;33:197-205.
90. Melo JB, Agostinho P, Oliveira CR. Amyloid beta-peptide 25-35 reduces [^{3}H]acetylcholine release in retinal neurons. Involvement of metabolic dysfunction. Amyloid 2002;9:221-8.
91. Satoh Y, Hirakura Y, Shibayama S, et al. Beta-amyloid peptides inhibit acetylcholine release from cholinergic nerve endings isolated from an electric ray. Neurosci Lett 2001;302:97-100.
92. Vaucher E, Amount N, Rowe W, et al. Amyloid β peptide levels and its effects on hippocampal acetylcholine release in aged, cognitively-impaired and -unimpaired rats. J Chem Neuroanat 2001;21:323-9.
93. Farr SA, Banks WA, Uezy K, et al. Antibody to β-amyloid protein increases acetylcholine in the hippocampus of 12 month SAMP8 male mice. Life Sci 2003;73:555-62.
94. Lee TF, Shiao YJ, Chen CF, et al. Effect of ginseng saponins on beta-amyloid-suppressed acetylcholine release from rat hippocampal slices. Planta Med 2001;67:634-7.
95. Lee T, Chen C, Wang AC. Effect of Ginkolides on β-amyloid-suppressed acetylcholine release from rat hippocampal slices. Phytother Res 2004;18: 556-60.
96. Wecker L. The synthesis and release of acetylcholine by depolarized hippocampal slices is increased by increased choline available in vitro prior to stimulation. J Neurochem 1991;57:1119-27.
97. Zambrzycka A, Alberghina M, Strosznajder JB. Effects of aging and amyloid-beta peptides on choline acetyltransferase activity in rat brain. Neurochem Res 2002;27:277-81.
98. Dobransky T, Brewer D, Lajoie G, et al. Phosphorylation of 69-kDa choline acetyltransferase at threonine 456 in response to amyloid-beta peptide 1-42. J Biol Chem 2003;278:5883-93.
99. Kristofikova Z, Tekalova H, Klaschka J. Amyloid beta peptide1-40 and the function of rat hippocampal hemicholinium-3 sensitive choline carriers: effects of a proteolytic degradation *in vitro*. Neurochem Res 2001;26:203-12.
100. Pedersen WA, Kloczewiak MA, Blusztajn JK. Amyloid β-protein reduces acetylcholine synthesis in a cell line derived from cholinergic neurones of the basal forebrain. Proc Natl Acad Sci USA 1996;93:8068-71.
101. Pedersen WA, Blusztajn JK. Characterization of the acetylcholine reducing effect of the amyloid-beta peptide in mouse SN56 cells. Neurosci Lett 1997;239:77-80.
102. Hoshi M, Takashima A, Murayama M, et al. Nontoxic amyloid β peptide$_{1-42}$ supresses acetylcholine synthesis. J Biol Chem 1997;272:2038-41.
103. Jhamandas JH, Harris KH, Cho C, et al. Human amylin actions on rat cholinergic basal forebrain neurons: antagonism of beta-amyloid effects. J Neurophysiol 2003;89:2923-30.
104. Joslin G, Krause JE, Hershey AD, et al. Amyloid-β peptide, substance, and bombesin bind to the serpin-enzyme complex receptor. J Biol Chem 1991; 266:21897-902.
105. El Khoury J, Hickman SE, Thomas CA, et al. Scavenger receptor-mediated adhesion of microglia to β-amyloid fibrils. Nature 1996;382:716-9.
106. Yan SD, Chen X, Fu J, et al. RAGE and amyloid-β peptide neurotoxicity in Alzheimer's disease. Nature 1996;382:685-91.
107. Kuner P, Schubenel R, Hertel C. β-amyloid binds to $p75^{NTR}$ and activates NFκB in human neuroblastoma cells. J Neurosci Res 1998;54:798-804.
108. Wang HY, Lee DHS, D'Andrea MR, et al. β-amyloid1-42 binds α_7 nicotinic acetylcholine receptor with high affinity: implications for Alzheimer's disease pathology. J Biol Chem 2000; 275: 5626-32.
109. Wang HY, Lee DHS, Davis CB, et al. Amyloid peptide Aβ1-42 binds selectively and with picomolar affinity to α_7 nicotinic acetylcholine receptors. J Neurochem 2000;75:1155-61.
110. Liu Q, Kawai H, Berg DK. β-amyloid peptide blocks the response of α_7-containing nicotinic receptors on hippocampal neurons. Proc Natl Acad Sci USA 2001;98:4734-9.
111. Pettit DL, Shao Z, Yakel JL. β-amyloid$_{1-42}$ peptide directly modulates nicotinic receptors in the rat hippocampal slice. J Neurosci 2001;21:RC120-24.
112. Tozaki H, Matsumoto A, Kanno T, et al. The inhibitory and facilitatory actions of amyloid-beta peptides on nicotinic ACh receptors and AMPA receptors. Biochem Biophys Res Commun 2002; 294:42-5.
113. Grassi F, Palma E, Tonini R, et al. Amyloid beta(1-42) peptide alters the gating of human and mouse alpha-bungarotoxin-sensitive nicotinic receptors. J Physiol 2003;547:147-57.
114. Wu J, Kuo YP, George AA, et al. β-amyloid directly inhibits α4β2-nicotinic acetylcholine receptors heterologously expressed in human SH-EP1 cells. J Biol Chem 2004;279:37842-51.
115. Dineley KT, Bell K, Bui D, et al. beta-Amyloid peptide activates alpha 7 nicotinic acetylcholine receptors expressed in Xenopus oocytes. J Biol Chem 2002;277:25056-61.

116. Fu W, Jhamandas JH. Beta-amyloid peptide activates non-alpha7 nicotinic acetylcholine receptors in rat basal forebrain neurons. J Neurophysiol 2003;90:3130-6.
117. Nagele RG, D'Andrea MR, Anderson WJ, et al. Intracellular accumulation of beta-amyloid(1-42) in neurons is facilitated by the alpha 7 nicotinic acetylcholine receptor in Alzheimer's disease. Neuroscience 2002;110:199-211
118. Wang HY, Li W, Benedetti NJ, et al. α7 nicotinic acetylcholine receptors mediate β-amyloid peptide-induced tau protein phosphorylation. J Biol Chem 2003;278:31547-53.
119. Lee DHS, Wang HY. Differential physiologic responses of α_7 nicotinic acetylcholine receptors to β-amyloid$_{1-40}$ and β-amyloid$_{1-42}$. J Neurobiol 2003;55:25-30.
120. Kelly JF, Furukawa K, Barger SW, et al. Amyloid β-peptide disrupts carbachol-induced muscarinic cholinergic signal transduction in cortical neurons. Proc Natl Acad Sci USA 1996;93:6753-8.
121. Huang HM, Ou HC, Hsieh SJ. Amyloid beta peptide impaired carbachol but not glutamate-mediated phosphoinositide pathways in cultured rat cortical neurons. Neurochem Res 2000;25:303-12.
122. Selkoe DJ, Schenk D. Alzheimer's disease: molecular understanding predicts amyloid-based therapeutics. Annu Rev Pharmacol Toxicol 2003;43: 545-84.
123. Pike CJ, Cotman CW. Cultured GABA-immunoreactive neurons are resistant to toxicity induced by β-amyloid. Neuroscience 1993;56:269-74.
124. Olesen OF, Dago L, Mikkelsen JD. Amyloid β neurotoxicity in the cholinergic but not in the serotonergic phenotype of RN46A cells. Mol Brain Res 1998;57:266-74.
125. Le W, Xie WJ, Kong R, et al. β-amyloid-induced neurotoxicity of a hybrid septal cell line associated with increased tau phosphorylation and expression β-amyloid precursor protein. J Neurochem 1997; 69:978-85.
126. Zheng WH, Bastianetto S, Mennicken F, et al. Amyloid β peptide induces tau phosphorylation and neuronal degeneration in rat primary septal cultured neurons. Neuroscience 2002;115:201-11.
127. Jhamandas JH, MacTavish D. Antagonist of the amylin receptor blocks beta-amyloid toxicity in rat cholinergic basal forebrain neurons. J Neurosci 2004;24:5579-84.
128. Jhamandas JH, Wie MB, Harris K, et al. Fucoidan inhibits cellular and neurotoxic effects of beta-amyloid (A beta) in rat cholinergic basal forebrain neurons. Eur J Neurosci 2005;21:2649-59.
129. Behl C, Cole GM, Schubert D. Vitamin E protects nerve cells from amyloid β protein toxicity. Biochem Biophys Res Commun 1992;186:944-50.
130. Mattson MP, Cheng B, Davis D, et al. β-amyloid peptides destabilize calcium homeostasis and render human cortical neurons vulnerable to excitotoxicity. J Neurosci 1992;12:376-89.
131. Hensley K, Carney JM, Mattson MP, et al. A model for β-amyloid aggregation and neurotoxicity based on free radical generation by the peptide: relevance to Alzheimer's disease. Proc Natl Acad Sci USA 1994;91:3270-74.
132. Perini G, Della-Bianca V, Politi V, et al. Role of p75 neurotrophin receptor in the neurotoxicity by beta-amyloid peptides and synergistic effect of inflammatory cytokines. J Exp Med 2002;195:907-18
133. ZhangY, Hong Y, Bounhar Y, et al., p75 neurotrophin receptor protects primary cultures of human neurons against extracellular amyloid beta peptide cytotoxicity. J Neurosci 2003;23:7385-94.
134. Busciglio J, Lorenzo A, Yeh J, et al. β-amyloid fibrils induce tau phosphorylation and loss of microtubule binding. Neuron 1995;14:879-88.
135. Shea TB, Prabhakar S, Ekinci FJ. β-amyloid and ionophore A23187 evoke tau hyper-phosphorylation by distinct intracellular pathways: differential involvement of the calpain/protein kinase C system. J Neurosci Res 1997;49:759-68.
136. Alvarez G, Munoz-Montano JR, et al. Lithium protects cultured neurons against β-amyloid-induced neurodegeneration. FEBS Lett 1999;453:260-64.
137. Mark RJ, Lovell MA, Markesbery WR, et al. A role for 4-hydroxynonenal, an aldehydic product of lipid peroxidation, in disruption of ion homeostasis and neuronal death induced by amyloid β-peptide. J Neurochem 1997;68:255-64.
138. Greenberg SM, Kosik KS. Secreted β-APP stimulates MAP kinase and phosphorylation of tau in neurons. Neurobiol Aging 1995;16:403-8.
139. Takashima A, Honda T, Yasutake K, et al. Activation of tau protein kinase I/glycogen synthase kinase-3 beta by amyloid beta peptide (25-35) enhances phosphorylation of tau in hippocampal neurons. Neurosci Res 1998;4:317-23.
140. Alvarez A, Toro R, Caceres A, et al. Inhibition of tau phosphorylating protein kinase cdk5 prevents β-amyloid-induced neuronal death. FEBS Lett 2001;459:421-6.
141. Hong M, Chen DCR, Klein PS, et al. Lithium reduces tau phosphorylation by inhibition of glycogen synthase kinase-3. J Biol Chem 1997;40:25326-32.
142. Sadot E, Gurwitz D, Barg J, et al. Activation of m1 muscarinic acetylcholine receptor regulates τ phosphorylation in transfected PC12 cells. J Neurochem 1996;66:877-80.
143. Hellstrom-Lindahl E, Moore H, Nordberg A. Increased levels of tau protein in SH-SY5Y cells

after treatment with cholinesterase inhibitors and nicotinic agonists. J Neurochem 2000;74:777-84.
144. Yamaguchi Y, Kawashima S, Effects of amyloid-beta-(25-35) on passive avoidance, radial-arm maze learning and choline acetyltransferase activity in the rat. Eur J Pharmacol 2001;412:265-72.
145. Gonzalo-Ruiz A, Sang JM, Arevalo J, Amyloid beta peptide-induced cholinergic fibres loss in the cerebral cortex of the rat is modified by diet high in lipids and by age. J Chem Neuroanat 2005;29:31-48.
146. Abe E, Casamenti F, Giovannelli L, et al. Administration of amyloid β-peptides into the medial septum of rats decreases acetylcholine release from hippocampus in vivo. Brain Res 1994; 636:162-4.
147. Giovannini MG, Scali C, Prosperi C, et al. Beta-amyloid-induced inflammation and cholinergic hypofunction in the rat brain in vivo: involvement of the p38MAPK pathway. Neurobiol Dis 2002;11:257-74.
148. Boncristiano S, Calhoun ME, Kelly PH, et al. Cholinergic changes in the APP23 transgenic mouse model of cerebral amyloidosis. J Neurosci 2002;22:3234-43.
149. Bronfman FC, Moechars D, Van Leuven F. Acetylcholinesterase-positive fiber deafferentation and cell shrinkage in the septohippocampal pathway of aged amyloid precursor protein london mutant transgenic mice. Neurobiol Dis 2000;7:152-68.
150. Wong TP, Debeir T, Duff K, et al. Reorganization of cholinergic terminals in the cerebral cortex and hippocampus in transgenic mice carrying mutated presenilin-1 and amyloid precursor protein transgenes. J Neurosci 1999;19:2706-16.
151. Jaffar S, Counts SE, Ma SY, et al. Neuropathology of mice carrying mutant APPswe and/or $PS1_{M146L}$ transgenes: alterations in the p^{75NTR} cholinergic basal forebrain septohippocampal pathway. Exp Neurol 2001;170:227-43.
152. Gau JT, Steinhilb ML, Kao TC, et al. Stable β-secretase activity and presynaptic cholinergic markers during progressive central nervous system amyloidogenesis in Tg2576 mice. Am J Pathol 2002;160:731-8.
153. Vaucher E, Fluit P, Chishti MA, et al. Alteration in working memory but not cholinergic receptor binding sites in transgenic mice expressing human presenilin 1 transgenes. Exp Neurol 2002;21:323-9.
154. Dineley KT, Xia X, Bui D, et al. Accelerated plaque accumulation, associative learning deficits and upregulation of α_7 nicotinic receptor protein in transgenic mice co-expressing mutant human presenilin 1 and amyloid precursor proteins. J Biol Chem 2002;227:22768-80.
155. Chishti MA, Yang DS, Janus C, et al. Early-onset amyloid deposition and cognitive deficits in transgenic mice expressing a double mutant form of amyloid precursor protein 695. J Biol Chem 2001;276:21562-70.
156. Slowikowski SPM, Chishti MA, Zheng WH, et al. Alterations in cholinergic parameters in the hippocampus of transgenic mice expressing mutated amyloid precursor protein and/or presenilin-1 transgenes. Soc Neurosci Abs 2002;295.18.
157. Apelt J, Kumar A, Schliebs R. Impairment of cholinergic neurotransmission in adult and aged transgenic Tg2576 mouse brain expressing the Swedish mutation of human beta-amyloid precursor protein. Brain Res 2002;953:17-30.
158. Hernandez D, Sugaya K, Qu T, et al. Survival and plasticity of basal forebrain cholinergic system in mice transgenic for presenilin-1 and amyloid precursor protein mutant genes. NeuroReport 2001; 12:1377-84.
159. German DC, Yazdani U, Speciale SG, et al. Cholinergic neuropathology in a mouse model of Alzheimer's disease. J Comp Neurol 2003;462:371-81.
160. Hu L, Wong TP, Cote SL, et al. The impact of Aβ-plaques on cortical cholinergic and non-cholinergic presynaptic boutons in Alzheimer's disease-like transgenic mice. Neuroscience 2003;121:421-32.
161. Feng Z, Chang Y, Cheng Y, et al. Melatonin alleviates behavioral deficits associated with apoptosis and cholinergic system dysfunction in the APP 695 transgenic mouse model of Alzheimer's disease. J Pineal Res 2004;37:129-36.
162. Klingner M, Apelt J, Kumar A, Alterations in cholinergic and non-cholinergic neurotransmitter receptor densities in transgenic Tg2576 mouse brain with beta-amyloid plaque pathology. Int J Dev Neurosci 2003;21:357-69.
163. Oddo S, Caccamo A, Green KN, et al. Chronic nicotine administration exacerbates tau pathology in a transgenic model of Alzheimer's disease. Proc Natl Acad Sci USA 2005;102:3046-3051.
164. Hartmann J, Erb C, Ebert U, et al., Central cholinergic functions in human amyloid precursor protein knock-in/presenilin-1 transgenic mice. Neuroscience 2004;125:1009-17.
165. Haass C, Schlossmacher MG, Hung AY, et al. Amyloid β-peptide is produced by cultured cells during normal metabolism. Nature 1992;359:322-5.
166. Seubert P, Vigo-Pelfrey C, Esch F, et al. Isolation and quantification of soluble Alzheimer's β-peptide from biological fluids. Nature 1992;359:325-7.
167. Shoji M, Golde TE, Ghiso J, et al. Production of Alzheimer's β protein by normal proteolytic processing. Science 1992;258:126-9.

168. Iverfeldt K, Walaas SI, Greengard P. Altered processing of Alzheimer's amyloid precursor protein in response to neuronal degeneration. Proc Natl Acad Sci USA 1993;90:4146-50.
169. Wallace W, Ahlers ST, Gotlib J, et al. Amyloid precursor protein in the cerebral cortex is rapidly and persistently induced by loss of subcortical innervation. Proc Natl Acad Sci USA 1993;90:8712-6.
170. Lin L, Georgievska B, Mattsson A, et al. Cognitive changes and modified processing of amyloid precursor protein in the cortical and hippocampal system after cholinergic synapse loss and muscarinic receptor activation. Proc Natl Acad. Sci USA 1999;96:12108-13.
171. Buxbaum JD, Oishi M, Chen HI, et al. Cholinergic agonists and interleukin 1 regulate processing and secretion of the β/A_4 amyloid precursor protein. Proc Natl Acad Sci USA 1992;89:10075-8.
172. Allen DD, Galdzicki Z, Brining SK, et al. Beta-amyloid induced increase in choline flux across PC12 cell membranes. Neurosci Lett 1997;234:71-3.
173. Wurtman R. Choline metabolism as a basis for the selective vulnerability of cholinergic neurones. Trends Neurosci 1992;15:117-22.

11 Physiologic and Neurotoxic Properties of Aβ Peptides

Gillian C. Gregory, Claire E. Shepherd, and Glenda M. Halliday

11.1 Introduction

Alzheimer's disease (AD) is characterized by a gradual decline of numerous cognitive processes, culminating in dementia and neurodegeneration. It is the most common form of dementia and a significant cause of death in the elderly. Definitive diagnosis of AD requires the presence of the extracellular accumulation of Aβ peptides in senile plaques in the cortex of the brain (Fig. 11.1) [1]. β-Amyloid (Aβ) peptides are ~4-kDa polypeptides with the main alloforms consisting of 40 and 42 amino acids. Analysis of the insoluble protein fraction has identified the longer $A\beta_{42}$ alloform as the predominant peptide species in the neuropathologic accumulations (see [2]), although Aβ peptides of variable length accumulate within plaques [3–8]. The association between the abnormal accumulation of Aβ peptides in the brain and dementia is strong evidence that Aβ peptides are vital for normal brain functioning.

Some of our understanding about Aβ and brain function has occurred after the identification of genetic mutations in the amyloid precursor protein (APP) that cause AD [9, 10] and the subsequent use of molecular biology to study the cellular mechanisms involved in Aβ production and clearance. Initial reports using human APP695 mice and PDAPP mice with the APP717 mutation revealed that these mutations caused Aβ levels to increase two to three times over control mice with Aβ deposition only occurring at these levels of production [11–13]. Subsequent studies revealed that these genetic mutations increase the amount of the $A\beta_{42}$ alloform over other Aβ species [14–16]. The study of these abnormalities in Aβ processing has led to a better understanding of the role Aβ peptides play within the brain.

11.2 Production of Aβ Peptides

The Aβ peptides are derived from the proteolytic processing of APP [17]. APP belongs to a heterogenous group of ubiquitously expressed polypeptides, with the heterogeneity arising from alternative splicing and post-translational modifications [18]. The pre-mRNA is spliced to produce three major isoforms APP_{770}, APP_{751}, and APP_{695} with the APP_{695} isoform expressed at high levels in neurons (APP 770:751:695 mRNA ratio is 1:10:20 in the cortex [19]). APP is a single membrane–spanning protein with a large extracellular N-terminal and small intracellular C-terminal domain and is localized to numerous membranous structures in the cell; the endoplasmic reticulum, Golgi compartments, and cell membrane [18]. In the axonal membrane, APP acts as a receptor for kinesin 1 during the fast axoplasmic transport of vesicles containing numerous proteins [20]. In addition to its possible role in membrane functions, APP undergoes considerable post-translational modifications including glycosylation and specific proteolytic cleavage to produce fragments that are believed to be extensively involved in adhesion, neurotrophic and neuroproliferative activity, intercellular communication, and membrane-to-nucleus signaling [21].

Proteolytic cleavage of APP occurs via at least two pathways involving three secretases (α, β,

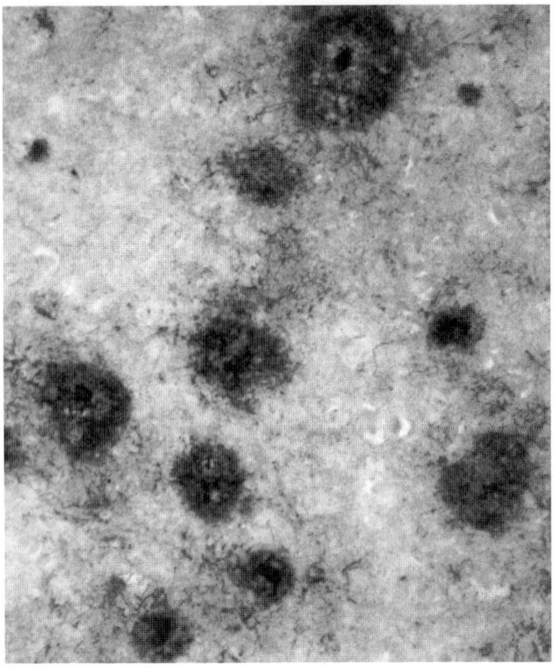

FIGURE 11.1. Tissue section from the temporal lobe of an early-onset AD case immunohistochemically stained for $A\beta_{42}$. Initially, Aβ deposits in diffuse plaques that are typically 10–200 μm in diameter with ill-defined boundaries. Over time, the accumulating Aβ becomes fibrillar acquiring a β-pleated sheet structure, and neuritic plaques develop. These plaques are associated with axonal and dendritic injury of pyramidal cells, known as dystrophic neurites, which occur both within this amyloid deposit and immediately surrounding it. The accumulating Aβ in neuritic plaques develops further into the classic senile plaques that have a distinct concentrated Aβ core surrounded by a ring or "corona" of neuritic pathology.

and γ), with only one pathway generating full-length Aβ peptide [18]. The α- and β-secretase cleavages are seen as mutually exclusive events, each releasing a large extracellular domain of the APP protein, soluble APP (sAPP). α-Secretase cleavage precludes the formation of Aβ, instead producing a shortened fragment, together with γ-secretase cleavage, called p3 [22]. Production of these non-amyloidogenic sAPP and p3 fragments occurs within the endoplasmic reticulum, the trans-Golgi apparatus, and at the cell membrane [23].

The Aβ peptides are generated early in the secretory trafficking of APP and at the cell surface. APP not cleaved at the cell surface by α-secretase is reinternalized for processing in the endosome/lysosome system by β-secretase [24, 25]. β-secretase, an aspartyl protease known as BACE (β-site APP cleavage enzyme) [26], cleaves APP both within the endocytic and secretory pathways of the endoplasmic reticulum and the Golgi [27]. The remaining APP fragment, the C-terminal fragment, is secured to the membrane. γ-Secretase cleavage occurs in the hydrophobic transmembrane domain, after the α- or β-secretase cleavage events, and creates the carboxyl terminus of the Aβ peptide. Studies suggest that Aβ peptides produced in the endoplasmic reticulum may not be secreted and are instead retained and catabolized inside the cell [27]. Most Aβ, however, is believed to be secreted into the extracellular space [18].

The γ-secretase consists of a complex of proteins made up of presenilin 1 and 2 (PS1 and PS2), nicastrin [28, 29], Aph-1 [30, 31], and pen-2 [31], though recent data suggest that different combinations of these proteins may exist [32]. This cleavage event occurs at different sites in the C-terminal fragment producing the predominant $A\beta_{1\text{-}40}$ and $A\beta_{1\text{-}42}$ fragments as well as $A\beta_{1\text{-}39}$ and $A\beta_{1\text{-}43}$. It is not clearly understood how the γ-secretase determines its particular cleavage site in the C-terminal fragment and what regulates the production of one peptide length over another. Such regulation is likely to have a substantial effect on overall Aβ function due to the different physicochemical properties of the peptides.

11.3 Detection and Tissue Location of Aβ Peptides

The Aβ peptides can be detected in numerous biological milieus, such as the CSF, plasma, and brain. Many studies have determined the concentrations of the peptides in these different locations, predominantly in the plasma and CSF because availability and access to these areas is markedly easier than brain tissue [33–46]. Comparisons and quantification of Aβ in plasma and CSF between control and AD samples have been performed for the development of biomarkers or objective predictors of cognitive dysfunction [47]. However, conflicting results have precluded any advances in this area because Aβ peptide concentrations in both CSF and plasma are highly variable [33, 35, 45, 48, 49].

The CSF bathes and drains from the brain, which implies that CSF Aβ mainly arises from brain tissue and in nondiseased states reflects brain tissue concentrations of these peptides. In control CSF, $A\beta_{40}$ is the dominant species, with concentrations consistently higher than $A\beta_{42}$ [33–36]. This suggests that the dominant Aβ peptide secreted by the cells of the brain is $A\beta_{40}$ and that γ-secretase cleavage preferentially produces this shorter Aβ peptide. It has been shown that CSF Aβ levels follow a natural U-shaped course in normal aging (Fig. 11.2). Proportionately higher concentrations of both $A\beta_{40}$ and $A\beta_{42}$ are detected in children compared with adults between 30 and 60 years of age [36, 37]. Concentrations then increase proportionately with further aging [36]. Low levels of Aβ during adulthood suggests that equilibrium has been reached between the cellular synthesis and extracellular

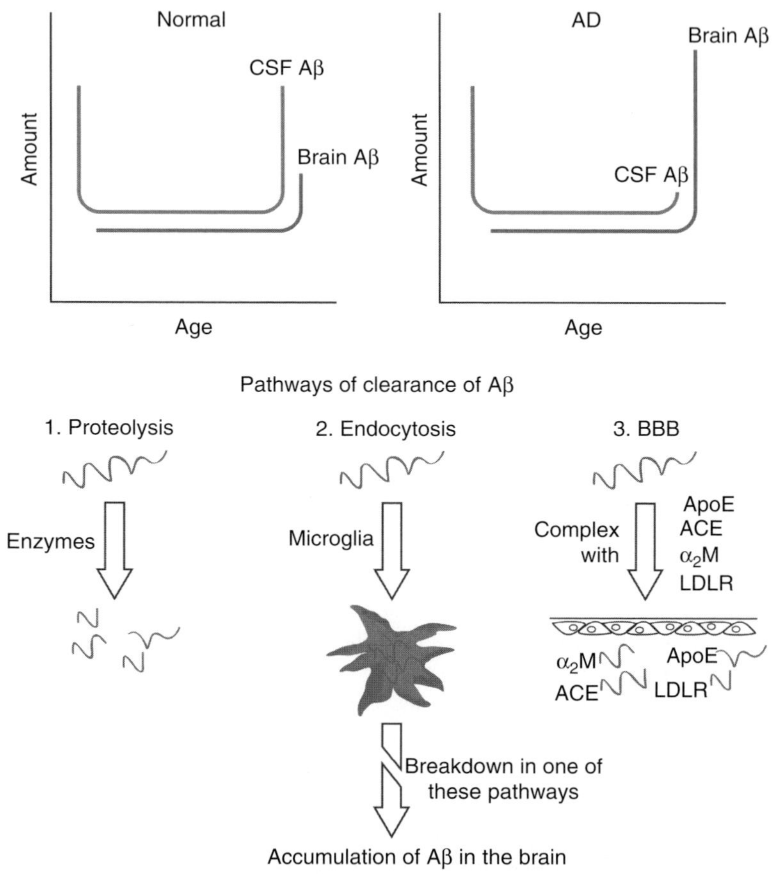

FIGURE 11.2. Graphs depicting normal (left) and abnormal (right) Aβ brain levels and a diagram depicting the mechanisms of Aβ clearance from the brain. The left-hand graph shows the natural U-shaped course of CSF Aβ during normal aging. Proportionately high concentrations of both $A\beta_{40}$ and $A\beta_{42}$ occur in childhood and are then downregulated between the ages of 30 and 60 years. Aβ peptide levels then proportionately increase with subsequent aging. Low levels of Aβ during adulthood suggests that equilibrium has been reached between the cellular synthesis and extracellular clearance of these peptides, and that with older age this equilibrium is changed. The diagrams in the lower part of the figure depict Aβ clearance mechanisms. Normal removal of Aβ from the brain occurs via extracellular proteolysis, receptor-mediated endocytosis, and transport across the blood brain barrier (BBB) via angiotensin-converting enzyme (ACE) and α_2-macroglobulin (α_2M) through interactions with LDL-receptor–related protein (LDLR) and apolipoproteins (ApoE). The right-hand graph shows that breakdown in one of the clearance pathways, and failure to clear the Aβ peptide, leads to increased brain Aβ and, hence, AD.

clearance of these peptides and that with older age this equilibrium is changed (Fig. 11.2).

Numerous studies of CSF Aβ in AD show a consistent decrease in $Aβ_{42}$ concentrations compared with controls [33–35, 37–44] and a negative correlation between $Aβ_{42}$ levels and disease severity [40, 50]. $Aβ_{40}$ levels in AD CSF remain the same [33–35, 37, 39] or decrease [40, 43] compared with controls. The lower $Aβ_{42}$ CSF levels in AD are thought to be due to reduced $Aβ_{42}$ clearance consistent with the preferential deposition of $Aβ_{42}$ in AD brain [51]. However, there is an overlap in CSF $Aβ_{42}$ values between AD and control groups [35] with the clearance problem occurring primarily in early disease [50]. More intriguing are studies that show low CSF $Aβ_{42}$ levels in patients with a variety of other disorders, some of which do not deposit Aβ in the brain. These include major depression [40, 50] and Creutzfeldt-Jakob disease [52], suggesting a possible dissociation between Aβ clearance and deposition. In addition, the same deficit occurs in patients with dementia with Lewy bodies [53] limiting the role of this measurement as a specific diagnostic marker for AD. Overall, these findings suggest that $Aβ_{40}$ is preferentially cleared through the CSF at all ages and in all brain disorders compared with $Aβ_{42}$.

In the plasma of normal elderly, the $Aβ_{40}$ peptide is the dominant species, with average concentrations of $Aβ_{40}$ well above those of $Aβ_{42}$ [35, 45, 46]. Plasma Aβ originates from many sources, but particularly blood-borne platelets, which preferentially produce $Aβ_{40}$ [54]. Platelet activation releases Aβ, and in patients with AD there is an increase in the plasma concentrations of Aβ, particularly $Aβ_{42}$ [45, 55, 56]. The binding of platelet-activating factor to platelets in AD has been used to measure platelet activation. This measure correlates with the degree of cognitive impairment in patients with AD [57], with decreasing platelet APP predicting conversion to dementia [58]. This raises the possibility that increased platelet activation and plasma Aβ may play some role in the dementing process.

Aβ peptides complex with apolipoprotein E (ApoE) and apolipoprotein J (ApoJ) to cross the blood-brain barrier (BBB) [59]. In primates, infused $Aβ_{40}$ readily crosses the BBB compared with other peptides, with the rate of Aβ sequestration into the brain parenchyma after a single exposure increasing with age [60]. In rats, infusions of $Aβ_{40}$ or $Aβ_{42}$ increase BBB permeability [61]. Enhancement of Aβ transport across the BBB along with reduced CSF clearance is thought to contribute to the increased brain deposition of Aβ in a transgenic model of AD [62]. Alternatively, intravenous administration of anti-Aβ antibody promotes a rapid efflux of Aβ from the CNS into plasma [63]. These studies show considerable flux of Aβ peptide across the BBB and suggest that a proportion of brain Aβ could originate from the circulating pool found in plasma.

The "amyloid cascade" hypothesis proposes that the increased burden of Aβ in the brain is the primary intrinsic pathogenic event in AD [64]. Consequently, most studies analyzing brain Aβ peptide levels have concentrated on AD tissue with few studies focusing on Aβ levels in normal (disease free) brain tissue [3, 15, 65–79]. In contrast with the results obtained in CSF and plasma, a large number of these studies show that $Aβ_{40}$ levels in elderly controls are low compared with the levels of $Aβ_{42}$ (for review, see [2]). This suggests that $Aβ_{40}$ is preferentially cleared from the brain, consistent with higher levels in the CSF. Despite these consistent findings, the literature commonly states that $Aβ_{40}$ is the dominant peptide species in the normal brain (for review, see [2]). This misconception is consistent with measurements from peripheral tissues and supernatant from cell lines (equivalent of CSF) [80] but is not supported by data from nondiseased human brain tissue. Unfortunately, this has also influenced research into AD pathogenesis to focus on changes in the production from the more "normal" $Aβ_{40}$ peptide to the $Aβ_{42}$ peptide that has been wrongly thought to only associate with AD.

11.4 Structure of Aβ Peptides

Aβ peptides exist as monomers, dimers, and higher oligomers, with aggregation producing protofibrils and eventually fibrils, in a β-pleated sheet conformation. The Aβ oligomers are believed to play a key role in AD neurotoxicity [81–85]. The formation of Aβ oligomers by the different alloforms occurs through different pathways. $Aβ_{40}$ aggregates as monomers, dimers, trimers, and tetramers in rapid equilibrium, whereas $Aβ_{42}$ preferentially forms pentamer/hexamer units that are able to assemble further to form early protofibril structures [86, 87]. These differences suggest different peptide functions.

Recent experiments have established that the major secondary structure adopted by Aβ depends on the environment [88]. The Aβ monomer contains an amphipathic sequence that favors an α-helix structure (Fig. 11.3) in a membrane or membrane-mimicking environment [89, 90], whereas in an aqueous solution, a nontoxic random coil configuration with few components of α-helix and/or β-sheet conformations is preferred [91–94]. The highly hydrophobic C-terminus of Aβ is embedded in the lipid membrane with its hydrophilic N-terminus protruding extracellularly [95]. Two lipophilic regions (Lys16 to Ala21 and Lys28 to Val40) are believed to be the main functional areas. The first region has an α-helical structure and the second a β-pleated sheet structure, which is able to form hydrophobic forces with other β-sheets of Aβ peptides [91]. The two lipophilic helical regions are separated by a flexible hinge or kink region (Fig. 11.3), which may be important for its membrane-inserting properties and conformational rearrangements [89, 95, 96].

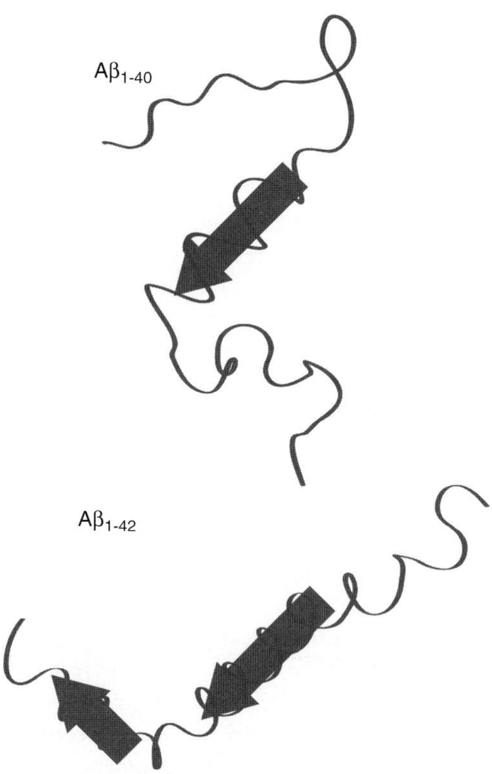

FIGURE 11.3. Membrane-bound structure of the main Aβ peptides, $A\beta_{40}$ and $A\beta_{42}$. Both peptides exhibit α-helical conformations (shown as large arrows) in conditions mimicking lipid membranes (in the presence of organic modifiers such as SDS). $A\beta_{42}$ has two α-helices, on either side of the "kink" region, in contrast with $A\beta_{40}$, which has only one α-helical domain.

The different lengths and structure of the Aβ peptides contribute to their different oligomeric states. Aβ aggregation into oligomers occurs when the dominant structure of Aβ is converted from an α-helix or random coil to a β-sheet conformation [97, 98] through intermediates of mixed helices and β-sheets [88, 92]. In contrast with $A\beta_{42}$, $A\beta_{40}$ has a tendency to move out of the lipid environment [88], possibly contributing to the smaller and more soluble oligomers formed by this peptide.

In disease conditions, when Aβ fibrillogenesis occurs, the structure of the Aβ peptides changes substantially due to increased concentrations and conformational effects. Over time, the helical Aβ residues 29–40 that are embedded into the stabilizing cell membrane leave the lipid bilayer and enter the extracellular environment where they have a high tendency to form short β-sheets in a concentration-dependent fashion thereby precipitating polymers [88, 92]. During the "lag phase" prior to the development of Aβ fibrils, no Aβ precipitates are detectable in brain tissue, suggesting that nucleation of a different structure is required, like seeding a crystallization process. The lag phase can be removed by seeding Aβ monomers with preaggregated Aβ fibrils [99]. Using kinetic studies, $A\beta_{42}$ has been shown to form precipitated fibrils significantly faster than $A\beta_{40}$, leading to the frequently coined phrase that $A\beta_{42}$ is more amyloidogenic than $A\beta_{40}$ [99]. This is probably due to its greater propensity for helical structures and lipid association. In fact, $A\beta_{40}$ has been shown to be comparatively neuroprotective against $A\beta_{42}$-induced neurotoxicity in vitro and in vivo. The mechanism for this neuroprotection may involve the $A\beta_{40}$ peptide inhibiting the β-sheet transformation and fibril formation of $A\beta_{42}$ [100].

Comparison between the concentrations of soluble and insoluble Aβ peptides in control brain tissue [3, 66, 69, 72, 74, 78] suggests that $A\beta_{40}$ is greater in the soluble fraction, whereas $A\beta_{42}$ is the predominant species in the insoluble fraction [2], as may be expected based on the physiochemical properties of the two peptides. There is a significant change in the Aβ levels in the brain tissue of AD cases (both sporadic [3, 15, 65–67, 70, 72, 73, 75, 78, 79, 101, 102] and familial [15, 65, 70, 77, 101, 103, 104]), with significant increases in the amount and insolubility of $A\beta_{42}$ in AD compared with controls (Fig. 11.2), in agreement with the dominant hypothesis that it is the pathogenic species in AD. In addition to the changes in $A\beta_{42}$, $A\beta_{40}$ levels are also increased in AD cases (Fig. 11.2), with greater increases in the amount of insoluble $A\beta_{40}$ than insoluble $A\beta_{42}$ in sporadic AD (for review, see [2]). These studies support the concept that increases in Aβ peptide levels promote significant changes in their structure and therefore their solubility and that these structural changes produce less soluble Aβ peptides and have significant pathogenic effects.

11.5 Other Aβ Binding Partners

Apart from concentration-dependent self-aggregation, Aβ peptides readily bind to other molecules, including lipids, proteins, and metal ions. Three histidine residues in the N-terminal hydrophilic region provide primary metal binding sites on the Aβ peptides. The binding of certain metal ions to Aβ can promote aggregation. Zn^{2+} induces Aβ aggregation at acidic to neutral pH and is the most powerful metal inducer of Aβ aggregation [105]. Cu^{2+} induces aggregation at mildly acidic pH comparable with the pH-dependent effect of Cu^{2+} on insulin aggregation [105]. Under normal physiologic conditions, Cu^{2+} protects Aβ against Zn^{2+}-induced aggregation by competing with Zn^{2+} for the histidine residues of Aβ [106]. A mildly acidic environment together with increased Zn^{2+} and Cu^{2+} are common features of inflammation, which suggests that Aβ aggregation by these factors may be a response to local injury [105].

Lipid membranes are important binding partners for Aβ as the peptide plays a role in the regulation of lipid membrane function, metabolism, and homeostasis [107]. The binding efficacy of lipids to Aβ increases when Aβ forms polymers [108] with the lipids binding to the hydrophobic areas of aggregated Aβ. Cholesterol is a key component of membranes and interacts with Aβ in a reciprocal manner [107]. Aggregated $A\beta_{40}$ in particular has a high affinity for cholesterol with oligomeric Aβ peptides promoting the normal release of lipid from neurons [109]. These Aβ-lipid particles have a very low binding affinity for neurons, reducing lipid internalization and thereby affecting intracellular lipid metabolism. Gangliosides (sialylated glycosphingolipids) are the predominant glycans on neuronal plasma membranes and are concentrated into membrane rafts by cholesterol where they mediate important physiological functions. These lipid rafts (made of cholesterol, sphingomyelin, and glycosphingolipids such as GM1 ganglioside) play an essential role in cell-cell communications and signal transduction across membranes [110]. GM1 ganglioside associates with cholesterol and binds to Aβ peptides, with GM1 ganglisoside–bound Aβ acting as a seed for Aβ fibrillogenesis [111].

In addition to the binding of Aβ to lipids, Aβ also binds to lipid-trafficking lipoproteins. Aβ complexes with ApoJ, a universal lipoprotein expressed in many cells throughout the body. Soluble Aβ also binds to normal human plasma high-density lipoprotein (HDL), including apolipoprotein A (ApoA)-I, ApoA-II, ApoE, and ApoJ [112]. Aβ. binding with ApoE, alleles E2 and E3, form stable membrane-bound complexes that are more abundant than ApoE4-Aβ. complexes [113]. In contrast with neurons, Aβ-ApoE lipid particles are internalized mainly by glia and vascular cells presenting a clearance pathway through which parenchymal Aβ is modulated [114]. Exogenous ApoE3 but not ApoE4 prevents Aβ-induced neurotoxicity by a process requiring ApoE receptors [113].

A subset of plasma membrane proteins and receptors also bind Aβ (for review, see [115]). Heparan sulfate proteoglycans are cell-surface binding sites for Aβ. The serpin-enzyme complex receptor and the insulin receptor can bind monomeric forms of Aβ peptides. The alpha7nicotinic acetylcholine receptor, integrins, RAGE (receptor for advanced glycosylation end-products), and formyl peptide receptor-like 1 are able to bind monomeric and fibrillar forms of Aβ peptides. In addition, APP, collagen-like Alzheimer's amyloid plaque component precursor/collagen XXV, the NMDA (*N*-methyl-D-aspartate) receptor, P75 neurotrophin receptor, scavenger receptors A, BI, and CD36 and complexes bind fibrillar forms of Aβ peptides. It is therefore likely that the function of Aβ differs depending on the associated binding partners, which are modulated by its structure and solubility.

11.6 Function of the Aβ Peptides

The functional properties of the Aβ peptides have not been completely elucidated to date, though numerous studies suggest that the peptides possess a number of neurotrophic and neurotoxic properties. As stated above, the divergent roles of Aβ seem dependent on their physicochemical properties, aggregation state, and binding partners, with $A\beta_{40}$ function primarily studied (both neurotoxic and trophic) due to its greater solubility. Recent studies suggest that soluble Aβ plays important roles in the facilitation of neuronal growth and survival, in the modulation of synaptic function, and in neurotoxic surveillance and defense against oxidative stress [116, 117], whereas oligomeric and fibrillar Aβ have less trophic and greater toxic properties.

11.6.1 Neurotrophic Functions

Recent studies have shown that Aβ peptides may be vital for neuronal development, plasticity, and survival due to its integral membrane interactions [118]. Neuronal viability appears to be dependent on Aβ [117] with the peptide possessing neurogenic properties [119]. Despite some controversy [120, 121], there is increased differentiation of hippocampal neural stem cells treated with $A\beta_{42}$, with no change to the rate of cell death or proliferation. Interestingly, this effect is only seen with soluble oligomeric $A\beta_{42}$ peptide, as neither monomeric $A\beta_{42}$, $A\beta_{25\text{-}35}$, nor $A\beta_{40}$ (aggregated or not) increased the percentage of neurons [119]. This may suggest that the formation of new neurons is induced by the more "soluble" forms of $A\beta_{42}$ that form larger pentamer/hexamer subunits and membrane channels.

11.6.2 Physiologic Functions

Because Aβ binds to the plasma membranes in both soluble and fibrillar forms, it changes the structure and function of the membranes by modifying the fluidity or forming ion channels [115]. Soluble $A\beta_{40}$ increases voltage-gated K(+) channel currents in cerebellar granular neurons without neurotoxic consequences [122]. Neuronally released soluble Aβ selectively depresses excitatory synaptic transmission through interactions with NMDA receptors [116]. The modification of membrane channels in vascular smooth cells causes vasoconstriction, with $A\beta_{40}$ having significantly greater vasoconstrictive effects compared with $A\beta_{42}$ [123]. The negative feedback after synaptic excitation coupled with an ability to reduce local blood flow and oxygen and glucose delivery would keep neuronal hyperactivity in check [116]. This suggests that the nonpathologic soluble forms of Aβ are important synaptic protectors through their ability to change ionic channel functions within cell membranes [122].

Monomeric Aβ peptide is also thought to have an antioxidant function through its metal-binding capabilities, particularly capturing Zn, Cu, and Fe ions and preventing them from participating in redox cycling with other ligands [124]. Aβ production increases with oxidative stress [125–127], and the peptides may be involved in altering ion fluxes by chelating metal ions in an attempt to prevent oxidation [128]. This suggests that Aβ production, in conjunction with its neuroprotective and neurotrophic properties, may be a normal stress response to minimize oxidative damage [129]. The formation of diffuse Aβ plaques in AD may be a compensatory event for the removal of reactive oxygen species.

11.6.3 Neurotoxic Properties

The key to Aβ cellular toxicity appears to be its aggregation state [130]. Aβ appears to promote neuron degeneration only when the peptide assumes a particular β-pleated structure either in oligomeric and/or fibrillar forms. Yankner and colleagues first showed that synthetic $A\beta_{1\text{-}40}$ was neurotoxic in primary rat hippocampal cell cultures [131]. Roher et al. reported that Aβ isolated from AD brains inhibited neurite sprouting and caused cell death in cultured sympathetic neurons [132]. Further studies then demonstrated that the toxicity of the peptide was strongly correlated with its propensity to form fibrillar aggregates [130, 133–137]. However, more recent work has indicated that oligomeric Aβ, the Aβ form required prior to fibrillization, may be the most toxic species involved in neuronal death [81–85]. Studies have shown that oligomeric Aβ induces greater cell death and apoptosis than soluble or fibrillar forms [138, 139], confirming that the structural conformation of the peptide is important in determining its physiological action.

A change in the binding properties of Aβ peptides may induce significant toxicity. In particular, the interaction between oligomeric Aβ and lipids may be an important cause of neuronal degeneration and would certainly impact on lipid homeostasis and function [109]. Michikawa and colleagues propose that the stimulation of lipid release from neurons by the increase in oligomeric Aβ in AD induces a disruption of cholesterol homeostasis and membrane raft maintenance in the brain, with the consequent neurotoxic changes such as an increase in tau phosphorylation [109, 140].

A change in the neurotrophic properties of Aβ peptides may also induce considerable toxicity. Physiological levels of Aβ can interfere with functions critical for neuronal plasticity [141]. Pretreatment of neurons with sublethal concentrations of the more amyloidogenic $A\beta_{1\text{-}42}$ suppresses the phosphorylation of cAMP-response element binding protein (CREB) and the downstream activation of brain-derived neurotrophic factor (BDNF). As both CREB and BDNF play critical roles in neuronal plasticity, an increase in the $A\beta_{1\text{-}42}$ suppression of this function may play a role in the cognitive deficits associated with AD [141].

Significant toxicity may also be induced by a change in the regulation of synaptic feedback and local blood flow by Aβ peptides. Increased release of Aβ from neurons significantly downregulates synaptic activity [116], and increased Aβ binding to vascular smooth muscle cells increases vasoconstriction and decreases local blood flow [123]. These changes would reduce synaptic function and therefore affect cognition. Aβ aggregation also changes synaptic properties due to downstream increases in intracellular free Ca^{2+} and decreased transmitter manufacturing through lower enzyme activities [142].

Changes in metal binding to Aβ peptides may also induce significant toxicity due to increased oxidation [143–146] leading to mitochondrial dysfunction [147]. The methionine residue 35 (met-35) of Aβ is critical to its oxidative stress and neurotoxic properties, with its removal abolishing the neurotoxic properties of $A\beta_{1\text{-}42}$ [148]. Although Zn^{2+} binding induces the greatest Aβ aggregation, the oxidative toxicity of Aβ in cell culture is mediated through its interaction with Cu^{2+} and Fe^{3+} [149, 150]. Aβ catalyzes the reduction of Cu^{2+} to Cu^{+} and Fe^{3+} to Fe^{2+}, generating H_2O_2 from molecular oxygen and available biological reducing agents such as vitamin C, cholesterol, and catecholamines [150]. Any reduced activity of the detoxifying enzymes, such as cytosolic Cu/Zn superoxide dismutase (SOD1), catalase, and/or glutathione peroxidase, allows H_2O_2 to further react with reduced Fe^{2+} and Cu^{+} to generate toxic hydroxyl radicals. $A\beta_{42}$ has greater oxidative toxicity than $A\beta_{40}$ [149] due to their relative Cu^{2+} and Fe^{3+} reducing potentials and the ability to catalytically generate H_2O_2 from biological reducing agents [150].

11.7 Clearance of Aβ Peptides from the Brain

Aβ clearance occurs through at least three pathways (Fig. 11.2): extracellular proteolysis by degrading enzymes [151], transport across the BBB [152], and receptor-mediated endocytosis [152]. Several proteolytic enzymes have been implicated in the degradation of Aβ. Two metalloproteinases; insulin-degrading enzyme (IDE) and endothelin-converting enzyme (ECE) 1 and 2 [153], the plasmin system, and a neutral endopeptidase known as neprilysin are involved in the extracellular degradation of Aβ [154–156]. IDE acts on soluble monomeric and particularly intracellular Aβ [157, 158], whereas plasmin is capable of degrading aggregated Aβ [156]. The ECE zinc metallopeptidases are a class of type II integral membrane protein named for their ability to hydrolyze a family of biologically inactive intermediate endothelins [159]. ECE-1 has been shown to cleave Aβ at multiple sites within the peptide sequence, with ECE inhibitors significantly increasing the accumulation of Aβ in culture, indicating a role for this protease in Aβ catabolism [153]. Neprilysin plays a major role in $A\beta_{42}$ degradation [160] with this enzyme concentrating in the brain regions most vulnerable to AD [161]. A loss of such clearance mechanisms may be responsible for the accumulation of Aβ with recent work showing that the degrading activity of neprilysin is insufficient to clear brain Aβ accumulation in either AD or pathologic aging [162].

Aβ transport across the BBB is less well understood. Aβ is thought to be able to move from the extracellular spaces into the perivascular pathways, along the small and large intracranial artery walls, possibly draining to the lymph nodes in the neck [163]. This mechanism of clearance occurs via the endothelium, mediated by the enzymes angiotensin-converting enzyme and α_2-macroglobulin through interactions with LDL-receptor–related protein and apolipoproteins [164, 165]. Microglia and astrocytes also take up Aβ through receptor-mediated mechanisms [166, 167]. Aβ-ApoJ complexes are transported over the BBB through the ApoJ receptor megalin [59]. The high affinity of aggregated $A\beta_{40}$ with cholesterol suggests that cholesterol bound peptide trafficking may also play a role in its removal from the extracellular space [108]. $A\beta_{40}$ transport across the BBB is faster than $A\beta_{42}$ [168] with $A\beta_{40}$ the predominant constituent of abnormal Aβ peptide deposits in blood vessel walls [169]. There is some evidence that age-associated changes in BBB transport stops the efflux of $A\beta_{42}$ via this route [168].

Although still poorly understood, it appears that a number of regulatory mechanisms are important for modulating Aβ levels in the brain (Fig. 11.2). Under normal circumstances, local catabolism or clearance mechanisms efficiently prevent accumulation of these amyloidogenic peptides in the brain [170]. In AD, the considerable build-up of Aβ peptides suggests difficulties with Aβ clearance even if other production pathways are affected. In the absence of knowing any common initiating event or mechanism for AD, modification of clearance pathways provides the most obvious therapeutic targets for this disease.

11.8 Potential Therapeutic Strategies for Aβ Toxicity

Genetic and animal models of AD have provided an important basis for the design and testing of therapeutic strategies to alter Aβ production, aggregation, and/or accumulation. Strategies for lowering Aβ production include secretase inhibitors [171]. Strategies for reducing Aβ aggregation include metal chelators [172], and strategies for ameliorating Aβ accumulation include Aβ immunization, nonsteroidal anti-inflammatory drugs (NSAIDs), peroxisome proliferator-activated receptor-γ (PPAR) agonists, and statin medication [173].

11.8.1 Secretase Inhibitors

Since identifying the importance of β- and γ-secretase in the production of the Aβ alloforms, therapeutics aimed at inhibiting these enzymes have been the focus of a great deal of research. Initial studies of BACE1 therapy in mouse models appeared promising as, despite their role in normal physiological functioning, BACE1/BACE2 double knockout animals do not show any phenotypic problems (for review, see [174]). To date, no BACE inhibitors have been trialed in the literature, although significant numbers have been patented [175]. In contrast, models knocking out γ-secretase have been more problematic behaviorally due to the importance of PS1 in the γ-secretase protein complex and Notch signaling [176]. Fortunately, specific γ-secretase inhibitors have recently shown promising results with a shift toward the production of the less toxic $A\beta_{38}$ alloform and a reduction in $A\beta_{40}$ and $A\beta_{42}$ both in vitro and in transgenic mice [177, 178]. Importantly, these effects were achieved without affecting other components of the γ-secretase complex, although clinical trials have not yet been carried out. Unfortunately, clinical trials of 70 AD patients with the γ-secretase inhibitor LY450139, which showed promising results in animal models, have failed to show a marked reduction in CSF $A\beta_{42}$ [179]. Although there is still great promise for the development of specific and efficacious γ-secretase inhibitors, many researchers are calling on the development of BACE1 inhibitors as a safer alternative.

11.8.2 Metal Chelators

Given the interaction between Aβ and metal ions, and the suggestion that they may mediate Aβ aggregation and toxicity, therapeutic strategies have focused on disrupting this interaction. Many of these studies have generated promising data with the demonstration that specific chelators of Zn and

Cu ions can solubilize Aβ plaques from Alzheimer's disease postmortem brain tissue [180]. The compound used, cloquinol, also substantially decreased Aβ deposition in the brains of transgenic mice after just 9 weeks of treatment [181]. This drug also slowed the rate of cognitive decline in a clinical trial of AD and controls and appeared to be well-tolerated among patients [182]. Interestingly, this improvement was only reported as evident in individuals who were more severely impaired and scored over 25 on the Alzheimer's Dementia Assessment Scale–cognition subscale (ADAS-cog), although this could have been a type I error and greater sample numbers need to be assessed. In contrast, while no significant effect on cognition was seen in individuals who scored below 25 (the authors suggest a lack of sensitivity in this measure [182]), their plasma $A\beta_{42}$ levels were significantly decreased. These discrepant results warrant further experimental studies in this area, although given the heterogeneous roles of Aβ and the potential antioxidant roles arising from an interaction between Aβ and metal ions, great caution is required when trialing such therapies.

11.8.3 Aβ Immunization

Recent evidence suggests that reducing Aβ deposition in the brain by way of immunotherapy can reverse disease-associated functional deficits [183, 184]. The immunization of transgenic APP mice with $A\beta_{42}$ appears to prevent the formation of Aβ-containing plaques and subsequent AD-related neuropathologic changes in animals as young as 6 weeks to 11 months [184]. This reduction in Aβ is associated with reductions in memory impairment [185]. Similar results occur with the administration of other Aβ alloforms [186] and shorter peptide fragments [187], as well as with peripheral immunization with Aβ antibodies [188]. Clinical trials using active $A\beta_{42}$ immunization, however, caused severe central nervous system inflammation in a small but significant number of subjects [189]. Although no definitive data exists, it is generally agreed that these side effects were attributable to a cytotoxic T-cell–mediated response against Aβ, raising questions about immunizing against a self-protein and the effect of such a reaction on normal peptide function [190]. An additional safety concern arises with the use of Aβ alloforms that are capable of forming toxic fibrils and seeding plaque formation [191]. Despite this data, neuropathologic studies of patients treated with the AB vaccine showed low levels of cortical Aβ [192]. In addition, those subjects who developed robust antibody titers did show some clinical improvement [193]. These data provide support for the continued development of immunization strategies in the treatment of AD.

Active immunization with nontoxic Aβ fragments may be more effective in clinical trials as they have been shown to have reduced fibrillogenic properties while maintaining immunogenicity in transgenic mice [187]. More recent studies have also shown promising results from intracerbroventricular immunization of Aβ fragments in transgenic mice [194], thereby avoiding perivascular hemmorhage concerns associated with intravenous administration. Despite promising results using transgenic murine models, these animals still express endogenous APP and are therefore less likely to reflect the autoimmune problems that may be associated with human Aβ vaccines. With this in mind, the serious adverse immune reactions seen in clinical trials highlights the need to test potential therapies in large primate cohorts [195] prior to clinical testing in patients.

11.8.4 NSAIDs and PPAR-γ Agonists

Epidemiological evidence indicates that NSAIDs may lower the risk of developing AD [196, 197]. Although a direct effect on reducing the damaging Aβ-stimulated inflammation has been postulated, recent studies have demonstrated that NSAIDs are capable of directly affecting Aβ production via several mechanisms. Ibuprofen, indomethacin, and sulindac sulfide are capable of reducing $A\beta_{42}$ production, and increasing the less toxic $A\beta_{38}$ alloform, in cultured cells [198]. These effects have also been reported in transgenic mice and are proposed to occur by shifting γ-secretase activity [199]. Unfortunately, clinical trials of NSAIDs have been less fruitful [200], possibly due to the fact that most trials have been carried out in AD patients where the disease is too advanced for NSAID therapy to be effective. However, recent reports suggest that the doses required to lower Aβ in patients may be toxic [201] and better results may be achieved through the development of more specific inhibitors of $A\beta_{42}$.

A subset of NSAIDs can also bind to and activate the nuclear hormone receptor, PPAR-γ [202, 203]. Given that the principal effect of PPAR-γ is to transcriptionally silence proinflammatory gene expression [204, 205], it was argued that the anti-inflammatory effects of NSAIDs may be partially mediated through this pathway. Recent studies have demonstrated a decrease in focal $A\beta_{42}$-positive amyloid deposits and soluble $A\beta_{42}$ levels in transgenic mice treated with ibuprofen and the PPAR-γ agonist pioglitazone [206]. Whether these effects on Aβ occur directly or via inflammation-mediated mechanisms remains to be seen, but decreased BACE1 mRNA and protein levels were also evident. These studies suggest that combination therapies may be valuable in the treatment of AD to treat both the Aβ accumulation and downstream events.

11.8.5 Cholesterol and Statins

As described above, several findings suggest a link between cholesterol metabolism, Aβ levels, and the development of AD [107]. Indeed, reduction of cholesterol using specific inhibitors of 3-hydroxy-3-methyl-glutaryl coenzyme A (HMG CoA) reductase (statins) have been supported in the possible treatment of AD [207]. Direct links between cholesterol and Aβ processing are supported by studies showing that cholesterol-rich diets increase the production of Aβ [208], and statins decrease Aβ deposition in transgenic mice [209]. These effects are thought to be mediated by shifting APP processing to a non-amyloidogenic route, possibly via changes in membrane fluidity and cholesterol gradients [210]. However, immunomodulatory properties of statins have also been identified and are thought to act by reducing leukocyte migration into the CNS and by inhibiting a number of proinflammatory factors [211]. In this regard, statins may have roles similar to NSAIDs in the treatment of AD.

Data from clinical trials of AD patients have reported lower serum cholesterol and lower CSF APP fragments after treatment with simvastatin for 12 weeks [212]. Despite this, patients continued to show cognitive decline during the study. However, this effect is difficult to assess after such a short period of treatment, and a more recent double-blind, placebo-controlled study has shown significant improvements in cognition in AD patients after 6 months, and a trend toward significance at 1-year, of treatment with atorvastatin [213]. Unfortunately, epidemiological studies have been less useful in determining whether statins are protective against AD. A recent large study of 2798 older adults reported a reduced incidence of AD in current statin users versus never-users [214], consistent with other case-control studies [207]. However, an increase risk of dementia was seen among individuals who had previously used statins compared with never-users [214]. Although this study involved a large number of patients required to trial such therapies, only prospective case-control studies can answer whether statins can prevent AD. Fortunately, such studies are currently in progress.

References

1. Glenner GG, Wong CW. Alzheimer's disease: initial report of the purification and characterization of a novel cerebrovascular amyloid protein. Biochem Biophys Res Commun 1984;120:885-90.
2. Gregory GC, Halliday GM. What is the dominant Abeta species in human brain tissue? A review. Neurotox Res 2005;7:29-41.
3. Funato H, Yoshimura M, Kusui K, et al. Quantitation of amyloid beta-protein (A beta) in the cortex during aging and in Alzheimer's disease. Am J Pathol 1998;152:1633-40.
4. Fukumoto H, Asami-Odaka A, Suzuki N, et al. Association of A beta 40-positive senile plaques with microglial cells in the brains of patients with Alzheimer's disease and in non-demented aged individuals. Neurodegeneration 1996;5:13-7.
5. Iwatsubo T, Odaka A, Suzuki N, et al. Visualization of A beta 42(43) and A beta 40 in senile plaques with end-specific A beta monoclonals: evidence that an initially deposited species is A beta 42(43). Neuron 1994;13:45-53.
6. Roher AE, Lowenson JD, Clarke S, et al. beta-Amyloid-(1-42) is a major component of cerebrovascular amyloid deposits: implications for the pathology of Alzheimer's disease. Proc Natl Acad Sci U S A 1993;90:10836-40.
7. Mann DM, Iwatsubo T, Fukumoto H, et al. Microglial cells and amyloid beta protein [A beta] deposition; association with A beta 40-containing plaques. Acta Neuropathol (Berlin) 1995;90:472-7.

8. Mann DM, Iwatsubo T, Pickering-Brown SM, et al. Preferential deposition of amyloid beta protein (Abeta) in the form Abeta40 in Alzheimer's disease is associated with a gene dosage effect of the apolipoprotein E E4 allele. Neurosci Lett 1997;221:81-4.
9. Goate A, Chartier-Harlin MC, Mullan M, et al. Segregation of a missense mutation in the amyloid precursor protein gene with familial Alzheimer's disease. Nature 1991;349:704-6.
10. Van Broeckhoven C, Haan J, Bakker E, et al. Amyloid beta protein precursor gene and hereditary cerebral hemorrhage with amyloidosis (Dutch). Science 1990;248:1120-2.
11. Borchelt DR, Ratovitski T, van Lare J, et al. Accelerated amyloid deposition in the brains of transgenic mice coexpressing mutant presenilin 1 and amyloid precursor proteins. Neuron 1997;19:939-45.
12. Games D, Adams D, Alessandrini R, et al. Alzheimer's-type neuropathology in transgenic mice overexpressing V717F beta-amyloid precursor protein. Nature 1995;373:523-7.
13. Hsiao K, Chapman P, Nilsen S, et al. Correlative memory deficits, Abeta elevation, and amyloid plaques in transgenic mice. Science 1996;274: 99-102.
14. Suzuki N, Cheung TT, Cai XD, et al. An increased percentage of long amyloid beta protein secreted by familial amyloid beta protein precursor (beta APP717) mutants. Science 1994;264:1336-40.
15. Tamaoka A, Odaka A, Ishibashi Y, et al. APP717 missense mutation affects the ratio of amyloid beta protein species (A beta 1-42/43 and a beta 1-40) in familial Alzheimer's disease brain. J Biol Chem 1994;269:32721-4.
16. Scheuner D, Eckman C, Jensen M, et al. Secreted amyloid beta-protein similar to that in the senile plaques of Alzheimer's disease is increased in vivo by the presenilin 1 and 2 and APP mutations linked to familial Alzheimer's disease. Nat Med 1996;2:864-70.
17. Haass C, Schlossmacher MG, Hung AY, et al. Amyloid beta-peptide is produced by cultured cells during normal metabolism. Nature 1992;359:322-5.
18. Selkoe DJ. Alzheimer's disease: genes, proteins, and therapy. Physiol Rev 2001;81:741-66.
19. Tanaka S, Shiojiri S, Takahashi Y, et al. Tissue-specific expression of three types of beta-protein precursor mRNA: enhancement of protease inhibitor-harboring types in Alzheimer's disease brain. Biochem Biophys Res Commun 1989;165: 1406-14.
20. Kamal A, Almenar-Queralt A, LeBlanc JF, et al. Kinesin-mediated axonal transport of a membrane compartment containing beta-secretase and presenilin-1 requires APP. Nature 2001;414:643-8.
21. Turner PR, O'Connor K, Tate WP, et al. Roles of amyloid precursor protein and its fragments in regulating neural activity, plasticity and memory. Prog Neurobiol 2003;70:1-32.
22. Haass C, Hung AY, Schlossmacher MG, et al. beta-Amyloid peptide and a 3-kDa fragment are derived by distinct cellular mechanisms. J Biol Chem 1993;268:3021-4.
23. Nunan J, Small DH. Regulation of APP cleavage by alpha-, beta- and gamma-secretases. FEBS Lett 2000;483:6-10.
24. Brown MS, Ye J, Rawson RB, et al. Regulated intramembrane proteolysis: a control mechanism conserved from bacteria to humans. Cell 2000;100:391-8.
25. Ulery PG, Beers J, Mikhailenko I, et al. Modulation of beta-amyloid precursor protein processing by the low density lipoprotein receptor-related protein (LRP). Evidence that LRP contributes to the pathogenesis of Alzheimer's disease. J Biol Chem 2000;275:7410-5.
26. Vassar R, Bennett BD, Babu-Khan S, et al. Beta-secretase cleavage of Alzheimer's amyloid precursor protein by the transmembrane aspartic protease BACE. Science 1999;286:735-41.
27. Cook DG, Forman MS, Sung JC, et al. Alzheimer's A beta(1-42) is generated in the endoplasmic reticulum/intermediate compartment of NT2N cells. Nat Med 1997;3:1021-3.
28. Yu G, Nishimura M, Arawaka S, et al. Nicastrin modulates presenilin-mediated notch/glp-1 signal transduction and betaAPP processing. Nature 2000;407:48-54.
29. Esler WP, Kimberly WT, Ostaszewski BL, et al. Activity-dependent isolation of the presenilin-gamma-secretase complex reveals nicastrin and a gamma substrate. Proc Natl Acad Sci U S A 2002;99:2720-5.
30. Goutte C, Tsunozaki M, Hale VA, et al. APH-1 is a multipass membrane protein essential for the Notch signaling pathway in Caenorhabditis elegans embryos. Proc Natl Acad Sci U S A 2002;99:775-9.
31. Francis R, McGrath G, Zhang J, et al. aph-1 and pen-2 are required for Notch pathway signaling, gamma-secretase cleavage of betaAPP, and presenilin protein accumulation. Dev Cell 2002;3:85-97.
32. Gu Y, Sanjo N, Chen F, et al. The presenilin proteins are components of multiple membrane-bound complexes that have different biological activities. J Biol Chem 2004;279:31329-36.
33. Ida N, Hartmann T, Pantel J, et al. Analysis of heterogeneous A4 peptides in human cerebrospinal

fluid and blood by a newly developed sensitive Western blot assay. J Biol Chem 1996;271:22908-14.
34. Tamaoka A, Sawamura N, Fukushima T, et al. Amyloid beta protein 42(43) in cerebrospinal fluid of patients with Alzheimer's disease. J Neurol Sci 1997;148:41-5.
35. Mehta PD, Pirttila T, Mehta SP, et al. Plasma and cerebrospinal fluid levels of amyloid beta proteins 1-40 and 1-42 in Alzheimer's disease. Arch Neurol 2000;57:100-5.
36. Shoji M. Cerebrospinal fluid Abeta40 and Abeta42: natural course and clinical usefulness. Front Biosci 2002;7:d997-1006.
37. Nakamura T, Shoji M, Harigaya Y, et al. Amyloid beta protein levels in cerebrospinal fluid are elevated in early-onset Alzheimer's disease. Ann Neurol 1994;36:903-11.
38. Motter R, Vigo-Pelfrey C, Kholodenko D, et al. Reduction of beta-amyloid peptide42 in the cerebrospinal fluid of patients with Alzheimer's disease. Ann Neurol 1995;38:643-8.
39. Shoji M, Matsubara E, Kanai M, et al. Combination assay of CSF tau, A beta 1-40 and A beta 1-42(43) as a biochemical marker of Alzheimer's disease. J Neurol Sci 1998;158:134-40.
40. Samuels SC, Silverman JM, Marin DB, et al. CSF beta-amyloid, cognition, and APOE genotype in Alzheimer's disease. Neurology 1999;52:547-51.
41. Andreasen N, Hesse C, Davidsson P, et al. Cerebrospinal fluid beta-amyloid(1-42) in Alzheimer's disease: differences between early- and late-onset Alzheimer's disease and stability during the course of disease. Arch Neurol 1999;56:673-80.
42. Tapiola T, Pirttila T, Mehta PD, et al. Relationship between apoE genotype and CSF beta-amyloid (1-42) and tau in patients with probable and definite Alzheimer's disease. Neurobiol Aging 2000;21:735-40.
43. Tapiola T, Pirttila T, Mikkonen M, et al. Three-year follow-up of cerebrospinal fluid tau, beta-amyloid 42 and 40 concentrations in Alzheimer's disease. Neurosci Lett 2000;280:119-22.
44. Kanai M, Matsubara E, Isoe K, et al. Longitudinal study of cerebrospinal fluid levels of tau, A beta1-40, and A beta1-42(43) in Alzheimer's disease: a study in Japan. Ann Neurol 1998;44:17-26.
45. Mayeux R, Tang MX, Jacobs DM, et al. Plasma amyloid beta-peptide 1-42 and incipient Alzheimer's disease. Ann Neurol 1999;46:412-6.
46. Hoglund K, Wiklund O, Vanderstichele H, et al. Plasma levels of beta-amyloid(1-40), beta-amyloid(1-42), and total beta-amyloid remain unaffected in adult patients with hypercholesterolemia after treatment with statins. Arch Neurol 2004;61:333-7.
47. Hampel H, Mitchell A, Blennow K, et al. Core biological marker candidates of Alzheimer's disease—perspectives for diagnosis, prediction of outcome and reflection of biological activity. J Neural Transm 2004;111:247-72.
48. Tamaoka A, Fukushima T, Sawamura N, et al. Amyloid beta protein in plasma from patients with sporadic Alzheimer's disease. J Neurol Sci 1996;141:65-8.
49. Kuo YM, Emmerling MR, Lampert HC, et al. High levels of circulating Abeta42 are sequestered by plasma proteins in Alzheimer's disease. Biochem Biophys Res Commun 1999;257:787-91.
50. Andreasen N, Minthon L, Vanmechelen E, et al. Cerebrospinal fluid tau and Abeta42 as predictors of development of Alzheimer's disease in patients with mild cognitive impairment. Neurosci Lett 1999;273:5-8.
51. Shoji M, Kanai M, Matsubara E, et al. Taps to Alzheimer's patients: a continuous Japanese study of cerebrospinal fluid biomarkers. Ann Neurol 2000;48:402.
52. Otto M, Esselmann H, Schulz-Shaeffer W, et al. Decreased beta-amyloid1-42 in cerebrospinal fluid of patients with Creutzfeldt-Jakob disease. Neurology 2000;54:1099-102.
53. Kanemaru K, Kameda N, Yamanouchi H. Decreased CSF amyloid beta42 and normal tau levels in dementia with Lewy bodies. Neurology 2000;54:1875-6.
54. Li QX, Fuller SJ, Beyreuther K, et al. The amyloid precursor protein of Alzheimer's disease in human brain and blood. J Leukoc Biol 1999;66:567-74.
55. Wild-Bode C, Yamazaki T, Capell A, et al. Intracellular generation and accumulation of amyloid beta-peptide terminating at amino acid 42. J Biol Chem 1997;272:16085-8.
56. Hartmann T, Bieger SC, Bruhl B, et al. Distinct sites of intracellular production for Alzheimer's disease A beta40/42 amyloid peptides. Nat Med 1997;3:1016-20.
57. Hershkowitz M, Adunsky A. Binding of platelet-activating factor to platelets of Alzheimer's disease and multiinfarct dementia patients. Neurobiol Aging 1996;17:865-8.
58. Borroni B, Colciaghi F, Caltagirone C, et al. Platelet amyloid precursor protein abnormalities in mild cognitive impairment predict conversion to dementia of Alzheimer's type: a 2-year follow-up study. Arch Neurol 2003;60:1740-4.
59. Zlokovic BV. Cerebrovascular transport of Alzheimer's amyloid beta and apolipoproteins J and E: possible anti-amyloidogenic role of the blood-brain barrier. Life Sci 1996;59:1483-97.

60. Mackic J, Ghiso J, Frangione B, et al. Differential cerebrovascular sequestration and enhanced blood-brain barrier permeability to circulating Alzheimer's amyloid-β peptide in aged Rhesus vs. aged Squirrel monkey. Vascular Pharmacol 2002;18:303-13.
61. Rhodin JA, Thomas TN, Clark L, et al. In vivo cerebrovascular actions of amyloid beta-peptides and the protective effect of conjugated estrogens. J Alzheimers Dis 2003;5:275-86.
62. Kawarabayashi T, Younkin LH, Saido TC, et al. Age-dependent changes in brain, CSF, and plasma amyloid (beta) protein in the Tg2576 transgenic mouse model of Alzheimer's disease. J Neurosci 2001;21:372-81.
63. DeMattos RB, Bales KR, Cummins DJ, et al. Brain to plasma amyloid-beta efflux: a measure of brain amyloid burden in a mouse model of Alzheimer's disease. Science 2002;295:2264-7.
64. Hardy JA, Higgins GA. Alzheimer's disease: the amyloid cascade hypothesis. Science 1992;256:184-5.
65. Naslund J, Schierhorn A, Hellman U, et al. Relative abundance of Alzheimer A beta amyloid peptide variants in Alzheimer's disease and normal aging. Proc Natl Acad Sci U S A 1994;91:8378-82.
66. Tamaoka A, Kondo T, Odaka A, et al. Biochemical evidence for the long-tail form (A beta 1-42/43) of amyloid beta protein as a seed molecule in cerebral deposits of Alzheimer's disease. Biochem Biophys Res Commun 1994;205:834-42.
67. Gravina SA, Ho L, Eckman CB, et al. Amyloid beta protein (A beta) in Alzheimer's disease brain. Biochemical and immunocytochemical analysis with antibodies specific for forms ending at A beta 40 or A beta 42(43). J Biol Chem 1995;270:7013-6.
68. Kuo YM, Emmerling MR, Vigo-Pelfrey C, et al. Water-soluble Abeta (N-40, N-42) oligomers in normal and Alzheimer's disease brains. J Biol Chem 1996;271:4077-81.
69. Shinkai Y, Yoshimura M, Morishima-Kawashima M, et al. Amyloid beta-protein deposition in the leptomeninges and cerebral cortex. Ann Neurol 1997;42:899-908.
70. Tamaoka A, Fraser PE, Ishii K, et al. Amyloid-beta-protein isoforms in brain of subjects with PS1-linked, beta APP-linked and sporadic Alzheimer's disease. Brain Res Mol Brain Res 1998;56:178-85.
71. Kuo YM, Emmerling MR, Bisgaier CL, et al. Elevated low-density lipoprotein in Alzheimer's disease correlates with brain abeta 1-42 levels. Biochem Biophys Res Commun 1998;252:711-5.
72. Lue LF, Kuo YM, Roher AE, et al. Soluble amyloid beta peptide concentration as a predictor of synaptic change in Alzheimer's disease. Am J Pathol 1999;155:853-62.
73. Beffert U, Cohn JS, Petit-Turcotte C, et al. Apolipoprotein E and beta-amyloid levels in the hippocampus and frontal cortex of Alzheimer's disease subjects are disease-related and apolipoprotein E genotype dependent. Brain Res 1999;843:87-94.
74. Wang J, Dickson DW, Trojanowski JQ, et al. The levels of soluble versus insoluble brain Abeta distinguish Alzheimer's disease from normal and pathologic aging. Exp Neurol 1999;158:328-37.
75. Naslund J, Haroutunian V, Mohs R, et al. Correlation between elevated levels of amyloid beta-peptide in the brain and cognitive decline. JAMA 2000;283:1571-7.
76. Morishima-Kawashima M, Oshima N, Ogata H, et al. Effect of apolipoprotein E allele epsilon4 on the initial phase of amyloid beta-protein accumulation in the human brain. Am J Pathol 2000;157:2093-9.
77. Miklossy J, Taddei K, Suva D, et al. Two novel presenilin-1 mutations (Y256S and Q222H) are associated with early-onset Alzheimer's disease. Neurobiol Aging 2003;24:655-62.
78. Ingelsson M, Fukumoto H, Newell KL, et al. Early Abeta accumulation and progressive synaptic loss, gliosis, and tangle formation in AD brain. Neurology 2004;62:925-31.
79. Li R, Lindholm K, Yang LB, et al. Amyloid beta peptide load is correlated with increased beta-secretase activity in sporadic Alzheimer's disease patients. Proc Natl Acad Sci U S A 2004;101:3632-7.
80. Turner RS, Suzuki N, Chyung AS, et al. Amyloids beta40 and beta42 are generated intracellularly in cultured human neurons and their secretion increases with maturation. J Biol Chem 1996;271:8966-70.
81. Klein WL, Krafft GA, Finch CE. Targeting small Abeta oligomers: the solution to an Alzheimer's disease conundrum? Trends Neurosci 2001;24:219-24.
82. Walsh DM, Lomakin A, Benedek GB, et al. Amyloid beta-protein fibrillogenesis. Detection of a protofibrillar intermediate. J Biol Chem 1997;272:22364-72.
83. Hartley DM, Walsh DM, Ye CP, et al. Protofibrillar intermediates of amyloid beta-protein induce acute electrophysiological changes and progressive neurotoxicity in cortical neurons. J Neurosci 1999;19:8876-84.
84. Nilsberth C, Westlind-Danielsson A, Eckman CB, et al. The 'Arctic' APP mutation [E693G] causes Alzheimer's disease by enhanced Abeta protofibril formation. Nat Neurosci 2001;4:887-93.
85. Walsh DM, Klyubin I, Fadeeva JV, et al. Naturally secreted oligomers of amyloid beta protein potently inhibit hippocampal long-term potentiation in vivo. Nature 2002;416:535-9.

86. Bitan G, Vollers SS, Teplow DB. Elucidation of primary structure elements controlling early amyloid beta-protein oligomerization. J Biol Chem 2003;278:34882-9.
87. Bitan G, Kirkitadze MD, Lomakin A, et al. Amyloid beta-protein [Abeta] assembly: Abeta 40 and Abeta 42 oligomerize through distinct pathways. Proc Natl Acad Sci U S A 2003;100:330-5.
88. Xu Y, Shen J, Luo X, et al. Conformational transition of amyloid beta-peptide. Proc Natl Acad Sci U S A 2005;102:5403-7.
89. Coles M, Bicknell W, Watson AA, et al. Solution structure of amyloid beta-peptide(1-40) in a water-micelle environment. Is the membrane-spanning domain where we think it is? Biochemistry 1998;37:11064-77.
90. Soto C, Castano EM, Frangione B, et al. The alpha-helical to beta-strand transition in the amino-terminal fragment of the amyloid beta-peptide modulates amyloid formation. J Biol Chem 1995;270:3063-7.
91. Mager PP. Molecular simulation of the primary and secondary structures of the Abeta(1-42)-peptide of Alzheimer's disease. Med Res Rev 1998;18:403-30.
92. Kirkitadze MD, Condron MM, Teplow DB. Identification and characterization of key kinetic intermediates in amyloid beta-protein fibrillogenesis. J Mol Biol 2001;312:1103-19.
93. Serpell LC. Alzheimer's amyloid fibrils: structure and assembly. Biochim Biophys Acta 2000;1502:16-30.
94. Good TA, Murphy RM. Aggregation state-dependent binding of beta-amyloid peptide to protein and lipid components of rat cortical homogenates. Biochem Biophys Res Commun 1995;207:209-15.
95. Temussi PA, Masino L, Pastore A. From Alzheimer to Huntington: why is a structural understanding so difficult? Embo J 2003;22:355-61.
96. Crescenzi O, Tomaselli S, Guerrini R, et al. Solution structure of the Alzheimer amyloid beta-peptide (1-42) in an apolar microenvironment. Similarity with a virus fusion domain. Eur J Biochem 2002;269:5642-8.
97. Pike CJ, Overman MJ, Cotman CW. Amino-terminal deletions enhance aggregation of beta-amyloid peptides in vitro. J Biol Chem 1995;270:23895-8.
98. Teplow DB. Structural and kinetic features of amyloid beta-protein fibrillogenesis. Amyloid 1998;5:121-42.
99. Jarrett JT, Berger EP, Lansbury PT, Jr. The carboxy terminus of the beta amyloid protein is critical for the seeding of amyloid formation: implications for the pathogenesis of Alzheimer's disease. Biochemistry 1993;32:4693-7.
100. Zou K, Kim D, Kakio A, et al. Amyloid beta-protein 1-40 protects neurons from damage induced by Abeta1-42 in culture and in rat brain. J Neurochem 2003;87:609-19.
101. Tamaoka A, Sawamura N, Odaka A, et al. Amyloid beta protein 1-42/43 (A beta 1-42/43) in cerebellar diffuse plaques: enzyme-linked immunosorbent assay and immunocytochemical study. Brain Res 1995;679:151-6.
102. Hosoda R, Saido TC, Otvos L, Jr., et al. Quantification of modified amyloid beta peptides in Alzheimer's disease and Down syndrome brains. J Neuropathol Exp Neurol 1998;57:1089-95.
103. Houlden H, Baker M, McGowan E, et al. Variant Alzheimer's disease with spastic paraparesis and cotton wool plaques is caused by PS-1 mutations that lead to exceptionally high amyloid-beta concentrations. Ann Neurol 2000;48:806-8.
104. Verdile G, Gnjec A, Miklossy J, et al. Protein markers for Alzheimer's disease in the frontal cortex and cerebellum. Neurology 2004;63:1385-92.
105. Atwood CS, Moir RD, Huang X, et al. Dramatic aggregation of Alzheimer's abeta by Cu[II] is induced by conditions representing physiological acidosis. J Biol Chem 1998;273:12817-26.
106. Miura T, Suzuki K, Kohata N, et al. Metal binding modes of Alzheimer's amyloid beta-peptide in insoluble aggregates and soluble complexes. Biochemistry 2000;39:7024-31.
107. Gibson Wood W, Eckert GP, Igbavboa U, et al. Amyloid beta-protein interactions with membranes and cholesterol: causes or casualties of Alzheimer's disease. Biochim Biophys Acta 2003;1610:281-90.
108. Avdulov NA, Chochina SV, Igbavboa U, et al. Lipid binding to amyloid beta-peptide aggregates: preferential binding of cholesterol as compared with phosphatidylcholine and fatty acids. J Neurochem 1997;69:1746-52.
109. Michikawa M, Gong JS, Fan QW, et al. A novel action of Alzheimer's amyloid beta-protein [Abeta]: oligomeric Abeta promotes lipid release. J Neurosci 2001;21:7226-35.
110. Tsui-Pierchala BA, Encinas M, Milbrandt J, et al. Lipid rafts in neuronal signaling and function. Trends Neurosci 2002;25:412-7.
111. Kakio A, Nishimoto S, Yanagisawa K, et al. Interactions of amyloid beta-protein with various gangliosides in raft-like membranes: importance of GM1 ganglioside-bound form as an endogenous seed for Alzheimer amyloid. Biochemistry 2002;41:7385-90.
112. Koudinov AR, Berezov TT, Koudinova NV. Alzheimer's amyloid beta and lipid metabolism: a missing link? Faseb J 1998;12:1097-9.
113. Manelli AM, Stine WB, Van Eldik LJ, et al. ApoE and Abeta1-42 interactions: effects of isoform and

conformation on structure and function. J Mol Neurosci 2004;23:235-46.
114. Holtzman DM, Bales KR, Wu S, et al. Expression of human apolipoprotein E reduces amyloid-beta deposition in a mouse model of Alzheimer's disease. J Clin Invest 1999;103:R15-R21.
115. Verdier Y, Zarandi M, Penke B. Amyloid beta-peptide interactions with neuronal and glial cell plasma membrane: binding sites and implications for Alzheimer's disease. J Pept Sci 2004;10:229-48.
116. Kamenetz F, Tomita T, Hsieh H, et al. APP processing and synaptic function. Neuron 2003;37:925-37.
117. Plant LD, Boyle JP, Smith IF, et al. The production of amyloid [beta] peptide is a critical requirement for the viability of central neurons. J Neurosci 2003;23:5531-5.
118. De Ferrari GV, Inestrosa NC. Wnt signaling function in Alzheimer's disease. Brain Res Brain Res Rev 2000;33:1-12.
119. Lopez-Toledano MA, Shelanski ML. Neurogenic effect of {beta}-amyloid peptide in the development of neural stem cells. J Neurosci 2004;24:5439-44.
120. Haughey NJ, Liu D, Nath A, et al. Disruption of neurogenesis in the subventricular zone of adult mice, and in human cortical neuronal precursor cells in culture, by amyloid beta-peptide: implications for the pathogenesis of Alzheimer's disease. Neuromolecular Med 2002;1:125-35.
121. Haughey NJ, Nath A, Chan SL, et al. Disruption of neurogenesis by amyloid beta-peptide, and perturbed neural progenitor cell homeostasis, in models of Alzheimer's disease. J Neurochem 2002;83:1509-24.
122. Ramsden M, Plant LD, Webster NJ, et al. Differential effects of unaggregated and aggregated amyloid beta protein (1-40) on K(+) channel currents in primary cultures of rat cerebellar granule and cortical neurones. J Neurochem 2001;79: 699-712.
123. Crawford F, Suo Z, Fang C, et al. Characteristics of the in vitro vasoactivity of beta-amyloid peptides. Exp Neurol 1998;150:159-68.
124. Zou K, Gong J-S, Yanagisawa K, et al. A novel function of monomeric amyloid beta -protein serving as an antioxidant molecule against metal-induced oxidative damage. J Neurosci 2002;22: 4833-41.
125. Misonou H, Morishima-Kawashima M, Ihara Y. Oxidative stress induces intracellular accumulation of amyloid beta-protein (Abeta) in human neuroblastoma cells. Biochemistry 2000;39:6951-9.
126. Zhang L, Zhao B, Yew DT, et al. Processing of Alzheimer's amyloid precursor protein during H2O2-induced apoptosis in human neuronal cells. Biochem Biophys Res Commun 1997;235:845-8.
127. Paola D, Domenicotti C, Nitti M, et al. Oxidative stress induces increase in intracellular amyloid beta-protein production and selective activation of betaI and betaII PKCs in NT2 cells. Biochem Biophys Res Commun 2000;268:642-6.
128. Cuajungco MP, Goldstein LE, Nunomura A, et al. Evidence that the beta-amyloid plaques of Alzheimer's disease represent the redox-silencing and entombment of abeta by zinc. J Biol Chem 2000;275:19439-42.
129. Maynard CJ, Bush AI, Masters CL, et al. Metals and amyloid-beta in Alzheimer's disease. Int J Exp Pathol 2005;86:147-59.
130. Pike CJ, Burdick D, Walencewicz AJ, et al. Neurodegeneration induced by beta-amyloid peptides in vitro: the role of peptide assembly state. J Neurosci 1993;13:1676-87.
131. Yankner BA, Duffy LK, Kirschner DA. Neurotrophic and neurotoxic effects of amyloid beta protein: reversal by tachykinin neuropeptides. Science 1990;250:279-82.
132. Roher AE, Ball MJ, Bhave SV, et al. Beta-amyloid from Alzheimer's disease brains inhibits sprouting and survival of sympathetic neurons. Biochem Biophys Res Commun 1991;174:572-9.
133. Pike CJ, Walencewicz AJ, Glabe CG, et al. In vitro aging of beta-amyloid protein causes peptide aggregation and neurotoxicity. Brain Res 1991;563:311-4.
134. Mattson MP, Tomaselli KJ, Rydel RE. Calcium-destabilizing and neurodegenerative effects of aggregated beta-amyloid peptide are attenuated by basic FGF. Brain Res 1993;621:35-49.
135. Mattson MP, Rydel RE. beta-Amyloid precursor protein and Alzheimer's disease: the peptide plot thickens. Neurobiol Aging 1992;13:617-21.
136. Lorenzo A, Yankner BA. Beta-amyloid neurotoxicity requires fibril formation and is inhibited by congo red. Proc Natl Acad Sci U S A 1994;91: 12243-7.
137. Busciglio J, Lorenzo A, Yankner BA. Methodological variables in the assessment of beta amyloid neurotoxicity. Neurobiol Aging 1992;13:609-12.
138. Gong Y, Chang L, Viola KL, et al. Alzheimer's disease-affected brain: presence of oligomeric A beta ligands (ADDLs) suggests a molecular basis for reversible memory loss. Proc Natl Acad Sci U S A 2003;100:10417-22.
139. Lacor PN, Buniel MC, Chang L, et al. Synaptic targeting by Alzheimer's-related amyloid beta oligomers. J Neurosci 2004;24:10191-200.

140. Fan QW, Yu W, Senda T, et al. Cholesterol-dependent modulation of tau phosphorylation in cultured neurons. J Neurochem 2001;76:391-400.
141. Tong L, Thornton PL, Balazs R, et al. Beta-amyloid (1-42) impairs activity-dependent cAMP-response element-binding protein signaling in neurons at concentrations in which cell survival Is not compromised. J Biol Chem 2001;276:17301-6.
142. Zheng WH, Bastianetto S, Mennicken F, et al. Amyloid beta peptide induces tau phosphorylation and loss of cholinergic neurons in rat primary septal cultures. Neuroscience 2002;115:201-11.
143. Butterfield DA, Bush AI. Alzheimer's amyloid (beta)-peptide (1-42): involvement of methionine residue 35 in the oxidative stress and neurotoxicity properties of this peptide. Neurobiol Aging 2004;25:563-8.
144. Yatin SM, Varadarajan S, Link CD, et al. In vitro and in vivo oxidative stress associated with Alzheimer's amyloid beta-peptide (1-42). Neurobiol Aging 1999;20:325-30; discussion 39-42.
145. Schubert D, Behl C, Lesley R, et al. Amyloid peptides are toxic via a common oxidative mechanism. Proc Natl Acad Sci U S A 1995;92:1989-93.
146. Mark RJ, Blanc EM, Mattson MP. Amyloid beta-peptide and oxidative cellular injury in Alzheimer's disease. Mol Neurobiol 1996;12:211-24.
147. Lustbader JW, Cirilli M, Lin C, et al. ABAD directly links Abeta to mitochondrial toxicity in Alzheimer's disease. Science 2004;304:448-52.
148. Butterfield DA. Amyloid beta-peptide (1-42)-induced oxidative stress and neurotoxicity: implications for neurodegeneration in Alzheimer's disease brain. A review. Free Radic Res 2002;36:1307-13.
149. Huang X, Atwood CS, Hartshorn MA, et al. The A beta peptide of Alzheimer's disease directly produces hydrogen peroxide through metal ion reduction. Biochemistry 1999;38:7609-16.
150. Opazo C, Huang X, Cherny RA, et al. Metalloenzyme-like activity of Alzheimer's disease beta-amyloid. Cu-dependent catalytic conversion of dopamine, cholesterol, and biological reducing agents to neurotoxic H(2)O(2). J Biol Chem 2002;277:40302-8.
151. Selkoe DJ. Clearing the brain's amyloid cobwebs. Neuron 2001;32:177-80.
152. Zlokovic BV. Clearing amyloid through the blood-brain barrier. J Neurochem 2004;89:807-11.
153. Eckman EA, Reed DK, Eckman CB. Degradation of the Alzheimer's amyloid beta peptide by endothelin-converting enzyme. J Biol Chem 2001;276:24540-8.
154. Iwata N, Tsubuki S, Takaki Y, et al. Identification of the major Abeta1-42-degrading catabolic pathway in brain parenchyma: suppression leads to biochemical and pathological deposition. Nat Med 2000;6:143-50.
155. Qiu WQ, Walsh DM, Ye Z, et al. Insulin-degrading enzyme regulates extracellular levels of amyloid beta-protein by degradation. J Biol Chem 1998;273:32730-8.
156. Tucker HM, Kihiko M, Caldwell JN, et al. The plasmin system is induced by and degrades amyloid-beta aggregates. J Neurosci 2000;20:3937-46.
157. Sudoh S, Frosch MP, Wolf BA. Differential effects of proteases involved in intracellular degradation of amyloid beta-protein between detergent-soluble and -insoluble pools in CHO-695 cells. Biochemistry 2002;41:1091-9.
158. Morelli L, Llovera R, Gonzalez SA, et al. Differential degradation of amyloid beta genetic variants associated with hereditary dementia or stroke by insulin-degrading enzyme. J Biol Chem 2003;278:23221-6.
159. Turner AJ, Murphy LJ. Molecular pharmacology of endothelin converting enzymes. Biochem Pharmacol 1996;51:91-102.
160. Iwata N, Tsubuki S, Takaki Y, et al. Metabolic regulation of brain Abeta by neprilysin. Science 2001;292:1550-2.
161. Yasojima K, McGeer EG, McGeer PL. Relationship between beta amyloid peptide generating molecules and neprilysin in Alzheimer's disease and normal brain. Brain Res 2001;919:115-21.
162. Wang DS, Lipton RB, Katz MJ, et al. Decreased neprilysin immunoreactivity in Alzheimer's disease, but not in pathological aging. J Neuropathol Exp Neurol 2005;64:378-85.
163. Weller RO. Pathology of cerebrospinal fluid and interstitial fluid of the CNS: significance for Alzheimer's disease, prion disorders and multiple sclerosis. J Neuropathol Exp Neurol 1998;57:885-94.
164. Lauer D, Reichenbach A, Birkenmeier G. Alpha 2-macroglobulin-mediated degradation of amyloid beta 1-42: a mechanism to enhance amyloid beta catabolism. Exp Neurol 2001;167:385-92.
165. Qiu Z, Strickland DK, Hyman BT, et al. Alpha2-macroglobulin enhances the clearance of endogenous soluble beta-amyloid peptide via low-density lipoprotein receptor-related protein in cortical neurons. J Neurochem 1999;73:1393-8.
166. Shaffer LM, Dority MD, Gupta-Bansal R, et al. Amyloid beta protein (A beta) removal by neuroglial cells in culture. Neurobiol Aging 1995;16:737-45.
167. Kakimura J, Kitamura Y, Taniguchi T, et al. Bip/GRP78-induced production of cytokines and

uptake of amyloid-beta(1-42) peptide in microglia. Biochem Biophys Res Commun 2001;281:6-10.
168. Banks WA, Ronbinson SM, Verma S, et al. Efflux of human and mouse amyloid β proteins 1-40 and 1-42 from brain: impairment in a mouse model of Alzheimer's disease. Neuroscience 2003;121:487-92.
169. Iwatsubo T, Mann DM, Odaka A, et al. Amyloid beta protein (A beta) deposition: a beta 42(43) precedes A beta 40 in Down syndrome. Ann Neurol 1995;37:294-9.
170. Calero M, Rostagno A, Matsubara E, et al. Apolipoprotein J (clusterin) and Alzheimer's disease. Microsc Res Tech 2000;50:305-15.
171. Pollack SJ, Lewis H. Secretase inhibitors for Alzheimer's disease: challenges of a promiscuous protease. Curr Opin Investig Drugs 2005;6:35-47.
172. Doraiswamy PM, Finefrock AE. Metals in our minds: therapeutic implications for neurodegenerative disorders. Lancet Neurol 2004;3:431-4.
173. Lahiri DK, Farlow MR, Sambamurti K, et al. A critical analysis of new molecular targets and strategies for drug developments in Alzheimer's disease. Curr Drug Targets 2003;4:97-112.
174. Pietrzik C, Behl C. Concepts for the treatment of Alzheimer's disease: molecular mechanisms and clinical application. Int J Exp Pathol 2005;86:173-85.
175. Cumming JN, Iserloh U, Kennedy ME. Design and development of BACE-1 inhibitors. Curr Opin Drug Discov Devel 2004;7:536-56.
176. Kobayashi DT, Chen KS. Behavioral phenotypes of amyloid-based genetically modified mouse models of Alzheimer's disease. Genes Brain Behav 2005;4:173-96.
177. Anderson JJ, Holtz G, Baskin PP, et al. Reductions in beta-amyloid concentrations in vivo by the gamma-secretase inhibitors BMS-289948 and BMS-299897. Biochem Pharmacol 2005;69:689-98.
178. Barten DM, Guss VL, Corsa JA, et al. Dynamics of β-amyloid reductions in brain, cerebrospinal fluid, and plasma of β-amyloid precursor protein transgenic mice treated with a γ-secretase inhibitor. J Pharmacol Exp Ther 2005;312:635-43.
179. Siemers E, Skinner M, Dean RA, et al. Safety, tolerability, and changes in amyloid beta concentrations after administration of a gamma-secretase inhibitor in volunteers. Clin Neuropharmacol 2005;28:126-32.
180. Cherny RA, Legg JT, McLean CA, et al. Aqueous dissolution of Alzheimer's disease Abeta amyloid deposits by biometal depletion. J Biol Chem 1999;274:23223-8.
181. Cherny RA, Atwood CS, Xilinas ME, et al. Treatment with a copper-zinc chelator markedly and rapidly inhibits beta-amyloid accumulation in Alzheimer's disease transgenic mice. Neuron 2001;30:665-76.
182. Ritchie CW, Bush AI, Mackinnon A, et al. Metal-protein attenuation with iodochlorhydroxyquin (clioquinol) targeting Abeta amyloid deposition and toxicity in Alzheimer's disease: a pilot phase 2 clinical trial. Arch Neurol 2003;60:1685-91.
183. DeMattos RB, Bales KR, Cummins DJ, et al. Peripheral anti-A beta antibody alters CNS and plasma A beta clearance and decreases brain A beta burden in a mouse model of Alzheimer's disease. Proc Natl Acad Sci U S A 2001;98:8850-5.
184. Schenk D, Barbour R, Dunn W, et al. Immunization with amyloid-beta attenuates Alzheimer-disease-like pathology in the PDAPP mouse. Nature 1999;400:173-7.
185. Gotz J, Streffer JR, David D, et al. Transgenic animal models of Alzheimer's disease and related disorders: histopathology, behavior and therapy. Mol Psychiatry 2004;9:664-83.
186. Weiner HL, Lemere CA, Maron R, et al. Nasal administration of amyloid-beta peptide decreases cerebral amyloid burden in a mouse model of Alzheimer's disease. Ann Neurol 2000;48:567-79.
187. Sigurdsson EM, Scholtzova H, Mehta PD, et al. Immunization with a nontoxic/nonfibrillar amyloid-beta homologous peptide reduces Alzheimer's disease-associated pathology in transgenic mice. Am J Pathol 2001;159:439-47.
188. Bard F, Cannon C, Barbour R, et al. Peripherally administered antibodies against amyloid beta-peptide enter the central nervous system and reduce pathology in a mouse model of Alzheimer's disease. Nat Med 2000;6:916-9.
189. Orgogozo JM, Gilman S, Dartigues JF, et al. Subacute meningoencephalitis in a subset of patients with AD after Abeta42 immunization. Neurology 2003;61:46-54.
190. Das P, Golde TE. Open peer commentary regarding Abeta immunization and CNS inflammation by Pasinetti et al. Neurobiol Aging 2002;23:671-4; discussion 83-4.
191. Sigurdsson E, Wisniewski, T., Frangione, B. Infectivity of amyloid diseases. Trends Mol Med 2002.
192. Nicoll JA, Wilkinson D, Holmes C, et al. Neuropathology of human Alzheimer's disease after immunization with amyloid-beta peptide: a case report. Nat Med 2003.

193. Hock C, Konietzko U, Streffer JR, et al. Antibodies against beta-amyloid slow cognitive decline in Alzheimer's disease. Neuron 2003;38:547-54.
194. Chauhan NB, Siegel GJ. Efficacy of anti-Abeta antibody isotypes used for intracerebroventricular immunization in TgCRND8. Neurosci Lett 2005;375:143-7.
195. Li SB, Wang HQ, Lin X, et al. Specific humoral immune responses in rhesus monkeys vaccinated with the Alzheimer's disease-associated beta-amyloid 1-15 peptide vaccine. Chin Med J (Engl) 2005;118:660-4.
196. Etminan M, Gill S, Samii A. Effect of non-steroidal anti-inflammatory drugs on risk of Alzheimer's disease: systematic review and meta-analysis of observational studies. BMJ 2003;327:128.
197. Szekely CA, Thorne JE, Zandi PP, et al. Nonsteroidal anti-inflammatory drugs for the prevention of Alzheimer's disease: a systematic review. Neuroepidemiology 2004;23:159-69.
198. Weggen S, Eriksen JL, Das P, et al. A subset of NSAIDs lower amyloidogenic Abeta42 independently of cyclooxygenase activity. Nature 2001;414:212-6.
199. Weggen S, Eriksen JL, Sagi SA, et al. Evidence that nonsteroidal anti-inflammatory drugs decrease amyloid beta 42 production by direct modulation of gamma-secretase activity. J Biol Chem 2003;278:31831-7.
200. Aisen PS, Schafer KA, Grundman M, et al. Effects of rofecoxib or naproxen vs placebo on Alzheimer's disease progression: a randomized controlled trial. JAMA 2003;289:2819-26.
201. Cole GM, Morihara T, Lim GP, et al. NSAID and antioxidant prevention of Alzheimer's disease: lessons from in vitro and animal models. Ann N Y Acad Sci 2004;1035:68-84.
202. Lehmann JM, Kliewer SA, Moore LB, et al. Activation of the nuclear receptor LXR by oxysterols defines a new hormone response pathway. J Biol Chem 1997;272:3137-40.
203. Jaradat MS, Wongsud B, Phornchirasilp S, et al. Activation of peroxisome proliferator-activated receptor isoforms and inhibition of prostaglandin H(2) synthases by ibuprofen, naproxen, and indomethacin. Biochem Pharmacol 2001;62:1587-95.
204. Jiang C, Ting AT, Seed B. PPAR-gamma agonists inhibit production of monocyte inflammatory cytokines. Nature 1998;391:82-6.
205. Ricote M, Li AC, Willson TM, et al. The peroxisome proliferator-activated receptor-gamma is a negative regulator of macrophage activation. Nature 1998;391:79-82.
206. Heneka MT, Sastre M, Dumitrescu-Ozimek L, et al. Acute treatment with the PPARgamma agonist pioglitazone and ibuprofen reduces glial inflammation and Abeta1-42 levels in APPV717I transgenic mice. Brain 2005;128:1442-53.
207. Wolozin B. Cholesterol and Alzheimer's disease. Biochem Soc Trans 2002;30:525-9.
208. Refolo LM, Malester B, LaFrancois J, et al. Hypercholesterolemia accelerates the Alzheimer's amyloid pathology in a transgenic mouse model. Neurobiol Dis 2000;7:321-31.
209. Refolo LM, Pappolla MA, LaFrancois J, et al. A cholesterol-lowering drug reduces beta-amyloid pathology in a transgenic mouse model of Alzheimer's disease. Neurobiol Dis 2001;8:890-9.
210. Sidera C, Parsons R, Austen B. The regulation of beta-secretase by cholesterol and statins in Alzheimer's disease. J Neurol Sci 2005;229-230:269-73.
211. Stuve O, Youssef S, Steinman L, et al. Statins as potential therapeutic agents in neuroinflammatory disorders. Curr Opin Neurol 2003;16:393-401.
212. Sjogren M, Gustafsson K, Syversen S, et al. Treatment with simvastatin in patients with Alzheimer's disease lowers both alpha- and beta-cleaved amyloid precursor protein. Dement Geriatr Cogn Disord 2003;16:25-30.
213. Sparks DL, Sabbagh MN, Connor DJ, et al. Atorvastatin therapy lowers circulating cholesterol but not free radical activity in advance of identifiable clinical benefit in the treatment of mild-to-moderate AD. Curr Alzheimer Res 2005;2:343-53.
214. Rea TD, Breitner JC, Psaty BM, et al. Statin use and the risk of incident dementia: the Cardiovascular Health Study. Arch Neurol 2005;62:1047-51.

12
Impact of β-Amyloid on the Tau Pathology in Tau Transgenic Mouse and Tissue Culture Models

Jürgen Götz, Della C. David, and Lars M. Ittner

12.1 Introduction

Dementia is a generic term that describes chronic or progressive dysfunction of cortical and subcortical functions that result in complex cognitive decline. These cognitive changes are commonly accompanied by disturbances of mood, behavior, and personality. In developed countries with an increasingly aging population, the prevalence of dementia is currently at around 1.5% at 65 years of age, which doubles every 4 years and reaches about 30% at the age of 80 [1].

Of all age-related neurodegenerative disorders, Alzheimer's disease (AD) is the most prevalent. It is characterized histopathologically by β-amyloid (Aβ)-containing plaques, tau-containing neurofibrillary tangles (NFTs), reduced synaptic density and neuronal loss in selected brain areas [2]. In familial forms of AD (FAD), pathogenic mutations have been identified in both the gene encoding the precursor of the Aβ peptide, APP, itself and in the presenilin genes, which encode part of the protease complex involved in processing APP. This genetic evidence supports the amyloid cascade hypothesis, which claims that Aβ causes or enhances the NFT pathology.

Frontotemporal dementia (FTD) is the preferred term for a spectrum of non-Alzheimer dementias characterized by focal atrophy of frontal and anterior temporal regions and NFTs in the absence of Aβ deposition. Recent epidemiological studies suggest that FTD is the second most common cause of dementia in persons younger than 65 years [3]. In familial forms of FTD (frontotemporal dementia with parkinsonism linked to chromosome 17; FTDP-17), pathogenic mutations have been identified in tau proving that tau dysfunction in itself can lead to neurodegeneration and dementia.

AD and FTD have a distinct neuropathological profile, but histopathological studies have shown that mixed states (with people presenting with features of more than one type of dementia) are probably more frequent than pure dementia syndromes [1, 4, 5]. Here, we discuss how aspects of the human pathology have been modeled in animals, with a special emphasis on tau transgenic mice. Furthermore, we present experimental evidence obtained in tau transgenic mouse and tissue-culture models that to some extent support the amyloid cascade hypothesis in mice.

12.2 Alzheimer's Disease

The clinical presentation of AD is dominated by early memory deficits, followed by gradual erosion of other cognitive functions such as judgment, verbal fluency, or orientation. Although this sequential order may vary, memory impairment is normally the first and dominating feature.

In addition to a reduced synaptic density and neuronal loss in selected brain areas, AD is characterized by two forms of insoluble protein aggregates, the extracellular Aβ-containing plaques and the intracellular NFTs. The major component of the plaques is a 40–42 amino acid aggregated polypeptide termed β-amyloid (Aβ; $A\beta_{40}$ and $A\beta_{42}$), which is derived by proteolysis from the larger amyloid precursor protein, APP (Fig. 12.1) [6, 7]. APP can be proteolytically cleaved by the

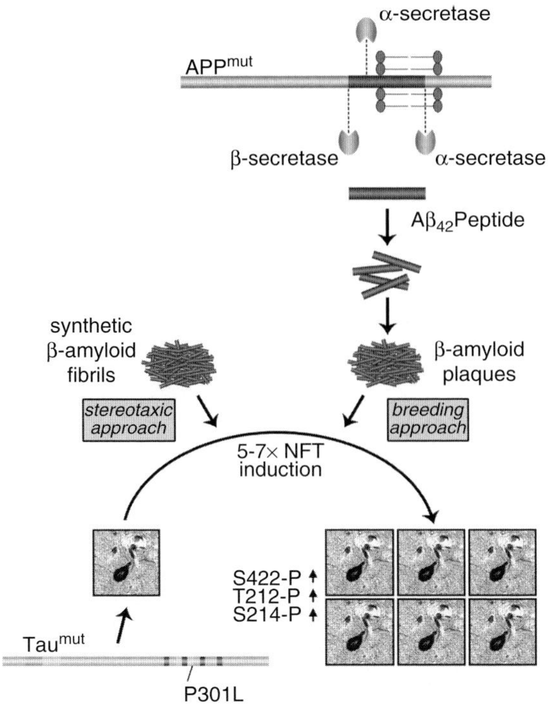

FIGURE 12.1. Cleavage of the amyloid precursor protein (APP) by the membrane-associated α-secretase is within the Aβ domain and thus precludes the formation of Aβ. Therefore, this pathway is non-amyloidogenic. Alternatively, cleavage may occur in the endosomal-lysosomal pathway, first by β-secretase and then by γ-secretase generating the Aβ peptide. Aβ is deposited around meningeal and cerebral vessels and in the gray matter as β-amyloid plaques. To determine the relationship between Aβ and the NFT/tau pathology in AD, two alternative approaches were pursued. One involved the intercrossing of APP and tau mutant mice with a plaque and NFT pathology ("breeding approach"), the other the stereotaxic injection of fibrillar preparations of $A\beta_{42}$ into mutant tau transgenic brains ("stereotaxic approach"). These approaches resulted in five- to sevenfold increased NFT formation, which was associated with phosphorylation of tau at the phospho-epitopes Thr212/Ser214 and Ser422. Together, these studies provide evidence for the amyloid cascade hypothesis in mice. The finding that $A\beta_{42}$ was not capable of inducing NFT formation in non-NFT-forming wild-type tau transgenic mice may reflect species differences between mice and men. Alternatively, it may imply that, at least in mice, $A\beta_{42}$ cannot induce NFT formation *de novo*.

membrane-associated α-secretase, which cleaves APP within the Aβ domain. This pathway is non-amyloidogenic, as this cleavage precludes the formation of Aβ. Alternatively, cleavage may occur in the endosomal-lysosomal pathway, first by β-secretase and then by γ-secretase, which together generate the Aβ peptide. β-Secretase activity has been attributed to a single protein, BACE, whereas γ-secretase activity was shown to depend on the presence of a total of four components: presenilin, nicastrin, APH-1 and PEN-2 [8, 9] (Fig. 12.1).

The second histopathological hallmark of AD are the neurofibrillary lesions that are found in cell bodies and apical dendrites as NFTs, in distal dendrites as neuropil threads, and in the abnormal neurites that are associated with some Aβ plaques (neuritic plaques). NFTs develop in specific sites and spread in a predictable, nonrandom manner across the brain. This sequence of the tau pathology is subjected to little inter-individual variation and provides a basis for distinguishing six stages in the progression of the disease [10, 11].

The major component of NFTs are abnormal filaments [12, 13]. The core protein of these filaments is tau, a microtubule-associated protein [14]. In the course of the disease, tau becomes abnormally phosphorylated, it adopts an altered conformation and is relocalized from axonal to somatodendritic compartments. Phosphorylation tends to dissociate tau from microtubules. Because this increases the soluble pool of tau, it might be an important first step in the assembly of tau filaments [5, 15–21]. Tau filaments have a clear β-cross structure, which is the defining feature of amyloid fibers [22]. They share this structure with the extracellular deposits present in the systemic and organ-specific amyloid diseases. It is therefore appropriate to consider the diseases with filamentous tau aggregates, the so-called tauopathies, a form of brain amyloidosis [23].

Physiological functions of tau include the assembly and stabilization of microtubules. Microtubules are hollow, 25-nm-wide cylindrical polymers, assembled primarily from heterodimers of α- and β-tubulin and a collection of microtubule-associated proteins (MAPs). Microtubules have two general functions, as the primary structural component of the mitotic spindle and in organizing the cytoplasm. Microtubules isolated from cell extracts by multiple cycles of assembly/disassembly and differential centrifugation yield a final microtubule preparation of which about 80% is tubulin, while the remaining 20% are MAPs.

Initially isolated from mammalian neurons, MAPs were named according to the three major size classes of polypeptides: MAP1 (>250 kDa), MAP2 (~200 kDa), and tau protein (50–70 kDa). MAP2 and tau are expressed together in most neurons, where they localize to separate subcellular compartments. MAP2 is largely found in dendrites, whereas tau is concentrated in axons. Tau has also been found in astrocytes and oligodendrocytes, although, under physiological conditions, levels are relatively low [24]. Additional roles have been assigned to tau in signal transduction, the organization of the actin cytoskeleton, intracellular vesicle transport, and anchoring of phosphatases and kinases [25–34]. In the adult human brain, six tau isoforms are produced by alternative mRNA splicing of exons 2, 3, and 10 (Fig. 12.2). They differ by the presence or absence of one or two short inserts in the amino-terminal half and have either three or four microtubule-binding repeat motifs in the carboxy-terminal half (3R and 4R). All six brain tau isoforms are found in the neurofibrillary lesions of AD brains [35].

In early-onset familial forms of AD (FAD), mutations were identified in three genes: in the APP gene itself and in the genes encoding presenilin 1 and 2 [36, 37]. Expression of FAD mutant forms of APP in transgenic mice by several research groups caused Aβ-plaque formation and concomitant memory deficits that progressed with age (reviewed in Ref. 5). These were more pronounced in transgenic mice coexpressing mutant forms of presenilin and APP, yet, NFT formation could not be reproduced [5].

For late-onset sporadic AD (SAD), around two dozen risk-conferring genes have been identified until today, but of these only the apolipoprotein E (APOE) gene has been confirmed unanimously and found to be associated with SAD [38]. When FAD is compared with SAD, the histopathological hallmarks are indistinguishable. This implies that lessons learned from the familial forms of AD may be applicable also to the sporadic forms.

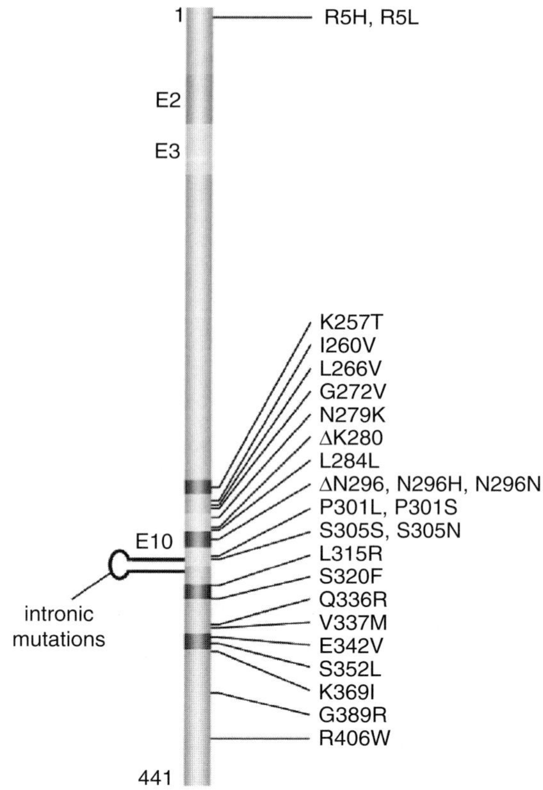

FIGURE 12.2. By alternative mRNA splicing of exons E2, E3, and E10, six tau isoforms are produced in the adult human brain. They differ by the presence or absence of one or two short inserts in the amino-terminal half (0N, 1N, and 2N, respectively) and have either three or four microtubule-binding repeat motifs in the carboxy-terminal half (3R and 4R). The microtubule-binding motifs are indicated in black. All six brain tau isoforms are found in the neurofibrillary lesions of AD patients. In FTDP-17, the majority of the exonic mutations in tau are clustered around the microtubule binding domain, whereas the intronic mutations (indicated by the stem loop) result in a shift of 3R to 4R tau isoforms.

12.3 Frontotemporal Dementias

Although AD is the most frequent form of dementia at high age, NFTs are, in the absence of β-amyloid plaques, also abundant in additional neurodegenerative diseases. The preferred term for this spectrum of non-Alzheimer dementias is "frontotemporal dementia" (FTD) [39]. FTD is characterized by focal atrophy of frontal and anterior temporal regions. Three broad subdivisions have been recognized, depending on the profile of immunohistochemical staining and the pattern of intracellular inclusions [39–42]: one with tau-positive aggregates (Pick disease [PiD], progressive supranuclear palsy [PSP], corticobasal degeneration [CBD], argyrophilic grain disease [AgD], and frontotemporal dementia with parkinsonism linked to chromosome 17 [FTDP-17]), a second with tau-negative and ubiquitin-positive inclusions (FTD with motor neuron inclusions; FTD-MND), and a third category named dementia lacking distinctive histology (DLDH) [39].

The tau field experienced significant advances with the identification of both exonic and intronic tau mutations in FTDP-17; this established that dysfunction of tau in itself can cause neurodegeneration and lead to dementia (Fig. 12.2). Initially, three missense ("exonic") mutations were identified in exons 9, 10, and 13 (G272V, P301L, and R406W) and three ("intronic") mutations in the 5′ splice site of the alternatively spliced exon 10 [43]. At the same time, the V279M mutation and a G to A mutation in the nucleotide adjacent to the exon 10 splice-donor site of the tau gene, were identified [45]. The intronic mutations all destabilize a potential stem-loop structure, which is probably involved in regulating the alternative splicing of exon 10. This causes a more frequent use of the 5′ splice site and an increased proportion of tau transcripts that include exon 10. This increase in exon 10–containing mRNAs results in an increased proportion of tau with four microtubule-binding repeats (4R > 3R). Together, these findings indicate that either an altered ratio of 4R to 3R tau isoforms or a missense mutation can lead to the formation of abnormal tau filaments. The majority of the tau mutations identified so far are in the carboxy-terminal half of the tau protein, suggesting that this is a hot spot for disease-causing mutations [21] (Fig. 12.2). In the amino-terminus, two mutations have been identified at position R5, which may affect the conformation of tau. Mutations in exons 9, 12, and 13 (such as G272V) affect all six tau isoforms. By contrast, mutations in the alternatively spliced exon 10 (such as P301L) only affect 4R tau isoforms. The silent mutations L284L (CTT to CTC) and N296N (AAT to AAC) in exon 10 are believed to disrupt an exon 10 splicing silencer sequence, which causes an increased production of exon 10–containing 4R tau mRNAs [46–60]. Until today, a total of 32 mutations have been described in more than 100 families with FTDP-17 [23].

All frontotemporal dementias with tau mutations that have been examined to date have a filamentous tau pathology. The morphology of these tau filaments and their isoform composition appears to be determined by whether tau mutations affect mRNA splicing of exon 10 or whether they are missense mutations located inside or outside of exon 10 [61]. The major component of NFTs in AD are straight (SF) and paired helical filaments (PHFs) [12, 13]. The Pick bodies found in PiD ultrastructurally consist of random coiled and straight tau filaments. There are reports showing that only 3R tau isoforms aggregate into Pick bodies [62]. One recent study showed that cases containing predominantly 3R tau were classic PiD (100%), cases with predominantly 4R tau were either CBD (71%) or PSP (29%), cases with both 3R and 4R tau were either a combination of PiD and AD (67%) or NFTD (neurofibrillary tangle dementia, 33%) [63]. Aggregated tau proteins in PiD are not reactive with the monoclonal antibody 12E8 directed against the phosphorylated tau epitope Ser262/Ser356 (for a map of tau phospho-epitopes, see Ref. 21). In contrast, this phosphorylation site is readily detected in other tauopathies [62].

Although tau is mainly a neuronal protein, it has also been found albeit at low levels in astrocytes and oligodendrocytes [24]. In PSP and CBD, tau forms aggregates in these cell-types, much in contrast with AD [20, 64]. In PSP, the neuritic and glial changes are composed of straight filaments and tubules, and in CBD of twisted filaments, which are different from the PHFs [65–67]. Although the filament morphologies and their tau isoform composition vary between diseases, it is the repeat region that forms the core of the filament, with the amino- and carboxy-terminal regions forming a fuzzy coat around the filament [68]. During the course of the disease, the fuzzy coat is frequently proteolysed, such that filaments may comprise only the repeat region of tau [69]. However, it is the full-length protein that assembles into filaments in the first place [35].

To which extent do the familial forms of FTD model other tauopathies such as PSP or CBD? Interestingly, nine of the missense mutations in tau found in FTDP-17 (K257T, L266V, G272V, L315R, S320F, Q336R, E342V, K369I, and G389R) gave rise to a clinical and neuropathological phenotype reminiscent of PiD [48, 58, 70–76], cases with four exonic (R5L, N279K, ΔN296 and S305S) and one intronic (+16) mutation presented a clinical picture similar to PSP [49, 59, 77–79], and some patients with mutations N296N and P301S presented a disease resembling CBD [80, 81].

12.4 Pathogenic Relationship of Plaques and NFTS

The pathogenic relationship of the two major lesions of AD, plaques and NFTs, and their relative contribution to the clinical features of the disease are a long-standing matter of debate, especially when sporadic forms of AD are considered, which comprise the majority of all cases. Human carriers of pathogenic mutations in the APP gene ultimately develop both Aβ plaques and NFTs. This finding led to the proposition of the amyloid cascade hypothesis, which claims that β-amyloid causes or enhances the NFT pathology in AD. Although this concept at first sight seems intriguing, it is difficult to reconcile with the anatomical distribution of plaques and NFTs.

The NFTs develop in specific predilection sites and spread in a predictable, nonrandom manner across the brain. This sequence of the tau pathology provides a basis for distinguishing six stages of disease progression [10, 11]: the transentorhinal stages I–II representing clinically silent cases; the limbic stages III–IV of incipient AD; and the neocortical stages V–VI of fully developed AD. A comparative study of the Aβ-associated pathology defined five phases. These differ markedly from the stages, which define the spreading of NFTs: The neocortical phase 1 is followed by the allocortical phase 2. In phase 3, the diencephalic nuclei, the striatum, and the cholinergic nuclei of the basal forebrain develop Aβ deposits, and in phase 4, several brain-stem nuclei become additionally involved. Finally, phase 5 is characterized by cerebellar Aβ-deposition. These findings suggest that Aβ deposition expands anterogradely into regions that receive neuronal projections from regions already exhibiting Aβ [82].

Numerous studies failed to demonstrate a clear relationship between the severity of dementia and Aβ deposition (that is, Aβ plaques) in human AD brain, whereas a correlation between NFT numbers and severity of dementia has been reported [83–86]. It was shown that total NFT counts in specific brain areas such as the entorhinal and frontal cortex, as well as neuron numbers in the CA1 region of the hippocampus were the best predictors of cognitive deficits in brain aging and AD [87]. Recently, however, Delacourte and co-workers proposed a synergistic interaction between the APP- and tau-related pathology, despite a different spatiotemporal distribution of plaques and NFTs [88, 89]. They also found that whenever Aβ aggregates were detected, a tau pathology was found, at least in the entorhinal cortex. The opposite was not true as cases were found with an advanced tau pathology and no trace of Aβ aggregates [89]. As far as Aβ is concerned, the focus has recently shifted from plaques and their fibrillar Aβ constituent to mono- and oligomeric Aβ with the latter possibly being the more toxic species [90]. This implies that to correlate with dementia, Aβ levels may need to be measured rather than merely counting plaque numbers.

A relationship has been postulated between neuronal loss and NFT formation in AD [91], yet only part of the neuronal loss can be explained by NFT formation as demonstrated for brain areas such as the visual cortex, the superior temporal sulcus, the entorhinal cortex, and area 9 [87, 92, 93]. For the CA1 region, the number of extracellular NFTs accounted for less than 20% (2.2–17.2%, mean 8.1%) of neurons lost in all cases [94]. These calculations were based on the assumption that NFTs persist until the end of the life, once they have formed. The findings imply that non-NFT-related mechanisms of neurodegeneration may also compromise vulnerable subsets of neurons. Alternatively, tau-related neuronal dysfunction may lead to cell death long before sufficient numbers of tau filaments accumulate and become visible as NFTs at the light microscopic level using silver impregnation techniques. A quantitative analysis of NFTs in human brain revealed that a substantial number of pyramidal cells may persist either unaffected or in a transitional stage of NFT formation. Whereas it is not possible to assess whether such transitional neurons are fully functional, these affected neurons might respond positively to therapeutic strategies aimed at protecting the cells that are prone to neurofibrillary degeneration [95].

As plaques and NFTs are the histopathological hallmarks of both FAD and SAD, it will be important to know what triggers their formation and how they are functionally related. Some insight may be gained by the analysis of adult lifestyle risk factors combined with the evidence of a genetic predisposition (as determined by the inheritance of risk alleles of susceptibility genes), which together may cause SAD [96]. Although the etiology of FAD and SAD differ, the clinical picture and the morphological end stage in the brain appear to be the same.

12.5 Tau Transgenic Mice: Requirements for and Role of NFT Formation

To better understand the role of β-amyloid plaques and NFTs in AD and related disorders, experimental animal models have been developed that reproduce aspects of the neuropathological characteristics of these diseases (reviewed in Ref. 5). Their suitability largely depends on the purpose a model has to suit. If one wants to model histopathological features, one has to discriminate between the precise anatomical "reproduction" of the pathology and modeling at the cellular level. This is important when the animals (in particular transgenic mice) are employed in behavioral studies intended to correlate the histopathology with dementia. These animal models may either offer a general proof of principle or reproduce more specific aspects of the human disease. Animal models may be used to identify disease modifiers, components of pathocascades, and susceptibility genes [97]. Furthermore, they may be employed in drug screenings [5]. Finally, insight gained from these models can be translated to human disease and assist in the development of treatment therapies [99].

After the very first APP transgenic animals had failed to show an extensive AD-like neuropathology, in 1995 Games and co-workers successfully expressed high levels of the disease-linked V717F mutant form of APP, under control of the platelet-derived growth factor (PDGF) mini-promoter. These PDAPP mice showed many of the pathological features of AD, including extensive deposition of extracellular amyloid plaques, astrocytosis, and neuritic dystrophy [100]. Similar features were observed in a second transgenic model by Hsiao and co-workers that expressed the APP^{sw} mutation inserted into a hamster prion protein (PrP) cosmid vector [101]. Then, by expressing the Swedish double APP mutation under control of the mThy1.2 promoter, a research group at Novartis established the APP23 mouse model with a sevenfold overexpression of APP [102, 103]. Subsequently, many more models have been developed by both academic and industrial research groups (such as the TgCRND8 [104] or J20 mice [105]). Using these mice, aspects of Aβ toxicity have been addressed and therapies have been tested. The APP transgenic mice were also crossed with presenilin, BACE, ApoE, and TGF-β1 transgenic and/or knockout strains (reviewed in Ref. 5).

The first tau transgenic models were established by us in 1995 (Table 12.1) and expressed the longest human 4R brain tau isoform (2N4R), without a pathogenic mutation, in mice using the hThy1 promoter for neuronal expression [106]. Despite the lack of NFT pathology, these mice modeled aspects of human AD, such as the somatodendritic localization of hyperphosphorylated tau and, therefore, represented an early pre-NFT phenotype. The subsequent use of stronger promoters caused a more pronounced phenotype in transgenic mice [107–109] (Table 12.1). In some strains, high expression levels of the transgene in motor neurons caused the formation of large numbers of pathologically enlarged axons with neurofilament- and tau-immunoreactive spheroids, a neuropathological characteristic of most cases of amyotrophic lateral sclerosis (ALS), where they are believed to impair slow axonal transport [110–112]. Tau protein extracted from transgenic brain and spinal cord was shown to be increasingly insoluble as the mice became older. Despite the decreased solubility of tau, NFTs did not form with the exception of one study where they were reported to be present at low numbers when the mice had reached a very old age [113]. Taken together, these findings demonstrate that overexpression of human tau can lead to an axonopathy resulting in nerve cell dysfunction and amyotrophy [5, 20].

When the first pathogenic FTDP-17 mutations had been identified in the tau gene in 1998, several groups achieved NFT formation both in neurons [114–118] and in glial cells of transgenic mice [119–122] (Table 12.1).

TABLE 12.1. List of currently available tau transgenic mice.

Promoter	Tau isoform	Mutation	Strain name	Reference
hThy1	4R-tau (2N)	Wild-type	ALZ7	106
mHMG-CoAR	3R-tau (0N)	Wild-type	TG23	154
mPrP	3R-tau (0N)	Wild-type	htau44	107
mThy1.2	4R-tau (2N)	Wild-type	htau40	108
mThy1.2	4R-tau (2N)	Wild-type	ALZ17	109
mPrP	3R-tau (2N)	P301L	JNPL3	114
mThy1.2	4R-tau (2N)	P301L	pR5	115
mPrP-TA	4R-tau (2N)	G272V	pR3	119
mPDGF	4R-tau (2N)	V337M	Tg214	116
CaMKII	4R-tau (2N)	R406W		117
mThy1.2	4R-tau (0N)	P301S		118
Tα1 α-tubulin	3R-tau (0,1,2N)	Wild-type	Tα1-3RT	120
mThy1.2	4R-tau (0N)	P301L	3×Tg-AD	121*
mThy1.2	4R-tau (2N)	P301L	tau-P301L	122†
mThy1.2	4R-tau (2N)	Wild-type	tau-4R/2N	122†
KOKI	4R-tau (2N)	Wild-type	KOKI	122†

*Triple transgenic approach: PS1 M146V knock-in oocytes microinjected with APP^{sw} and P301L tau transgenes.

†Three transgenic approaches in parallel: P301L tau transgenic mice were compared with wild-type tau transgenic mice of comparable expression levels. A third strain contained a single copy of a wild-type tau transgene (under the control of the mThy1.2 promoter) inserted into the endogenous murine tau locus.

The P301L mutation was one of the first FTDP-17 mutations that had been identified in human patients [43]; it is quite frequent [123] and was the first mutation to be expressed in transgenic mice. Expression of a human tau isoform lacking the two amino-terminal inserts (0N4R) together with the P301L mutation under the control of the murine PrP promoter [114] caused severe motor and behavioral disturbances in 90% of the mice by 10 months of age (Table 12.1). These were more pronounced than in the previously published wild-type tau transgenic mouse models [107–109]. Importantly, NFTs were identified by Gallyas silver stainings and thioflavin S-fluorescent microscopy both in brain and spinal cord, and motor neurons were reduced twofold in the spinal cord [114]. We expressed the same mutation using the longest human tau isoform containing both amino-terminal inserts (2N4R). The mThy1.2 promoter was chosen instead of the PrP promoter, which may account for different expression patterns [115]. Again, NFTs were identified and tau filaments were revealed by immuno-electron microscopy of sarkosyl extracts using phospho-tau-specific antibodies. No motor phenotype was observed, possibly due to low expression levels of the transgene in motor neurons of the spinal cord.

The P301S mutation is an aggressive mutation that causes clinical signs of FTDP-17 already in the third decade of life [80]. When P301S mutant tau was expressed under control of the mThy1.2 promoter, massive NFT formation was observed [118]. To address the role of distinct tau phospho-epitopes in tau filament formation, tau was analyzed in both the soluble and insoluble fraction. Perchloric-acid soluble tau was phosphorylated at many phospho-epitopes of tau, with the exception of the AT100 phospho-epitope S214, whereas sarkosyl-insoluble tau was strongly immunoreactive with all antibodies including AT100. Interestingly, this site has been shown, together with S422, to be linked to NFT formation in P301L mice (see below) [124]. Together, this indicates that immunoreactivity for phospho-S214 closely mirrors the presence of tau filaments, suggesting that phosphorylation of this site occurs in the course of, or after, filament assembly.

To address the tau pathology in glial cells, G272V mutant tau was expressed by combining a PrP-driven expression system with an autoregulatory transactivator loop that resulted in high expression in a subset of both neurons and oligodendrocytes. Electron microscopy established filament formation associated with hyperphosphorylation of tau. Thioflavin S-positive fibrillary inclusions were identified in oligodendrocytes and motor neurons in spinal cord [119]. The clinical phenotype of these mice was subtle. In contrast, when human wild-type tau was overexpressed in neurons and glial cells using the mouse Tα1 α-tubulin promoter, a glial pathology was found resembling the astrocytic plaques in CBD and the coiled bodies in CBD and PSP [120].

To reproduce the plaque and NFT pathology in one single animal model, triple-transgenic mice were developed harboring PS1 M146V, the APP^{Swe} and P301L tau transgenes. Instead of crossing independent lines, the APP and tau transgenes were microinjected into transgenes embryos derived from homozygous PS1 M146V knock-in mice, generating mice with the same genetic background. In the triple transgenic mice, synaptic dysfunction, including LTP deficits, manifested in an age-related manner, but before plaque and NFT pathology [121].

To allow a better side-by-side comparison of wild-type and P301L mutant mice, a total of three strains were generated by another research group and analyzed in parallel [122]. First, they compared two strains, both expressing the longest human tau isoform, one bearing the P301L mutation and one without mutations, at similar, moderate levels [122]. The two strains developed very different phenotypes. Nonmutant mice became motor-impaired already around at 6–8 weeks of age, accompanied by axonopathy, but no tau aggregates, and survived normally. In contrast, the mutant mice developed NFTs from 6 months of age, without axonal dilatations and, despite displaying only minor motor problems, all succumbed before the age of 13 months. The authors concluded that excessive binding of wild-type human tau as opposed to reduced binding of P301L mutant tau to microtubules may be responsible for the development of axonopathy and tauopathy, respectively, in the two strains and that the conformational change of P301L tau is a major determinant in triggering the tauopathy. The third strain (a tau knock-in of human wild-type tau-4R/2N aimed to inactivate the endogenous murine tau gene and to replace it with a single copy of the *thy1*-tau-4R/2N expression construct) survived normally with minor motor problems late in life and without any obvious pathology [122]. When these findings are compared with those obtained by other research groups, it becomes obvious that the different strains show a range of phenotypes, possibly due to the use of different promoters for transgene expression, the integration site of the transgene, expression levels, and the mouse strain used for transgenesis [5].

In light of the neuropathological findings in humans that only a subset of the neuronal loss can be explained by NFTs, an important question arises, namely whether NFTs are an incidental marker for the neurotoxic cascade in AD or rather represent a protective neuronal response, allowing sequestration of neurotoxic species into a less harmful stable form [125]. To address this question, P301L mice were generated where the transgene can be turned off (or at least reduced from very high to only high overexpression levels). It was found that mice expressing doxycycline-repressible human P301L mutant tau developed progressive age-related NFTs, a remarkable neuron loss, and behavioral impairment. After the suppression of transgenic tau from 13- to 2.5-fold overexpression, memory function recovered, and neuron numbers stabilized, but NFTs continued to accumulate. These data convincingly show that tau dysfunction impairs memory, when massively overexpressed. The data further imply that NFTs per se (as entities of fibrillar accumulation that are visible by light microscopy) are not sufficient to cause cognitive decline or neuronal death in this model of tauopathy [125]. Not surprisingly, cognitive impairment in a second P301L tau transgenic mouse strain was shown to occur in the absence of NFT formation [126, 127]. As NFTs make up only a small percentage of all neurons in any animal model published so far, and as they are by far exceeded by dysfunctional neurons with tau aggregates but lacking NFTs, it is not surprising that, considering the limited life-span of mice compared with humans, NFT numbers do not correlate with functional impairment in these mice but rather the high number of cells that display tau aggregates.

12.6 Tau Transgenic Mice: Correlation of Histopathology and Behavioral Impairment

Similar to the APP transgenic models, the tau transgenic mouse models have been assessed using a wide range of behavioral tasks. Our mThy1.2 promoter-driven P301L mice accumulate tau in many brain areas but develop NFTs mainly in the amygdala. This brain area is involved in mediating effects of emotion and stress on learning and memory [124, 128, 129]. Therefore, behavioral alterations and cognitive deficits of the P301L mice were investigated using an amygdala-specific test battery for anxiety-related and cognitive behavior. These included an open-field, a light-dark box, fear conditioning, and a conditioned taste aversion (CTA) test [126]. The P301L mice showed an increased exploratory behavior but normal anxiety levels and no impairment in fear conditioning. In the P301L mice, fear conditioning was unaffected probably due to the absence of tau aggregates in the central and lateral nucleus of the amygdala. In the CTA test, the mice learn to associate a novel taste with nausea and, as a consequence, avoid consumption of this specific taste at the next presentation. We found that acquisition and consolidation of CTA memory was not significantly affected by the P301L transgene. However, transgenic mice extinguished the CTA memory more rapidly than did wild-type mice [126]. This rapid extinction may be due to the presence of tau aggregates in the basolateral nucleus of the amygdala, which has been shown to be essential for the extinction of CTA memory, whereas acquisition is dependent on an intact central nucleus, where no tau aggregates were found. When the P301L mice were assessed in hippocampus-dependent behavioral tests, the Morris water maze and Y-maze revealed intact spatial working memory but impairment in spatial reference memory at 6 and 11 months of age. In addition, a modest disinhibition of exploratory behavior at 6 months of age was confirmed in the open-field and the elevated O-maze and was more pronounced during aging [127].

The PrP promoter-driven P301L tau transgenic mice strongly overexpress mutant tau in several neuronal cell-types, including motor neurons. Therefore, they develop a progressive motor phenotype [114]. The V337M tau mutant mice show a very confined expression pattern as mutant tau was detected only in the hippocampus. These mice show an increased locomotor activity and memory deficits in the elevated plus maze, increased spontaneous locomotion in the open-field, but no significant impairment in the Morris water maze [130]. R406W tau mutant mice express tau at highest levels in the hippocampus and, to a lesser extent, in other cortical and subcortical brain areas. However, in the amygdala, only a few cells strongly express mutant tau, even in old animals [117]. These mice show a slight decrease in locomotor activity during the first minutes of the open-field test and a significant impairment in the contextual and cued fear-conditioning test.

When triple-transgenic mice (PS1 M146V knock-in microinjected with APP^{sw} and P301L tau transgenes) were analyzed, 2-month-old mice were cognitively unimpaired. The earliest cognitive impairment manifested at 4 months as a deficit in long-term retention and correlated with the accumulation of intraneuronal Aβ in the hippocampus and amygdala. Plaque or NFT pathology was not apparent at this age, suggesting that they contribute to cognitive dysfunction at later time points [131].

In summary, these findings demonstrate that tau aggregation in distinct brain areas directly affects the performance in memory tests controlled by these brain areas. They also show that tau aggregation per se, in the absence of NFT formation, is sufficient to cause behavioral deficits.

12.7 Cross-Talk of β-Amyloid and Tau in Experimental Model Systems

Before NFT formation had been achieved in tau transgenic mice, the interaction of plaques and NFTs has been addressed in different non-transgenic species such as rats and monkeys [132]. Intracerebral injection of plaque-equivalent concentrations of fibrillar, but not soluble, Aβ resulted in profound neuronal loss, tau phosphorylation, and microglial proliferation in the aged rhesus monkey cerebral cortex. In contrast, the same preparations were not toxic in the young adult rhesus brain, indicating a role for age in Aβ toxicity. This toxicity was also highly species-specific as it was neither observed in young nor in aged rats [132]. These results suggested that Aβ neurotoxicity in vivo is a pathological response of the aging brain, which is most pronounced in higher order primates. Thus, longevity may contribute to the unique susceptibility of humans to AD by rendering the brain vulnerable to Aβ neurotoxicity.

In transgenic mice, the presence of the P301L mutation appeared to accelerate tau filament formation as transgenic mice with high expression levels of human tau developed NFTs only at a high age [113–115]. P301L mutant mice are therefore suitable models to determine whether Aβ affects the tau pathology in these mice. Synthetic preparations of fibrillar $Aβ_{42}$ were stereotaxically injected into the somatosensory cortex and the hippocampal CA1 region of P301L and wild-type human tau transgenic mice and non-transgenic littermate controls, causing a fivefold increase of NFTs in the amygdala of P301L transgenic, but not wild-type tau transgenic or control mice, 18 days after the injections [124]. In contrast, when the non-fibrillogenic reversed peptide $Aβ_{42\text{-}1}$ was injected, levels of NFTs were not affected (Fig. 12.1). NFT formation in the $Aβ_{42}$-injected P301L mice was tightly correlated with the pathological phosphorylation of tau at S422 and the epitope AT100 (T212/S214), but not AT8 (S202/T205). The finding that $Aβ_{42}$ was not capable of inducing NFT formation in non-NFT-forming wild-type tau transgenic mice may reflect species differences between mice and men. Alternatively, it may imply that, at least in mice, $Aβ_{42}$ cannot induce NFT formation *de novo*, which would be in disagreement with the amyloid cascade hypothesis. Interestingly, in cultured murine hippocampal neurons, toxicity of $Aβ_{42}$ has been shown to be dependent on the presence of tau [133].

An alternative approach was chosen by Lewis and co-workers who crossed Aβ-producing APP-mutant Tg2576 mice with their PrP promoter-driven P301L tau mutant mice [134]. Double transgenic mice showed a more than sevenfold increase in NFT numbers in the olfactory bulb, the entorhinal cortex, and the amygdala compared with P301L single transgenic mice, whereas Aβ plaque formation was unaffected by the presence of the tau lesions (Fig. 12.1).

When both approaches are taken together, they imply that not all brain areas are similarly susceptible to Aβ-mediated NFT induction. In both studies, the amygdala is a hot spot of NFT induction. Unless tau levels are particularly high in the amygdala compared with other brain areas such as the hippocampus or cortical areas, a different mRNA/protein profile may account for the observed differences. A recent study of amygdala-specific gene expression provided a list of genes, some of which may confer an increased tau-related vulnerability of amygdaloid neurons to $Aβ_{42}$ [135]. Alternatively, it may be the nerve terminals, which are susceptible to $Aβ_{42}$, whereas direct exposure of the cell body or neurites may not pose a risk to the tau-expressing neuron. Whether Aβ is taken up by receptor-mediated mechanisms or whether it forms pores is still a matter of debate [136, 137] (Fig. 12.3).

Antibody-directed approaches were pursued in a recent study to dissect the cross-talk of Aβ and tau. When triple transgenic mice (PS1 M146V knock-in microinjected with APP^{sw} and P301L tau transgenes) were intracerebrally injected with anti-Aβ antibodies or a γ-secretase inhibitor, this resulted in

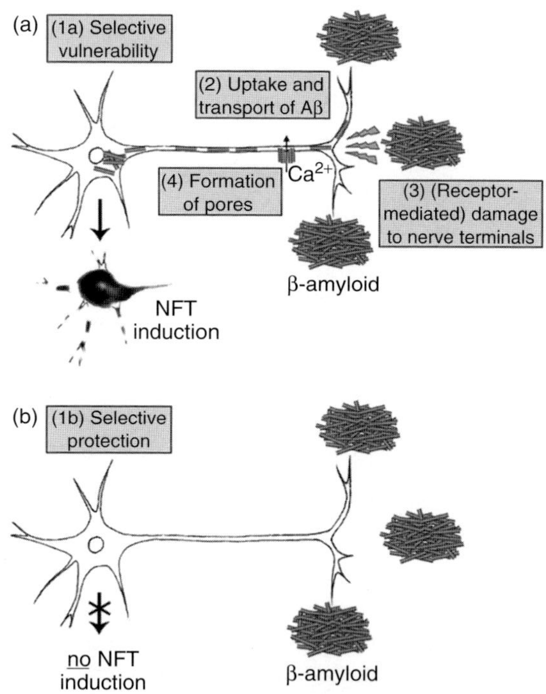

FIGURE 12.3. The mechanism of Aβ-mediated neurotoxicity is not understood at all. Whereas some neurons are particularly vulnerable already early in disease (A, 1a), others are relatively spared (B, 1b). Possible mechanisms of Aβ neurotoxicity and downstream NFT formation include uptake and transport of Aβ (2), (receptor-mediated) damage to nerve terminals (3), and the formation of pores (4). The receptors may have a selective specificity for Aβ or may, alternatively, bind peptides with a β-cross structure as the defining feature of amyloid fibers such as Aβ.

the disappearance of somatodendritic tau staining in young, but not old, mice [138]. It thus appears that extracellular Aβ deposits can exacerbate the intraneuronal pathology caused by the expression of mutant human tau protein [23].

An interaction between Aβ and tau was also demonstrated after the functional validation of proteomics findings in P301L tau transgenic mice [139]. Here, mainly mitochondrial proteins, antioxidant enzymes, and synaptic proteins were identified as modified in the proteome pattern of P301L tau mice. Significantly, the reduction in mitochondrial complex V levels in the P301L tau mice found by using proteomics was also confirmed as decreased in brains derived from human carriers of the P301L mutation of tau. Functional analysis demonstrated a mitochondrial dysfunction in P301L tau mice together with reduced NADH-ubiquinone oxidoreductase activity and, with age, impaired mitochondrial respiration and ATP synthesis. Mitochondrial dysfunction was associated with higher levels of reactive oxygen species in aged transgenic mice. Increased tau pathology as in aged homozygous P301L tau mice revealed modified lipid peroxidation levels and the upregulation of antioxidant enzymes in response to oxidative stress. To investigate whether brain cells from P301L tau mice are more susceptible to Aβ, we measured the mitochondrial membrane potential of isolated cortical brain cells with and without Aβ treatment [139]. Previous experiments using PC12 cells had shown that extracellular Aβ treatment lead to a significant decrease in mitochondrial membrane potential [140]. We found that, interestingly, the basal mitochondrial membrane potential was still conserved in cerebral cells from P301L tau mice. However a secondary insult with $A\beta_{42}$ resulted in a higher reduction in membrane potential in P301L tau mitochondria than in wild-type controls. Importantly, this effect was brain region–specific and therefore probably dependent on the presence of P301L tau because cells from the cerebellum with very low P301L tau expression levels were not vulnerable to this damage whereas cells from the cerebrum with high P301L tau expression levels were. These data suggests a synergistic action of Aβ and tau pathology on mitochondrial function. Moreover, it can be concluded that the tau pathology involves a mitochondrial and oxidative stress disorder distinct from that caused by Aβ [139].

The interaction between Aβ and tau has also been addressed in cell lines. Several studies have shown that tau-expressing cell lines are responsive to different forms of pathogenic stimuli. For example, when human SH-SY5Y neuroblastoma cells were incubated with okadaic acid (OA), a potent phosphatase inhibitor, together with HNE, a product of lipid oxidation found to be associated with NFTs in vivo [141–143], this resulted in the assembly of tau into aberrant polymers [144]. Most of them had a diameter of 2–3 nm and were straight, whereas PHFs have a diameter of 20 nm and are twisted. Fibrillar aggregates of tau were also observed in Chinese hamster ovary (CHO) cells that have been transfected with mutant tau expression constructs [145]. For example, Δ280K, but not several other single tau mutants (such as V337M, P301L, and R406W), developed insoluble amorphous and fibrillar aggregates, whereas a triple tau mutant containing V337M, P301L, and R406W substitutions (VPR) also formed similar aggregates. Furthermore, the aggregates increased in size over time. The formation of aggregated Δ280K and VPR tau protein correlated with their reduced affinity to bind microtubules. Reduced phosphorylation and altered proteolysis was also observed in R406W and Δ280K tau mutants. Thus, distinct pathological phenotypes, including the formation of insoluble filamentous tau aggregates, result from the expression of different FTDP-17 tau mutants in transfected CHO cells suggesting that these missense mutations cause diverse neurodegenerative FTDP-17 syndromes by multiple mechanisms.

As mentioned above, in human tauopathies other than AD, tau-positive inclusions are not restricted to neurons. They are found in oligodendrocytes and are a consistent neuropathological feature of CBD, PSP, and some forms of FTDP-17. When an oligodendroglial cell line was engineered to stably express high levels of the longest human tau isoform, treatment with OA caused tau hyperphosphorylation and a decreased binding of tau to microtubules. Transiently, tau-positive aggregates formed that could be stained with the amyloid-binding dye thioflavin-S. However, when the proteasome was inhibited by MG-132 after OA treatment, the aggregates were stabilized and were still detectable after 18 h in the absence of OA. Incubation with MG-132 alone did not induce the formation of thioflavin-S-positive aggregates. Hence, although tau hyperphosphorylation induced by protein phosphatase inhibition contributed to pathological aggregate formation, only hyperphosphorylation of tau followed by proteasome inhibition led to stable fibrillar deposits of tau similar to those observed in human tauopathies [146]. Together, these studies demonstrate that tau is capable of forming filamentous aggregates under specific experimental conditions.

Previous stereotaxic injection experiments have demonstrated principal differences between mice and men: Whereas Aβ induced NFT formation in human P301L mutant mice, it failed to do so in human wild-type tau transgenic mice. This is different from the situation in human AD, where Aβ aggregation and NFT formation occur in the absence of pathogenic tau mutations. Therefore, to address the role of Aβ in tau fibrillogenesis in a tissue culture system, we chose the human SH-SY5Y neuroblastoma instead of a murine cell line. SH-SY5Y cells can be neuronally differentiated by the sequential treatment with retinoic acid and brain-derived neurotrophic factor (BDNF) [147] (Fig. 12.4). They can be transplanted into mouse brain where they persist for a couple of days.

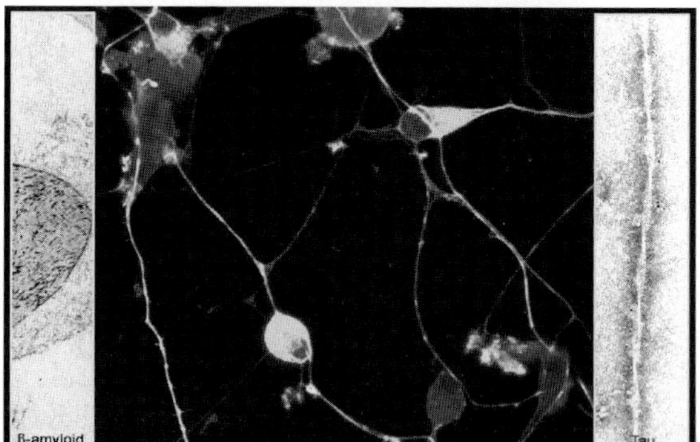

FIGURE 12.4. The formation of PHFs in tissue culture was reproduced by stably expressing human tau (both wild-type and P301L mutant) in neuronally differentiated human SH-SY5Y cells and exposing them for 5 days to aggregated synthetic $A\beta_{42}$. An electron micrograph of the fibrillar preparations of $A\beta_{42}$ is included (on the left). This incubation caused the generation of PHF-like tau containing filaments that were 20 nm wide and had periodicities of 130 to 140 nm in the presence of P301L mutant tau or 150 to 160 nm in the presence of wild-type tau (on the right).

Moreover, they anatomically integrate into organotypic hippocampal slices where they express synaptic markers and fire action potentials after 20 days in culture [O. Rainteau, A. Ferrari, and J. Götz, unpublished observations]. We stably expressed human tau with and without pathogenic mutations in these cells and exposed them for 5 days to aggregated synthetic $A\beta_{42}$ (Fig. 12.4) [148]. This caused a decreased solubility of tau along with the generation of PHF-like tau containing filaments, which were 20 nm wide and had periodicities of 130 to 140 nm in the presence of P301L mutant tau or 150 to 160 nm in the presence of wild-type tau (Fig. 12.4). As the stereotaxic $A\beta_{42}$ injection experiments had linked the S422 epitope of tau to NFT formation, we mutagenized serine 422 into alanine (which was intended to abrogate phosphorylation) and glutamic acid (intended to mimic phosphorylation). To our surprise, both mutations prevented the $A\beta_{42}$-mediated decrease in solubility and the generation of PHF-like filaments suggesting a role of S422 or its phosphorylation in tau filament formation. S422 is located next to a putative caspase-3 cleavage site at position 421, and altered caspase cleavage has been shown to be involved in the rates of tau filament formation [149–151]. Together, these data underscore a role of $A\beta_{42}$ in the formation of PHF-like filaments. These data are consistent with our previous results of $A\beta_{42}$-induced PHF-like tau filament formation in P301L tau transgenic mice [124] but in contrast to the transgenic mice $A\beta_{42}$-induced PHF formation in tissue culture also occurred with wild-type mice. This may be related to the species difference and points to the possibility that human cells in culture may be more susceptible to the formation of abnormal tau filaments than are murine cells in vivo.

The tissue culture system has since been used to map additional phospho-epitopes of tau involved in PHF formation and revealed that mutagenesis of some sites is even inhibitory to tau filament formation of endogenous, non-mutant tau [152]. Further adaptation of the system may allow the screening and validation of compounds designed to prevent PHF formation.

In summary, the above experiments demonstrate pathological interactions between Aβ and tau that led to increased NFT formation. Moreover, the region-specific induction of Aβ-mediated NFT formation in P301L tau transgenic mice mirrors, to some extent, the regional vulnerability observed in AD brains. Finally, besides their major advantages for an understanding of the pathophysiology of NFT formation, these models may assist in the development of therapies designed to reduce NFT formation and tau-related dysfunction, be they Aβ-mediated or not.

12.8 Outlook

The recent advent of transcriptomic and proteomic technology and its application to transgenic mouse models and tissue culture systems is likely to assist in the dissection of the pathocascade of AD and FTD [153]. Transcriptomics and proteomics identify individual, differentially regulated mRNAs and proteins and are in addition employed to dissect signaling pathways and reveal networks by using an integrated approach. This will undoubtedly lead to a redefinition and subdivision of disease entities based on biochemical criteria rather than the clinical presentation. Moreover, it will determine whether the pathogenesis of FAD and SAD are shared. Whether this can be reconciled with a unifying theory for AD remains to be determined. In any case, the new knowledge will have important implications for treatment strategies [97, 98].

Acknowledgments J.G. is a Fellow of the Medical Foundation. This work was supported by grants from the University of Sydney and the Medical Foundation (University of Sydney) to J.G.

References

1. Ritchie K, Lovestone S. The dementias. Lancet 2002; 360:1759-66.
2. Arnold SE, Hyman BT, Flory J, et al. The topographical and neuroanatomical distribution of neurofibrillary tangles and neuritic plaques in the cerebral cortex of patients with Alzheimer's disease. Cereb Cortex 1991; 1:103-16.
3. Ratnavalli E, Brayne C, Dawson K, Hodges JR. The prevalence of frontotemporal dementia. Neurology 2002; 58:1615-21.
4. Kurosinski P, Guggisberg M, Gotz J. Alzheimer's and Parkinson's disease—Overlapping or synergistic pathologies? Trends Mol Med 2002; 8:3-5.
5. Gotz J, Streffer JR, David D, et al. Transgenic animal models of Alzheimer's disease and related disorders: Histopathology, behavior and therapy. Mol Psychiatry 2004; 9:664-683.
6. Glenner GG, Wong CW. Alzheimer's disease: initial report of the purification and characterization of a novel cerebrovascular amyloid protein. Biochem Biophys Res Commun 1984; 120:885-90.
7. Masters CL, Simms G, Weinman NA, et al. Amyloid plaque core protein in Alzheimer's disease and Down syndrome. Proc Natl Acad Sci U S A 1985; 82:4245-9.
8. Vassar R, Bennett BD, Babu-Khan S, et al. Beta-secretase cleavage of Alzheimer's amyloid precursor protein by the transmembrane aspartic protease BACE. Science 1999; 286:735-41.
9. Edbauer D, Winkler E, Regula JT, et al. Reconstitution of gamma-secretase activity. Nat Cell Biol 2003; 5:486-8.
10. Braak H, Braak E. Neuropathological stageing of Alzheimer-related changes. Acta Neuropathol (Berlin) 1991; 82:239-59.
11. Braak H, Braak E. Staging of Alzheimer's disease-related neurofibrillary changes. Neurobiol Aging 1995; 16:271-8; discussion 278-84.
12. Crowther RA, Wischik CM. Image reconstruction of the Alzheimer paired helical filament. EMBO J 1985; 4:3661-5.
13. Wischik CM, Crowther RA, Stewart M, Roth M. Subunit structure of paired helical filaments in Alzheimer's disease. J Cell Biol 1985; 100:1905-12.
14. Goedert M, Wischik CM, Crowther RA, et al. Cloning and sequencing of the cDNA encoding a core protein of the paired helical filament of Alzheimer's disease: identification as the microtubule-associated protein tau. Proc Natl Acad Sci U S A 1988; 85:4051-5.
15. Lichtenberg B, Mandelkow EM, Hagestedt T, Mandelkow E. Structure and elasticity of microtubule-associated protein tau. Nature 1988; 334:359-62.
16. Schweers O, Schonbrunn-Hanebeck E, Marx A, Mandelkow E. Structural studies of tau protein and Alzheimer paired helical filaments show no evidence for beta-structure. J Biol Chem 1994; 269:24290-7.
17. Goedert M, Spillantini MG, Jakes R, et al. Molecular dissection of the paired helical filament. Neurobiol Aging 1995; 16:325-34.
18. Buee L, Bussiere T, Buee-Scherrer V, et al. Tau protein isoforms, phosphorylation and role in neurodegenerative disorders. Brain Res Brain Res Rev 2000; 33:95-130.
19. Lee VM, Goedert M, Trojanowski JQ. Neurodegenerative tauopathies. Annu Rev Neurosci 2001; 24:1121-59.

20. Gotz J. Tau and transgenic animal models. Brain Res Brain Res Rev 2001; 35:266-86.
21. Chen F, David D, Ferrari A, Gotz J. Posttranslational modifications of tau—Role in human tauopathies and modeling in transgenic animals. Curr Drug Targets 2004; 5:503-15.
22. Berriman J, Serpell LC, Oberg KA, et al. Tau filaments from human brain and from in vitro assembly of recombinant protein show cross-beta structure. Proc Natl Acad Sci U S A 2003; 100:9034-8.
23. Goedert M, Jakes R. Mutations causing neurodegenerative tauopathies. Biochim Biophys Acta 2005; 1739:240-50.
24. Tashiro K, Hasegawa M, Ihara Y, Iwatsubo T. Somatodendritic localization of phosphorylated tau in neonatal and adult rat cerebral cortex. Neuroreport 1997; 8:2797-801.
25. Lee G, Rook SL. Expression of tau protein in non-neuronal cells: Microtubule binding and stabilization. J Cell Sci 1992; 102:227-37.
26. Reszka AA, Seger R, Diltz CD, et al. Association of mitogen-activated protein kinase with the microtubule cytoskeleton. Proc Natl Acad Sci U S A 1995; 92:8881-5.
27. Morishima-Kawashima M, Kosik KS. The pool of map kinase associated with microtubules is small but constitutively active. Mol Biol Cell 1996; 7:893-905.
28. Flanagan LA, Cunningham CC, Chen J, et al. The structure of divalent cation-induced aggregates of PIP2 and their alteration by gelsolin and tau. Biophys J 1997; 73:1440-7.
29. Ebneth A, Godemann R, Stamer K, et al. Overexpression of tau protein inhibits kinesin-dependent trafficking of vesicles, mitochondria, and endoplasmic reticulum: Implications for Alzheimer's disease. J Cell Biol 1998; 143:777-94.
30. Jenkins SM, Johnson GV. Tau complexes with phospholipase C-gamma in situ. Neuroreport 1998; 9:67-71.
31. Sontag E, Nunbhakdi-Craig V, Lee G, et al. Molecular interactions among protein phosphatase 2A, tau, and microtubules. Implications for the regulation of tau phosphorylation and the development of tauopathies. J Biol Chem 1999; 274:25490-8.
32. Anderton BH, Dayanandan R, Killick R, Lovestone S. Does dysregulation of the Notch and wingless/Wnt pathways underlie the pathogenesis of Alzheimer's disease? Mol Med Today 2000; 6:54-9.
33. De F, Gv, Inestrosa NC. Wnt signaling function in Alzheimer's disease [In Process Citation]. Brain Res Brain Res Rev 2000; 33:1-12.
34. Maas T, Eidenmuller J, Brandt R. Interaction of tau with the neural membrane cortex is regulated by phosphorylation at sites that are modified in paired helical filaments. J Biol Chem 2000; 275:15733-40.
35. Goedert M, Spillantini MG, Cairns NJ, Crowther RA. Tau proteins of Alzheimer paired helical filaments: abnormal phosphorylation of all six brain isoforms. Neuron 1992; 8:159-68.
36. Sherrington R, Rogaev EI, Liang Y, et al. Cloning of a gene bearing missense mutations in early-onset familial Alzheimer's disease. Nature 1995; 375:754-60.
37. Van Broeckhoven C, Backhovens H, Cruts M, et al. Mapping of a gene predisposing to early-onset Alzheimer's disease to chromosome 14q24.3. Nat Genet 1992; 2:335-9.
38. Rocchi A, Pellegrini S, Siciliano G, Murri L. Causative and susceptibility genes for Alzheimer's disease: A review. Brain Res Bull 2003; 61:1-24.
39. Hodges JR, Davies RR, Xuereb JH, et al. Clinicopathological correlates in frontotemporal dementia. Ann Neurol 2004; 56:399-406.
40. Dickson DW. Pick's disease: A modern approach. Brain Pathol 1998; 8:339-54.
41. McKhann GM, Albert MS, Grossman M, et al. Clinical and pathological diagnosis of frontotemporal dementia: Report of the Work Group on Frontotemporal Dementia and Pick's Disease. Arch Neurol 2001; 58:1803-9.
42. Hodges JR, Davies R, Xuereb J, et al. Survival in frontotemporal dementia. Neurology 2003; 61:349-54.
43. Hutton M, Lendon CL, Rizzu P, et al. Association of missense and 5′-splice-site mutations in tau with the inherited dementia FTDP-17. Nature 1998; 393:702-5.
44. Poorkaj P, Bird TD, Wijsman E, et al. Tau is a candidate gene for chromosome 17 frontotemporal dementia. Ann Neurol 1998; 43:815-25.
45. Spillantini MG, Murrell JR, Goedert M, et al. Mutation in the tau gene in familial multiple system tauopathy with presenile dementia. Proc Natl Acad Sci U S A 1998; 95:7737-41.
46. Clark LN, Poorkaj P, Wszolek Z, et al. Pathogenic implications of mutations in the tau gene in pallido-ponto-nigral degeneration and related neurodegenerative disorders linked to chromosome 17. Proc Natl Acad Sci U S A 1998; 95:13103-7.
47. Dumanchin C, Camuzat A, Campion D, et al. Segregation of a missense mutation in the microtubule-associated protein tau gene with familial frontotemporal dementia and parkinsonism. Hum Mol Genet 1998; 7:1825-9.
48. Spillantini MG, Crowther RA, Kamphorst W, et al. Tau pathology in two Dutch families with mutations in the microtubule-binding region of tau. Am J Pathol 1998; 153:1359-63.
49. Delisle MB, Murrell JR, Richardson R, et al. A mutation at codon 279 (N279K) in exon 10 of the

Tau gene causes a tauopathy with dementia and supranuclear palsy. Acta Neuropathol (Berlin) 1999; 98:62-77.
50. D'Souza I, Poorkaj P, Hong M, et al. Missense and silent tau gene mutations cause frontotemporal dementia with parkinsonism-chromosome 17 type, by affecting multiple alternative RNA splicing regulatory elements. Proc Natl Acad Sci U S A 1999; 96:5598-603.
51. Goedert M, Spillantini MG, Crowther RA, et al. Tau gene mutation in familial progressive subcortical gliosis. Nat Med 1999; 5:454-7.
52. Mirra SS, Murrell JR, Gearing M, et al. Tau pathology in a family with dementia and a P301L mutation in tau. J Neuropathol Exp Neurol 1999; 58:335-45.
53. Rizzu P, Van Swieten JC, Joosse M, et al. High prevalence of mutations in the microtubule-associated protein tau in a population study of frontotemporal dementia in the Netherlands. Am J Hum Genet 1999; 64:414-21.
54. Sperfeld AD, Collatz MB, Baier H, et al. FTDP-17: An early-onset phenotype with parkinsonism and epileptic seizures caused by a novel mutation [see comments]. Ann Neurol 1999; 46:708-15.
55. Arima K, Kowalska A, Hasegawa M, et al. Two brothers with frontotemporal dementia and parkinsonism with an N279K mutation of the tau gene. Neurology 2000; 54:1787-95.
56. Delisle MB, Uro-Coste E, Murrell JR, et al. Neurodegenerative disease associated with a mutation of codon 279(N279K) in exon 10 of Tau protein. Bull Acad Natl Med 2000; 184:799-809.
57. Tolnay M, Grazia Spillantini M, Rizzini C, et al. A new case of frontotemporal dementia and parkinsonism resulting from an intron 10 +3-splice site mutation in the tau gene: Clinical and pathological features. Neuropathol Appl Neurobiol 2000; 26:368-78.
58. Lippa CF, Zhukareva V, Kawarai T, et al. Frontotemporal dementia with novel tau pathology and a Glu342Val tau mutation. Ann Neurol 2000; 48:850-8.
59. Pastor P, Pastor E, Carnero C, et al. Familial atypical progressive supranuclear palsy associated with homozigosity for the delN296 mutation in the tau gene. Ann Neurol 2001; 49:263-7.
60. Hayashi S, Toyoshima Y, Hasegawa M, et al. Late-onset frontotemporal dementia with a novel exon 1 (Arg5His) tau gene mutation. Ann Neurol 2002; 51:525-30.
61. Goedert M, Crowther RA, Spillantini MG. Tau mutations cause frontotemporal dementias. Neuron 1998; 21:955-8.
62. Delacourte A, Sergeant N, Wattez A, et al. Vulnerable neuronal subsets in Alzheimer's and Pick's disease are distinguished by their tau isoform distribution and phosphorylation. Ann Neurol 1998; 43:193-204.
63. Mott RT, Dickson DW, Trojanowski JQ, et al. Neuropathologic, biochemical, and molecular characterization of the frontotemporal dementias. J Neuropathol Exp Neurol 2005; 64:420-8.
64. Kurosinski P, Gotz J. Glial cells under physiologic and pathological conditions. Arch Neurol 2002; 59: 1524-8.
65. Cervos-Navarro J, Schumacher K. Neurofibrillary pathology in progressive supranuclear palsy (PSP). J Neural Transm Suppl 1994; 42:153-64.
66. Delacourte A. Pathological Tau proteins of Alzheimer's disease as a biochemical marker of neurofibrillary degeneration. Biomed Pharmacother 1994; 48:287-95.
67. Ksiezak-Reding H, Morgan K, Mattiace LA, et al. Ultrastructure and biochemical composition of paired helical filaments in corticobasal degeneration. Am J Pathol 1994; 145:1496-508.
68. Wischik CM, Novak M, Thogersen HC, et al. Isolation of a fragment of tau derived from the core of the paired helical filament of Alzheimer's disease. Proc Natl Acad Sci U S A 1988; 85:4506-10.
69. Bondareff W, Wischik CM, Novak M, et al. Molecular analysis of neurofibrillary degeneration in Alzheimer's disease. An immunohistochemical study. Am J Pathol 1990; 137:711-23.
70. Murrell JR, Spillantini MG, Zolo P, et al. Tau gene mutation G389R causes a tauopathy with abundant pick body-like inclusions and axonal deposits. J Neuropathol Exp Neurol 1999; 58:1207-26.
71. Rizzini C, Goedert M, Hodges JR, et al. Tau gene mutation K257T causes a tauopathy similar to Pick's disease. J Neuropathol Exp Neurol 2000; 59:990-1001.
72. Neumann M, Schulz-Schaeffer W, Crowther RA, et al. Pick's disease associated with the novel Tau gene mutation K369I. Ann Neurol 2001; 50: 503-13.
73. Rosso SM, van Herpen E, Deelen W, et al. A novel tau mutation, S320F, causes a tauopathy with inclusions similar to those in Pick's disease. Ann Neurol 2002; 51:373-6.
74. Kobayashi T, Ota S, Tanaka K, et al. A novel L266V mutation of the tau gene causes frontotemporal dementia with a unique tau pathology. Ann Neurol 2003; 53:133-7.
75. van Herpen E, Rosso SM, Serverijnen LA, et al. Variable phenotypic expression and extensive tau pathology in two families with the novel tau mutation L315R. Ann Neurol 2003; 54:573-81.
76. Pickering-Brown SM, Baker M, Nonaka T, et al. Frontotemporal dementia with Pick-type histology associated with Q336R mutation in the tau gene. Brain 2004; 127:1415-26.

77. Stanford PM, Halliday GM, Brooks WS, et al. Progressive supranuclear palsy pathology caused by a novel silent mutation in exon 10 of the tau gene: expansion of the disease phenotype caused by tau gene mutations. Brain 2000; 123(Pt 5):880-93.
78. Poorkaj P, Muma NA, Zhukareva V, et al. An R5L tau mutation in a subject with a progressive supranuclear palsy phenotype. Ann Neurol 2002; 52:511-6.
79. Morris HR, Osaki Y, Holton J, et al. Tau exon 10 +16 mutation FTDP-17 presenting clinically as sporadic young onset PSP. Neurology 2003; 61:102-4.
80. Bugiani O, Murrell JR, Giaccone G, et al. Frontotemporal dementia and corticobasal degeneration in a family with a P301S mutation in tau. J Neuropathol Exp Neurol 1999; 58:667-77.
81. Spillantini MG, Yoshida H, Rizzini C, et al. A novel tau mutation (N296N) in familial dementia with swollen achromatic neurons and corticobasal inclusion bodies. Ann Neurol 2000; 48:939-43.
82. Thal DR, Rub U, Orantes M, Braak H. Phases of A beta-deposition in the human brain and its relevance for the development of AD. Neurology 2002; 58:1791-800.
83. Crystal H, Dickson D, Fuld P, et al. Clinico-pathologic studies in dementia: nondemented subjects with pathologically confirmed Alzheimer's disease. Neurology 1988; 38:1682-7.
84. Arriagada PV, Growdon JH, Hedley-Whyte ET, Hyman BT. Neurofibrillary tangles but not senile plaques parallel duration and severity of Alzheimer's disease. Neurology 1992; 42:631-9.
85. Bierer LM, Hof PR, Purohit DP, et al. Neocortical neurofibrillary tangles correlate with dementia severity in Alzheimer's disease. Arch Neurol 1995; 52:81-8.
86. Nagy Z, Jobst KA, Esiri MM, et al. Hippocampal pathology reflects memory deficit and brain imaging measurements in Alzheimer's disease: clinicopathologic correlations using three sets of pathologic diagnostic criteria. Dementia 1996; 7:76-81.
87. Giannakopoulos P, Herrmann FR, Bussiere T, et al. Tangle and neuron numbers, but not amyloid load, predict cognitive status in Alzheimer's disease. Neurology 2003; 60:1495-500.
88. Delacourte A, David JP, Sergeant N, et al. The biochemical pathway of neurofibrillary degeneration in aging and Alzheimer's disease [see comments]. Neurology 1999; 52:1158-65.
89. Delacourte A, Sergeant N, Champain D, et al. Nonoverlapping but synergetic tau and APP pathologies in sporadic Alzheimer's disease. Neurology 2002; 59:398-407.
90. Caughey B, Lansbury PT. Protofibrils, pores, fibrils, and neurodegeneration: separating the responsible protein aggregates from the innocent bystanders. Annu Rev Neurosci 2003; 26:267-98.
91. Fukutani Y, Cairns NJ, Shiozawa M, et al. Neuronal loss and neurofibrillary degeneration in the hippocampal cortex in late-onset sporadic Alzheimer's disease. Psychiatry Clin Neurosci 2000; 54:523-9.
92. Leuba G, Kraftsik R. Visual cortex in Alzheimer's disease: occurrence of neuronal death and glial proliferation, and correlation with pathological hallmarks. Neurobiol Aging 1994; 15:29-43.
93. Gomez-Isla T, Hollister R, West H, et al. Neuronal loss correlates with but exceeds neurofibrillary tangles in Alzheimer's disease. Ann Neurol 1997; 41:17-24.
94. Kril JJ, Patel S, Harding AJ, Halliday GM. Neuron loss from the hippocampus of Alzheimer's disease exceeds extracellular neurofibrillary tangle formation. Acta Neuropathol (Berlin) 2002; 103:370-6.
95. Bussiere T, Gold G, Kovari E, et al. Stereologic analysis of neurofibrillary tangle formation in prefrontal cortex area 9 in aging and Alzheimer's disease. Neuroscience 2003; 117:577-92.
96. Holness MJ, Langdown ML, Sugden MC. Early-life programming of susceptibility to dysregulation of glucose metabolism and the development of Type 2 diabetes mellitus. Biochem J 2000; 349 Pt 3:657-65.
97. David D, Hoerndli, F., Gotz, J. Functional Genomics meets neurodegenerative disorders Part I: Transcriptomic and proteomic technology. Prog Neurobiol 2005:1-16.
98. Hoerndli D, David, D., Gotz, J. Functional Genomics meets neurodegenerative disorders Part II: Transcriptomic and proteomic technology. Prog Neurobiol 2005:1-21.
99. Gotz J, Schild A, Hoerndli F, Pennanen L. Amyloid-induced neurofibrillary tangle formation in Alzheimer's disease: Insight from transgenic mouse and tissue-culture models. Int J Dev Neurosci 2004; 22:453-65.
100. Games D, Adams D, Alessandrini R, et al. Alzheimer-type neuropathology in transgenic mice overexpressing V717F beta-amyloid precursor protein [see comments]. Nature 1995; 373:523-7.
101. Hsiao K, Chapman P, Nilsen S, et al. Correlative memory deficits, Abeta elevation, and amyloid plaques in transgenic mice [see comments]. Science 1996; 274:99-102.
102. Sturchler-Pierrat C, Abramowski D, Duke M, et al. Two amyloid precursor protein transgenic mouse models with Alzheimer's disease-like pathology. Proc Natl Acad Sci U S A 1997; 94:13287-92.
103. Stalder M, Phinney A, Probst A, et al. Association of microglia with amyloid plaques in brains of APP23 transgenic mice. Am J Pathol 1999; 154:1673-84.

104. Janus C, Pearson J, McLaurin J, et al. A beta peptide immunization reduces behavioural impairment and plaques in a model of Alzheimer's disease. Nature 2000; 408:979-82.
105. Mucke L, Masliah E, Yu GQ, et al. High-level neuronal expression of abeta 1-42 in wild-type human amyloid protein precursor transgenic mice: Synaptotoxicity without plaque formation. J Neurosci 2000; 20:4050-8.
106. Gotz J, Probst A, Spillantini MG, et al. Somatodendritic localization and hyperphosphorylation of tau protein in transgenic mice expressing the longest human brain tau isoform. EMBO J 1995; 14:1304-13.
107. Ishihara T, Hong M, Zhang B, et al. Age-dependent emergence and progression of a tauopathy in transgenic mice overexpressing the shortest human tau isoform. Neuron 1999; 24:751-62.
108. Spittaels K, Van den Haute C, Van Dorpe J, et al. Prominent axonopathy in the brain and spinal cord of transgenic mice overexpressing four-repeat human tau protein. Am J Pathol 1999; 155:2153-65.
109. Probst A, Gotz J, Wiederhold KH, et al. Axonopathy and amyotrophy in mice transgenic for human four-repeat tau protein. Acta Neuropathol (Berlin) 2000; 99:469-81.
110. Hirano A, Nakano I, Kurland LT, et al. Fine structural study of neurofibrillary changes in a family with amyotrophic lateral sclerosis. J Neuropathol Exp Neurol 1984; 43:471-80.
111. Munoz DG, Greene C, Perl DP, Selkoe DJ. Accumulation of phosphorylated neurofilaments in anterior horn motoneurons of amyotrophic lateral sclerosis patients. J Neuropathol Exp Neurol 1988; 47:9-18.
112. Rouleau GA, Clark AW, Rooke K, et al. SOD1 mutation is associated with accumulation of neurofilaments in amyotrophic lateral sclerosis. Ann Neurol 1996; 39:128-31.
113. Ishihara T, Zhang B, Higuchi M, et al. Age-dependent induction of congophilic neurofibrillary tau inclusions in tau transgenic mice. Am J Pathol 2001; 158:555-62.
114. Lewis J, McGowan E, Rockwood J, et al. Neurofibrillary tangles, amyotrophy and progressive motor disturbance in mice expressing mutant (P301L) tau protein. Nat Genet 2000; 25:402-5.
115. Gotz J, Chen F, Barmettler R, Nitsch RM. Tau filament formation in transgenic mice expressing P301L tau. J Biol Chem 2001; 276:529-34.
116. Tanemura K, Akagi T, Murayama M, et al. Formation of filamentous tau aggregations in transgenic mice expressing V337M human tau. Neurobiol Dis 2001; 8:1036-45.
117. Tatebayashi Y, Miyasaka T, Chui DH, et al. Tau filament formation and associative memory deficit in aged mice expressing mutant (R406W) human tau. Proc Natl Acad Sci U S A 2002; 99:13896-901.
118. Allen B, Ingram E, Takao M, et al. Abundant tau filaments and nonapoptotic neurodegeneration in transgenic mice expressing human P301S tau protein. J Neurosci 2002; 22:9340-51.
119. Gotz J, Tolnay M, Barmettler R, et al. Oligodendroglial tau filament formation in transgenic mice expressing G272V tau. Eur J Neurosci 2001; 13:2131-40.
120. Higuchi M, Ishihara T, Zhang B, et al. Transgenic mouse model of tauopathies with glial pathology and nervous system degeneration. Neuron 2002; 35:433-46.
121. Oddo S, Caccamo A, Shepherd JD, et al. Triple-transgenic model of Alzheimer's disease with plaques and tangles. Intracellular abeta and synaptic dysfunction. Neuron 2003; 39:409-21.
122. Terwel D, Lasrado R, Snauwaert J, et al. Changed conformation of mutant Tau-P301L underlies the moribund tauopathy, absent in progressive, nonlethal axonopathy of Tau-4R/2N transgenic mice. J Biol Chem 2005; 280:3963-73.
123. Sobrido MJ, Miller BL, Havlioglu N, et al. Novel tau polymorphisms, tau haplotypes, and splicing in familial and sporadic frontotemporal dementia. Arch Neurol 2003; 60:698-702.
124. Gotz J, Chen F, van Dorpe J, Nitsch RM. Formation of neurofibrillary tangles in P301L tau transgenic mice induced by Abeta 42 fibrils. Science 2001; 293:1491-5.
125. Santacruz K, Lewis J, Spires T, et al. Tau suppression in a neurodegenerative mouse model improves memory function. Science 2005; 309:476-81.
126. Pennanen L, Welzl H, D'Adamo P, et al. Accelerated extinction of conditioned taste aversion in P301L tau transgenic mice. Neurobiol Dis 2004; 15:500-9.
127. Pennanen L, Wolfer D, Nitsch RM, Gotz J. Impaired spatial reference memory in P301L tau transgenic mice. Genes, Brain and Behavior 2006; 5:369-379.
128. LeDoux JE. Emotion circuits in the brain. Annu Rev Neurosci 2000; 23:155-84.
129. Welzl H, D'Adamo P, Lipp HP. Conditioned taste aversion as a learning and memory paradigm. Behav Brain Res 2001; 125:205-13.
130. Tanemura K, Murayama M, Akagi T, et al. Neurodegeneration with tau accumulation in a transgenic mouse expressing V337M human tau. J Neurosci 2002; 22:133-41.

131. Billings LM, Oddo S, Green KN, et al. Intraneuronal Abeta causes the onset of early Alzheimer's disease-related cognitive deficits in transgenic mice. Neuron 2005; 45:675-88.
132. Geula C, Wu CK, Saroff D, et al. Aging renders the brain vulnerable to amyloid beta-protein neurotoxicity [see comments]. Nat Med 1998; 4:827-31.
133. Rapoport M, Dawson HN, Binder LI, et al. Tau is essential to beta-amyloid-induced neurotoxicity. Proc Natl Acad Sci U S A 2002; 99:6364-9.
134. Lewis J, Dickson DW, Lin W-L, et al. Enhanced neurofibrillary degeneration in transgenic mice expressing mutant Tau and APP. Science 2001; 293:1487-91.
135. Zirlinger M, Kreiman G, Anderson DJ. Amygdala-enriched genes identified by microarray technology are restricted to specific amygdaloid subnuclei. Proc Natl Acad Sci U S A 2001; 98:5270-5.
136. Verdier Y, Penke B. Binding sites of amyloid beta-peptide in cell plasma membrane and implications for Alzheimer's disease. Curr Protein Pept Sci 2004; 5:19-31.
137. Lashuel HA, Hartley D, Petre BM, et al. Neurodegenerative disease: amyloid pores from pathogenic mutations. Nature 2002; 418:291.
138. Oddo S, Billings L, Kesslak JP, et al. Abeta immunotherapy leads to clearance of early, but not late, hyperphosphorylated tau aggregates via the proteasome. Neuron 2004; 43:321-32.
139. David DC, Hauptmann S, Scherping I, et al. Proteomic and functional analysis reveal a mitochondrial dysfunction in P301L tau transgenic mice. J Biol Chem 2005; 280:23802-14.
140. Keil U, Bonert A, Marques CA, et al. Amyloid beta-induced changes in nitric oxide production and mitochondrial activity lead to apoptosis. J Biol Chem 2004; 279:50310-20.
141. Sayre LM, Zelasko DA, Harris PL, et al. 4-Hydroxynonenal-derived advanced lipid peroxidation end products are increased in Alzheimer's disease. J Neurochem 1997; 68:2092-7.
142. Perez M, Cuadros R, Smith MA, et al. Phosphorylated, but not native, tau protein assembles following reaction with the lipid peroxidation product, 4-hydroxy-2-nonenal. FEBS Lett 2000; 486:270-4.
143. Takeda A, Smith MA, Avila J, et al. In Alzheimer's disease, heme oxygenase is coincident with Alz50, an epitope of tau induced by 4-hydroxy-2-nonenal modification. J Neurochem 2000; 75:1234-41.
144. Perez M, Hernandez F, Gomez-Ramos A, et al. Formation of aberrant phosphotau fibrillar polymers in neural cultured cells. Eur J Biochem 2002; 269:1484-9.
145. Vogelsberg-Ragaglia V, Bruce J, Richter-Landsberg C, et al. Distinct FTDP-17 missense mutations in tau produce tau aggregates and other pathological phenotypes in transfected CHO cells. Mol Biol Cell 2000; 11:4093-104.
146. Goldbaum O, Oppermann M, Handschuh M, et al. Proteasome inhibition stabilizes tau inclusions in oligodendroglial cells that occur after treatment with okadaic acid. J Neurosci 2003; 23:8872-80.
147. Encinas M, Iglesias M, Liu Y, et al. Sequential treatment of SH-SY5Y cells with retinoic acid and brain-derived neurotrophic factor gives rise to fully differentiated, neurotrophic factor-dependent, human neuron-like cells. J Neurochem 2000; 75:991-1003.
148. Ferrari A, Hoerndli F, Baechi T, et al. Beta-amyloid induces PHF-like tau filaments in tissue culture. J Biol Chem 2003; 278:40162-8.
149. Abraha A, Ghoshal N, Gamblin TC, et al. C-terminal inhibition of tau assembly in vitro and in Alzheimer's disease. J Cell Sci 2000; 113:3737-45.
150. Fasulo L, Ugolini G, Visintin M, et al. The neuronal microtubule-associated protein tau is a substrate for caspase-3 and an effector of apoptosis. J Neurochem 2000; 75:624-33.
151. Berry RW, Abraha A, Lagalwar S, et al. Inhibition of tau polymerization by its carboxy-terminal caspase cleavage fragment. Biochemistry 2003; 42:8325-31.
152. Pennanen L, Gotz J. Different tau epitopes define Abeta(42)-mediated tau insolubility. Biochem Biophys Res Commun 2005; 337:1097-101.
153. Chen F, Wollmer MA, Hoerndli F, et al. Role for glyoxalase I in Alzheimer's disease. Proc Natl Acad Sci U S A 2004; 101:7687-92.
154. Brion JP, Tremp G, Octave JN. Transgenic expression of the shortest human tau affects its compartmentalization and its phosphorylation as in the pretangle stage of Alzheimer's disease [see comments]. Am J Pathol 1999; 154:255-70.

13
Glial Cells and Aβ Peptides in Alzheimer's Disease Pathogenesis

Gilbert Siu, Peter Clifford, Mary Kosciuk, Venkat Venkataraman, and Robert G. Nagele

13.1 Introduction

Alzheimer's disease (AD) is a tragic neurodegenerative disorder that targets the elderly and ultimately ends in dementia. Unfortunately, the ever increasing length of the human life span in the United States and throughout the world is now being paralleled by corresponding increases in the incidence of AD as well as in the duration of this disease in individual patients. AD is characterized symptomatically by progressive cognitive and memory loss, language deficits, impairment of judgment, deficient problem solving, and reduced abstract thought. At the root of these symptoms is widespread loss of neurons and their synapses primarily in the cerebral cortex, entorhinal area, hippocampus, ventral striatum, and basal forebrain [1–5]. Other pathological features that make their appearance in the brain tissue include a variety of different kinds of amyloid deposits collectively called amyloid plaques (Fig. 13.1), persistent accumulations of abnormal tau filaments referred to as neurofibrillary tangles, dense focal deposits of fibrillar amyloid in the walls of certain blood vessels (mostly small arterioles), intraneuronal accumulation of amyloid, reactive gliosis, and inflammation [1, 2, 6–9].

The presence of numerous amyloid plaques in AD brains has attracted great interest because they appear relatively early in the course of the disease and thus provide a potential early therapeutic target. These plaques consist of amyloid deposits, microglial cells, dystrophic neurites, and bundles of astrocytic processes. A principal component of plaques in human brain is amyloid β (Aβ) peptide, especially Aβ (1-42) (Aβ42), a 42-amino-acid peptide fragment derived from the sequential proteolytic cleavage of the amyloid precursor protein by beta- and gamma-secretases [1, 10]. An enormous number of studies have implicated Aβ42 as a key player in the observed neurodegenerative cascade, and many investigators believe that it may be directly responsible for the rampant synaptic and neuronal loss observed during the course of this disease [11, 12]. Exactly how the accumulation of this "toxic" peptide is linked to the observed cognitive and memory decline remains to be elucidated, but this is an area of intense research interest with the hope of changing the long-term outcome of this disease or, better yet, eradicating the disease altogether.

It is now well-recognized that glial cells (especially astrocytes and microglia) play a critical, dynamic role in inflammatory and neurodegenerative events that occur in the brain during the course of AD. Traditionally, astrocytes were assigned the role of filling tissue voids caused by degenerative events, a process called glial scar formation, whereas microglia were presumed to function primarily as brain phagocytes, responsible for the removal of Aβ deposits and debris from degenerating neurons and their processes. More recently, as will be discussed here, it has become apparent that there may be some sharing of phagocytic responsibility among these cell types and that their contribution to events occuring in the brain is considerably more complex than previously thought. In this chapter, we highlight the responses of astrocytes and microglia to intraneuronal Aβ accumulation, neuronal and synaptic degeneration, and amyloid plaque formation and focus on how their responses

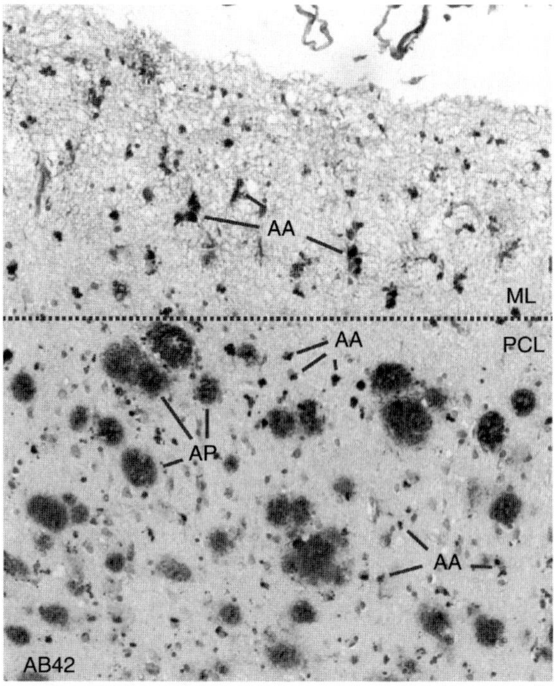

FIGURE 13.1. Section through entorhinal cortex of AD brain immunostained with anti-Aβ42 antibody (Chemicon International), which does not show appreciable reactivity with Aβ40 in ELISA or APP, showing amyloid plaques (AP) confined to the pyramidal cell layer (PCL). Activated astrocytes (AA) in both the molecular layer (ML) and PCL contain substantial quantities of Aβ42-positive material. The Aβ42-positive material in ML astrocytes is presumed to be derived from their active role in clearing debris associated with local synaptic and dendritic loss, which is rampant in this layer. Dendritic and synaptic loss in the ML appears to be temporally linked to the accumulation of Aβ42-positive material by the parent neurons in the underlying PCLs.

are intimately and irrevocably integrated into the fate of Aβ peptides and evolving pathology in AD brains.

13.2 Astrocytes and the Fate of Aβ in AD Brains

13.2.1 Astrocytes: Structure and Function in Normal Healthy Brain

Astrocytes, the predominant glial cell type found in the gray matter of the human CNS, extend numerous cytoplasmic processes that contain abundant bundles of intermediate filaments composed mainly of glial fibrillary acidic protein (GFAP) (Figs. 13.2A and 13.2B) [13]. In each astrocyte, the fine, highly branched tips of these cytoplasmic processes generally lack GFAP and can come into contact with thousands of local synapses [14]. In addition to structurally and functionally isolating synapses from events in the surrounding brain tissue, astrocytic processes are now thought to play an active role in synaptogenesis, the construction of neuronal circuits during development, synaptic stability and plasticity in the adult brain, ensuring normal neuronal excitability by maintaining extracellular ion homeostasis, and in clearing potassium from the region of synapses [15–19]. In addition, astrocytes are able to take up the excitatory amino acid glutamate from the synaptic cleft to levels up to 10,000 times higher than that in the extracellular space, a function that is pivotal for optimal gluataminergic neurotransmission and avoiding neuronal excitotoxicity [20–24]. The interchange of metabolites between astrocytes and between astrocytes and neurons is complex and is not well-understood, but gap junctions are now thought to be critical for this function [17].

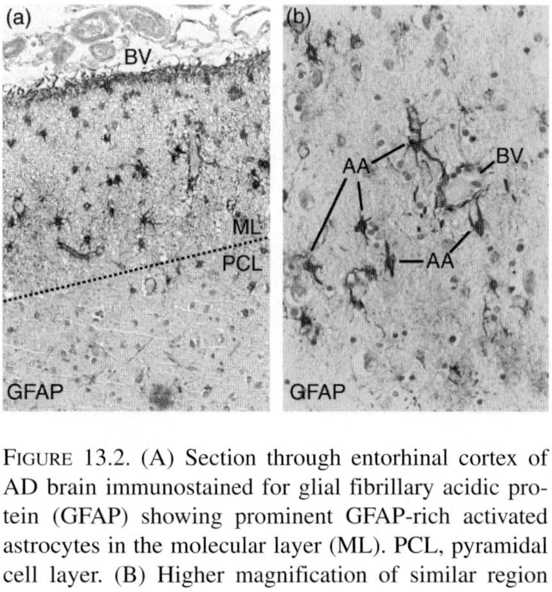

FIGURE 13.2. (A) Section through entorhinal cortex of AD brain immunostained for glial fibrillary acidic protein (GFAP) showing prominent GFAP-rich activated astrocytes in the molecular layer (ML). PCL, pyramidal cell layer. (B) Higher magnification of similar region showing the GFAP-rich processes of activated astrocytes (AA) with their associated end-feet in contact with the wall of a blood vessel (BV).

Lastly, the end feet of astrocytic processes encapsulate brain capillaries that pass through the brain tissue (Fig. 13.2B), most likely providing additional structural support for the blood-brain barrier and participating in regulation of the exchange between the smaller blood vessels and the surrounding brain tissue [25–27].

13.2.2 "Activation" of Astrocytes in Response to Local AD Pathology Compromises Astrocytic Function

In addition to playing a critical role in the functions described above, all of which are ultimately devoted to the maintenance of normal neuronal activity, astrocytes are capable of responding to pathological situations, where they engage in a series of structural and functional changes collectively referred to as "activation," "reactive astrogliosis," or "astrocytosis" [28–32]. These "activated astrocytes" exhibit a pronounced enlargement of their cell bodies and a dramatic thickening and lengthening of their cytoplasmic processes. They are readily identifiable in regions of CNS trauma, hypoxia, and in many neurodegenerative conditions by virtue of the dramatically elevated expression of glial fibrillary acidic protein (GFAP), vimentin, and nestin in their cell bodies and in the main trunks and branches of their cytoplasmic processes, compared with their more quiescent counterparts [13, 14, 28, 33] (Figs. 13.2A and 13.2B). Unfortunately, these changes come with a price—"activation" forces astrocytes to give up many of the activities mentioned above that were essential for normal neuronal function. Physiological functions such as the buffering of neuronally released potassium and glutamate from the extracellular space may be impaired, favoring local nerve cell depolarization, excessive Ca^{2+} influx, and excitotoxic damage to neurons [18, 34–35]. In addition, retraction of astrocytic end-feet and processes from synapses and the walls of local blood vessels may jeopardize the integrity of synapses and the local blood-brain barrier. Thus, although astrocyte activation no doubt is intended to be a protective response in the normal day to day activities in the brain, the intense and widespread astrocyte activation seen throughout AD brains may also exacerbate the extent of neuronal damage and even accelerate the rate of disease progression [36].

13.2.3 Astrocyte "Activation" Compromises the Blood-Brain Barrier, Leading to Leakage of Blood-Borne Substances, Including Soluble Aβ42, into Brain Tissue

The blood-brain barrier (BBB) is a diffusion barrier that blocks the movement of blood-borne substances into the brain parenchyma [37]. The three main components of the BBB are endothelial cells, the end-feet of astrocytes, and pericytes. Tight junctions between the endothelial cells in cerebral vessels are thought to provide the structural basis for the seal. Astrocyte end-feet tightly ensheath the vessel wall and most likely lend additional stability to the integrity of the barrier (Fig. 13.2B). Activation of astrocytes causes them to pull away many of their processes from the walls of blood vessels. The loss of astrocyte-endothelial cell contact can lead to breakdown of the BBB, resulting in an efflux of serum components into the brain tissue. Studies have shown that a significant pool of Aβ exists in the peripheral circulation [38–40]. Because the breakdown of the BBB is unlikely to occur uniformly throughout the brain, regions showing such leaks also exhibit increased levels of plasma components including serum immunoglobulin, complement, and Aβ42 [R. Nagele, unpublished observations]. Leakage of these components into AD brains can often be detected in AD brain as immunopositive perivascular "leak clouds" that, unexpectedly, are most often associated with small arterioles rather than capillaries within the brain parenchyma (Fig. 13.3). Elevated levels of these substances in the brain tissue may play an important role in the development of AD pathology as described in more detail below and could conceivably explain the frequent observation of "hot spots" of AD pathology, especially in the brains of patients that are early in the course of the disease.

13.2.4 Activated Astrocytes Accumulate Aβ42 in AD Brains

In early AD pathology, activated astrocytes are conspicuous in two regions: in the molecular layer of the cerebral cortex and in the immediate vicinity of amyloid plaques in the underlying pyramidal cell layers (Figs. 13.1 and 13.4A). What triggers these cells to become activated in response to AD-

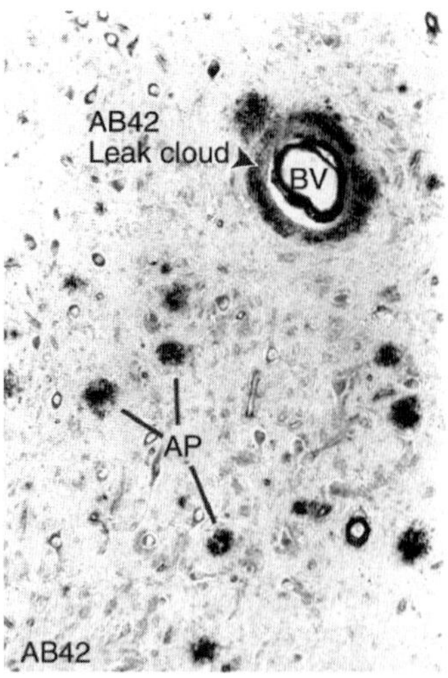

FIGURE 13.3. Section through the entorhinal cortex of an AD brain immunostained for Aβ42 showing and Aβ42-rich perivascular "leak cloud" surrounding a small blood vessel (BV) (arteriole). These leak clouds are observed preferentially around small arterioles and are only seen around brain capillaries in regions showing advanced pathology and well-developed inflammation. AP, amyloid plaque.

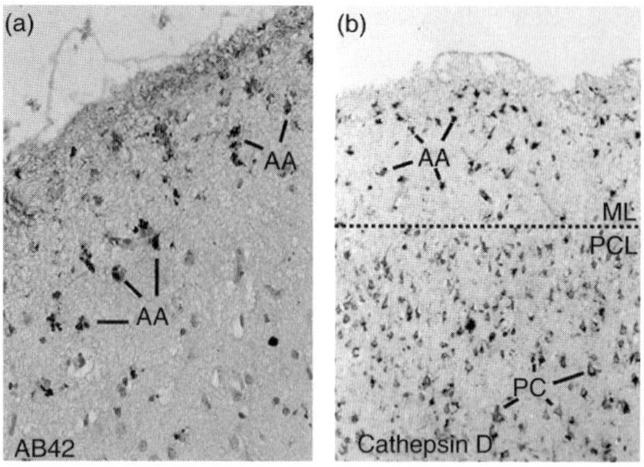

FIGURE 13.4. (A) Section through AD cortex immunostained with anti-Aβ42 antibodies showing large Aβ42-rich deposits in activated astrocytes (AA) in the molecular layer (ML). These same cells also exhibit intense cathepsin D (Sigma) immunoreactivity, suggesting increased activity of their lysosomal compartment. PC, pyramidal cells; PCL, pyramidal cell layer.

related pathological changes is not clear, but in vitro studies have shown that aggregated Aβ and the cores of amyloid plaques isolated from human AD brain tissue are effective in stimulating astrocyte activation [41]. Once activated, these cells are capable of internalizing and degrading Aβ42, suggesting that they may play a direct role in its clearance from the brain parenchyma. In support of this possibility, activated astrocytes in AD brains positioned in the cortical molecular layer as well as those closely association with neuritic or dense-core plaques in the underlying pyramidal cell layers can accumulate substantial amounts of Aβ42 (Figs. 13.1 and 13.4A) [42–46]. In the cortical pyramidal cell layers, astrocytes stationed outside of amyloid plaques, just beyond the outer edge of the Aβ42-rich corona, extend thick, intensely GFAP-immunopositive, cytoplasmic processes that envelop the amyloid plaque and thinner (mostly GFAP negative) branches from these processes that infiltrate deep into the plaque interior. In addition to intense GFAP immunostaining, these cells often show impressive intracellular accumulations of Aβ42-immunopositive material, suggesting that they are capable of internalizing Aβ42 via their processes and transporting it back to the cell body, presumably for degradation within the lysosomal compartment. In fact, most Aβ42-immunopositive material within astrocytes localizes to prominent granules in the perinuclear cytoplasm, and these granules have the same distribution and size as those that immunostain with antibodies specific for cathepsin D (Fig. 13.4B) [44].

13.2.5 The Amount of Aβ42 in Activated Astrocytes Is Linked to the Local Abundance of Neurons Containing Substantial Intracellular Aβ42 Deposits

The amount of Aβ42-positive material contained within activated astrocytes is not uniform throughout the cerebral cortex of AD brains but rather appears to be both spatially and temporally correlated with the extent of local AD pathology [44]. In the pyramidal cell layers, the Aβ42 content within individual astrocytes is proportional to the amount of intracellular Aβ42-positive material within nearby neurons as well as the presence and local density of plaques (Fig. 13.1). By contrast, cortical molecular layer astrocytes contain abundant Aβ42-positive material despite the fact that this layer generally lacks Aβ42-burdened neurons and plaques, especially in the early stages of AD pathogenesis

(Fig. 13.1). Interestingly, the amount of Aβ42-positive material within these astrocytes correlates closely with the severity of pathology exhibited by the pyramidal cell layers lying directly under this layer. In brain regions where pyramidal cells lack significant intracellular Aβ42 deposits, most of the overlying molecular layer astrocytes are quiescent and generally devoid of Aβ42-positive material [44]. Taken together, these observations emphasize the temporal and spatial link between Aβ42 accumulation in pyramidal neurons and the appearance of similar intracellular deposits in the overlying molecular layer astrocytes.

13.2.6 Activated Astrocytes Accumulate Aβ42 While Clearing the Products of Neuronal and Synaptic Degeneration and Loss

The source of the Aβ42 and the mechanism by which it accumulates selectively in activated astrocytes and not in their more quiescent counterparts remains to be determined. Expression of the amyloid precursor protein is either extremely low or nonexistent in astrocytes, thus internal production is unlikely to be a major source of the Aβ42 that accumulates in these cells. By contrast, exogenous (soluble) Aβ42 from the surrounding extracellular fluid is a much more likely source, and its accumulation in astrocytes could occur via receptor-mediated endocytosis and/or phagocytosis. In support of this possibility, the phagocytic capability of activated astrocytes has already been demonstrated and includes the removal of local synaptic material [47]. In addition, our previous study has provided strong evidence that most (possibly all) of the accumulated Aβ42 within activated astrocytes positioned in the cortical molecular layer is of neuronal origin and is derived from internalization of degenerating synapses and dendrites belonging to neurons in the underlying pyramidal cell layers [44]. Further evidence for this mode of astrocytic Aβ42 accumulation comes from the fact that Aβ42 in activated astrocytes colocalizes with other neuron-specific proteins, including choline acetyltransferase (ChAT) and the alpha7 nicotinic acetylcholine receptor (α7nAChR) (Fig. 13.5A), neither of which is synthesized by astrocytes [44]. The selective accumulation of these neuronal proteins and Aβ42

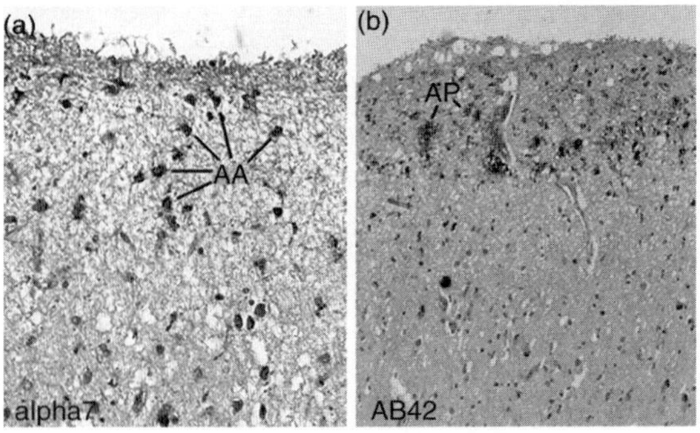

FIGURE 13.5. (A) Section through entorhinal cortex of AD brain immunostained with rabbit polyclonal antibodies directed against the alpha7 nicotinic acetylcholine receptor (alpha7) (Santa Cruz Biotechnology, sc-1447, raised against amino acids 367–502 mapping at the C-terminus of human a7nAChR). Activated astrocytes (AA) in the cortical molecular layer are strongly immunopositive for alpha7. Alpha7 accumulation in these cells is a by-product of their action in clearing local dendritic and synaptic debris. Confirmation of the specificity of this antibody was obtained by Western blot analysis and deletion of staining by preabsorption with the immunogen peptide. (B) Death and lysis of Aβ42-overburdened, activated astrocytes leads to the formation of small astrocytic amyloid plaques (AP) in the cortical molecular layer that are both Aβ42- and alpha7-immunopositive.

in activated astrocytes is an expected consequence of their debris-clearing activity in response to elevated levels of AD-related degeneration of local dendrites and synapses. The fact that accumulated ChAT- and α7nAChR-immunopositive material is most prominent in astrocytes populating the cortical molecular layer is a reflection of the abundance of synapses containing these proteins in this region [44]. Studies using electron microscopy have shown that the corona of dense-core amyloid plaques in the pyramidal cell layers and the amyloid aggregates associated with capillaries are extensively infiltrated with astrocytic processes in both human AD and APP tg mouse brains [48–50]. Aβ deposits can apparently be degraded by metalloproteases, including neprilysin and insulysin [51, 52], and neprilysin has been localized in astrocytes closely associated with amyloid plaques, suggesting that they possess the requisite elements for Aβ degradation [53]. In view of the above, the idea that astrocytes may not become phagocytic until the phagocytic capacity of brain microglia has become saturated [54] may have to be discarded. In fact, the reverse seems more likely—that microglia are not activated until after the phagocytic activity of astrocytes is overwhelmed or, at least, sufficiently taxed above some unknown threshold level.

13.2.7 Effects of Intracellular Aβ42 Accumulation on the Functional Activity of Astrocytes

It is not known whether Aβ42-burdened, activated astrocytes are capable of clearing internalized and accumulated Aβ42. The fact that the total astrocytic amyloid burden seems to increase in AD brains with the degree of AD pathology suggests that astrocytes are either not capable of clearing internalized Aβ42 or that their clearance mechanism may be deficient. The effects of gradual intracellular Aβ42 accumulation on the functional activity of astrocytes is unknown, but it is likely to have a progressively deleterious effect on these cells throughout the accumulation process, eventually ending in cell death and lysis. As mentioned above, astrocytes are known to make contacts with multiple neurons in their immediate vicinity. This position between neurons allows astrocytes to facilitate information transfer between neighboring neurons and other astrocytes, maintain neuronal excitability by keeping close control over ion homeostasis, and may contribute to synaptic plasticity [15, 16, 20, 55–57]. Recent work has led to a new appreciation of the active role of astrocytes and astrocyte-derived cytokines in the response to injury and repair and their influence on the integrity of the blood brain barrier [53, 58, 59]. Degeneration of cortical dendrites and synapses in AD brains may stimulate the conversion of "quiescent" to "activated" astrocytes [31]. Such degeneration would result in a severing of astrocyte-neuron contacts, which may itself provide a signal for activation of astrocytes, the clearing of local neuronal debris, and, thus, drive the accumulation of neuron-derived materials, including Aβ42, in these cells. Another consequence is impairment of astrocyte-maintained extracellular ion homeostasis, which favors excitotoxic neuronal damage [32]. It is possible that, as in many otherwise protective processes, this may get out of hand by favoring oxidative neuronal damage and enhanced Aβ toxicity, thus providing a therapeutic target to possibly slow it down [31].

13.2.8 Aβ42-Overburdened Astrocytes Can Undergo Lysis to Form Astrocyte-Derived Amyloid Plaques

The progressive and extensive synaptic loss in the cortical molecular layer appears to gradually increase the intracellular load of Aβ42-immunopositive material that has accumulated in local activated astrocytes (Fig. 13.1). We have shown that this increased load is eventually accompanied by the appearance of a new population of amyloid plaques within the cortical molecular layer (Fig. 13.5B). This new population of plaques appears to be derived from the death and lysis of Aβ42-overburdened astrocytes [44]. Upon lysis, cytoplasmic material from ruptured astrocytes is dispersed somewhat radially, including their content of accumulated Aβ42. This dispersion may initially be facilitated by the action of lysosomal enzymes that are also released at that time. Cell lysis leaves in its wake a persistent, roughly spherical, Aβ42-rich residue that takes the form of a distinctive population of amyloid plaques. That these plaques are derived from the lysis of astrocytes is bolstered by the fact that they first appear in the subpial portion of the cortical molecular layer and are observed only in regions where nearby astrocytes contain large intracellular deposits of Aβ42-positive material (Fig. 13.1B) [44]. This proposed mode of "astrocytic" plaque formation is nearly identical to that which has been described previously for the larger, spherical, neuron-derived plaques that populate the underlying pyramidal cell layers, many of which appear to be the lysis remnants of Aβ42-overburdened neurons [10, 60]. Although both types of plaque are Aβ42-immunopositive, astrocytic plaques are readily distinguished from neuron-derived plaques because of their location, much smaller size, and particularly intense GFAP-immunoreactivity. The consistent spherical shape of most plaques (Fig. 13.1) and the close relationship between the size of both neuron- and astrocyte-derived amyloid plaques and the cells from which they are presumably derived argue strongly against proposed mechanisms for amyloid plaque formation that describe the gradual growth of plaques from a seeding site or "nidus," at least for this morphological subset of plaques.

13.2.9 Astrocyte Activation May Be Triggered by the Intraneuronal Accumulation of Aβ42 in AD Brains

The formation of large intracellular deposits of Aβ42 have been reported in several types of neurons in the cerebral cortex and cerebellum of AD and Down syndrome brains (Fig. 13.6) [8, 44, 61–64]. Our recent studies suggest that their ability to do so may be linked to neuronal expression of the alpha7 nicotinic acetylcholine receptor (α7nAChR) [60]. Previous studies have shown that Aβ42 binds with exceptionally high affinity to α7nAChRs on neuronal surfaces [63–64]. As described above, the leak of serum Aβ42 into the brain parenchyma through local breaches in the BBB (cf. Fig. 13.3) would be expected to provide a constant source of exogenous Aβ42 to local neurons. Thus, neurons that are particularly well-endowed with α7nAChRs (e.g., cortical pyramidal cells) would form relatively high levels of Aβ42/α7nAChR complex on their surfaces. It follows then that any membrane recycling or endocytic activity on the part of the neurons would tend to drive the internalization of Aβ42/α7nAChR complex into neurons and target this complex to the lysosomal compartment. Consistent with this mechanism, Aβ42 and the α7nAChR are invariably colocalized within intraneuronal deposits in AD brains, and these deposits are also immunopositive for cathepsins, confirming that this accumulation occurs within the lysosomal compartment [60]. We have suggested that the binding of "exogenous" Aβ42 to the α7nAChR-bearing dendrite trees of neurons may not only facilitate internalization and accumulation of Aβ42 in these cells via endocytosis but also provides a plausible explanation for the well-known selective vulnerability of cholinergic and cholinoceptive neurons to AD pathogenesis [60].

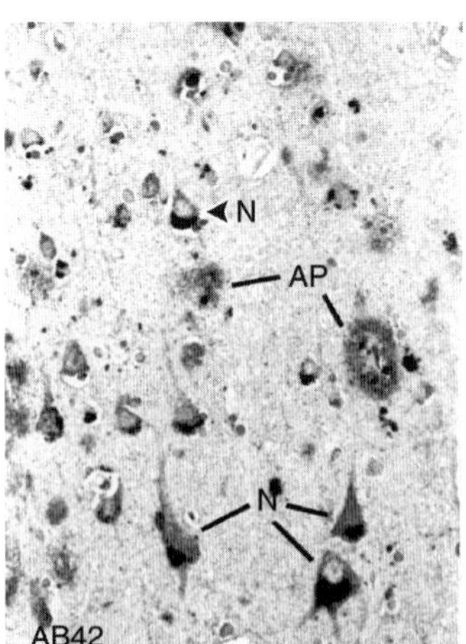

FIGURE 13.6. Section through the entorhinal cortex of an AD brain immunostained with anti-Aβ42 antibodies showing Aβ42 localized to amyloid plaques (AP) and large intracellular deposits within pyramidal neurons (N).

The accumulation of Aβ42/α7nAChR complex in cortical pyramidal neurons is one of the earliest signs of developing AD pathology, and work on transgenic mice has temporally linked this event with early synaptic degeneration and loss [44, 66–68]. It is likely that these events are also directly linked to the observed early activation of astrocytes in the cortical molecular layer. This layer is densely packed with the fine, α7nAChR-rich dendrite branches that extend from the main dendrite trunks of neurons positioned in the pyramidal cell layers lying directly below. We have suggested that excessive accumulation of Aβ42 in neurons (Fig. 13.6) impairs the ability of these cells to maintain their extensive dendritic arbors. If this is the case, then the most distal dendrite branches and their associated synapses, located in the cortical molecular layer, would be most vulnerable to degeneration and loss, which is consistent with what is observed. If Aβ42/α7nAChR complex is present on degenerating dendrites and synapses, clearing of this debris by local astrocytes via phagocytosis/endocytosis and the targeting of this material to the lysosomal compartment would explain the source of Aβ42 seen in these cells (Fig. 13.5A). In addition, it would explain why other neuron-specific proteins, such as α7nAChR and choline acetyl transferase (ChAT), are also colocalized within Aβ42-immunopositive deposits of astrocytes [44]. Thus, "activated" astrocytes are capable of internalizing neuron-derived materials, including surface-bound Aβ42, presumably through their endocytic/phagocytic activity and, as in neurons, this activity is paralleled by a dramatic elevation of lysosomal cathepsin D levels [44]. The great affinity of Aβ42, but not Aβ40, for the α7nAChR also provides a straightforward explanation for Aβ42 as the dominant Aβ peptide species in astrocytic intracellular deposits and in amyloid plaques throughout AD brains [60]. The proposed mechanism described above pinpoints a few variables that may dictate variations in both the nature of the pathology and rate at which it evolves in individual AD patients. These variables could include the serum levels of Aβ42, the location(s) of the breach in the BBB, whether the breach is focal or global, and whether the breach is sufficient to allow passage of materials from the blood into the brain that could contribute to AD pathology (e.g., Aβ42, immunoglobulin, and complement).

13.3 Microglia and the Fate of Aβ in AD Brains

Microglia are resident cells of monocyte-phagocyte lineage in the brain that, when activated, are capable of phagocytosis and participating in immune responses by presenting antigens to invading immune cells. In the normal healthy brain, they are referred to as "resting microglia" and are widely scattered, seeming to occupy their own individual defined territory within the brain parenchyma. The function of these cells in the resting state is unknown. However, in response to pathological changes in the brain tissue, microglia can rapidly transition to an activated state (Fig. 13.7A). In the activated state, these cells take on a more amoeboid character and migrate to the site of injury, where they can proliferate, launch a phagocytic attack on the offending material including tissue debris, and release inflammatory mediators such as cytokines into the surrounding tissue [69–74]. Much of what we know about the activity of microglia has been derived from studies on the actions of these cells in the culture environment. In cell cultures, microglia show increased cell surface expression of MHCII [75], a classic marker for activated microglia, as well as an increased secretion of inflammatory cytokines such as interleukin-1B (Il-1B), interleukin-6 (IL-6), and tumor necrosis factor-α (TNF-α), and chemokines such as interleukin-8 (IL-8), macrophage inflammatory protein-1α (MIP-1α), and monocyte chemoattractant peptide-1 (MCP-1) [76]. In addition, mRNAs encoding C1q, C3, C4, IL-1 receptor antagonist, and transforming growth factor-β (TGF-β) have been detected in AD microglia [77–80]. Where the tissue devastation is particularly great, brain microglia intermingle with additional monocytic cells that appear to migrate into the brain tissue from the blood (Fig. 13.7B). At this point, it is often difficult to distinguish microglia from these immigrant macrophages and, for this reason, it is probably best to refer to them as microglia/macrophages. The precise identity and nature of the signals that cause the initial activation of microglia that are resident in the brain are unknown.

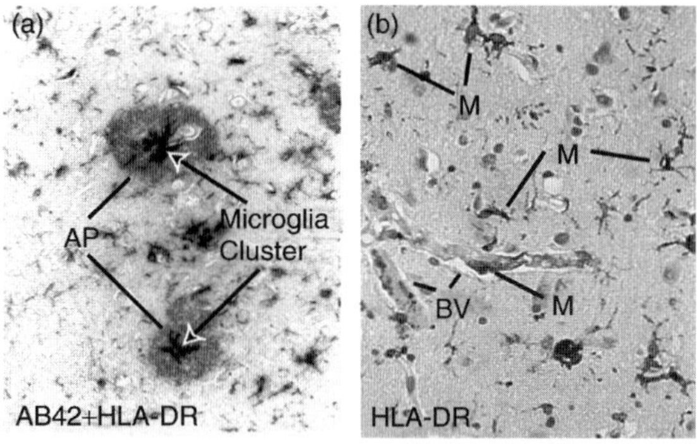

FIGURE 13.7. (A) Section through entorhinal cortex of AD brain double-immunostained with anti-Aβ42 antibodies and HLA-DR antibodies to immunolabel activated microglia. Note the strong tendency for activated microglia to congregate at the exact center of the amyloid plaque (AP), a region known to contain a neuronal nuclear remnant and other debris associated with neuronal lysis. (B) Section through the pyramidal cell layer immunostained with anti-HLA-DR antibodies showing microglia/macrophages (M), some of which appear to be in the process of entering into the brain tissue from local small blood vessels (BV).

13.3.1 Relationship Between the Phagocytic Activity of Microglia and Aβ in AD Brains

In AD brains, activated microglia are widely distributed throughout the brain parenchyma but are also focally concentrated within amyloid plaques where they are generally thought to be actively engaged in the clearance of Aβ from the plaque interior via phagocytosis [30, 55, 56, 70, 71, 81–89]. In culture, microglia derived from AD brains are not only able to congregate around aggregated Aβ deposits, but they appear to be able to remove these deposits over a period of 2–4 weeks [90]. In addition, the intracellular accumulation of Aβ occurs more rapidly and to a greater extent in these cells when serum is added to the culture medium, suggesting that serum contains some factor(s) that facilitates Aβ endocytosis [55]. Microglia applied to unfixed brain sections in culture reportedly phagocytose Aβ deposits when anti-Aβ antibodies are included in the culture medium, suggesting that opsonization of the Aβ facilitates this activity [55, 82]. The Aβ is subsequently found in phagosome-like intracellular vesicles [82].

Unfortunately, studies on the activities of microglia in the context of the AD brain have been less revealing. Although ultrastructural studies have reported that microglia in the AD cortex contain some intracytoplasmic Aβ fibrils, it is not dramatic, and there have been reports to the contrary [85, 91, 92]. One possible explanation for this apparent discrepancy is that microglia might process internalized Aβ so rapidly that little of this material can be demonstrated in a cell at any particular time. Of course, another possibility is that Aβ internalization by microglia is a culture anomaly and that they do not internalize Aβ at all in the brain. If the latter proves to be true, we are still left without assigning a definitive function to the microglia that are stationed within amyloid plaques. In contrast to a role in the clearance of Aβ from plaques and the brain, it has also been suggested that microglia participate in the conversion of soluble or oligomeric Aβ into polymerized amyloid fibrils in the parenchyma, within plaques and in the walls of blood vessels [85]. This idea is based on the observation that plaque-associated microglia display dilated intracellular channels of endoplasmic reticulum that appear to contain amyloid fibers [91, 92]. Also, largely because of their location within plaques, the actions of plaque-associated microglia have been postulated to play a role in the reported transformation of diffuse amyloid plaques into neuritic or dense-core Aβ plaques. However, this role seems to be unlikely in view of the fact that microglia are generally not found in association with diffuse plaques but rather clearly prefer to congregate at the central portions of dense-core plaques in both AD brains and the brains of APP-overexpressing transgenic mice [93] (Fig. 13.7A). In addition, it has not yet been determined whether one morphological type of plaque can evolve into another or whether they represent different plaque types with unique origins. The general lack of an obvious, well-defined function for plaque-associated microglia that is related to either Aβ clearance or deposition inevitably leads one to consider the possibility that their presence within plaques may have nothing at all to do with Aβ clearance in the brain.

13.3.2 Microglial Chemotaxis: Aβ or DNA Fragments as Chemoattractants

What lures microglia to amyloid plaques is unknown, but their preferential association with dense-core plaques as well as the tendency for them to be positioned at or near the dense core of plaques suggests that there is something either at or emanating from the plaque core that is strongly chemotactic to microglia. In elegant studies carried out by Rogers and co-workers on cultured microglia originally isolated from the brains of both AD and non-demented patients, these cells exhibited obvious chemotaxis to preaggregated Aβ42 deposits that were adherent to the culture substratum [94, 95] but were not attracted to Aβ42 scrambled sequence [96]. It has been reported that Aβ can bind to several types of microglial cell surface receptors, including RAGE [97]. Although they have provided a wealth of information on the phagocytic actions of microglia, cultured microglia models also have some limitations that raise questions about how accurately and directly the actions of these cells in culture reflect those of their counterparts in the context of the brain. One obvious limitation is that the responses of cultured microglia to various test agents or conditions are occurring in an artificial environment that lacks their usual interactions with neurons, neuronal processes, astrocytes, and elements of the local blood vasculature. Another limitation is that the culture environment alone is sufficient to activate microglia, which makes it difficult to determine the identity of factors that can either induce or influence the activation state. Lastly, compared with what happens in a slowly evolving disease state such as AD, studies on cultured microglia are of very short duration.

Direct extrapolation of the results on chemotaxis obtained from studies on cultured microglia to the actions of microglia in vivo does not seem to fit well with the apparent behavior of microglia and their response to Aβ peptides in the AD brain. For example, if Aβ42 is chemotactic to microglia in AD brains, one would expect to see abundant microglia near and within all types of Aβ42-containing plaques. Contrary to this expectation, microglia are generally not found either within or associated with diffuse plaques, which contain abundant Aβ42. In addition, these cells apparently pass through the Aβ42-rich outer corona of dense-core plaques and take up residence preferentially at the plaque core (Fig. 13.7A), which is also rich in Aβ42. Together, these observations suggest that, in AD brains, something at or within the core of dense-core plaques is highly chemotactic to activated microglia. One likely chemotactic factor is DNA fragments. Microglia have been shown to accumulate damaged DNA fragments in AD brain, and fragmented DNA has been suggested as a potent promoter of microglial activation [98]. In support of this possibility, our previous studies have provided strong evidence that many (possibly all) dense-core plaques in the pyramidal cell layers of the cerebral cortex are derived from the lysis of Aβ42-overburdened neurons. Neuronal lysis releases the contents of the neuronal perikaryon, including Aβ42 and lysosomes. The local release of lysosomal enzymes probably facilitates the radial diffusion of neuron-derived Aβ peptide, which explains both the generally spherical shape of all plaques as well as the fact that their individual sizes seems to correlate with the size of local neurons (Fig. 13.6) [8, 44, 60]. Another consequence of neuronal lysis is the persistent presence of a nuclear remnant at the core of the dense-core plaque [8]. Here, we propose that the gradual degradation of this nuclear material releases DNA fragments that diffuse out from the plaque core into the surrounding brain parenchyma. Because microglia are capable of responding to DNA fragments, it is reasonable to suppose that the release of these fragments is chemotactic to microglia, drawing them ever closer to the source of the DNA positioned at the plaque core (Fig. 13.7A). In addition, peripheral monocytes are often observed emigrating from local small blood vessels into regions where dense-core plaques are nearby or adjacent, which is not observed in brain regions containing only diffuse plaques [66, 99, 100] (Fig. 13.7B). In fact, it is entirely possible that most of the so-called microglia/macrophages seen within dense-core amyloid plaques in AD brains are immigrants from the blood and that the involvement of resting/resident microglia in the formation/evolution and eventual clearance of Aβ42 and plaques is minimal. The practicality of DNA fragments serving as the principal chemotactic signal attracting local microglia and moncytic cells from local blood vessels is obvious because its release into the local milieu can only occur via local cell death, thus making it an unambiguous marker indicating that local cellular degeneration and death has actually occurred.

13.3.3 Mediators of Microglial Phagocytosis

There are likely to be multiple mediators of microglial activation, chemotaxis, and phagocytic activity in the brain, and some of these may depend on the nature of the pathology that develops in association with specific brain diseases. The formyl peptide receptor (FPR), the macrophage scavenger receptors (MSR) [101], and the receptor for advanced glycation end products (RAGE) are expressed by microglia, have opsin-independent activity, and appear to have Aβ as a ligand [102, 103]. Microglia also express the complement opsonin receptors CR3 and CR4 and the anaphylatoxins C3a and C5a [104–107]. Complement is well-known to facilitate the phagocytosis of tissue debris, and there is some evidence that complement can opsonize Aβ fibrils, facilitating their removal by microglial phagocytosis. The well-known pathway for complement activation is initiated with the attachment of C1q to a target, its interaction with a number of proteases (including C1r, C1s, C4, C2, C3) followed by the attachment of C4b and C3b, which act as ligands for complement receptors on microglia and other phagocytic cells [108]. When completed, complement terminal components (C5b–C9) are assembled into the membrane attack complex. Complement activation and opsonization of fibrillar Aβ by C1q in amyloid plaques has been demonstrated in AD brains [109–111]. The difficulties mentioned above in detecting significant amounts of phagocytosed Aβ within brain microglia raise a question as to the relevance of opsonization of Aβ fibrils within plaques. If this were, in fact, a driving influence for Aβ-mediated microglial chemotaxis and the phagocytic activity of these cells, it fails to explain why such microglia are not found in association with Aβ42-rich diffuse plaques. Perhaps the idea of opsonization-enhanced phagocytosis is correct except for what is being opsonized. The lack of microglia in diffuse plaques and the preferential localization of microglia at the core of dense-core plaques suggest that the opsonized material is located exclusively at the core of dense-core plaques.

13.3.4 Positive and Negative Aspects of Microglial Activity in AD Brain

The intent of inflammation is to allow a series of specific cellular events to occur that will ultimately result in the removal of the offending agent and its associated cell and tissue debris from the affected tissue, leaving the way open for either tissue repair or replacement (scar formation). The process seems to work well in instances where there are clear limits to the amount of offending agent and the extent of tissue destruction caused by this agent and in situations where the vascularity of the tissue can be restored. In this case, elimination of the offending agent and tissue debris largely by phagocytic activity can then be followed by a period of tissue repair or replacement without additional insults. On the other hand, this process does not seem to work well in cases of chronic diseases such as AD, where the offending agent (presumably Aβ) is constantly supplied throughout the course of the disease, leaving little opportunity for repair to occur in an environment free of additional insults and progressive tissue destruction. Unfortunately, AD seems to be one of those diseases where the rate of tissue destruction exceeds the capacity of local cells (astrocytes and microglia) to resolve it. Inevitably, such conditions lead to the recruitment of additional cells (e.g., blood-borne monocytic cells) to the site of damage. When the brain tissue becomes heavily populated with inflammatory cells (Figs. 13.8A and 13.8B), the additional production of unusually high levels of inflammatory mediators and the excessive phagocytic activity of these immigrant cells becomes more destructive than beneficial. Thus, in AD, the chronic and progressive nature of the disease eventually tips the balance of the resulting inflammation to the destructive side, leading to the loss of irreplaceable neurons.

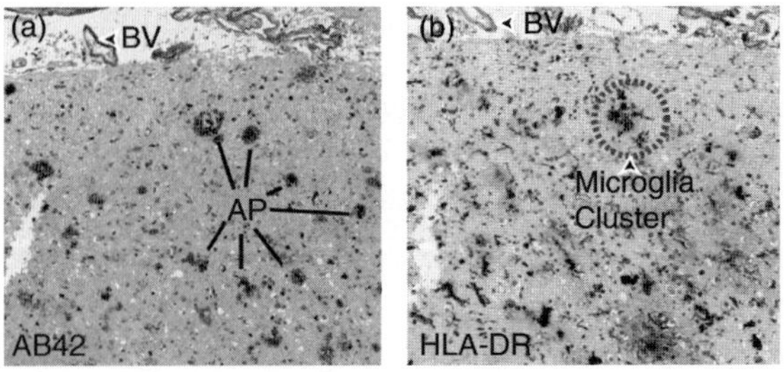

FIGURE 13.8. Consecutive sections through the entorhinal cortex of an AD brain immunostained with anti-Aβ42 (A) and anti-HLA-DR (B) antibodies. The brain tissue shows considerable inflammation with microglia/macrophages occurring both individually (in the space between plaques) and in clusters (within plaques).

13.3.5 Microglia as Therapeutic Targets

As detailed above, microglia have been assigned a role in the inflammatory response associated with AD pathology and also possibly with the processing and/or clearance of Aβ from the brain. The concept that runaway inflammation in the brain may actually precipitate some of the observed neurodegeneration in AD has raised the possibility that at least some of this damage may be avoided or alleviated through the use of nonsteroidal anti-inflammatory drugs (NSAIDs). The results of a number of clinical trails using NSAIDs, with some claiming a reduced incidence of AD, have been somewhat less than convincing [112–120]. Part of the problem may be that the levels of brain inflammation at the time the patient enters into the clinical trial may be too advanced. Another possibility proposed by Streit and co-workers is that microglia in the AD brain show a loss or deterioration of function that may represent a type of cellular senescence [121, 122]. If this is the case, then the collective phagocytic capability of microglia/macrophages in the brain both before and after treatment would be insufficient to keep up with the rate of tissue destruction. This could explain the marginal benefit of NSAIDs for AD.

In the past few years, great attention has been given to the possibility that immune stimulation by vaccination with Aβ peptides (especially Aβ42) would lead to the production of anti-Aβ peptide antibodies. From the therapeutic standpoint, the hope is that this vaccination will ultimately result in microglia/macrophages becoming more efficient at phagocytosing amyloid deposits, which are considered by many to be the direct or indirect cause of the neurodegeneration that is associated with this disease. Some success with this approach has been reported in animal models of AD, where antibodies generated against Aβ42 caused a reduction in the amyloid load in the brain of transgenic mice [123–125]. In these experiments, the clearance of amyloid fibrils from the brain parenchyma was determined to occur by the binding of Aβ42/immunoglobulin complexes to immunoglobulin Fc receptors on microglia/macrophages, which enhanced the rate and extent of phagocytosis of these complexes. On the other hand, results of clinical trials with humans have not been encouraging, and the development of encephalitis has been problematic. Several potential problems with this approach are predictable and noteworthy. First, the ability of anti-Aβ antibody to bind to anything in the brain requires that the BBB not be intact, so that the induced immunoglobulin can enter into the brain from the blood. A question arises as to whether the long-term, global breach in the BBB can ever be repaired in AD brains, even if the amyloid load of the brain is successfully lowered. Second, as mentioned above, Aβ42 has great affinity for the α7nAChR, which is abundantly present on the surfaces of many types of neurons throughout the brain. Thus, because Aβ42 is also able to enter into the brain from the blood, many neurons in AD brains at the time of treatment will possess Aβ42/α7nAChR complexes on their surfaces, which, of course, will be immunoreactive to the incoming anti-Aβ42 antibodies. In addition to inducing the formation of cell surface patches of aggregated anti-Aβ42-Aβ42/α7nAChR complex, this may prompt stripping of these complexes from the cell surface via endocytosis. The net effect is that binding of the anti-Aβ42 antibody to neuronal surfaces could actually accelerate the rate of Aβ42 internalization and accumulation within neurons. Another potential negative effect of the binding of anti-Aβ42 antibodies to neuronal surfaces is that it attracts complement (including the membrane attack complex), which can promote neuron degeneration and death. Because accelerated neuronal degeneration and death would be expected to elicit an enhanced inflammatory response, it is not surprising that the vaccination approach runs the risk of global brain inflammation.

13.4 Perspectives

The combined activities of astrocytes and microglia/macrophages eventually become deleterious and make a major and direct contribution to evolution of AD pathology in the brain. Evaluation of recent data in the context of what is already known about these two important cell types and the formation of amyloid plaques has allowed us to construct a proposed pathological sequence that highlights the entangled interactions of Aβ and these cells and their involvement in the pathogenesis of AD (Fig. 13.9). A key starting point for AD appears to be the focal or global compromise of the BBB. Of course, this can happen in association with any head or brain trauma but can also evolve as a result of aging-associated changes in the walls of blood vessels. The requirement for this step may explain why aging seems to be a prerequisite for one to express AD symptoms and pathology. The chronic leak of serum-bound Aβ42 into the brain tissue through the defective BBB provides a constant supply of exogenous Aβ42 that can bind with high affinity to neurons (especially cortical pyramidal cells) abundantly endowed with α7nAChR. For unknown reasons, neurons begin to internalize Aβ42/α7nAChR complex via endocytosis. Once neurons have accumulated sufficient Aβ42-positive material to elicit distal synaptic and dendritic loss, first in the cortical molecular layer, local astrocytes are activated and begin to internalize the resulting neuronal debris, which includes neuron-specific proteins such as α7nAChR, ChAT and Aβ42 [44]. Aβ42-overburdened neurons and astrocytes eventually die and undergo lysis, releasing their content of Aβ42-positive material [8, 44, 60]. The material released by cell lysis is dispersed radially with the aid of the activity of released lysosomal enzymes, leading to the formation of a spherical deposition of cell residue in the form of a plaque. Both smaller astrocytic plaques and larger neuron-derived plaques are rapidly infiltrated with macrophages/microglia, many of which are derived from blood monocytes that immigrate into the brain parenchyma from local capillaries. The lack of microglia/macrophages in diffuse plaques and their direct migration through the Aβ42-rich corona and into the cores of dense-core plaques suggest that DNA fragments gradually released from the nuclear

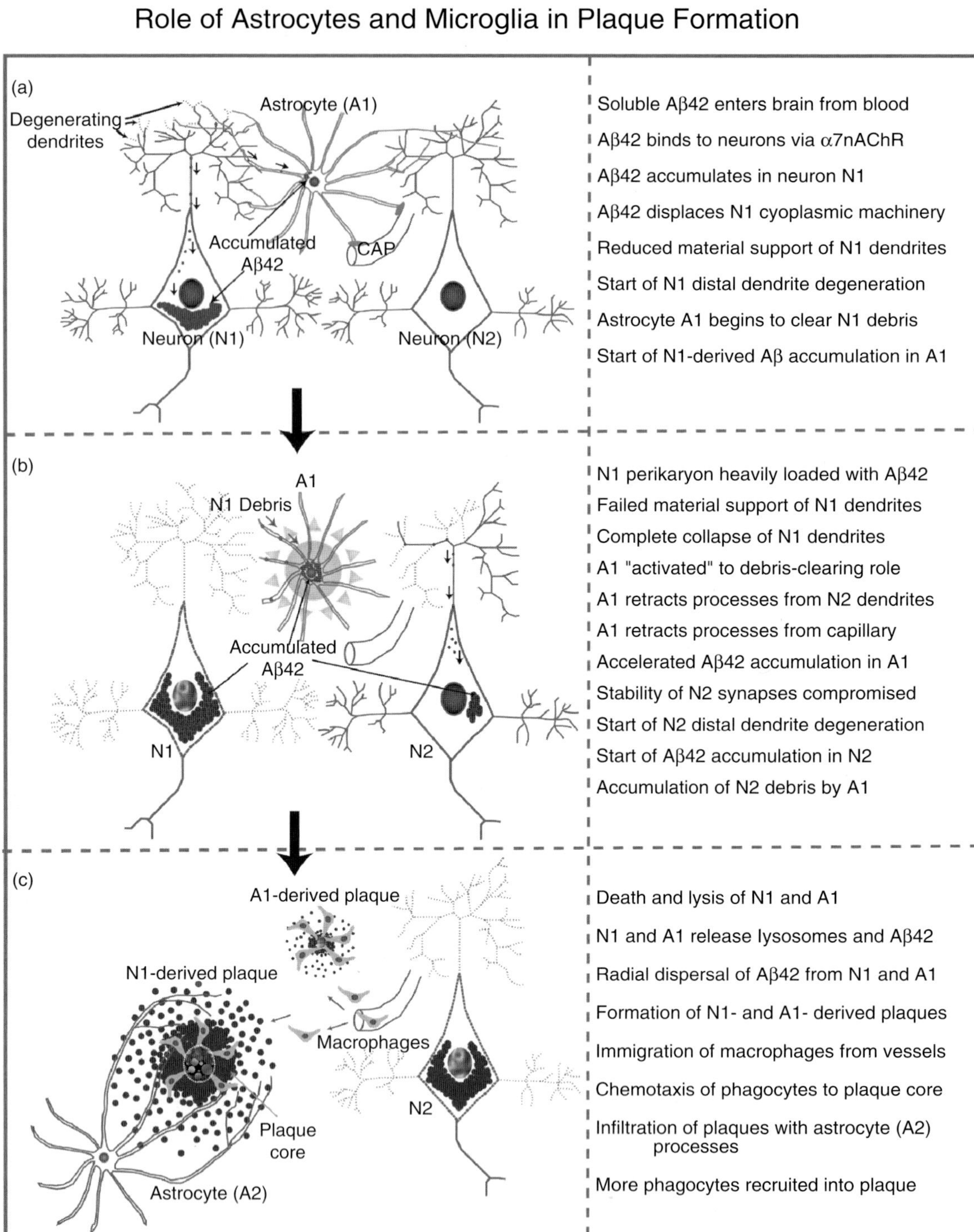

FIGURE 13.9. Proposed scenario for the involvement of astrocytes and microglia/macrophages in AD pathogenesis in the context of developing neuronal pathology.

remnant at the plaque core, and not Aβ peptides, may be chemotactic to microglia/macrophages. While at the plaque core, it is not clear if microglia/macrophages ingest Aβ in AD brains. It is more likely that their role is to clear remaining nuclear debris from the plaque core. Local activated astrocytes that are positioned just outside the plaque margin extend long, GFAP immunopositive cytoplasmic processes toward the plaque and both encapsulate it and infiltrate it with finer GFAP-negative processes. In addition, plaque-associated astrocytes clearly are able to internalize Aβ42-immunopositive material which accumulates in their cell bodies.

The suggested ability of different cell types to independently give rise to amyloid plaques (especially neurons and astrocytes) can account, at least in part, for the broad spectrum of plaque morphologies observed in AD brains. The proposed pathological sequence described in Figure 13.9 highlights the link between the loss of BBB integrity and the initiation of AD pathological changes. Equally important is the dramatic intraneuronal Aβ42 accumulation of Aβ42. The trigger for this phenomenon is unknown, but the possibilities include one or more of the following; binding of serum-derived, exogenous Aβ42 to α7nAChR on neuronal surfaces, oxidative damage, reduced delivery of materials to distal dendrites, impaired neuronal Aβ42 clearance, or binding of neuron-specific immunoglobulins and complement that have gained entry into the brain parenchyma via local or global breaches in the BBB. Regardless of the cause of neuronal Aβ42 accumulation, the fact that it leads to degeneration of distal dendrites and synapses in the cortical molecular layer provides a plausible explanation for the early telltale signs of AD progression (i.e., cognitive and memory decline), even prior to the appearance of amyloid plaques within the brain tissue. From a therapeutic perspective, maintaining or restoring BBB integrity could be a first line of defense against AD, and blocking the initial accumulation of Aβ42 in neurons is an obvious and early target. Success on either or both fronts would provide an opportunity to block or at least slow the progression of AD pathology in the brains of the elderly.

Acknowledgments The authors wish to thank Drs. Michael D'Andrea and Hoau-Yan Wang for their many helpful discussions and Alison Rigby, Jennifer Dubay, Seth Vatsky, Emily Sim, and James Novak for their technical assistance. This work is supported by grants from the National Institute on Aging (AG00925), the Alzheimer's Association., the New Jersey Gerontological Institute, and NJ Governor's Council on Autism.

References

1. Selkoe D. Alzheimer's disease: genes, proteins, and therapy. Physiol Rev 2001; 81:741-66.
2. Wisniewski KE, Wisniewski HM, Wen GY. Occurrence of neuropathological changes and dementia of Alzheimer's disease in Down's syndrome. Ann Neurol 1985; 17:278-82.
3. Selkoe DJ. Alzheimer's disease: genotypes, phenotypes, and treatments. Science 1997; 275:630-1.
4. Terry RD, Masliah E, Slmon DP, et al. Physical basis of cognitive alterations in Alzheimer's disease: synapse loss is the major correlate of cognitive impairment. Ann Neurol 1991; 30:572-80.
5. Cummings JL, Vinters HV, Cole GM, et al. Alzheimer's disease: etiologies, pathophysiology, cognitive reserve, and treatment opportunities. Neurology 1998;51(1 Suppl 1):S2-17; discussion S65-7.
6. Dickson DW. The pathogenesis of senile plaques. J Neuropathol Exp Neurol 1997; 56:321-39.
7. Scheff SW and Price DA. Synaptic density in the inner molecular layer of the hippocampal dentate gyrus in Alzheimer's disease. J Neuropathol Exp Neurol 1998; 57:1146-53.
8. D'Andrea MR, Nagele RG, Wang HY, et al. Evidence that neurones accumulating amyloid can undergo lysis to form amyloid plaques in Alzheimer's disease. Histopathology 2001; 38:120-34.
9. D'Andrea MR, Nagele RG, Wang HY, et al. Consistent immunohistochemical detection of intracellular beta-amyloid42 in pyramidal neurons of Alzheimer's disease entorhinal cortex. Neurosci Lett 2002; 333:163-6.
10. Citron M, Diehl TS, Gordon G, et al. Evidence that the 42- and 40-amino acid forms of amyloid β protein are generated from the β-amyloid precursor protein by different protease activities. Proc Natl Acad Sci U S A 1996; 93:13170-5.
11. Gendelman HE, Folks DG. Innate and acquired immunity in neurodegenerative disorders. J Leukoc Biol 1999; 65:407-8.
12. Cotter RL, Burke WJ, Thomas VS, et al. Insights into the neurodegenerative process of Alzheimer's disease: a role for mononuclear phagocyte-associated inflammation and neurotoxicity. J Leukoc Biol 1999; 65:416-27.

13. Eng LF, Ghirnikar RS, Lee YL. Glial fibrillary acidic protein: GFAP -thirty-one years (1969-2000). Neurochem Res 2000; 25:1439-51.
14. Bushong EA, Martone ME, Jones YZ, Ellisman MH. Protoplasmic astrocytes in CA1 stratum radiatum occupy separate anatomical domains. J Neurosci 2002; 22(1):183-92.
15. Ullian EM, Sapperstein SK, Christopherson KS, et al. Control of synapse number by glia. Science 2001; 291:657-61.
16. Martin ED, Araque A, Buno W. Synaptic regulation of the slow Ca2+-activated K+ current in hippocampal CA1 pyramidal neurons: implication in epileptogenesis. J Neurophysiol 2001; 86:2878-86.
17. Fields RD, Stevens-Graham B. New insights into neuron-glia communication. Science 2002; 298:556-62.
18. Hansson E, Ronnback L. Glial neuronal signaling in the central nervous system. FASEB J 2003; 17:341-8.
19. Vernadakis A. Glia-neuron intercommunications and synaptic plasticity. Prog Neurobiol 1996; 49:185-214.
20. Barres BA. New roles for glia. J Neurosci 1991; 11:3685-94.
21. Danbolt NC. The high affinity uptake system for excitatory amino acids in the brain. Prog Neurobiol 1994; 44:377-96.
22. Rothstein JD, Martin L, Levey AI, et al. Localization of neuronal and glial glutamate transporters. Neuron 1994; 13:713-25.
23. Lehre KP, Levy LM, Ottersen OP, et al. Differential expression of two glial glutamate transporters in the rat brain: quantitative and immunocytochemical observations. J Neurosci 1995; 15:1835-53.
24. Hertz L, Zielke HR. Astrocytic control of glutamatergic activity: astrocytes as stars of the show. Trends Neurosci 2004; 27:735-43.
25. Dong Y, Benveniste EN. Immune function of astrocytes. Glia 2001; 36:180-90.
26. Hatten ME, Liem RK, Shelanski ML, et al. Astroglia in CNS injury. Glia 1991; 4:233-43.
27. Rubin LL, Staddon JM. The cell biology of the blood-brain barrier. Annu Rev Neurosci 1999; 22:11-28.
28. Pekny M, Nilsson M. Astrocyte activation and reactive gliosis. Glia 2005; 50:427-34.
29. Eng LF, Ghirnikar RS. GFAP and astrogliosis. Brain Pathol 1994; 4:229-37.
30. Itagaki S, McGeer PL, Akiyama H, et al. Relationship of microglia and astrocytes to amyloid deposits of Alzheimer's disease. J Neuroimmunol 1989; 24: 173-82.
31. Schubert P, Ogata T, Marchini C, et al. Glia-related pathomechanisms in Alzheimer's disease: a therapeutic target? Mech. Ageing Dev 2001; 123:47-57.
32. Panickar KS, Norenberg MD. Astrocytes in cerebral ischemic injury: morphological and general considerations. Glia 2005; 50:287-98.
33. Wilhelmsson U, Li L, Pekna M, et al. Absence of glial fibrillary acidic protein and vimentin prevents hypertrophy of astrocytic processes and improves post-traumatic regeneration. J Neurosci 2004;24:5016-21.
34. Anderson CM, Swanson RA. Astrocyte glutamate transport: review of properties, regulation, and physiological functions. Glia 2000; 32:1-14.
35. Parpura V, Scemes E, Spray DC. Mechanisms of glutamate release from astrocytes: gap junction "hemichannels," purinergic receptors and exocytotic release. Neurochem Int 2004; 45:259-64.
36. Norenberg MD. The reactive astrocyte. In: Aschner M, editor. The role of glia in neurotoxicity. Boca Raton, FL: CRC Press, 2005:73-92.
37. Ge S, Song L, Pachter JS. Where is the blood-brain barrier . . .really? J Neurosci Res 2005; 79:421-7.
38. Bush AI, Beyreuther K, Masters CL. The beta A4 amyloid protein precursor in human circulation. Ann N Y Acad Sci 1993; 695:175-82.
39. Haas C, Hung AY, Citron M, et al. beta-Amyloid, protein processing and Alzheimer's disease. Arzneimittelforschung 1995; 45:398-402.
40. Younkin SG. The role of A beta 42 in Alzheimer's disease. J Physiol (Paris) 1998; 92:289-92.
41. Dewitt DA, Perry G, Cohen M, et al. Astrocytes regulate microglial phagocytosis of senile plaque cores of Alzheimer's disease. Exp Neurol 1998; 149: 329-40.
42. Akiyama H, Mori H, Saido T, et al. Occurrence of the diffuse amyloid beta-protein (Abeta) deposits with numerous Abeta-containing glial cells in the cerebral cortex of patients with Alzheimer's disease. Glia 1999; 25:324-31.
43. Kurt MA, Davies DC, Kidd MM. beta-Amyloid immunoreactivity in astrocytes in Alzheimer's disease brain biopsies: an electron microscope study. Exp Neurol 1999; 158:221-8.
44. Nagele RG, D'Andrea MR, Lee H, Venkataraman V, et al. Astrocytes accumulate Amyloid beta 42 and give rise to astrocytic amyloid plaques in Alzheimer's disease brains. Brain Res 2003; 971:197-209.
45. Thal DR, Schultz C, Dehghani F, et al. Amyloid beta-protein (Abeta)-containing astrocytes are located preferentially near N-terminal-truncated Abeta deposits in the human entorhinal cortex. Acta Neuropathol 2000; 100:608-17.
46. Thal DR, Hartig W, Schober R. Diffuse plaques in the molecular layer show intracellular AB8-17-immunoreactive deposits in subpial astrocytes. Clin Neuropathol 1999; 18:226-31.

47. Aldskogius H, Liu L, Svensson M. Glial responses to synaptic damage and plasticity. J Neurosci Res 1999; 58:33-41.
48. Wegiel J, Wang K-C, Tarnawski M, et al. Microglial cells are the driving force in fibrillar plaque formation whereas astrocytes are a leading factor in plaque degradation. Acta Neuropathol 2000; 100:356-64.
49. Wegiel J, Wang K-C, Imaki H, et al. The role of microglial cells and astrocytes in fibrillar plaque evolution in transgenic APPsw mice. Neurobiol Aging 2001; 22:49-61.
50. Wegiel J, Imaki H, Wang K-C, et al. Origin and turnover of microglial cells in fibrillar plaques of APPsw transgenic mice. Acta Neuropathol 2003; 105:393-402.
51. Kurochkin IV, Goto S. Alzheimer's beta-amyloid peptide specifically interacts with and is degraded by insulin degrading enzymes. FEBS Lett 1994; 345:33-7.
52. Qiu WQ, Walsh DM, Ye Z, et al. Insulin-degrading enzyme regulates extracellular levels of amyloid β-protein by degradation. J Biol Chem 1998; 273: 32730-8.
53. Apelt J, Ach K, Schliebs R. Aging-related down-regulation of neprilysin, a putative β-amyloid-degrading enzyme, in transgenic Tg2576 Alzheimer-like mouse brain is accompanied by an astroglial upregulation in the vicinity of β-amyloid plaques. Neurosci Lett 2003; 339:183-6.
54. Magnus T, Chan A, Linker RA, et al. Astrocytes are less efficient in the removal of apoptotic lymphocytes than microglia cells: implications for the role of glial cells in the inflamed central nervous system. J Neuropathol Exp Neurol 2002; 61:760-6.
55. Ard MD, Cole GM, Wei J, et al. Scavenging of Alzheimer's amyloid β-protein by microglia in culture. J Neurosci Res 1996; 43:190-202.
56. Wisniewski HM, Wegiel J, Wang K-C, et al. Ultrastructural studies of the cells forming amyloid fibers in classical plaques. Can J Neurol Sci 1989; 16:535-42.
57. Wisniewski HM, Wegiel J. Spatial relationships between astrocytes and classical plaque components. Neurobiol Aging 1991; 12:593-600.
58. Perry VH, Gordon S. Macrophages in the nervous system. Int Rev Cytol 1991; 125:203-44.
59. Biernacki K, Prat A, Blain M, et al. Regulation of Th1 and Th2 lymphocyte migration by human adult brain endothelial cells. J Neuropathol Exp Neurol 2001; 60:1127-36.
60. Nagele RG, D'Andrea MR, Anderson WJ, et al. Intracellular accumulation of B-amyloid in neurons is facilitated by the a7 nicotinic acetylcholine receptor in Alzheimer's disease. Neuroscience 2002; 110:199-211.
61. Gouras GK, Tsaiu J, Nalund J, et al. Intraneuronal Aβ42 accumulation in human brain. Am J Pathol 2000; 156:15-20.
62. Gyure KA, Durham R, Stewart WF, et al. Intraneuronal abeta-amyloid precedes development of amyloid plaques in Down syndrome. Arch Pathol Lab Med 2001; 125:489-92.
63. Wang H-Y, Lee DHS, D'Andrea MR, et al. β-amyloid1-42 binds to α7 nicotinic acetylcholine receptor with high affinity: implications for Alzheimer's disease pathology. J Biol Chem 2000; 275:5626-32.
64. Wang H-Y, D'Andrea MR, Nagele RG. Cerebellar diffuse amyloid plaques are derived from dendritic Aβ42 accumulations in Purkinje cells. Neurobiol Aging 2002; 23:213-23.
65. Lee DH, Wang HY. Differential physiologic responses of alpha7 nicotinic acetylcholine receptors to beta-amyloid1-40 and beta-amyloid1-42. J Neurobiol 2003; 55:25-30.
66. Nagele RG, Wegiel J, Venkataraman V, et al. Contribution of glial cells to the development of amyloid plaques in Alzheimer's disease. Neurobiol Aging 2004; 25:663-74.
67. Oddo S, Caccamo A, Sheppard JD, et al. Triple-transgenic model of Alzheimer's disease with plaques and tangles: intracellular Abeta and synaptic dysfunction. Neuron. 2003; 39:409-21.
68. Mori C, Spooner ET, Wisniewsk KE, et al. Intraneuronal Abeta42 accumulation in Down syndrome brain. Amyloid 2002; 9:88-102.
69. Streit WJ, Walter SA, Pennell NA. Reactive microgliosis. Prog Neurobiol 1999; 57:563-81.
70. Griffin WS, Sheng JG, Roberts GW, et al. Interleukin-1 expression in different plaque types in Alzheimer's disease: significance in plaque evolution. J Neuropathol Exp Neurol 1995; 54:276-81.
71. Griffin WS, Stanley LC, Ling C, et al. Brain interleukin 1 and S-100 immunoreactivity are elevated in Down syndrome and Alzheimer's disease. Proc Natl Acad Sci U S A 1989; 86:7611-5.
72. Summers WK. Alzheimer's disease, oxidative injury, and cytokines. J Alzheimers Dis 2004; 6:651-7.
73. Perry VH, Andersson PB, Gordon S. Macrophages and inflammation in the central nervous system. Trends Neurosci 1993; 16:268-73.
74. Streit WJ, Graeber MB, Kreutzberg GW. Functional plasticity of microglia: a review. Glia 1988; 1:301-7.
75. Rogers J, Lue LF. Microglial chemotaxis, activation, and phagocytosis of amyloid beta-peptide as linked phenomena in Alzheimer's disease. Neurochem Int 2001; 39:333-40.
76. Burudi EM, Regnier-Vigouroux A. Regional and cellular expression of the mannose receptor in the

post-natal developing mouse brain. Cell Tissue Res 2001; 303:307-17.
77. Walker DG, Kim SU, McGeer PL. Complement and cytokine gene expression in cultured microglical derived from postmortem human brains. J Neurosci Res 1995; 40:478-93.
78. Strohmeyer R, Shen Y, Rogers J. Detection of complement alternative pathway mRNA and proteins in the Alzheimer's disease brain. Brain Res Mol Brain Res 2000; 81:7-18.
79. Walker DG, Lue LF, Beach TG. Gene expression profiling of amyloid beta peptide-stimulated human post-mortem brain microglia. Neurobiol Aging 2001; 22:957-66.
80. Shen Y, Li R, McGeer EG, et al. Neuronal expression of mRNAs for complement proteins of the classical pathway in Alzheimer brain. Brain Res 1997; 769:391-5.
81. Schenk D, Barbour R, Dunn W, et al. Immunization with amyloid-beta attenuates Alzheimer-disease-like pathology in the PDAPP mouse. Nature 1999; 400:173-7.
82. Bard F, Cannon C, Barbour R, et al. Peripherally administered antibodies against amyloid beta-peptide enter the central nervous system and reduce pathology in a mouse model of Alzheimer's disease. Nat Med 2000; 6:916-9.
83. Haga S, Akai K, Ishii T. Demonstration of microglial cells in and around senile (neuritic) plaques in the Alzheimer brain. An immunohistochemical study using a novel monoclonal antibody. Acta Neuropathol (Berlin) 1989; 77:569-75.
84. Perlmutter LS, Scott SA, Barron E, et al. MHC class II-positive microglia in human brain: association with Alzheimer lesions. J Neurosci Res 1992; 33:549-58.
85. Wisniewski HM, Wegiel J, Wang KC, et al. Ultrastructural studies of the cells forming amyloid in the cortical vessel wall in Alzheimer's disease. Acta Neuropathol (Berlin) 1992; 84:117-27.
86. Frautschy SA, Cole GM, Baird A. Phagocytosis and deposition of vascular beta-amyloid in rat brains injected with Alzheimer beta-amyloid. Am J Pathol 1992;140:1389-99.
87. Kopec KK, Carroll RT. Alzheimer's beta-amyloid peptide 1-42 induces a phagocytic response in murine microglia. J Neurochem 1998; 71:2123-31.
88. Weldon DT, Rogers SD, Ghilardi JR, et al. Fibrillar beta-amyloid induces microglial phagocytosis, expression of inducible nitric oxide synthase, and loss of a select population of neurons in the rat CNS in vivo. J Neurosci 1998; 18:2161-73.
89. Paresce DM, Chung H, Maxfield FR. Slow degradation of aggregates of the Alzheimer's disease amyloid beta-protein by microglial cells. J Biol Chem 1997; 272:29390-7.
90. Rogers J, Lue LF, Walker DG, et al. Elucidating molecular mechanisms of Alzheimer's disease in microglial cultures. Ernst Schering Res Found Workshop 2002; 39:25-44.
91. Frackowiak J, Wisniewski HM, Wegiel J, et al. Ultrastructure of the microglia that phagocytose amyloid and the microglia that produce β-amyloid fibrils. Acta Neuropathol 1992; 84:225-33.
92. Wisniewski HM, Weigel J. Migration of perivascular cells into the neuropil and their involvement in beta-amyloid plaque formation. Acta Neuropathol (Berlin) 1993; 85:586-95.
93. Stalder M, Phinney A, Probst A, et al. Association of microglia with amyloid plaques in brains of APP23 transgenic mice. Am J Pathol 1999; 154:1673-84.
94. Lue LF, Walker DG, Rogers J. Modeling microglial activation in Alzheimer's disease with human postmortem microglial cultures. Neurobiol Aging 2001; 22:945-56.
95. Rogers J, Lue LF, Walker DG, et al. Elucidating molecular mechanisms of Alzheimer's disease in microglial cultures. Ernst Schering Res Found Workshop 2002; 39:25-44.
96. Davis JB, McMurray HF, Schubert D. The amyloid beta-protein of Alzheimer's disease is chemotactic for mononuclear phagocytes. Biochem Biophys Res Commun 1992; 189:1096-100.
97. Yan SD, Chen X, Fu J, et al. RAGE and amyloid-beta peptide neurotoxicity in Alzheimer's disease. Nature 1996; 382:685-91.
98. Kato K, Suzuki F, Morishita R, et al. Selective increase in S-100 beta protein by aging in rat cerebral cortex. J Neurochem 1990; 54:1269-74.
99. D'Andrea MR, Nagele RG. MAP-2 immunolabeling can distinguish diffuse from dense-core amyloid plaques in brains with Alzheimer's disease. Biotech Histochem 2002; 77:95-103.
100. Fiala M, Zhang L, Gan X, et al. Amyloid-beta induces chemokine secretion and monocyte migration across a human blood–brain barrier model. Mol Med 1998; 4:480-9.
101. Loike JD, el Khoury J, Cao L, et al. Fibrin regulates neutrophil migration in response to interleukin 8, leukotriene B4, tumor necrosis factor, and formyl-methionyl-leucyl-phenylalanine. J Exp Med 1995; 181:1763-72.
102. Lorton D, Schaller J, Lala A, et al. Chemotactic-like receptors and Abeta peptide induced responses in Alzheimer's disease. Neurobiol Aging 2000; 21:463-73.
103. Du Yan S, Zhu H, Fu J, et al. Amyloid-beta peptide-receptor for advanced glycation endproduct interaction elicits neuronal expression of macrophage-

colony stimulating factor: a proinflammatory pathway in Alzheimer's disease. Proc Natl Acad Sci U S A 1997; 94:5296-301.
104. Gehrmann J, Schoen SW, Kreutzberg GW. Lesion of the rat entorhinal cortex leads to a rapid microglial reaction in the dentate gyrus. A light and electron microscopical study. Acta Neuropathol (Berlin) 1991; 82:442-55.
105. Lacy M, Jones J, Whittemore SR, et al. Expression of the receptors for the C5a anaphylatoxin, interleukin-8 and FMLP by human astrocytes and microglia. J Neuroimmunol 1995; 61:71-8.
106. Gasque P, Singhrao SK, Neal JW, et al. The receptor for complement anaphylatoxin C3a is expressed by myeloid cells and nonmyeloid cells in inflamed human central nervous system: analysis in multiple sclerosis and bacterial meningitis. J Immunol 1998; 160:3543-54.
107. Kaur C, Chan YG, Ling EA. Ultrastructural and immunocytochemical studies of macrophages in an excitotoxin induced lesion in the rat brain. J Hirnforsch 1992; 33:645-52.
108. Hauwel M, Furon E, Canova C, et al. Innate (inherent) control of brain infection, brain inflammation and brain repair: the role of microglia, astrocytes, "protective" glial stem cells and stromal ependymal cells. Brain Res Brain Res Rev 2005; 48:220-33.
109. Jiang H, Burdick D, Glabe CG, et al. beta-Amyloid activates complement by binding to a specific region of the collagen-like domain of the C1q A chain. J Immunol 1994; 152:5050-9.
110. McGeer PL, Akiyama H, Itagaki S, et al. Activation of the classical complement pathway in brain tissue of Alzheimer patients. Neurosci Lett 1989; 107:341-6.
111. Tacnet-Delorme P, Chevallier S, Arlaud GJ. Beta-amyloid fibrils activate the C1 complex of complement under physiological conditions: evidence for a binding site for A beta on the C1q globular regions. J Immunol 2001; 167:6374-81.
112. Akiyama H, Barger S, Barnum S, et al. Inflammation and Alzheimer's disease. Neurobiol Aging 2000; 21:383-421.
113. Anthony JC, Breitner JC, Zandi PP, et al. Reduced prevalence of AD in users of NSAIDs and H2 receptor antagonists: the Cache County study. Neurology 2000; 54:2066-71.
114. Colton CA, Chernyshev ON, Gilbert DL, et al. Microglial contribution to oxidative stress in Alzheimer's disease. Ann N Y Acad Sci 2000; 899:292-307.
115. Flynn BL, Theesen KA. Pharmacologic management of Alzheimer's disease part III: nonsteroidal antiinflammatory drugs—emerging protective evidence? Ann Pharmacother 1999; 33:840-9.
116. in 't Veld BA, Launer LJ, Hoes AW, et al. NSAIDs and incident Alzheimer's disease. The Rotterdam Study. Neurobiol Aging 1998; 19:607-11.
117. McGeer PL, McGeer EG. Inflammation, autotoxicity and Alzheimer's disease. Neurobiol Aging 2001; 22:799-809.
118. Mortimer JA. New findings consistent with Alzheimer's-NSAIDs link. Neurobiol Aging 1998; 19:615-6.
119. Pasinetti GM. Cyclooxygenase and inflammation in Alzheimer's disease: experimental approaches and clinical interventions. J Neurosci Res 1998; 54:1-6.
120. Schubert P, Ogata T, Marchini C, et al. Glia-related pathomechanisms in Alzheimer's disease: a therapeutic target? Mech Ageing Dev 2001; 123:47-57.
121. Flanary BE, Streit WJ. Progressive telomere shortening occurs in cultured rat microglia, but not astrocytes. Glia 2004; 45:75-88.
122. Streit WJ, Sammons NW, Kuhns AJ, et al. Dystrophic microglia in the aging human brain. Glia 2004; 45:208-12.
123. Wilcock DM, Gordon MN, Ugen KE, et al. Number of Abeta inoculations in APP+PS1 transgenic mice influences antibody titers, microglial activation, and congophilic plaque levels. DNA Cell Biol 2001; 20:731-6.
124. Schenk D, Barbour R, Dunn W, et al. Immunization with amyloid-beta attenuates Alzheimer's-disease-like pathology in the PDAPP mouse. Nature 1999; 400:173-7.
125. Bard F, Cannon C, Barbour R, et al. Peripherally administered antibodies against amyloid beta-peptide enter the central nervous system and reduce pathology in a mouse model of Alzheimer's disease. Nat Med 2000; 6:916-9.

14 The Role of Presenilins in Aβ-Induced Cell Death in Alzheimer's Disease

Maria Ankarcrona

14.1 Introduction

Neuronal death in specific brain regions is a common feature of neurodegenerative disorders. Alzheimer's disease (AD) is characterized by synaptic loss and a substantial amount of neuronal degeneration in regions involved in memory and learning processes (e.g., temporal, entorhinal and frontal cortex; hippocampus). The neuropathologic hallmarks of AD include the accumulation of amyloid plaques and hyperphosphorylated tau forming intracellular tangles. However, no correlation has been established between the number of plaques and the cognitive performance in AD patients [1, 2]. Instead, synaptic failure and intracellular production of amyloid beta (Aβ) appears to correlate well with the early cognitive dysfunction in AD patients [3, 4]. This has also been tested in a triple transgenic mouse model of AD where accumulation of intracellular $A\beta_{1-42}$ corresponded with the early cognitive impairment [5]. Interestingly, no extracellular deposits of $A\beta_{1-42}$ were detected in these mice at 4 months of age suggesting that $A\beta_{1-42}$ accumulate intracellulary early in the disease process. Moreover, intracellular accumulation of $A\beta_{1-42}$ was cleared with administration of anti-Aβ antibodies and rescued the retention deficits seen in young 3×Tg AD mice. Together, results from this and several other studies indicate that intracellular $A\beta_{1-42}$-generation causes the primary toxicity to neurons in AD [6].

In this chapter, the functions of presenilin (PS) in Aβ-generation and toxicity will be described. PS appears to play several roles in cell death mechanisms associated with AD: (i) functional PS is crucial for the generation of Aβ [7, 8], (ii) PS interacts with proteins involved in cell signaling, regulation of calcium homeostasis, and apoptosis [9], and (iii) PS mutations sensitize cells to different apoptotic stimuli in vitro [10] and increase the generation of $A\beta_{1-42}$. Whether it is the overproduction of $A\beta_{1-42}$ *per se* or other non-Aβ-related changes that cause the increased sensitivity of cells carrying PS mutations is not clear, and the different possibilities will be discussed here.

14.2 Cell Death in AD Brain

The mechanisms of cell death in the AD brain are not fully elucidated, however it is likely that several forms of cell death are involved. Loss of synapses is an early phenotypic manifestation in the pathology of AD, and synapse density is significantly decreased in AD. Synapse loss and impaired long-term potentiation also precede accumulation of plaques and tangles in 3×Tg mice [11]. Cytosolic extracts from synaptosomes exposed to Aβ induced chromatin condensation and fragmentation of isolated nuclei showing that apoptotic signals can be generated locally in synapses [12, 13]. Neurons that lose synapses and therefore also contact and communication with other cells are still alive but do not function as before and will not survive in the long run. Such cells could, however, stay in the tissue as "ghost cells" before they are cleared away by, for example, apoptosis. There are several evidences for apoptosis in AD. Postmortem analysis of AD brain showed TUNEL positive neurons and glia in hippocampus and cortex indicating DNA fragmentation [14–20]. Increased expression of Bcl-2 family members [21–25], as well as increased caspase activities and cleavage of caspase substrates have been detected in AD brain [26–32]. Cells that are triggered to die by apoptosis (e.g., have active caspase 8 and 9, which are initiator caspases), but fail to complete the process because executor caspases such as caspase-3 and -7 are not active, have also been detected in AD brain [33]. This phenomenon is called "abortosis" and is as an anti-apoptotic mechanism that might try to protect neurons from death. However, this process is probably finally overridden as many neurons still die in AD. There is also evidence for activation of cell-cycle proteins in AD brain [34, 35]. This may be a defense mechanism initiated to survive bad conditions or toxic stimuli. However the neurons do not go through mitosis, instead they are stuck in a cycle they cannot complete and eventually die. Postmitotic neurons do not normally divide, but it is possible that reentry of the cell cycle is necessary for the completion of apoptotis. Normally proliferating cells are regularly checked throughout the cell cycle and taken aside to die by apoptosis when damaged. Maybe also postmitotic cells have to take this way to death.

A cell dying by apoptosis leaves no traces in the tissue because it is silently disassembled and phagocytosed. Therefore, the main part of cells, which presumably have died by apoptosis during the course of AD, have already been cleared from the tissue at the time of autopsy. This is one of the difficulties with proving the impact of apoptotic cell death in AD. It has also been argued that the great difference in time spans between the disease process (approximately 20 years) and the apoptotic process (approximately 24 hours), rules out apoptosis as a mechanism for cell death in AD. However, if cell death is triggered at different times during the course of the disease, it is very likely that cells die by apoptosis in AD.

From a therapeutic point of view, it would of course be most attractive to target the early cognitive changes in AD presumably associated with intracellular accumulation of Aβ and synaptic failure. When the neuron is dead, it is too late. Therefore, it is of great importance to understand the mechanisms behind neuronal failure to be able to design the best neuroprotection. The treatment strategies are also highly dependent on diagnostic methods: the earlier a correct diagnosis can be given, the earlier a potential treatment could start.

14.3 Presenilins, γ-Secretase Activity, and APP Processing

Most AD cases are sporadic or have not so far been genetically linked. Only a minor number of AD cases have been associated with mutations in specific genes. These genes are presenilin 1 (PS1), presenilin 2 (PS2), and amyloid precursor protein (APP) [36]. All these mutations are autosomal dominant and fully penetrant. Generally, familial Alzheimer's disease (FAD) cases have a lower age of onset (PS1 mutation carriers 44 ± 7.8 years and PS2 carriers 58.6 ± 7.0 years [37]) and show a more aggressive form of the disease compared with sporadic cases. PS1 and PS2 are encoded on chromosomes 14 and 1, respectively [38–40] and show 63% homology. PS are membrane-bound proteins with eight transmembrane domains and localized to the endoplasmatic reticulum, Golgi apparatus [41–44], plasma membrane [45], nuclear envelope [46], lysosomes [47], and mitochondria [48]. Deficiency of PS1 inhibits Aβ generation from β-amyloid precursor protein (APP) suggesting that PS1 is involved in γ-secretase cleavage [49]. The γ-secretase complex consists of at least PS1/PS2, nicastrin (Nct), presenilin enhancer-2 (Pen-2), and anterior pharynx defective-1 (Aph-1), and γ-secretase activity has been reconstituted by expressing these four components in yeast [50] (Fig. 14.1). The γ-secretase complex is assembled in the ER and then trafficked to late secretory compartments including the plasma membrane where it exerts its biological function [51]. The four γ-secretase components are assembled stepwise. Nct and Aph-1

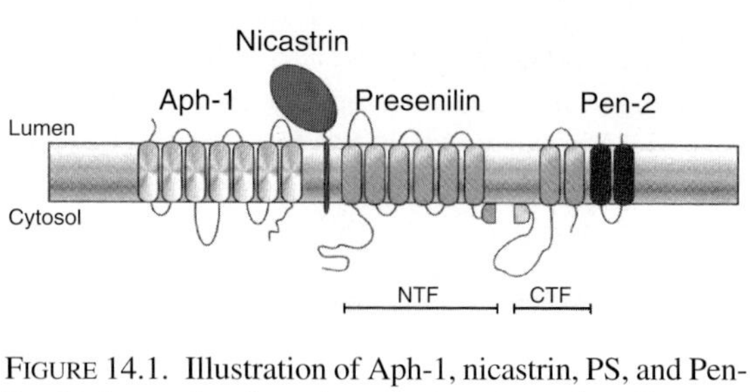

FIGURE 14.1. Illustration of Aph-1, nicastrin, PS, and Pen-2, which together form the γ-secretase complex. Courtesy of Dr. Jan Näslund, Karolinska Institutet, Sweden.

first form a stable subcomplex followed by the addition of full-length PS. Then Pen-2 is added to the complex and full-length PS is cleaved into C-terminal (CTF) and N-terminal (NTF) fragments forming the functional heterodimer of PS. Full-length PS, CTFs, and NTFs as well as Pen-2 are degraded by the proteasome when not incorporated into the γ-secretase complex [52–54]. The importance of PS for γ-secretase activity has been demonstrated in several ways: (i) in PS-deficient cells [7, 8], (ii) by the use of γ-secretase inhibitors that bind to PS [55, 56], (iii) by the substitution of either of two aspartyl residues in transmembrane domains 6 (Asp257) and 7 (Asp385) of PS1 [57]. All these studies showed inhibited γ-secretase activity and lower production of Aβ.

The γ-secretase complex cleaves APP and other type I membrane proteins [58]. Before γ-secretase cleavage, the N-terminal part of APP either facing the extracellular space or the lumen is cleaved by β-site APP cleaving enzyme (BACE), a process referred to as ectodomain shedding. BACE cleavage releases secreted sAPP-β and leaves a 99-amino-acid C-terminal fragment (C99) in the membrane. Subsequently, the γ-secretase complex cleaves C99 generating APP intracellular domain (AICD) and Aβ. In the non-amyloidogenic pathway, α-secretase cleaves APP in the middle of the Aβ region resulting in the formation of secreted sAPPα and an 83-amino acid C-terminal fragment (C83). γ-Secretase cleavage of C83 also results in AICD formation [36]. Other γ-secretase substrates than APP include Notch, ErbB4, E-cadherin, Delta/Jagged, nectin-1α, CD44, and LRP [59]. Many of these substrates function in intercellular communication or adhesion. Notch signaling is important during development as γ-secretase cleavage of this receptor generates a Notch intracellular domain (NICD). NICD translocates to the nucleus and activates transcription of the cell-fate determining HES (Hairy/Enhancer of split) genes, thus initiating a non-neuronal development of the cell [60, 61]. Similarly, AICD has been detected in the nucleus where it interacts with the nuclear adaptor protein Fe65 and the histone acetyltransferase and activates transcription [62–64]. AICD has also been implicated in the regulation of phosphoinositide-mediated calcium signaling [65].

Mice knocked-out for both PS1/ PS2 die before embryonic day 13.5 [49, 66]. PS1 can compensate for the loss of PS2 (PS1+/+ PS2–/– and PS1 +/– PS2–/– embryos survive), while PS2 cannot fully compensate for the loss of PS1 (PS1–/– PS2+/+ die at birth; PS1–/– PS2+/– embryos die during E9.5–E13.5). The results from these animal models emphasize the importance of PS1/γ-secretase activity during embryogenesis. In accordance, PS1–/– PS2+/+ and PS K/O mouse embryonic fibroblasts (MEFs) accumulate C83/C99 showing that PS1 is responsible for most of the γ-secretase activity cleaving APP (Fig. 14.2). The residual γ-secretase activity comes from PS2 that contributes to Aβ production to a lesser extent than PS1 [49].

14.4 PS Mutations, Aβ-Generation, and Apoptosis

To date, almost 150 mutations have been identified in PS1 and 11 mutations in PS2 (AD mutation database: http://www.molgen.ua.ac.be/ADmutations/default.cfm). All these are missense mutations that generate single amino acid substitutions in the protein primary structure, with the exception of PS1 exon 9 deletion splice mutation [67]. The different PS mutations lead to a similar phenotype: an increased ratio of $A\beta_{1\text{-}42}$ to $A\beta_{1\text{-}40}$, increased plaque deposition, and early age of onset [68]. Although the mutations are distributed all over the PS molecule, with a clustering of mutations in the transmembrane regions, the effect on Aβ-generation is similar indicating a common mechanism. Fluorescent lifetime imaging microscopy (FLIM) [69] studies have suggested that PS1 mutations, spread in different regions, all cause a conformational change in PS1. The proximity between the N- and C-terminus of PS1 was increased in the mutant PS1

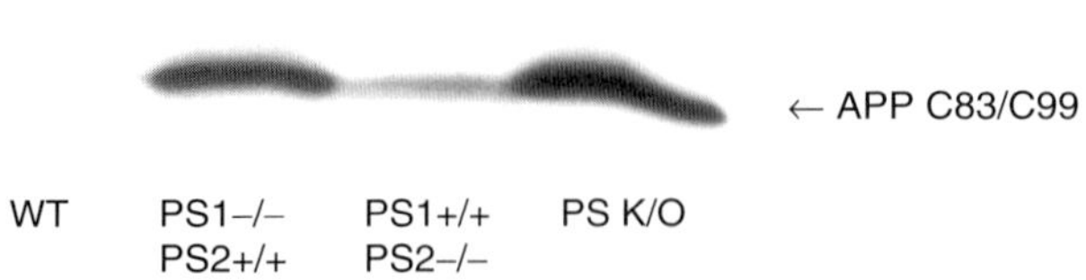

FIGURE 14.2. Western blot of cell lysates isolated from mouse embryonic fibroblasts (obtained from Prof. Bart de Strooper). Accumulation of APP C83/C99 fragments indicates lack of γ-secretase activity.

compared with PS1 wild-type. A consistent change was also detected in the configuration of the PS1-APP complex in PS1 mutants and could explain the common effect on Aβ generation [70]. In another study, using a random mutagenesis screen of PS1, five unique mutations that exclusively generated a high level of $A\beta_{1\text{-}43}$ were identified [71]. Together, these two studies show that PS1 mutations may change the activity and specificity of γ-secretase through a common mechanism.

PS1 is responsible for the major γ-secretase activity generating Aβ, and PS2 plays a minor role, still it has been shown that PS2 mutations also influence the ratio of Aβ42/40. Of the reported PS2 mutations T122P, N141I, M239V, and M239I significantly increased the Aβ42/40 ratio similar to very-early-onset PS1 FAD mutations [72]. The shift toward the production of longer and more amyloidogenic Aβ species induced both by PS1 and PS2 mutations suggest that this common alteration in APP processing by γ-secretase contributes to the increased neuronal death in FAD.

FAD mutant proteins are expressed from birth, but it takes decades for AD to manifest itself, and FAD mutant carriers start to develop the disease as adults. This suggests that FAD mutants do not induce neuronal cell death themselves but rather increase sensitivity to cell death stimuli [73]. Indeed, several in vitro studies have shown that presenilin mutations contribute to neuronal death and sensitize cells to apoptotic stimuli [10]. It has been reported that FAD-linked mutant PS1 enhances cell death in T lymphocytes [74], PC12 cells [75, 76], SH-SY5Y neuroblastoma cells [77, 78], and primary neurons [79]. However, another study failed to demonstrate that mutant PS1 increases sensitivity to cytotoxic insults in primary neurons [80].

Alterations in cells carrying PS1 mutations include higher caspase-3 activity [81], increased oxygen radical levels [82], induction of p53 and Bax upregulation of calpain, mitochondrial membrane depolarization [83], enhanced phospholipase C activity [84], and altered intracellular calcium regulation [75]. The PS2 mutant N141I-PS2 induces neuronal death in immortalized cell lines and primary neurons [85, 86]. The induction of apoptosis in PS2 mutant N141I-PS2 cells was accompanied by increased caspase-3 activity and decreased Bcl-2 expression after serum-deprivation [87].

Whether it is the increased Aβ42/40 ratio that causes the cellular alterations detected in PS mutants or vice versa is not known. One possibility is that PS mutations affect cellular functions independently of γ-secretase activity making such cells more vulnerable to Aβ and other cell death stimuli. Another possibility is that the high intracellular production of toxic Aβ species in PS mutant cells disturbs different cellular functions and thereby finally renders the cells more susceptible to cell death stimuli including Aβ.

14.5 PS Mutations, Aβ, and Intracellular Calcium Homeostasis

Many studies have shown dysregulation of intracellular calcium (Ca^{2+}) homeostasis in cells carrying PS mutations. Mutant forms of PS1 have been shown to enhance Ca^{2+} transients in several different cell systems including transfected PC12 cells [75, 88], fibroblasts from human FAD patients [89, 90], mutant knock-in mouse fibroblasts [91], cultured hippocampal neurons [92], and oocytes overexpressing mutant PS1 [93]. The effect on intracellular Ca^{2+} might be mediated by inositol triphosphate (IP_3) as FAD-linked PS1 mutations potentiate IP_3-mediated Ca^{2+} release from the ER [93]. The number of IP_3 receptors are not increased in cortical homogenates of PS1 knock-in mice, instead it has been suggested that the exaggregated cytosolic Ca^{2+} signals result from increased store filling [94].

Increased intracellular calcium concentrations $[Ca^{2+}]_i$ result in enhanced Aβ generation [95] and at the same time cells treated with Aβ show increased $[Ca^{2+}]_i$ [96]. One mechanism by which Aβ could increase $[Ca^{2+}]_i$ is the formation of calcium-permeable pores in membranes [97, 98–100]. More recently, Kayed and collegues suggested that amyloid oligomers rather induce permeabilization of membranes, without forming pores or channels, and thereby enhance the ability of ions to move through the lipid bilayer [101].

Other APP fragments have been shown to stabilize $[Ca^{2+}]_i$ and protect from Aβ toxicity. sAPPα is formed when α-secretase cleaves APP in the non-amyloidogenic pathway (Fig. 14.2). sAPPα has been shown to stabilize calcium homeostasis and protect neurons against excitotoxic, metabolic, and oxidative insults including Aβ [102, 103]. The proapoptotic action of mutant PS1 was counteracted by sAPPα, which stabilized $[Ca^{2+}]_i$ and mitochondrial function and suppressed oxidative stress by a mechanism involving activation of NF-κB [104].

Sorcin, calmyrin, and calsenilin are all Ca^{2+}-binding proteins that have been shown to interact with PS. Sorcin and calmyrin interact with PS2, while calsenilin interacts with both PS1 and PS2. Sorcin is found in mammalian brain associated with ryanodine receptors [105] and co-expressed with *N*-methyl-D-aspartate receptors [106], both involved in Ca^{2+} signaling. Calsenilin is a neuronal calcium-binding protein that interacts with the C-terminus of PS1 and PS2 [107]. The interaction with PS promotes $A\beta_{1\text{-}42}$ production and apoptosis in a γ-secretase dependent manner [108–110]. Calsenilin knock-out mice show decreased levels of brain $A\beta_{1\text{-}42}$ [111], and co-expression of mutant PS1 and calsenilin reverse presenilin-mediated enhancement of calcium signalling in *Xenopus* oocytes [112]. It appears that calsenilin regulates $A\beta_{1\text{-}42}$ production and alterations in calcium signaling by interaction with PS1 C-terminus.

14.6 PS Processing and Aβ Generation During Apoptosis

As stated above, full-length PS is processed into a NTF and a CTF that form the functional heterodimer in the γ-secretase complex. In addition, presenilins are substrates for calpains [113] and caspases, two groups of proteases activated during apoptosis. Two caspase cleavage sites have been identified within the cytoplasmic loop of PS1 ($ENDD_{329}$ and $AQRD_{341}$) and one in PS2 ($DSYD_{329}$) [114]. The resulting fragments may have a regulatory role in apoptosis. Both the normally cleaved CTF of PS2 and the caspase-cleaved CTF of PS1 are anti-apoptotic and delay cell death in different experimental paradigms [115, 116]. The caspase-cleaved CTF of PS1 is degraded by a calpain-like cysteine proteinase, which may also influence the regulation of apoptosis [52]. On the contrary, overexpression of full-length PS2 is pro-apoptotic [85] and triggers p53-dependent apoptosis leading to downregulation of PS1 [117] and Bcl-2 [118]. Downregulation of PS1 seems to lead to increased cell death, and therefore the full-length PS1 is suggested to be anti-apoptotic [119, 120].

The mature γ-secretase complex is very stable, and protease activity has been detected in samples from frozen human brain [121]. Results from our laboratory also show that the γ-secretase complex is preserved and active in apoptotic cells [Hansson et al., unpublished data]. Brockhaus and collegues [122] have previously shown that caspase cleavage of PS does not change the production of Aβ. These data suggest that Aβ generation occurs in dying cells and that these cells contribute to the amyloid burden in AD brain. The early cognitive impairments in AD are caused by loss of synapses in regions of the brain critical for memory function (entorhinal cortex, hippocampus). As discussed above, neurons without synapses can survive even though they do not signal and have contacts with other neurons. Therefore, these neurons are present in the tissue for some time before they finally degenerate, and apparently they can produce Aβ during this time. Maybe dying neurons form seeds for the growing plaques. Indeed, LaFerla and collegues suggested several years ago that intracellular accumulation of Aβ triggers cell death. Aβ is then released from the dead cells leading to extracellular deposits of Aβ and the formations of plaques [123]. They detected DNA fragmentation in cells from AD brain and Aβ plaques containing numerous neuronal ghosts, indicating that neuronal death proceeds the formation of extracellular deposition of Aβ in AD brains.

14.7 Mitochondria Are Targets for Aβ-Induced Cell Death

Recent data suggest that it is the intracellular Aβ species, and not extracellular plaques, that are primarily toxic to cells [5]. Increasing evidence show that functional mitochondria play a significant part as targets or mediators of Aβ toxicity. Neurons are dependent on aerobic oxidative phosphorylation for their energy needs, and mitochondria are therefore essential for neuronal function. Mitochondria are abundant in presynaptic nerve terminals where they provide energy for sustained neurotransmittor release. Mitochondrial damage may lead to release of death factors (e.g., cytochrome c, Omi/HtrA2, Smac/Diablo) resulting in apoptosis. Dysfunctional mitochondria also lead to decreased ATP production and impaired calcium buffering capacity. Apoptosis can be triggered locally in synapses [12, 13], and loss of synapses correlates well with the impairment of cognitive functions early in AD. Local Aβ production in synapses may therefore damage mitochondria and cause synapse loss.

Aβ accumulates in mitochondria in AD brain and in APP transgenic mice [124] and has been shown to inhibit enzymes important for mitochondrial functions in vitro, for example, cytochrome c oxidase, β-ketoglutarate dehydrogenase, and pyruvate dehydrogenase [125–127]. Another intracellular target for Aβ is alcohol binding dehydrogenase (ABAD) [124] (for a review, see [128]). ABAD is located to mitochondria where it binds to Aβ and promotes Aβ-induced cell stress. ABAD is overexpressed in AD brain and in brains from transgenic APP mice.

Aβ-toxicity is dependent on a functional electron transport chain [129], and Aβ has been shown to induce oxidative stress [130, 131] and induction of permeability transition [132, 133] in different cell models. Aβ also induces p53 and Bax activation [134] associated with apoptosis signaling through the mitochondrial pathway. In addition, Aβ triggers the release of cytochrome c from mitochondria [135]. Taken together, it seems that Aβ induces cell death by affecting different mitochondrial functions and triggering apoptotic mechanisms. As discussed above, cells carrying PS mutations have increased production of Aβ and are sensitized to apoptotic stimuli. Mitochondria seem to be an important target for Aβ-induced cell death in agreement with the central role of mitochondria in apoptosis signaling.

At present, it is not clear whether Aβ is produced in mitochondria or imported into mitochondria. Two studies have shown the localization of APP to mitochondria. First, APP immunoreactivity was detected by electron microscopy in the outer membrane of mitochondria [136]. Second, APP was shown to be imported into the outer mitochondrial membrane. However, the import is arrested by an acidic domain that spans sequence 220–290 of APP leaving a 73-kDa portion of the C-terminal side of the protein facing the cytoplasm. According to this topology, the Aβ peptide region of APP is not located to the membrane making it impossible for β- and γ-secretases to cleave out Aβ from APP located to mitochondria [137]. We have shown that PS, nicastrin, Pen-2, and Aph-1 form active γ-secretase complexes in mitochondria [138]. So far, no γ-secretase substrate has been identified in mitochondria, and the function of the mitochondrial γ-secretase complex is not known. In conclusion, it is most likely that Aβ is taken up by mitochondria and that the mitochondrial γ-secretase complex cleaves other substrates than APP. Exactly how Aβ gains access to mitochondria is not known, and this issue has to be addressed in future studies. Aβ is secreted luminally and has been detected in ER/Golgi, lysosomes/endosomes, and multivesicular bodies. One possiblility is that, for example, ER-to-mitochondrial transfer might occur [139].

14.8 Conclusions

It has been established that PS is essential for γ-secretase activity, and PS is therefore mandatory for the generation of Aβ. Aβ is toxic and kills cells by mechanisms involving perturbed intracellular calcium homeostasis, oxidative stress, and impaired mitochondrial functions. PS mutations sensitize cells to various toxic stimuli in vitro and increase the production of Aβ. Whether it is the increased Aβ load that causes the sensitization of PS mutant cells or if PS mutations cause cellular alterations independent of Aβ production have not been elucidated. Further studies have to be performed to shed more light on these complicated mechanisms. Under all circumstances, it is becoming clear that it is the intracellular Aβ that is primarily toxic. Therefore, it is of great importance to decrease Aβ-generation and protect neurons from Aβ in order to block cell death in AD.

References

1. Arriagada PV, Growdon JH, Hedley-Whyte ET, Hyman BT. Neurofibrillary tangles but not senile plaques parallel duration and severity of Alzheimer's disease. Neurology 1992; 42:631-9.
2. Samuel W, Terry RD, DeTeresa R, et al. Clinical correlates of cortical and nucleus basalis pathology in Alzheimer dementia. Arch Neurol 1994; 51:772-8.
3. Terry RD, Masliah E, Salmon DP, et al. Physical basis of cognitive alterations in Alzheimer's disease: synapse loss is the major correlate of cognitive impairment. Ann Neurol 1991; 30:572-80.

4. Selkoe DJ. Alzheimer's disease is a synaptic failure. Science 2002; 298:789-91.
5. Billings LM, Oddo S, Green KN, et al. Intraneuronal Abeta causes the onset of early Alzheimer's disease-related cognitive deficits in transgenic mice. Neuron 2005; 45:675-88.
6. Tseng BP, Kitazawa M, LaFerla FM Amyloid β-peptide: The inside story. Curr Alzheimer Res 2004; 1:231-239.
7. Zhang Z, Nadeau P, Song W, et al. Presenilins are required for gamma-secretase cleavage of beta-APP and transmembrane cleavage of Notch-1. Nat Cell Biol 2000; 2:463-5.
8. Herreman A, Serneels L, Annaert W, et al. Total inactivation of gamma-secretase activity in presenilin-deficient embryonic stem cells. Nat Cell Biol 2000; 2:461-2.
9. Chen Q, Schubert D. Presenilin-interacting proteins. Expert Rev Mol Med 2002; 2002:1-18.
10. Popescu BO, Ankarcrona, M. Neurons bearing presenilins: weapons for defense or suicide. J Cell Mol Med 2000; 4:249-261.
11. Oddo S, Caccamo A, Shepherd JD, et al. Triple-transgenic model of Alzheimer's disease with plaques and tangles. Intracellular abeta and synaptic dysfunction. Neuron 2003; 39:409-21.
12. Mattson MP, Keller JN, Begley JG. Evidence for synaptic apoptosis. Exp Neurol 1998; 153:35-48.
13. Mattson MP, Partin J, Begley JG. Amyloid beta-peptide induces apoptosis-related events in synapses and dendrites. Brain Res 1998; 807:167-76.
14. Su JH, Anderson AJ, Cummings BJ, Cotman CW. Immunohistochemical evidence for apoptosis in Alzheimer's disease. Neuroreport 1994; 5:2529-33.
15. Lassmann H, Bancher C, Breitschopf H, et al. Cell death in Alzheimer's disease evaluated by DNA fragmentation in situ. Acta Neuropathol 1995; 89:35-41.
16. Smale G, Nichols NR, Brady DR, et al. Evidence for apoptotic cell death in Alzheimer's disease. Exp Neurol 1995; 133:225-30.
17. Dragunow M, Faull RL, Lawlor P, et al. In situ evidence for DNA fragmentation in Huntington's disease striatum and Alzheimer's disease temporal lobes. Neuroreport 1995; 6:1053-7.
18. Li WP, Chan WY, Lai HW, Yew DT. Terminal dUTP nick end labeling (TUNEL) positive cells in the different regions of the brain in normal aging and Alzheimer patients. J Mol Neurosci 1997; 8:75-82.
19. Lucassen PJ, Chung WC, Kamphorst W, Swaab DF. DNA damage distribution in the human brain as shown by in situ end labeling; area-specific differences in aging and Alzheimer's disease in the absence of apoptotic morphology. J Neuropathol Exp Neurol 1997; 56:887-900.
20. Sugaya K, Reeves M, McKinney M. Topographic associations between DNA fragmentation and Alzheimer's disease neuropathology in the hippocampus. Neurochem Int 1997; 31:275-81.
21. Kitamura Y, Shimohama S, Kamoshima W, et al. Alteration of proteins regulating apoptosis, Bcl-2, Bcl-x, Bax, Bak, Bad, ICH-1 and CPP32, in Alzheimer's disease. Brain Res 1998; 780:260-9.
22. Giannakopoulos P, Kovari E, Savioz A, et al. Differential distribution of presenilin-1, Bax, and Bcl-X(L) in Alzheimer's disease and frontotemporal dementia. Acta Neuropathol (Berlin) 1999; 98:141-9.
23. Su JH, Deng G, Cotman CW. Bax protein expression is increased in Alzheimer's brain: correlations with DNA damage, Bcl-2 expression, and brain pathology. J Neuropathol Exp Neurol 1997; 56:86-93.
24. Drache B, Diehl GE, Beyreuther K, et al. Bcl-xl-specific antibody labels activated microglia associated with Alzheimer's disease and other pathological states. J Neurosci Res 1997; 47:98-108.
25. MacGibbon GA, Lawlor PA, Sirimanne ES, et al. Bax expression in mammalian neurons undergoing apoptosis, and in Alzheimer's disease hippocampus. Brain Res 1997; 750:223-34.
26. Yang F, Sun X, Beech W, et al. Antibody to caspase-cleaved actin detects apoptosis in differentiated neuroblastoma and plaque-associated neurons and microglia in Alzheimer's disease. Am J Pathol 1998; 152:379-89.
27. Uetsuki T, Takemoto K, Nishimura I, et al. Activation of neuronal caspase-3 by intracellular accumulation of wild-type Alzheimer amyloid precursor protein. J Neurosci 1999; 19:6955-64.
28. LeBlanc A, Liu H, Goodyer C, et al. Caspase-6 role in apoptosis of human neurons, amyloidogenesis, and Alzheimer's disease. J Biol Chem 1999; 274:23426-36.
29. Chan SL, Griffin WS, Mattson MP. Evidence for caspase-mediated cleavage of AMPA receptor subunits in neuronal apoptosis and Alzheimer's disease. J Neurosci Res 1999; 57:315-23.
30. Stadelmann C, Deckwerth TL, Srinivasan A, et al. Activation of caspase-3 in single neurons and autophagic granules of granulovacuolar degeneration in Alzheimer's disease. Evidence for apoptotic cell death. Am J Pathol 1999; 155:1459-66.
31. Rohn TT, Head E, Nesse WH, et al. Activation of caspase-8 in the Alzheimer's disease brain. Neurobiol Dis 2001; 8:1006-16.
32. Pompl PN, Yemul S, Xiang Z, et al. Caspase gene expression in the brain as a function of the clinical progression of Alzheimer's disease. Arch Neurol 2003; 60:369-76.

33. Raina AK, Hochman A, Zhu X, et al. Abortive apoptosis in Alzheimer's disease. Acta Neuropathol (Berlin) 2001; 101:305-10.
34. Yang Y, Mufson EJ, Herrup K. Neuronal cell death is preceded by cell cycle events at all stages of Alzheimer's disease. J Neurosci 2003; 23:2557-63.
35. Copani A, Sortino MA, Nicoletti F, Giuffrida SA. Alzheimer's disease research enters a "new cycle": how significant? Neurochem Res 2002; 27:173-6.
36. Mattson MP. Pathways towards and away from Alzheimer's disease. Nature 2004; 430:631-9.
37. Lippa CF, Swearer JM, Kane KJ, et al. Familial Alzheimer's disease: site of mutation influences clinical phenotype. Ann Neurol 2000; 48:376-9.
38. Levy-Lahad E, Wasco W, Poorkaj P, et al. Candidate gene for the chromosome 1 familial Alzheimer's disease locus. Science 1995; 269:973-7.
39. Rogaev EI, Sherrington R, Rogaeva EA, et al. Familial Alzheimer's disease in kindreds with missense mutations in a gene on chromosome 1 related to the Alzheimer's disease type 3 gene. Nature 1995; 376:775-8.
40. Sherrington R, Rogaev EI, Liang Y, et al. Cloning of a gene bearing missense mutations in early-onset familial Alzheimer's disease. Nature 1995; 375:754-60.
41. Cook DG, Sung JC, Golde TE, et al. Expression and analysis of presenilin 1 in a human neuronal system: localization in cell bodies and dendrites. Proc Natl Acad Sci U S A 1996; 93:9223-8.
42. Kovacs DM, Fausett HJ, Page KJ, et al. Alzheimer-associated presenilins 1 and 2: neuronal expression in brain and localization to intracellular membranes in mammalian cells. Nat Med 1996; 2:224-9.
43. Walter J, Capell A, Grunberg J, et al. The Alzheimer's disease-associated presenilins are differentially phosphorylated proteins located predominantly within the endoplasmic reticulum. Mol Med 1996; 2:673-91.
44. De Strooper B, Beullens M, Contreras B, et al. Phosphorylation, subcellular localization, and membrane orientation of the Alzheimer's disease-associated presenilins. J Biol Chem 1997; 272:3590-8.
45. Dewji NN, Singer SJ. Cell surface expression of the Alzheimer's disease-related presenilin proteins. Proc Natl Acad Sci U S A 1997; 94:9926-31.
46. Li J, Xu M, Zhou H, Ma J, Potter H. Alzheimer presenilins in the nuclear membrane, interphase kinetochores, and centrosomes suggest a role in chromosome segregation. Cell 1997; 90:917-27.
47. Pasternak SH, Bagshaw RD, Guiral M, et al. Presenilin-1, nicastrin, amyloid precursor protein, and gamma-secretase activity are co-localized in the lysosomal membrane. J Biol Chem 2003; 278: 26687-94.
48. Ankarcrona M, Hultenby K. Presenilin-1 is located in rat mitochondria. Biochem Biophys Res Commun 2002; 295:766-70.
49. De Strooper B, Saftig P, Craessaerts K, et al. Deficiency of presenilin-1 inhibits the normal cleavage of amyloid precursor protein. Nature 1998; 391:387-90.
50. Edbauer D, Winkler E, Regula JT, et al. Reconstitution of gamma-secretase activity. Nat Cell Biol 2003; 5:486-8.
51. Capell A, Beher D, Prokop S, et al. Gamma-secretase complex assembly within the early secretory pathway. J Biol Chem 2005; 280:6471-8.
52. Steiner H, Capell A, Pesold B, et al. Expression of Alzheimer's disease-associated presenilin-1 is controlled by proteolytic degradation and complex formation. J Biol Chem 1998; 273:32322-31.
53. Van Gassen G, De Jonghe C, Pype S, et al. Alzheimer's disease associated presenilin 1 interacts with HC5 and ZETA, subunits of the catalytic 20S proteasome. Neurobiol Dis 1999; 6:376-91.
54. Bergman A, Hansson EM, Pursglove SE, et al. Pen-2 is sequestered in the endoplasmic reticulum and subjected to ubiquitylation and proteasome-mediated degradation in the absence of presenilin. J Biol Chem 2004; 279:16744-53. Epub 2004 Jan 14.
55. Li YM, Xu M, Lai MT, et al. Photoactivated gamma-secretase inhibitors directed to the active site covalently label presenilin 1. Nature 2000; 405:689-94.
56. Esler WP, Kimberly WT, Ostaszewski BL, et al. Transition-state analogue inhibitors of gamma-secretase bind directly to presenilin-1. Nat Cell Biol 2000; 2:428-34.
57. Wolfe MS, Xia W, Ostaszewski BL, et al. Two transmembrane aspartates in presenilin-1 required for presenilin endoproteolysis and gamma-secretase activity. Nature 1999; 398:513-7.
58. Xia W, Wolfe MS. Intramembrane proteolysis by presenilin and presenilin-like proteases. J Cell Sci 2003; 116:2839-44.
59. Selkoe D, Kopan R. Notch and presenilin: regulated intramembrane proteolysis links development and degeneration. Annu Rev Neurosci 2003; 26: 565-97.
60. Jarriault S, Brou C, Logeat F, et al. Signalling downstream of activated mammalian Notch. Nature 1995; 377:355-8.
61. de la Pompa JL, Wakeham A, Correia KM, et al. Conservation of the Notch signalling pathway in mammalian neurogenesis. Development 1997; 124:1139-48.
62. Kimberly WT, Zheng JB, Guenette SY, Selkoe DJ. The intracellular domain of the beta-amyloid precursor protein is stabilized by Fe65 and translocates to the nucleus in a notch-like manner. J Biol Chem 2001; 276:40288-92.
63. Kimberly WT, Zheng JB, Town T, et al. Physiological regulation of the beta-amyloid precur-

sor protein signaling domain by c-Jun N-terminal kinase JNK3 during neuronal differentiation. J Neurosci 2005; 25:5533-43.
64. Cao X, Sudhof TC. Dissection of amyloid-beta precursor protein-dependent transcriptional transactivation. J Biol Chem 2004; 279:24601-11.
65. Leissring MA, Murphy MP, Mead TR, et al. A physiologic signaling role for the gamma -secretase-derived intracellular fragment of APP. Proc Natl Acad Sci U S A 2002; 99:4697-702.
66. Donoviel DB, Hadjantonakis AK, Ikeda M, et al. Mice lacking both presenilin genes exhibit early embryonic patterning defects. Genes Dev 1999; 13:2801-10.
67. Perez-Tur J, Froelich S, Prihar G, et al. A mutation in Alzheimer's disease destroying a splice acceptor site in the presenilin-1 gene. Neuroreport 1995; 7:297-301.
68. Duering M, Grimm MO, Grimm HS, et al. Mean age of onset in familial Alzheimer's disease is determined by amyloid beta 42. Neurobiol Aging 2005; 26:785-8.
69. Lleo A, Berezovska O, Herl L, et al. Nonsteroidal anti-inflammatory drugs lower Abeta42 and change presenilin 1 conformation. Nat Med 2004; 10:1065-6.
70. Berezovska O, Lleo A, Herl LD, et al. Familial Alzheimer's disease presenilin 1 mutations cause alterations in the conformation of presenilin and interactions with amyloid precursor protein. J Neurosci 2005; 25:3009-17.
71. Nakaya Y, Yamane T, Shiraishi H, et al. Random mutagenesis of presenilin-1 identifies novel mutants exclusively generating long amyloid beta-peptides. J Biol Chem 2005; 280:19070-7.
72. Walker ES, Martinez M, Brunkan AL, Goate A. Presenilin 2 familial Alzheimer's disease mutations result in partial loss of function and dramatic changes in Abeta 42/40 ratios. J Neurochem 2005; 92:294-301.
73. Esposito L, Gan L, Yu GQ, Essrich C, Mucke L. Intracellularly generated amyloid-beta peptide counteracts the antiapoptotic function of its precursor protein and primes proapoptotic pathways for activation by other insults in neuroblastoma cells. J Neurochem 2004; 91:1260-74.
74. Wolozin B, Alexander P, Palacino J. Regulation of apoptosis by presenilin 1. Neurobiol Aging 1998; 19:S23-7.
75. Guo Q, Furukawa K, Sopher BL, et al. Alzheimer's PS-1 mutation perturbs calcium homeostasis and sensitizes PC12 cells to death induced by amyloid beta-peptide. Neuroreport 1996; 8:379-83.
76. Weihl CC, Ghadge GD, Kennedy SG, et al. Mutant presenilin-1 induces apoptosis and downregulates Akt/PKB. J Neurosci 1999; 19:5360-9.
77. Tanii H, Ankarcrona M, Flood F, et al. Alzheimer's disease presenilin-1 exon 9 deletion and L250S mutations sensitize SH-SY5Y neuroblastoma cells to hyperosmotic stress-induced apoptosis. Neuroscience. 2000; 95:593-601.
78. Popescu BO, Cedazo-Minguez A, Popescu LM, et al. Caspase cleavage of exon 9 deleted presenilin-1 is an early event in apoptosis induced by calcium ionophore A 23187 in SH-SY5Y neuroblastoma cells. J Neurosci Res 2001; 66:122-34.
79. Czech C, Lesort M, Tremp G, et al. Characterization of human presenilin 1 transgenic rats: increased sensitivity to apoptosis in primary neuronal cultures. Neuroscience 1998; 87:325-36.
80. Bursztajn S, DeSouza R, McPhie DL, et al. Overexpression in neurons of human presenilin-1 or a presenilin-1 familial Alzheimer's disease mutant does not enhance apoptosis. J Neurosci 1998; 18:9790-9.
81. Kovacs DM, Mancini R, Henderson J, et al. Staurosporine-induced activation of caspase-3 is potentiated by presenilin 1 familial Alzheimer's disease mutations in human neuroglioma cells. J Neurochem 1999; 73:2278-85.
82. Keller JN, Guo Q, Holtsberg FW, et al. Increased sensitivity to mitochondrial toxin-induced apoptosis in neural cells expressing mutant presenilin-1 is linked to perturbed calcium homeostasis and enhanced oxyradical production. J Neurosci 1998; 18:4439-50.
83. Chan SL, Culmsee C, Haughey N, et al. Presenilin-1 mutations sensitize neurons to DNA damage-induced death by a mechanism involving perturbed calcium homeostasis and activation of calpains and caspase-12. Neurobiol Dis 2002; 11:2-19.
84. Cedazo-Minguez A, Popescu BO, Ankarcrona M, et al. The presenilin 1 deltaE9 mutation gives enhanced basal phospholipase C activity and a resultant increase in intracellular calcium concentrations. J Biol Chem 2002; 277:36646-55.
85. Wolozin B, Iwasaki K, Vito P, et al. Participation of presenilin 2 in apoptosis: enhanced basal activity conferred by an Alzheimer mutation. Science 1996; 274:1710-3.
86. Araki W, Yuasa K, Takeda S, et al. Overexpression of presenilin-2 enhances apoptotic death of cultured cortical neurons. Ann N Y Acad Sci 2000; 920:241-4.
87. Mori M, Nakagami H, Morishita R, et al. N141I mutant presenilin-2 gene enhances neuronal cell death and decreases bcl-2 expression. Life Sci 2002; 70:2567-80.
88. Guo Q, Sopher BL, Furukawa K, et al. Alzheimer's presenilin mutation sensitizes neural cells to apoptosis induced by trophic factor withdrawal and amyloid beta-peptide: involvement of calcium and oxyradicals. J Neurosci 1997; 17:4212-22.
89. Ito E, Oka K, Etcheberrigaray R, et al. Internal Ca2+ mobilization is altered in fibroblasts from patients with Alzheimer's disease. Proc Natl Acad Sci U S A 1994; 91:534-8.

90. Hirashima N, Etcheberrigaray R, Bergamaschi S, et al. Calcium responses in human fibroblasts: a diagnostic molecular profile for Alzheimer's disease. Neurobiol Aging 1996; 17:549-55.
91. Leissring MA, Akbari Y, Fanger CM, et al. Capacitative calcium entry deficits and elevated luminal calcium content in mutant presenilin-1 knockin mice. J Cell Biol 2000; 149:793-8.
92. Guo Q, Fu W, Sopher BL, et al. Increased vulnerability of hippocampal neurons to excitotoxic necrosis in presenilin-1 mutant knock-in mice. Nat Med 1999; 5:101-6.
93. Leissring MA, Paul BA, Parker I, et al. Alzheimer's presenilin-1 mutation potentiates inositol 1,4,5-trisphosphate-mediated calcium signaling in Xenopus oocytes. J Neurochem 1999; 72:1061-8.
94. Stutzmann GE, Caccamo A, LaFerla FM, Parker I. Dysregulated IP3 signaling in cortical neurons of knock-in mice expressing an Alzheimer's-linked mutation in presenilin1 results in exaggerated Ca2+ signals and altered membrane excitability. J Neurosci 2004; 24:508-13.
95. Querfurth HW, Selkoe DJ. Calcium ionophore increases amyloid beta peptide production by cultured cells. Biochemistry 1994; 33:4550-61.
96. Mattson MP, Cheng B, Davis D, et al. beta-Amyloid peptides destabilize calcium homeostasis and render human cortical neurons vulnerable to excitotoxicity. J Neurosci 1992; 12:376-89.
97. Arispe N, Rojas E, Pollard HB. Alzheimer's disease amyloid beta protein forms calcium channels in bilayer membranes: blockade by tromethamine and aluminum. Proc Natl Acad Sci U S A 1993; 90:567-71.
98. Rhee SK, Quist AP, Lal R. Amyloid beta protein-(1-42) forms calcium-permeable, Zn2+-sensitive channel. J Biol Chem 1998; 273:13379-82.
99. Kagan BL, Hirakura Y, Azimov R, et al. The channel hypothesis of Alzheimer's disease: current status. Peptides 2002; 23:1311-5.
100. Bhatia R, Lin H, Lal R. Fresh and globular amyloid beta protein (1-42) induces rapid cellular degeneration: evidence for AbetaP channel-mediated cellular toxicity. FASEB J 2000; 14:1233-43.
101. Kayed R, Sokolov Y, Edmonds B, et al. Permeabilization of lipid bilayers is a common conformation-dependent activity of soluble amyloid oligomers in protein misfolding diseases. J Biol Chem 2004; 279:46363-6.
102. Mattson MP, Cheng B, Culwell AR, et al. Evidence for excitoprotective and intraneuronal calcium-regulating roles for secreted forms of the beta-amyloid precursor protein. Neuron 1993; 10:243-54.
103. Furukawa K, Sopher BL, Rydel RE, et al. Increased activity-regulating and neuroprotective efficacy of alpha-secretase-derived secreted amyloid precursor protein conferred by a C-terminal heparin-binding domain. J Neurochem 1996; 67:1882-96.
104. Guo Q, Robinson N, Mattson MP. Secreted beta-amyloid precursor protein counteracts the proapoptotic action of mutant presenilin-1 by activation of NF-kappaB and stabilization of calcium homeostasis. J Biol Chem 1998; 273:12341-51.
105. Pickel VM, Clarke CL, Meyers MB. Ultrastructural localization of sorcin, a 22 kDa calcium binding protein, in the rat caudate-putamen nucleus: association with ryanodine receptors and intracellular calcium release. J Comp Neurol 1997; 386:625-34.
106. Gracy KN, Clarke CL, Meyers MB, Pickel VM. N-methyl-D-aspartate receptor 1 in the caudate-putamen nucleus: ultrastructural localization and co-expression with sorcin, a 22,000 mol. wt calcium binding protein. Neuroscience 1999; 90: 107-17.
107. Buxbaum JD, Choi EK, Luo Y, et al. Calsenilin: a calcium-binding protein that interacts with the presenilins and regulates the levels of a presenilin fragment. Nat Med 1998; 4:1177-81.
108. Jo DG, Kim MJ, Choi YH, et al. Pro-apoptotic function of calsenilin/DREAM/KChIP3. FASEB J 2001; 15:589-91.
109. Jo DG, Chang JW, Hong HS, et al. Contribution of presenilin/gamma-secretase to calsenilin-mediated apoptosis. Biochem Biophys Res Commun 2003; 305:62-6.
110. Jo DG, Jang J, Kim BJ, Lundkvist J, Jung YK. Overexpression of calsenilin enhances gamma-secretase activity. Neurosci Lett 2005; 378:59-64.
111. Lilliehook C, Bozdagi O, Yao J, et al. Altered Abeta formation and long-term potentiation in a calsenilin knock-out. J Neurosci 2003; 23:9097-106.
112. Leissring MA, Yamasaki TR, Wasco W, et al. Calsenilin reverses presenilin-mediated enhancement of calcium signaling. Proc Natl Acad Sci U S A 2000; 97:8590-3.
113. Maruyama K, Usami M, Kametani F, et al. Molecular interactions between presenilin and calpain: inhibition of m-calpain protease activity by presenilin-1, 2 and cleavage of presenilin-1 by m-, mu-calpain. Int J Mol Med 2000; 5:269-73.
114. van de Craen M, de Jonghe C, van den Brande I, et al. Identification of caspases that cleave presenilin-1 and presenilin-2. Five presenilin-1 (PS1) mutations do not alter the sensitivity of PS1 to caspases. FEBS Lett 1999; 445:149-54.
115. Vito P, Ghayur T, D'Adamio L. Generation of anti-apoptotic presenilin-2 polypeptides by alternative transcription, proteolysis, and caspase-3 cleavage. J Biol Chem 1997; 272:28315-20.
116. Vezina J, Tschopp C, Andersen E, Muller K. Overexpression of a C-terminal fragment of

presenilin 1 delays anti-Fas induced apoptosis in Jurkat cells. Neurosci Lett 1999; 263:65-8.
117. Alves da Costa C, Paitel E, Mattson MP, et al. Wild-type and mutated presenilins 2 trigger p53-dependent apoptosis and down-regulate presenilin 1 expression in HEK293 human cells and in murine neurons. Proc Natl Acad Sci U S A 2002; 99:4043-8.
118. Araki W, Yuasa K, Takeda S, et al. Pro-apoptotic effect of presenilin 2 (PS2) overexpression is associated with down-regulation of Bcl-2 in cultured neurons. J Neurochem 2001; 79:1161-8.
119. Roperch JP, Alvaro V, Prieur S, et al. Inhibition of presenilin 1 expression is promoted by p53 and p21WAF-1 and results in apoptosis and tumor suppression. Nat Med 1998; 4:835-8.
120. Hong CS, Caromile L, Nomata Y, et al. Contrasting role of presenilin-1 and presenilin-2 in neuronal differentiation in vitro. J Neurosci 1999; 19:637-43.
121. Farmery MR, Tjernberg LO, Pursglove SE, et al. Partial purification and characterization of gamma-secretase from post-mortem human brain. J Biol Chem 2003; 278:24277-84.
122. Brockhaus M, Grunberg J, Rohrig S, et al. Caspase-mediated cleavage is not required for the activity of presenilins in amyloidogenesis and NOTCH signaling. Neuroreport 1998; 9:1481-6.
123. LaFerla FM, Troncoso JC, Strickland DK, Kawas CH, Jay G. Neuronal cell death in Alzheimer's disease correlates with apoE uptake and intracellular Abeta stabilization. J Clin Invest 1997; 100:310-20.
124. Lustbader JW, Cirilli M, Lin C, et al. ABAD directly links Abeta to mitochondrial toxicity in Alzheimer's disease. Science 2004; 304:448-52.
125. Crouch PJ, Blake R, Duce JA, et al. Copper-dependent inhibition of human cytochrome c oxidase by a dimeric conformer of amyloid-beta1-42. J Neurosci 2005; 25:672-9.
126. Casley CS, Canevari L, Land JM, et al. Beta-amyloid inhibits integrated mitochondrial respiration and key enzyme activities. J Neurochem 2002; 80: 91-100.
127. Canevari L, Clark JB, Bates TE. beta-Amyloid fragment 25-35 selectively decreases complex IV activity in isolated mitochondria. FEBS Lett 1999; 457:131-4.
128. Yan SD, Stern DM. Mitochondrial dysfunction and Alzheimer's disease: role of amyloid-beta peptide alcohol dehydrogenase (ABAD). Int J Exp Pathol 2005; 86:161-71.
129. Cardoso SM, Santos S, Swerdlow RH, Oliveira CR. Functional mitochondria are required for amyloid beta-mediated neurotoxicity. FASEB J 2001; 15: 1439-41.
130. Rodrigues CM, Sola S, Brito MA, et al. Amyloid beta-peptide disrupts mitochondrial membrane lipid and protein structure: protective role of tauroursodeoxycholate. Biochem Biophys Res Commun 2001; 281:468-74.
131. Abramov AY, Canevari L, Duchen MR. Beta-amyloid peptides induce mitochondrial dysfunction and oxidative stress in astrocytes and death of neurons through activation of NADPH oxidase. J Neurosci 2004; 24:565-75.
132. Moreira PI, Santos MS, Moreno A, et al. Effect of amyloid beta-peptide on permeability transition pore: a comparative study. J Neurosci Res 2002; 69:257-67.
133. Parks JK, Smith TS, Trimmer PA, et al. Neurotoxic Abeta peptides increase oxidative stress in vivo through NMDA-receptor and nitric-oxide-synthase mechanisms, and inhibit complex IV activity and induce a mitochondrial permeability transition in vitro. J Neurochem 2001; 76:1050-6.
134. Zhang Y, McLaughlin R, Goodyer C, LeBlanc A. Selective cytotoxicity of intracellular amyloid beta peptide1-42 through p53 and Bax in cultured primary human neurons. J Cell Biol 2002; 156:519-29.
135. Kim HS, Lee JH, Lee JP, et al. Amyloid beta peptide induces cytochrome C release from isolated mitochondria. Neuroreport 2002; 13:1989-93.
136. Yamaguchi H, Yamazaki T, Ishiguro K, et al. Ultrastructural localization of Alzheimer amyloid beta/A4 protein precursor in the cytoplasm of neurons and senile plaque-associated astrocytes. Acta Neuropathol (Berlin) 1992; 85:15-22.
137. Anandatheerthavarada HK, Biswas G, Robin MA, Avadhani NG. Mitochondrial targeting and a novel transmembrane arrest of Alzheimer's amyloid precursor protein impairs mitochondrial function in neuronal cells. J Cell Biol 2003; 161:41-54.
138. Hansson CA, Frykman S, Farmery MR, et al. Nicastrin, presenilin, APH-1, and PEN-2 form active gamma-secretase complexes in mitochondria. J Biol Chem 2004; 279:51654-60.
139. Caspersen C, Wang N, Yao J, et al. Mitochondrial Aβ: a potential focal point for neuronal metabolic dysfunction in Alzheimer's disease. FASEB J. 2005; 19: 2040-2041.

15 Immunotherapeutic Approaches to Alzheimer's Disease

Josef Karkos

15.1 Concept of Immunotherapy for Alzheimer's Disease

The concept of immunotherapy for Alzheimer's disease (AD) is based on the molecular findings that place AD within the group of disorders called "protein-misfolding diseases." These disorders are caused by conformational changes coupled with the aggregation of misfolded proteins outside of the cell [1–4]. The concept emerged after the research group of Salomon [5–8] demonstrated that the immunologic approach in vitro was successful in inducing conformational changes in both antigen and antibody. In particular, it was demonstrated that the monoclonal antibodies were capable of stabilizing the conformation of an antigen against incorrect folding and recognize an incompletely folded epitope, inducing native conformation in a partially unfolded protein.

Support for the in vivo relevance of the concept has been provided by experiments published by the Schenk's research group [9]. They found that vaccination of a transgenic mouse expressing the human β-amyloid protein with the β-amyloid peptide ($A\beta_{42}$) significantly decreased the β-amyloid burden in areas of the brain important for cognition and memory. Furthermore, the studies carried out by Schenk's group indicated that the effect of the $A\beta_{42}$ peptide was mediated by antibodies it induced [10].

The functional relevance of the findings reported by Schenk's group was demonstrated in separated, independent follow-up studies carried out by Janus and Morgan and their colleagues [11, 12]. They showed that the β-amyloid peptide vaccine was able to protect transgenic mice from the memory deficits they normally develop and to ameliorate the preexisting behavioral and memory deficits.

After promising preclinical studies in several species, clinical trials were initiated using $A\beta_{42}$ (vaccine's name: AN-1792) in conjunction with the adjuvant QS-21 [13]. Despite numerous adverse effects that occurred in some patients that led to suspension of the study, preliminary data demonstrated that vaccination can reduce AD pathology and mitigate progressive cognitive decline associated with the disease.

The experimental and clinical data obtained to date indicate that the induction of the systemic adaptive response to $A\beta_{42}$ is an effective way to induce its clearance [14–17], supporting the amyloid cascade hypothesis of AD and implying that $A\beta_{42}$ deposition is driving the disease pathogenesis [18, 19]. Consistent with this hypothesis is the recent finding that the accumulation of Aβ is able to induce the development of tau pathology [20]. Aβ immunotherapy reduces first Aβ deposits and subsequently clears aggregates of tau-protein [21].

15.2 Immune Responses to Aβ

15.2.1 Molecular Structure and Immunological Properties of Aβ

In the Aβ structure, two domains can be discriminated: the N-terminal domain that encompasses amino acids 1 to 28 and C-terminal domain from amino acids 29 to 42. In aqueous solution, the N-terminal region exhibits different conformations and solubility properties depending on environmental conditions [22, 23]. The hydrophobic region in the C-terminal domain forms a β-strand structure in aqueous solutions, independently of pH and temperature. The amino acids sequences in the N-terminal domain permit the existence of a dynamic equilibrium between the α-helix and the β-strand conformations. In addition, results of in vitro experiments indicate a steady-state equilibrium between Aβ in plaques and in solution [24]. The most important conclusion from experiments in vitro is that amyloid formation might be subjected to modulation in terms of changes in conformation.

The Aβ molecule exhibits antigenic and immunogenic properties. Most of the $A\beta_{42}$-antibody-producing epitopes were detected in the N-terminal region of the peptide $A\beta_{42}$. The predominance of T-cell epitopes lies in the central to carboxy-terminal region of the peptide. The reported differences in the location of epitopes within the Aβ peptide depend on the different length of the peptides used for the detection of epitopes. The effects of antibody binding to various epitopes may be different. As $A\beta_{42}$ exists both in soluble and fibrillar forms, antibodies generated against this antigen may recognize different immunogenic structures within it. It is important to identify within $A\beta_{42}$ antigenic determinants for B and T cells in order to design the most effective vaccine.

Because the dominant B-cell and T-cell epitopes have distinct location, the humoral and cellular immune responses may be modulated. The modulation can be achieved for instance by using an antigen and various adjuvant combinations. Because the type of immune response generated may be critical to the efficacy and safety of a potential vaccine, a careful examination of the overall immune response, especially of the T_h1 and T_h2 responses, is of great importance [25].

15.2.2 Innate Immunoresponses to Aβ

Naturally occurring anti-Aβ antibodies (autoantibodies) were found in plasma in the elderly population [26]. There were detectable but very low levels of anti-amyloid antibodies in just over 50% of all samples and modest levels in under 5% of all samples. However, neither the presence nor the level of anti-amyloid-β antibodies correlated with the likelihood of developing dementia or with plasma levels of amyloid-β peptide. These findings suggest that low levels of anti-amyloid-β autoantibodies are frequent in the elderly population but do not confer protection against developing dementia.

Another group detected anti-amyloid-β autoantibodies in the CSF of AD patients [27, 28]. The titers of the antibodies were significantly lower in AD patients than in age-matched controls. These data indicate an impaired or reduced ability to generate antibodies specific against AD. This hypothesis has been supported by the finding that treatment of individuals with intravenous immunoglobulin preparation containing anti-Aβ antibodies increase both CSF and serum levels of anti-Aβ antibodies and significantly lowered CSF levels, possibly by facilitating transport of Aβ from the CSF to the serum [29]. These findings suggest that human Aβ antibodies are able to lower the Aβ concentration in the CSF, which may reduce Aβ deposition in brain. It seems that Aβ is recognized in the CNS as a molecule that needs to be cleared and provokes activation of microglia and astrocytes. The innate immunoresponse is also supported by such findings in AD patients as activation of complement; secretion of proinflammatory cytokines such as interleukin (IL)-1β and tumor necrosis factor (TNF)-α; expression of chemokines MIP-1α, MIP-1β, and MCP-1; and the secretion of nitric oxide [30, 31].

Monsonego et al. [32] found that some healthy, elderly individuals, as well as individuals with AD, possess elevated baseline levels of Aβ-reactive T-cells. While the general trend was toward a diminished immune response with aging, this demonstrates a selective increase in Aβ-reactive T cells in older individuals with and without dementia. The reason for this selective expansion of Aβ-reactive cells in elderly individuals is unclear. T-cell reactivity may be considered as an endogenous reaction to Aβ deposition in the brain in the context of the local innate immune response that occurs in AD [32].

The epitopes for Aβ-reactive T cells in humans are primarily amino acids 16–42. As in studies of active immunization of humans and of mouse models of AD, the primary epitope to which antibodies are generated are residues 1–12 [33]. There exists the possibility to influence both epitopes separately.

The function of microglia in AD seems also to be impaired. The role of microglia cells as a principle immune effector and phagocytic cells in the CNS is established. These cells are associated with plaques containing fibrillar β-amyloid found in the brains of AD patients. The plaque-associated glia undergo a phenotypic conversion into an activated phenotype. It is believed that microglia are responsible for the development of a focal inflammatory response that exacerbates and accelerates disease process. However, despite the presence of abundant activated microglia in the brains of AD patients, these cells fail to mount a phagocytic response to Aβ deposits but can efficiently phagocytose Aβ fibrils and plaques in vitro. It remains unclear why the plaque-associated microglia in vivo are unable to effectively phagocytose the amyloid deposits despite their close physical vicinity to the plaques [34]. It could be assumed that other plaque constituents block the interaction of the microglia with the plaque, as has been suggested for C1q [35].

15.2.3 Adaptive Immune Response to Aβ

15.2.3.1 Experience in Transgenic Animals

Although AD is associated with local innate immune responses, they are not sufficient to protect against the development of the disease or to attenuate the disease progression. The induction of systemic adaptive immune responses to Aβ in mouse models of AD has been found to be beneficial for both the neuropathologic and behavioral changes that these mice develop.

Active immunization with synthetic Aβ peptide or passive transfer with Aβ antibodies has been shown to prevent and reduce the cerebral amyloid load [9, 36, 37]. Using similar experimental settings, improvements in cognitive deficits in APP and APP/PS1 transgenic mice were observed [11, 12, 38, 39]. Schenk et al. [9] reported for the first time that intraperitoneal injections of $A\beta_{42}$ peptide, with complete or incomplete Freund`s adjuvant, almost completely prevented plaque deposition when given before initiation of plaque formation and significantly lowered cerebral levels if given after the initiation of plaque deposition in PDAPP transgenic mice. Evidence has been provided that the antibodies generated by active immunization with Aβ peptide recognized an epitope within the amino-terminus of the Aβ protein [37, 40–44]. Active immunization was shown to be less effective in reducing cerebral Aβ levels in very old APP transgenic mice with abundant cerebral Aβ plaques [45].

Passive administration of selected Aβ antibodies achieved similar effects to active immunization [36]. Passive transfer with a monoclonal antibody directed at the midregion of Aβ (mAb 266, recognizing $A\beta_{13\text{-}28}$) has been shown to lower cerebral levels while increasing Aβ levels in the blood [46]. When a single dose of Aβ mAb 266 was passively administered to aged transgenic mice, no reduction in Aβ levels in brain was found, nevertheless improvements in cognitive deficits were observed [38].

Since the first report on the effect of immunotherapy in animals, several formulations of Aβ have been investigated, for example, genetically engineered filamentous phages displaying $A\beta_{3\text{-}6}$ (EFRH) [47], intranasal Aβ immunization [37, 41], a soluble non-amyloidogenic, nontoxic homologue of Aβ [48], microencapsulated Aβ [49], and recombinant adeno-associated virus Aβ vaccine expressing a fusion protein containing $A\beta_{42}$ and cholera toxin B subunit [50]. Irrespective of the way of administration and the animal species used (mice, rabbits, guinea-pigs), the immunization entailed reductions in cerebral amyloid load and improvements in behavior.

Lemere et al. [51] immunized for the first time a non-human primate, the vervet monkey, with a cocktail of human Aβ peptides ($A\beta_{40,}$ $A\beta_{42}$). This monkey species develops cerebral amyloid plaques with aging, and the amyloid deposits are associated with gliosis and neuritic dystrophy. Immunized animals generated anti-Aβ antibodies that labeled Aβ plaques in human, transgenic mouse, and vervet. Anti-Aβ antibodies bound to $A\beta_{1\text{-}7}$ epitope and recognized monomeric and oligomeric Aβ but not full-length APP or C-terminal fragments of APP. The Aβ levels in the CNS were reduced, whereas they were increased in plasma. This finding confirms that Aβ can be moved from the central to peripheral compartment where the anti-Aβ antibodies bind them, enhancing clearance of Aβ [46]. In an experiment by Lemere et al. [51], immunization did not elicit any side effects. In particular, no Aβ-reactive T-cell populations were detected.

Plaque clearance can be invoked only by antibodies against epitopes located in the N-terminal region of Aβ [52]. It has also been shown that the isotype of the antibody prominently influences the degree of plaque clearance. For example, IgG2a antibodies against Aβ were more efficient that IgG1 or IgG2b antibodies in reducing pathology. Moreover, it was shown that the high affinity of the antibody for Fc receptors on microglial cells seems to be more important than high affinity for Aβ itself and that complement activation is not required for plaque clearance.

It was reported [53] that after intracranial anti-Aβ antibody injections into APP transgenic mice, there is a rapid removal of diffuse amyloid deposits apparently independent of microglial activation and also a later removal of compact amyloid deposits, which appears to require microglial activation. After suppression of microglial activation with dexamethasone, administration of anti-Aβ antibody inhibited the removal of compact, thioflavine-S-positive amyloid deposits [54].

Wilcock et al. [55] using antibody 2286 (mouse monoclonal anti-human $A\beta_{28\text{-}40}$ IgG1) for passive immunization in a transgenic mouse model showed that the antibody is able to enter the brain and bind to the amyloid deposits, likely opsonizing the Aβ and resulting in Fcγ receptor-mediated phagocytosis. This group also showed that passive immunization improved behavioral performance. Such improvement might reflect rapid reduction of the Aβ pool, closely linked to memory impairments yet not easily detected by immunochemistry. A similar phenomenon was previously reported by Dodart et al. [38] and Kotilinek et al. [39]. They observed rapid reversal of memory deficits in transgenic mice after passive immunization without significant reduction in brain Aβ.

The clearance of various types of amyloid plaque depends on the isotype of the administered antibody [56]. It was shown that IgG2a antibodies are efficacious in clearing fibrillar, thio-S-positive plaque. The high efficacy of IgG2a antibodies is consistent with their ability to best stimulate microglial and peripheral macrophage phagocytosis. This finding also supports a crucial role for microglial Fc receptor-mediated phagocytosis in the clearance of at least fibrillar plaques. However, because Fc knockout mice show a reduction of plaque burden after Aβ immunotherapy [57], alternative clearing mechanisms should be taken in consideration.

Mechanisms by which antibodies act are not entirely understood. Suggested mechanisms include (i) microglial-mediated phagocytosis (Fc-dependent, Fc-independent, or combination of Fc-dependent and Fc-independent mechanisms [53–55, 58]), and β1 integrin-dependent [59]; (ii) direct interaction of antibodies with Aβ with subsequent disaggregation of amyloid deposits [8, 53, 55]; and (iii) removal of Aβ from the brain by binding circulating Aβ in plasma with the anti-Aβ antibodies (so-called peripheral sink hypothesis) [38, 46, 60].

All three proposed mechanisms of anti-Aβ antibody-mediated amyloid removal are not mutually exclusive. They are likely to be synergistic if multiple mechanisms are elicited by a single antibody or serum. Other possible mechanisms of amyloid removal would include activation of scavenger receptors [61, 62] or receptors for advanced glycation end products [63].

The effect of immunization on vascular Aβ deposits has recently been addressed [64]. This issue seems to be important in light of a study showing that passive immunization of APP23 transgenic mice, characterized by prominent vascular Aβ deposition, with anti-Aβ IgG1 antibody, resulted in a twofold increase in the rate of hemorrhages [65]. To better understand this potential side effect, Racke et al. [64] characterized the binding properties of several monoclonal anti-Aβ antibodies to deposited Aβ in brain parenchyma and cerebral vessels (CAA; cerebral amyloid angiopathy). They observed an increase in both the incidence and severity of CAA-associated microhemorrhages when PDAPP transgenic mice were treated with N-terminally directed 3D6 antibody, whereas mice treated with central domain antibody 266 were unaffected. In this context, the question arises whether the amyloid angiitis that has been recently reported [66] would augment the risk of such hemorrhages. Taken together, circulating antibodies elicited by active immunization or administered passively cross the blood-brain barrier [67, 68]. Moreover, administration to transgenic animals of monoclonal Aβ antibodies against defined Aβ epitopes reduces plaque burden and improves cognitive deficits to the same degree as active immunization [8].

Assessment of morphological and behavioral changes in animals is a very important issue for comparative purposes and for effectivity and safety measurements of investigated agents. Assessment of behavioral deficits observed in transgenic mice may be particularly difficult, because these deficits are only in part related to amyloid deposition. As histological analyses by Dodart et al. [69] indicate, the behavioral deficits are also related to neuroanatomical alterations secondary to overexpression of the APP transgene and are independent of amyloid deposition.

Gandy and Walker [70] suggest the use of nonhuman primates as adjunctive models for assessing the efficacy and safety of immunotherapeutics for AD. Use of this animal model could contribute to further clarification of potential damage caused by immunization to the cerebral vessels.

15.2.3.2 Clinical Experience: Human Trials of Aβ Vaccination

The finding that active and passive vaccination with Aβ exerts remarkable Aβ-reducing effects in animal models of AD led to clinical trials in which an $A\beta_{42}$ synthetic peptide was administered parenterally with a previously tested adjuvant (QS-21) to patients with mild to moderate AD.

In a long-term phase I clinical trial [71], the safety, tolerability, and immunogenicity of AN1792 (human aggregated Aβ42) and exploratory evidence of efficacy in patients with mild to moderate AD were evaluated. Twenty patients were enrolled into each of four dose groups and randomly assigned to receive intramuscularly AN1792 (50 or 225 μg with QS-21 adjuvant 50 or 100 μg) or QS-21 only (control) in a 4:1 active-control ratio on day 0 and at weeks 4, 12, and 24. Patients were allowed to receive up to four additional injections of polysorbate 80 modified formulation at weeks 36, 48, 60, and 72.

During the period of the first four injections, 23.4% of AN1792-treated patients had a positive anti-AN1792 antibody titer (an anti-AN1792 antibody titre of ≥1:1000). This increased to 58.8% after additional injections with the modified formulation. With regard to efficacy, Disability Assessment for Dementia scores showed less decline among active compared with control patients at week 84 ($p = 0.002$).

No treatment differences were observed in three other efficacy measures. Treatment-related side effects were reported in 19 (23.8%) patients, but no relationship was observed between AN1792 dose and their incidence. One patient developed meningoencephalitis 219 days after discontinuing from the study. Diagnostics of meningoencephalitis was made postmortem, and the cause of death was considered non-treatment related. Another five deaths occurred during the study follow-up, but none was deemed directly related to study treatment.

Although no severe side effects occurred during the course of the phase I trials, phase IIa trials were halted when 18 of 298 patients immunized with AN-1792 presented with symptoms consistent with meningoencephalitis [72]. The symptoms and signs of encephalitis included headache, confusion, and changes on magnetic resonance imaging scans. Of the 18 patients in the phase II study, 12 have returned to their baseline status and six have experienced some type of prolonged neurological deficit. The majority of patients had IgG responses to Aβ, and all patients mounted at least a small IgM response. There was no correlation of the severity of encephalitis with either the level or epitope specificity of the antibody response. Moreover, the vast majority of individuals who mounted the antibody response to Aβ did not develop encephalitis.

A cohort of 30 patients who participated in the phase IIa multicenter trial was followed up after suspension of treatment [73]. The group of patients who generated antibodies against β-amyloid showed a marked and long-lasting increase in serum antibodies against aggregated $A\beta_{42}$ in both IgG and IgM classes.

AD patients who generated antibodies against Aβ performed markedly better on the Mini Mental State Examination (MMSE) 8 months and 1 year after the immunization, as compared with control patients, and they remained unchanged after 1 year, as compared with baseline. Within this period, patients in the control group worsened significantly. Taken together, the patients who generated antibodies exhibited slower rates of cognitive decline 1 year after the last immunization.

The neuropathologic findings in 3 patients who received AN1792/QS21 were reported to date [74–76]. Nicoll et al. [76] found infiltrates of lymphocytes in the leptomeninges that were identified as being composed of T lymphocytes (CD3+ and CD45RO+); the majority were CD4+ and very few were CD8+. B lymphocytes were not present. The large areas of neocortex contained very few Aβ plaques or they were devoid of plaques. In some regions devoid of plaques, Aβ-immunoreactivity was associated with microglia immunoreactive for CD68 and human leukocyte antigen DR. Moreover, in the neocortical areas devoid of plaques, densities of tangles, neuropil threads, and cerebral amyloid angiopathy similar to unimmunized AD patients were found. The plaque-associated dystrophic neurites and astrocyte clusters were not seen. At immunohistochemistry, the plaques were surrounded by IgG and C3 complement. Interestingly, cerebral white matter showed marked reduction in the density of myelinated fibers and extensive infiltration with macrophages that were not immunostained for Aβ.

Neuropathological data reported by Ferrer et al. [74] showed some differences in comparison with the above described case. A focal depletion of diffuse and neuritic plaques was observed, but not of amyloid angiopathy. In the cerebral white matter, there was loss of myelin that was accompanied by moderate microgliosis and astrogliosis. Moreover, multinucleated giant cells filled with dense $A\beta_{42}$ and $A4\beta_{40}$ were seen.

Interestingly, severe small cerebral blood vessel lesion (lipohyalinosis) and multiple cortical hemorrhages, including acute lesions and lesions with macrophages filled with hemosyderin, were found. Focal inflammatory infiltrates were seen in the meninges as well as in the cerebrum and they were composed mostly of CD8+, less often of CD4+, CD3+, CD5+, and, rarely, CD7+ lymphocytes. B lymphocytes and the detected T cytotoxic markers were negative.

Masliah et al. [75] reported the results of neuropathologic examination of the patient without clinical symptoms and signs of meningoencephalitis. They found that vaccination with $A\beta_{42}$ resulted in a considerable reduction of plaque burden and promoted amyloid phagocytosis in the frontal cortex and to a lesser extent in the temporal lobe. Plaque associated neuritic dystrophy in the frontal cortex was undetectable. Neurofibrillary pathology and CAA were unchanged. Only minimal lymphocytic reaction was observed in the leptomeninges and the white matter was unaffected.

In summary, it can be said that the clinical and pathologic data of these two trials support the concept of using immunization in the treatment of AD. However, many questions remain unanswered. First, the responder population needs to be characterized. Indeed, assuming that the anti-Aβ antibodies mediate the reduction in the observed amyloid pathology, only about half of the patients benefit from the treatment. Second, the risk to benefit ratio cannot be determined until an analysis of the phase IIa trial data is completed and the pathogenesis of the side effects is definitively determined. Inflammatory response, demyelination, and intracerebral bleeding would be severe and intolerable side effects of the immunization. Current data indicate that the meningoencephalitis may be due to a T-cell response rather than the anti-Aβ antibodies.

Immunization with the full-length $A\beta_{42}$ peptide, containing both B- and T-cell epitiopes, appears not to be optimual, because it brings about an extensive T-cell activation. The cerebral bleeding is possibly due to cerebral amyloid angiopathy (CAA). The cerebral hemorrhages were reported after passive anti-Aβ immunotherapy in mice [65]. Investigation into the pathogenesis of meningoencephalitis induced by vaccination with amyloid-β peptide should now be possible using a recently constructed appropriate animal model [77].

It cannot be excluded that the differences in safety results obtained in transgenic animals and in clinical trials depend, at least to some extent, on the different adjuvants used in protocols. In the studies in mice, the adjuvants CFA (complete Freund's adjuvant) and IFA (incomplete Freund's adjuvant) were used, whereas in clinical trials the immunogen was formulated in adjuvant QS21, a saponine derivative. Moreover, in clinical trial a detergent (polysorbate-80) was added to aid the manufacturing and stability of the Aβ peptide [13].

15.3 Current Directions in Experimental and Clinical Research

The experimental evidence indicates that the clearance of Aβ from the brain is dependent on anti-Aβ antibody and not on T cell–mediated mechanisms. These mechanisms were probably responsible for side-effects observed in the first clinical trials. It is clear that alternative approaches must be developed that bias the immune response toward a T_h2-phenotype and/or replace the Aβ T-cell epitope with a foreign T-cell epitope.

These goals may be attained through modifications of the Aβ molecule, synthesis of new immunogens, and by choice of suitable adjuvants. The use of humanized monoclonal anti-Aβ antibodies will entirely eliminate a cellular response to Aβ, with comparable effectiveness to active immunization. The development of new delivery systems can also contribute to the improvement of efficacy and safety aspects of immunization. Some of the current approaches are discussed below.

15.3.1 Active Immunization

An immunization procedure was developed for the production of effective anti-aggregating Aβ monoclonal antibodies based on filamentous phages displaying only one epitope, the EFRH epitope, as a specific and nontoxic antigen. Effective autoimmune responses were obtained after phage administration as an antigen in guinea-pigs, in which the amino acids sequence in the Aβ molecule is identical to that in humans. Because of the high antigenicity of the phage, no adjuvant was required to obtain high affinity anti-aggregating IgG antibodies [7].

The development of immunoconjugates seems to be a very promising strategy. The immunoconjugates are typically composed of a fragment of the Aβ peptide derived from either the amino-terminal or central region linked to a carrier protein that provides T-cell help. An epitope vaccine has been engineered composed of the B-cell epitope from the immunodominant region of $Aβ_{42}$, $Aβ_{1-15}$ in tandem with a universal synthetic T-cell epitope, pan HLA DR-binding peptide (PADRE). Immunization of BALB/c mice with the PADRE-$Aβ_{1-15}$ epitope vaccine produced high titers of anti-Aβ antibodies [78].

Seabrook et al. [79] have designed two multiantigen peptides (MAP) composed of either 8 copies of $Aβ_{1-7}$ or 16 copies of $Aβ_{1-15}$ and investigated the immune response in B6D2F1 mice. The MAP were formulated with the adjuvant LT (R192G). As the mice receiving $Aβ_{1-15}$ MAP generated very high anti-Aβ antibody titers of the mainly IgG isotype, it was suggested that this MAP may have potential as an AD vaccine.

Immunization with $Aβ_{40}$ fibrils generated two conformation-specific monoclonal antibodies in BALB/c mice [80]. The monoclonal antibodies WO1 and WO2 bound to the amyloid fibril state of the $Aβ_{40}$ peptide but not to its soluble, monomeric state. This new class of antibodies appears to recognize a common conformational epitope with little apparent dependence on amino acid side-chain conformation. Reduction in brain levels of soluble $Aβ_{42}$ by 57% was detected after immunization with a soluble non-amyloidogenic, nontoxic Aβ homologous peptide in Tg2576 mice. The cortical and hippocampal brain amyloid burden was reduced by 89% and 81%, respectively [48].

Although compelling evidence has been provided that the reduction of plaque burden after immunization is mediated through anti-Aβ antibodies, Frenkel et al. [81] reported that nasal vaccination with a proteasome-based adjuvant (IVX-908) and glatiramer acetate, a synthetic copolymer used in the treatment of multiple sclerosis, clears β-amyloid in a mouse model of AD in an antibody-independent fashion. Vaccinated animals developed activated microglia (CD11b+ cells), and the extent of microglial activation correlated strongly with the decrease in Aβ fibrils. They also found a strong correlation between CD11b+ cells and IFN-γ secreting cells and increased numbers of T cells, which may play a role in promoting microglial activation.

15.3.2 Passive Immunization

Passive immunotherapy has advantages over active immunization from both efficacy and safety perspectives. Particularly, passive immunotherapy using a humanized monoclonal anti-Aβ antibody will entirely eliminate a cellular response to Aβ. The use of polyclonal anti-Aβ antibodies can be considered as a promising alternative. Polyclonal anti-Aβ antibodies can be delivered by healthy individuals because they have circulating autoantibodies against Aβ-peptide.

Bard et al. [52] determinated prerequisites for monoclonal antibodies to prevent neuropathologic lesions in transgenic mice. For this purpose, immune sera with reactivity against different Aβ epitopes and monoclonal antibodies with different isotypes were examined for efficacy *ex vivo* and in vivo. They found that only antibodies against the N-terminal regions of Aβ were able to invoke plaque clearance. Plaque binding correlated with a clearance response, whereas the ability of antibodies to capture soluble Aβ was not necessarily correlated with efficacy. The isotype of the antibody influenced the degree of plaque clearance. High affinity of the antibody for Fc receptors seemed more important that high affinity for Aβ itself.

High-affinity anti-aggregating monoclonal anti-Aβ antibodies were obtained in human APP transgenic mice after a short immunization time with phage-EFRH. A dose-response relationship was observed between antibody-titer and reduced amyloid load. High immunogenicity of the phage enables intranasal administration without use of adjuvant [40].

Rangan et al. [82] have identified recombinant antibody light-chain fragments with proteolytic activity, capable of hydrolyzing Aβ in vitro. Although these fragments currently demonstrate broad substrate specificity, they may prove therapeutically useful if the antibody could be engineered to specifically target pathogenic forms of Aβ, such as oligomers or protofibrils.

By screening a human single-chain antibody (scFv) library for Aβ immunoreactivity, Fukuchi et al. [83] have isolated a battery of scFvs that specifically react with amyloid plaques in the brain. The efficacy of human scFv was tested in a mouse model of AD. It was observed that relative to control mice, injections of the scFv into the brain of transgenic mice reduced Aβ deposits and improved spatial learning in Morris water maze. They concluded that human scFvs against Aβ may be useful to treat AD patients without eliciting brain inflammation because scFvs lack the Fc-portion of the immunoglobulin molecule.

Frenkel et al. [6] suggested a novel approach, where intracellular expression of a site-directed single-chain antibody, which has been shown to inhibit fibrillogenesis and cytotoxicity in vitro, could target Aβ before it is released from the cell.

Reducing the ability of an amyloidogenic protein to form partly unfolded species has been suggested as an effective method of preventing its aggregation [84]. It was shown that a single-domain fragment of a camelid antibody raised against wild-type human lysosyme inhibits the in vitro aggregation of its amyloidogenic variant, D67H. The binding of the antibody achieves its effect by restoring the structural cooperativity characteristic of the wild-type protein. This appeared to occur at least in part through the transmission of long-range conformational effects to the interface between the two structural domains of the protein.

Ultrastructural investigation into structure of human classical plaques in different stages of development showed that in the early plaque, the leading pathology is fibrillar Aβ deposition by microglial cells. In the late plaques, microglial cells retract and activation of astrocytes predominate [85]. In line with these findings, Wyss-Coray et al. [86] found that adult mouse astrocytes degrade amyloid-β in vitro and *in situ*. Furthermore, it was demonstrated [87] that a modest increase in astroglial production of transforming growth factor β1 (TGF-β1) in aged transgenic mice expressing the human APP (hAPP) results in a threefold reduction in the number of parenchymal amyloid plaques, a 50% reduction in the overall Aβ load in the hippocampus and neocortex, and a decrease in the number of dystrophic neurites. In mice expressing hAPP and TGF-β1, the reduction of parenchymal plaques was associated with a strong activation of microglia and an increase in inflammatory mediators. Taken together, the stimulation of astrocytes and/or microglia could be considered an alternative approach for the treatment of AD. However, it was found [88] that over-activation of microglia induces apoptosis. Interestingly, in the experiment reported by Weiner et al. [37], the lowering of Aβ burden was associated with decreased local microglial and astrocytic activation after nasal administration of $A\beta_{40}$ to PDAPP mice. In serum, anti-Aβ antibodies of the IgG1 and Ig2b classes were detected, both of which are characteristic of the T_h2-type immune response.

It is possible to generate anti-Aβ antibodies that are capable of exerting their selective effect on Aβ fibrils. In the study by McLaurin et al. [43], the TgCRND8 mice were vaccinated with protofibrillar/oligomeric assemblies of $A\beta_{42}$ that reduced cerebral Aβ deposits and cognitive impairments and induced immunoglobulins of IgG2b isotype against residues 4–10 of Aβ. The generated anti-Aβ antibodies were able to inhibit Aβ fibrils assembly and toxicity without activating microglial or other cellular inflammatory responses. In the light of the above-mentioned results, both stimulation and inhibition of either microglia or astrocytes might be of therapeutic relevance in dependence, among others, of the stage in classical plaque development. Schmechel et al. [89] suggest that monoclonal antibody recognizing $A\beta_{42}$ homodimers, which are potentially the earliest form of synaptotoxic Aβ oligomers, might be useful for Aβ amyloid related therapeutic approaches by impeding its precipitation into existing plaques. A multi-antibody based approach, with one antibody targeted against Aβ and one against tau, was suggested by Oddo et al. [21].

Specific polyclonal anti-Aβ-IgG in both the serum and the CNS from non-immunized humans were identified [27, 29]. The distribution of the different IgG subclasses in the Aβ antibody sample were as follows: IgG1, 63.8%; IgG2, 19.9%; IgG3, 9%; and IgG4, 7.3%. These antibodies were able to block fibril formation, disrupt formation of fibrillar structures, and prevent neurotoxicity of Aβ in vitro [90]. In another experiment [52], purified anti-Aβ antibodies could disaggregate both preformed $A\beta_{40}$ as well as active truncated $A\beta_{25\text{-}35}$ and also block neurotoxicity induced by both peptides. These results indicate that the investigated antibody fractions include antibodies not only against the N-terminal of Aβ but also against the middle portion of Aβ.

In a pilot study [91], IgG were administered intravenously (IVIgG) in patients with AD. Five patients with AD were enrolled and received monthly IVIgG (0.4 g intravenous IG per kg body weight) over a 6-month period. After IVIgG, total Aβ levels in the CSF decreased by 30.1% compared with baseline. Total Aβ increased in the serum by 233%. No effect on $A\beta_{42}$ levels was observed. In addition, stabilization or a mild improvement in cognitive function was observed in the patients as detected using ADAS-cog. (improvement of 3.7 ± 2.9 points). It was postulated that the effects of IVIg in the AD patients were due to altered cytokine production by microglial cells. However, the patient population included in this study was too small to make definite conclusions regarding the efficacy of IVIg in AD. From the safety point of view, it is important that polyclonal antibodies do not bind complement. Taken together, the available data indicate that administration of polyclonal human anti-Aβ antibodies isolated from plasma might be a potential therapeutic agent in AD.

15.3.3 Gentechnologic Approaches

It could be expected that efficacy and safety issues associated with immunotherapy for AD could be improved using DNA vaccines or viral vectors [92, 93]. Among the most important goals of the work being done in the field are (i) the limitation of extension of amyloid accumulation through generation of high titers of epitope-specific anti-Aβ antibodies with favorable isotype-profiles; (ii) reduction of side effects related to T_h1-responses; (iii) induction of T_h2-based immune response; and (iv) breaking of self-tolerance to Aβ. Some of these goals have already been achieved in animals. For example, Qu et al. [94] have demonstrated that gene-gun–mediated genetic immunization with $A\beta_{42}$ gene can efficiently elicit humoral immune responses against mouse $A\beta_{42}$ peptide in wild-type BALB/c mice as well as against human $A\beta_{42}$ in transgenic mice. It was shown that induction of the humoral immune response did not induce a significant cellular immune response. A study is underway to detect whether this novel immunization approach leads to reduction of Aβ burden in the brains of mice.

Dodart et al. [95] investigated whether gene delivery of the three common human apoE isoforms can directly alter the brain Aβ pathology in PADPP transgenic mice. They demonstrated that intracerebral gene delivery of the lentivirus encoding apoE-constructs resulted in efficient and sustained expression of human apoE in the hippocampus as well as in a significant isoform-dependent effect of human apoE on hippocampal Aβ burden and amyloid formation. This experimental data suggests that gene delivery of human apoE2 may prevent and/or reduce brain Aβ burden and the subsequent formation of neuritic plaques. It is possible that the use of gene technology could enable the construction of new transgenic animals models suitable for further investigating the efficacy and safety of immunotherapy [96].

15.3.4 Role of Adjuvant

The choice of appropriate adjuvant can strengthen the antibody response to $A\beta_{42}$ and shift the type of the immune response generated (T_h1 vs. T_h2). To investigate the role of adjuvant in the humoral and cell-mediated immune response to

$A\beta_{42}$, immunization with $A\beta_{42}$ formulated in four different adjuvants, complete Freund's adjuvant (CFA), incomplete Freund's adjuvant (IFA), saponine QS21, alum, and TitreMax Gold (TMG), was performed in BALB/c mice [25]. All adjuvants induced a strong anti-$A\beta_{42}$ antibody response after the first boost, and the antibody titers increased considerably after the second and third boosts with fibrillar $A\beta_{42}$. A significant difference in the magnitude of the antibody response to $A\beta_{42}$ immunization with the different adjuvants was observed. The highest titers of antibody were generated in mice immunized and boosted with $A\beta_{42}$ formulated in QS21 followed by CFA/IFA > alum > TMG.

To provide a relative measure of the contribution of T_h2- and T_h1-type humoral responses, the ratios of IgG1 to IgG2a antibody generated in response to Aβ immunization were examined. All mice immunized with $A\beta_{42}$ formulated in alum had IgG1:IgG2a ratios >1, indicating that this adjuvant induced primarily T_h2-type antibody response against $A\beta_{42}$. On the other hand, CFA, TMG, and QS21 shifted the humoral immune responses toward a T_h1 phenotype. Promising results in terms of antibody generation and their isotypes were obtained in B6D2F1 mice after immunization with Aβ formulated in adjuvants monophosphoryl lipid A (MPL)/trehalose dicorynomycolate (TDM), cholera toxin B subunit (CTB), and LT (R192G) [97].

15.4 Other Suggested Treatment Approaches Targeting Aβ

Amyloid binding ligands (ABL) has been suggested as an alternative, non-immunological therapeutic strategy to delay the onset or slow the progression of AD [98]. The ABL represent derivatives of known amyloid-binding molecules such as Congo red, chrysamine G (CG), and thioflavin S (TS). The generated derivatives of CG and TS specifically recognize fibrillar Aβ in vitro, arrest the formation of Aβ fibrils, and contrary to the parent substances, they cross the blood-brain barrier of transgenic mice after intravenous administration. It was demonstrated that CG derivative IMSB binds to amyloid plaques composed of $A\beta_{40}$ with much higher affinity than $A\beta_{42}$, whereas TS derivative TDZM shows the opposite affinity. Furthermore, IMSB but not TDZM bound selectively to neurofibrillary tangles.

As the microglia activated by Aβ exert their toxic effects through NMDA receptors in vitro, the blocking of these receptors may be an effective therapeutic approach [99]. It is possible that small, bifunctional molecules that reveal antifibrillogenic properties may be of relevance in vivo [100]. Zinc-copper chelation resulting in the solubilization of Aβ offers promise as a new therapeutic approach for AD [101, 102]. Curcumin, the unconventional NSAID/antioxidant, has multiple anti-amyloid actions. Curcumin, targeting directly Aβ, may act as a "peripheral sink" [103].

15.5 Conclusions

Although transgenic animals are not the most favorable models of AD in terms of morphologic and immunologic aspects, compelling evidence exists that immunotherapy can prevent or reduce neuropathology and improve cognitive performance. The preventative effects of immunization are mediated by anti-Aβ antibodies, with titer, isotype, and epitope specificity playing crucial roles in their effects. Experimentally, the anti-Aβ antibodies reduced or prevented plaque formation, acted against aggregation and neurotoxicity, favored disaggregation, and promoted recovery of neuronal damage. Compelling experimental evidence also indicates that Aβ immunization may be useful for clearing aggregates of tau protein, another hallmark lesion of AD neuropathology, on condition that the treatment occurs early in the disease progression. Clinically, the primary concern is the safety of immunotherapy, especially the cause of side effects, including subacute meningoencephalitis, microhemorrhages, and demyelination. With regard to efficacy, slowing down of cognitive deficits after suspension of vaccine administration in a cohort study was observed. Modifications of Aβ-antigen, synthesis of new immunogens, generation of epitope-specific monoclonal antibodies, development of new adjuvants and delivery systems may contribute to future favorable efficacy and safety profiles of immunotherapy. In this respect, gentechnology seems to be a particularly promising approach.

References

1. Dobson CM. Protein folding and misfolding. Nature 2003, 426, 884-890.
2. Holtzman DM. Aβ conformational change is central to Alzheimer's disease. Neurobiol Aging 2002, 23, 1085-1088.
3. Selkoe DJ. Folding proteins in fatal ways. Nature 2003, 426, 900-904.
4. Soto C. Unfolding the role of protein misfolding in neurodegenerative diseases. Nat Rev Neurosci 2003, 4, 49-60.
5. Frenkel D, Balass M, Solomon B. N-terminal EFRH sequence of Alzheimer's β-amyloid peptide represents the epitope of its anti-aggregating antibodies. J Neuroimmunol 1998, 88, 85-90.
6. Frenkel D, Solomon B, Benhar I. Modulation of Alzheimer's beta-amyloid neurotoxicity by site-directed single-chain antibody. J Neuroimmunol 2000, 106, 21-31.
7. Solomon B. Generation of anti-β-amyloid antibodies via phage display technology towards Alzheimer's disease vaccination. Vaccine 2005, 23, 2327-2330.
8. Solomon B. Immunotherapeutic strategies for prevention and treatment of Alzheimer's disease. DNA Cell Biol 2001, 20, 697-703.
9. Schenk D, Barbour W, Dunn W, et al. Immunization with amyloid-β attenuates Alzheimer's-disease-like pathology in the PDAPP mouse. Nature 1999, 400, 173-177.
10. Schenk D. Amyloid-β immunotherapy for Alzheimer's disease: The end of the beginning. Nat Rev Neurosci 2002, 3, 824-828.
11. Janus C, Pearson J, McLaurin J, et al. Aβ peptide immunization reduces behavioural impairment and plaques in a model of Alzheimer's disease. Nature 2000, 408, 979-982.
12. Morgan D, Diamond DM, Gottschall PE, et al. Aβ - peptide vaccination prevents memory loss in an animal model of Alzheimer's disease. Nature 2000, 408, 982-985.
13. Schenk D, Hagen M, Seubert P. Current progress in beta-amyloid immunotherapy. Curr Opin Immunol 2004, 16, 599-606.
14. Cirrito JR, Holtzman DM. Amyloid-β and Alzheimer's disease therapeutics: the devil may be in the details. J Clin Invest 2003, 112, 321-323.
15. Gelinas DS, DaSilva K, Fenili D, et al. Immunotherapy for Alzheimer's disease. Proc Natl Acad Sci U S A 2004, 101 (suppl 2), 14657-14662.
16. Karkos J. Immuntherapie der Alzheimer-Krankheit. Experimentelle Untersuchungsergebnisse und Behandlungsperspektiven. Fortschr Neurol Psychiat 2004, 72, 204-219.
17. Selkoe DJ, Schenk D. Alzheimer's disease: molecular understanding predicts amyloid-based therapeutics. Annu Rev Pharmacol Toxicol 2002, 43, 545-584.
18. Gandy S. Molecular basis for anti-amyloid therapy in the prevention and treatment of Alzheimer's disease. Neurobiol Aging 2002, 23, 1009-1016.
19. Selkoe DJ. Alzheimer's disease results from the cerebral accumulation and cytotoxicity of amyloid β-protein. J Alzheimer's Dis 2001, 3, 75-81.
20. Hutton M, McGowan E. Clearing tau pathology with Aβ immunotherapy-reversible and irreversible stages revealed. Neuron 2004, 43, 293-294.
21. Oddo S, Billings L, Kesslak JP, et al. Aβ immunotherapy leads to clearance of early, but not late, hyperphosphorylated tau aggregates via proteasome. Neuron 2004, 43, 321-332.
22. Barrow CJ, Zagorski MG. Solution structures of β-peptide and its constituent fragments: relation to amyloid deposition. Science 1991, 253, 179-182.
23. Hollosi M, Otvos L, Kaijtar J, et al. Is amyloid deposition in Alzheimer's disease preceded by an environment induced double conformational transition? Peptide Res 1989, 2, 109-113.
24. Maggio JE, Mantyh PW. Brain amyloid – a physicochemical perspective. Brain Pathol 1996, 6, 147-162.
25. Cribbs DH, Ghochikyan A, Vasilevko V, et al. Adjuvant-dependent modulation of T_h1 and T_h2 responses to immunization with β-amyloid. Int Immunol 2003, 15, 505-514.
26. Hyman BT, Smith C, Buldyrev I, et al. Autoantibodies to amyloid-β and Alzheimer's disease. Ann Neurol 2001, 49, 808-810.
27. Du Y, Dodel R, Hampel H, et al. Eastwood B. Reduced levels of amyloid beta-peptide antibody in Alzheimer's disease. Neurology 2001, 57, 801-805.
28. Weksler ME, Relkin N, Turkenich R, et al. Patients with Alzheimer's disease have lower levels of serum anti-amyloid peptide antibodies than healthy elderly individuals. Exp Gerontol 2002, 37, 943-948.
29. Dodel R, Hampel H, Depboylu C, et al. Human antibodies against amyloid β peptide: a potential treatment for Alzheimer's disease. Ann Neurol 2002, 52, 253-256.
30. McGeer EG, McGeer PL. Inflammatory process in Alzheimer's disease. Progr Neuropsychopharmacol Biol Psychiatry 2003, 27, 741-749.
31. Monsonego A, Weiner HL. Immunotherapeutic approaches to Alzheimer's disease. Science 2003, 302, 834-838.
32. Monsonego A, Zota V, Karni A, et al. Increased T cell reactivity to amyloid β protein in older humans and patients with Alzheimer's disease. J Clin Invest 2003, 112, 415-422.

33. Town T, Tan J, Sansone N, et al. Characterization of murine immunoglobulin G antibodies against human amyloid-β1-42. Neurosci Lett 2001, 307, 101-104.
34. Stalder M, Deller T, Staufenbiel M, Jucker M. 3D-Reconstruction of microglia and amyloid in APP23 transgenic mice: no evidence of intracellular amyloid. Neurobiol Aging 2001, 22, 427-434.
35. Webster SD, Yabg AJ, Margol L, et al. Complement component C1q modulates the phagocytosis of Abeta by microglia. Exp Neurol 2000, 161, 127-138.
36. Bard F, Cannon C, Barbour R, et al. Peripherally administered antibodies against amyloid β-peptide enter the central nervous system and reduce pathology in a mouse model of Alzheimer's disease. Nat Med 2000, 6, 916-919.
37. Weiner HL, Lemere CA, Maron R, et al. Nasal administration of amyloid-β peptide decreases cerebral amyloid burden in a mouse model of Alzheimer's disease. Ann Neurol 2000, 48, 567-579.
38. Dodart J-C, Bales KR, Gannon KS, et al. Immunization reverses memory deficits without reducing brain Abeta burden in Alzheimer's disease model. Nat Neurosci 2002, 5, 452-457.
39. Kotilinek LA, Bacskai B, Westerman M, et al. Reversible memory loss in a mouse transgenic model of Alzheimer's disease. J Neurosci 2002, 22, 6331-6335.
40. Frenkel D, Dewachter F, Van Leuven F, Solomon B. Reduction of beta-amyloid plaques in brain of transgenic mouse model of Alzheimer's disease by EFRH-phage immunization. Vaccine 2003, 21, 1060-1065.
41. Lemere CA, Maron R, Spooner ET, et al. Nasal Aβ treatment induces anti-Aβ antibody production and decreases cerebral amyloid burden in PD-APP mice. Ann N Y Acad Sci 2000, 920, 328-333.
42. Lemere CA, Spooner E, Leverone J, Mori C, Clements J. Intranasal immunotherapy for the treatment of Alzheimer's disease: Escherichia coli LT and LT (R192G) as mucosal adjuvants. Neurobiol Aging 2002, 23, 991-1000.
43. McLaurin J, Cecal R, Kierstead ME, et al. Therapeutically effective antibodies against amyloid-β peptide target amyloid-β residues 4-10 and inhibit cytotoxicity and fibrillogenesis. Nat Med 2002, 8, 1263-1269.
44. Spooner E, Desai R, Mori C, et al. The generation and characterization of potentially therapeutic Aβ antibodies in mice: differences according to strain and immunization protocol. Vaccine 2000, 21, 290-297.
45. Das P, Murphy M, Younkin L, et al. Reduced effectiveness of Aβ1-42 immunization in APP transgenic mice with significant amyloid deposition. Neurobiol Aging 2001, 22, 721-727.
46. DeMattos RB, Bales KR, Cummins DJ, et al. Peripheral anti-Aβ antibody alters CNS and plasma clearance and decreases brain Aβ burden in a mouse model of Alzheimer's disease. Proc Natl Acad Sci U S A 2001, 98, 8850-8855.
47. Frenkel D, Katz O, Solomon B. Immunization against Alzheimer's β-amyloid plaques via EFRH phage administration. Proc Natl Acad Sci U S A 2000, 97, 11455-11459.
48. Sigurdsson EM, Scholtzova H, Mehta PD, et al. Immunization with non-toxic/nonfibrillar amyloid-β homologous peptide reduces Alzheimer's disease-associated pathology in transgenic mice. Am J Pathol 2001, 159, 439-447.
49. Brayden D, Templeton L, McClean S, et al. Encapsulation in biodegradable microparticles enhances serum antibody response to parenterally delivered β-amyloid in mice. Vaccine 2001, 19, 4185-4193.
50. Zhang J, Wu S, Qin C, et al. A novel recombinant adeno-associated virus vaccine reduces behavioural impairment and β-amyloid plaques in a mouse model of Alzheimer's disease. Neurobiol Dis 2003, 14, 365-379.
51. Lemere CA, Beierschmitt A, Iglesias M, et al. Alzheimer's disease Aβ vaccine reduces central nervous system Aβ levels in a non-human primate, the Caribbean vervet. Am J Pathol 2004, 165, 283-297.
52. Bard F, Barbour R, Cannon C, et al. Epitope and isotype specificities of antibodies to beta-amyloid peptide for protection against Alzheimer's disease-like neuropathology. Proc Natl Acad Sci U S A 2003, 100, 2023-2028.
53. Wilcock DM, DiCarlo G, Henderson D, et al. Intracranially administered anti-Aβ antibodies reduce β-amyloid deposition by mechanisms both independent of and associated with microglial activation. J Neurosci 2003, 213, 3745-3751.
54. Wilcock DM, Munireddy SK, Rosenthal A, et al. Microglial activation facilitates Aβ plaque removal following intracranial anti-Aβ administration. Neurobiol Dis 2004, 15, 11-20.
55. Wilcock DM, Rojiani A, Rosenthal A, et al. Passive amyloid immunotherapy clears amyloid and transiently activates microglia in a transgenic mouse model of amyloid deposition. J Neurosci 2004, 24, 6144-6151.
56. Bussière T, Bard F, Barbour R, et al. Morphological characterization of thioflavin-S-positive amyloid plaques in transgenic Alzheimer mice and effect of passive Aβ immunotherapy on their clearance. Am J Pathol 2004, 165, 987-995.
57. Das P, Howard V, Loosbrock N, et al. Amyloid-β immunization effectively reduces amyloid deposition

in FcR $\gamma^{-/-}$ knock-out mice. J Neurosci 2003, 23, 8532-8538.
58. Bacskai BJ, Kajdasz S, McLellan ME, et al. Non-Fc-mediated mechanisms are involved in clearance of amyloid-β in vivo by immunotherapy. J Neurosci 2002, 22, 7873-7878.
59. Koenigsknecht J, Landreth G. Microglial phagocytosis of fibrillar β-amyloid through a^{β}_{1} integrin-dependent mechanism. J Neurosci 2004, 24, 9838-9846.
60. Lemere CA, Spooner E, LaFrancois J, et al. Evidence for peripheral clearance of cerebral Aβ protein following chronic, active Aβ immunization in PSAPP mice. Neurobiol Dis 2003, 14, 10-18.
61. Huang F, Buttini M, Wyss-Coray T, et al. Elimination of the class A scavenger receptor does not affect amyloid plaque formation or neurodegeneration in transgenic mice expressing human amyloid protein precursors. Am J Pathol 1999, 155, 1741-1747.
62. Husemann J, Loike JD, Anankov R, et al. Scavenger receptors in neurobiology and neuropathology: their role on microglia and other cells of the nervous system. Glia 2002, 40, 195-205.
63. Münch G, Thome J, Foley P, et al. Advanced glycation end products in ageing and Alzheimer's disease. Brain Res Rev 1997, 23, 134-143.
64. Racke MM, Boone LI, Hepburn DL, et al. Exacerbation of cerebral amyloid angiopathy-associated microhemorrhage in amyloid precursor protein transgenic mice by immunotherapy is dependent on antibody recognition of deposited forms of amyloid β. J Neurosci 2005, 25, 629-636.
65. Pfeifer M, Boncristiano S, Bondolfi L, et al. Cerebral hemorrhage after passive anti-Aβ immunotherapy. Science 2002, 298, 1379.
66. Scolding NJ, Joseph F, Kirby PA, et al. Aβ-related angiitis: primary angiitis of the central nervous system associated with cerebral amyloid angiopathy. Brain 2005, 128, 500-515.
67. Banks W.A., Terell B., Farr S.A., et al. Passage of amyloid β protein antibody across the blood-brain barrier in a mouse model of Alzheimer's disease. Peptides 2002, 23, 2223-2226.
68. Hock C, Konietzko U, Papassotiropoulos A, et al. Generation of antibodies specific for β-amyloid by vaccination of patients with Alzheimer's disease. Nat Med 2002, 8, 1270-1275.
69. Dodart J-C, Mathis C, Saura J, et al. Neuroanatomical abnormalities in behaviourally characterized APP^{V717F} transgenic mice. Neurobiol Dis 2000, 7, 71-85.
70. Gandy S, Walker L. Toward modelling hemorrhagic and encephalitic complications of Alzheimer amyloid-β vaccination in nonhuman primates. Curr Opin Immunol 2004, 16, 607-615.
71. Bayer AJ, Bullock R, Jones RW, et al. Evaluation of the safety and immunogenicity of synthetic A[beta]42 (AN1792) in patients with AD. Neurology 2005, 64, 94-101.
72. Orgogozo J-M, Gilman S, Dartigues J-F, et al. Subacute meningoencephalitis in a subset of patients with AD after A [beta] 42 immunization. Neurology 2003, 61, 46-54.
73. Hock C, Konietzko U, Streffer JR, et al. Antibodies against β-amyloid slow cognitive decline in Alzheimer's disease. Neuron 2003, 38, 547-554.
74. Ferrer I, Boada Rovira M, et al. Neuropathology and pathogenesis of encephalitis following amyloid-β immunization in Alzheimer's disease. Brain Pathol 2004, 14, 11-20.
75. Masliah E, Hansen L, Adame A, et al. A [beta] vaccination effects on plaque pathology in the absence of encephalitis in Alzheimer's disease. Neurology 2005, 64, 129-131.
76. Nicoll JAR, Wilkinson D, Holmes C, et al. Neuropathology of human Alzheimer's disease after immunization with amyloid-β peptide: a case report. Nat Med 2003, 9, 448-452.
77. Furlan R, Brambilla E, Sanvito F, et al. Vaccination with amyloid-β peptide induces autoimmune encephalomyelitis in C57/BL6 mice. Brain 2003, 126, 285-291.
78. Agadjanyan MG, Ghochikyan A, Petrushina I, et al. Prototype Alzheimer's disease vaccine using the immunodominant B cell epitope from β-amyloid and promiscuous T cell epitope Pan HLA DR-binding peptide. J Immunol 2005, 174, 1580-1586.
79. Seabroook TJ, Bloom JK, Spooner ET, Lemere CA. The use of Aβ1-15 multi-antigen peptide as a potential vaccine for Alzheimer's disease. Program No. 830.1. *2004 Abstract Viewer/Itinerary Planner*. Washington, DC: Society for Neuroscience.
80. O'Nuallain B, Wetzel R. Conformational Abs recognizing a generic amyloid fibril epitope. Proc Natl Acad Sci U S A 2002, 99, 1485-1490.
81. Frenkel D, Maron R, Burt DS, Weiner HL. Nasal vaccination with a proteasome-based adjuvant and glatiramer acetate clears β-amyloid in a mouse model of Alzheimer's disease in an antibody-independent fashion. Program No. 674.4. *2004 Abstract Viewer/itinerary Planner*. Washington, DC: Society for Neuroscience.
82. Rangan SK, Ruitian L, Brune D, et al. Degradation of β-amyloid by proteolytic antibody light chains. Biochemistry 2003, 42, 14328-14334.
83. Fukuchi K, Accaviti–Loper M., Kim H.D, et al. Human single chain antibodies for treatment of Alzheimer's disease. Program No. 675.13. *2004 Abstract Viewer/Itinerary Planner*. Washington, DC: Society for Neuroscience.

84. Dumoulin M., Last AM., Desmyther A., et al. A camelid antibody fragment inhibits the formation of amyloid fibrils by human lysozyme. Nature 2003, 424, 783-788.
85. Wegiel J, Wang K-C, Tarnawski M, Lach B. Microglial cells are the driving force in fibrillar plaque formation, whereas astrocytes are a leading factor in plaque degradation. Acta Neuropathol 2000, 100, 356-364.
86. Wyss-Coray T, Loike JD, Brionne TC, et al. Adult mouse astrocytes degrade amyloid-β in vitro and in situ. Nat Med 2003, 9, 453-458.
87. Wyss-Coray T, Lin C, Yan F, et al. TGF-β1 promotes microglial amyloid-β clearance and reduces plaque burden in transgenic mice. Nat Med 2001, 7, 612-618.
88. Liu B, Wang K, Gao HM, et al. Molecular consequences of activated microglia in the brain: overactivation induces apoptosis. J Neurochem 2001, 77, 182-189.
89. Schmechel A, Zentgraf H, Scheuermann S, et al. Alzheimer β-amyloid homodimers facilitate Aβ fibrillization and the generation of conformational antibodies. J Biol Chem 2003, 278, 35317-35324.
90. Du Y, Wei X, Dodel R, et al. Human anti-β-amyloid antibodies block β-amyloid fibril formation and prevent β-amyloid-induced neurotoxicity. Brain 2003, 126, 1935-1939.
91. Dodel RC, Du Y, Depboylu C, et al. Intravenous immunoglobulins containing antibodies against β-amyloid for the treatment of Alzheimer's disease. J Neurol Neurosurg Psychiatry 2004, 75, 1472-1474.
92. Ghochikyan A, Vasilevko V, Petrushina I, et al. Generation and characterization of the humoral immune response to DNA immunization with a chimeric beta-amyloid-interleukin-4 minigene. Eur J Immunol 2003, 33, 3232-3241.
93. Tsuji S. DNA vaccination may open up a new avenue for treatment of Alzheimer's disease. Arch Neurol 2004, 61, 1832-1832.
94. Qu B, Rosenberg RN, Li L, et al. Gene vaccination to bias the immune response to amyloid-β peptide as therapy for Alzheimer's disease. Arch Neurol 2004, 61, 1859-1864.
95. Dodart J-C, Marr RA, Koistinaho M, et al. Gene delivery of human apolipoprotein E alters brain Aβ burden in a mouse model of Alzheimer's disease. Proc Natl Acad Sci U S A 2005, 102, 1211-1216.
96. Chan AW, Chong KY, Martinovich C, et al. Transgenic monkeys produced by retroviral gene transfer into mature oocytes. Science 2001, 291, 309-312.
97. Maier M, Seabrook TJ, Bloom JK, Lemere CA. Immune responses to alternative adjuvants and Aβ immunogens for Aβ immunotherapy. Program No. 830.2. *2004 Abstract Viewer/Itinerary Planner*. Washington, DC: Society for Neuroscience.
98. Lee V M-Y. Amyloid binding ligands as Alzheimer's disease therapies. Neurobiol Aging 2002, 23, 1039-1042.
99. Takeuchi H, Mizuno T, Zhang G., et al. Neuritic beading induced by activated microglia is an early feature on neuronal dysfunction toward neuronal death by inhibition of mitochondrial respiration and axonal transport. J Biol Chem 2005, 280, 10444-10454.
100. Gestwicki JE, Crabtree GR, Graef IA. Harnessing chaperones to generate small-molecule inhibitors of amyloid β aggregation. Science 2004, 306, 865-869.
101. Bush AI. The metallobiology of Alzheimer's disease. Trends Neurosci 2003, 26, 207-214.
102. Rosenberg RN. Metal chelation therapy for Alzheimer's Disease. Arch Neurol 2003, 60, 1678-1679.
103. Cole GM, Morihara T, Lim GP, et al. NSAID and antioxidant prevention of Alzheimer's disease. Lessons from in vitro and animals models. Ann N Y Acad Sci 2004, 1035, 68-84.

16
Mouse Models of Alzheimer's Disease

Dwight C. German

16.1 Introduction

Transgenic mouse models have been created that mimic many of the neuropathologic and behavioral phenotypes of Alzheimer's disease (AD). Using mutations found in familial AD, the mouse models exhibit some of the cardinal features of the human disease. Wong et al. [1] and Higgins and Jacobsen [2] have written reviews of this topic. The current review extends a previous one [3] and will describe the similarities in the neuropathology of AD and the mouse models of the disease, specifically regarding neurodegeneration, and also describe treatments being developed using the mouse models.

16.2 Neuropathology of Alzheimer's Disease

AD is characterized by extensive cortical and hippocampal neuropathology [4], including extracellular neuritic plaques composed of β-amyloid (Aβ) protein. There are also neurofibrillary tangles (NFTs), which accumulate within neurons in cortical and subcortical regions. In addition, several subcortical nuclei degenerate in AD, and many of the affected nuclei have been shown to project to the cerebral cortex. The first subcortical nucleus found to degenerate in AD was the nucleus basalis of Meynert [5–8], which contains cholinergic neurons that project to cortical and hippocampal regions [9]. Further study indicated that there is also degeneration of other cortical-projecting subcortical nuclei: for example, the serotonergic dorsal raphe nucleus [10], the dopaminergic ventral tegmental area [11], and the noradrenergic locus coeruleus [12–14].

AD is also characterized by inflammation; microglia are located near neuritic plaques and undergo a phenotypic activation [15]. Microglial activation results in the expression of a wide range of proinflammatory molecules that may actively damage/destroy neurons. Astrocytes are also activated in AD.

Neurogenesis is abnormal in AD. Adult neurogenesis occurs in brain structures that have a high degree of neuronal plasticity, such as the hippocampus and olfactory bulb [16–20]. In the adult rat hippocampus, it is estimated that more than 9000 new neurons are born each day [21]. Although the number of newly born neurons is thought to be much lower in human and nonhuman primates [22–24], the presence of adult neurogenesis in a wide range of species suggests a role for new neurons in shaping the form and function of the adult brain [25]. Adult neurogenesis is regulated by myriad environmental and physiological stimuli [26, 27]. In vivo, chronic stress, aging, inflammation, and repeated exposure to drugs of abuse decrease adult hippocampal neurogenesis [28–31].

Neurogenesis takes place in the hippocampus of the adult primate brain [17, 22, 24]. The first report of neurogenesis in AD postmortem brain indicates that it is abnormal [32] and is also abnormal in other neurodegenerative diseases like Parkinson disease [33] and Huntington disease [34].

16.3 Mouse Models of Alzheimer's Disease

Several transgenic mouse models of AD have been developed. Although the various models exhibit some of the neuropathologic features of the human disease, so far none exhibits all of the features. Table 16.1 summarizes the gene mutations used to create nine AD mouse models and how the neuropathology in the mouse models compares with that in AD. The early mouse models of AD contained mutant genes such as APP_{717} (PDAPP mouse, [35]) and APP_{695} (Tg2576 mouse, [36]). Additional bigenic models have been developed that contain mutant APP and PS1 [37, 38], mutant APP and tau [39], and a triple transgenic mouse that carries mutant amyloid precursor protein (APP), presenilin-1 (PS1), and tau [40]. Mice lacking PS1 and PS2 function also exhibit some AD neuropathology [41].

16.3.1 Amyloid-β Plaques

All APP and PS1 mouse models exhibit diffuse and/or neuritic Aβ-plaques in the cortex and hippocampus (Table 16.1), as illustrated in Figure 16.1. Two APP mouse models have been shown to exhibit an age-related development of neuritic plaques in the cerebral cortex and hippocampus [35, 36].

One of the earliest AD mouse models was developed in 1995 by Games et al. [35]: the PDAPP mouse. This mouse was generated using the platelet-derived growth factor-β promoter driving a human APP minigene encoding the $APP_{717V\text{-}F}$ mutation associated with familial AD [42]. Between 6 and 9 months of age, hemizygous PDAPP mice exhibit thioflavin-S-positive Aβ deposits and neuritic plaques. The Aβ-containing plaques are directly associated with reactive gliosis and dystrophic neurites, suggesting that the plaques may induce neurodegenerative changes. Some of the Aβ plaque pathology in the dentate gyrus appears to originate from nerve terminals whose axons traverse the perforant pathway, as lesions of this pathway in mouse models of AD result in a reduction in hippocampal plaque pathology [43, 44].

Protofibrils are precursors to the formation of fibrilar neuritic plaques, and evidence suggests that they play a role in the neurodegenerative process. Protofibrils are short assemblies, 5–200 nm in length, that assemble into Aβ plaques. The protofibrils have been shown to be neurotoxic [45, 46]. The Aβ oligomers, but not monomers, inhibit hippocampal long-term potentiation in the rat [47, 48]. The homozygous PDAPP mouse contains very high levels of soluble Aβ in both CSF and plasma [49]. That there are region-specific amounts of the oligomers in APP mouse models is suggested by the regional differences in splice variants of β-secretase enzyme, which may explain why Aβ-extracellular plaques are formed only in certain brain regions in AD and in AD mouse models [50].

16.3.2 Neurofibrillary Tangles

In some of the AD mouse models that express APP and/or PS mutations, there is an age-related hyperphosphorylation of tau protein, which comes after the formation of Aβ-plaques [51–55]. However, none of these models exhibit NFTs as defined by the presence of paired helical filaments (PHF) (Table 16.1). Kurt et al. [54] found evidence of PHF-like structures in the 24-month-old APP/PS1 mouse but not in younger animals, however, whether they represent PHF or Hirano bodies is not clear. In hemizygous PDAPP animals up to 20 months of age, no PHFs were observed [35, 51]. Even in transgenic mice that express mutant APP, PS1, and tau [40], and in those expressing APP and tau [39], the NFTs that occur within neurons in the neocortex and hippocampus are defined solely by immunostaining with phospho-specific tau antibodies and not by the presence of PHF. In a study using conditional knock-out of PS1 in PS2 KO mice (PS cDKO mice), there is hyperphosphorylation of tau in the cortex of 9-month-old mice and marked cortical shrinkage [41]. These studies indicate that mouse models containing mutant APP, PS, and/or tau accumulate abnormally phosphorylated tau in an age-related manner, but whether there is progression to PHF formation in older animals must await further study.

16.3.3 Glial Activation

In APP transgenic mouse models of AD that exhibit neuritic plaques in the cortex and hip-

TABLE 16.1. Neuropathology in mouse models of Alzheimer's disease.

	Name (*Alternate name*)							
	NSEAPP	PDAPP (*APPlon*)	Tg2576 (*APPswe*)	APP23	TgCRND8	TAPP (*APP/tau*)	PSAPP	PS1	Aβ-Arc
Transgene or mutation	APP$_{75}$[124]	APP$_{V717F}$[35]	APP$_{695}$[36]	APP$_{751}$[53]	APP$_{695+V717F}$[130]	APP$_{695}$ × JNPL3[39]	APP$_{695}$ × PS1[70]	PS1[133] or cPS1[94]	Aβ$_{E22G+}$ APP$_{695/V717F}$[134]
Amyloid-β plaques	Y[124]	Y[35]	Y[36]	Y[53]	Y[130]	Y[39]	Y[70]	Y[133]	Y[134]
Neurofibrillary tangles (Paired helical filaments)		N[51]				Y[39]	N[54]		
Glial activation	Y[125, 126]	Y[35]	Y[128]	Y[53]	Y[130]	Y[39]	Y[70]	Y[67]	
Hippocampal and/or cortical cell loss		N[35,58,127]		Y[59,129]			N[131,132]	N[94]	
Cholinergic cell loss		N[69]	N[128]	N[68]			N[63]		
Noradrenergic cell loss		N[88]							
Abnormal adult hippocampal neurogenesis		Y[95]	Y[92]					Y[94]	

Abbreviations: Y, yes; N, no; blank, not determined.
Reference numbers appear as superscripts. Details on mutations are found in original publications (see references).

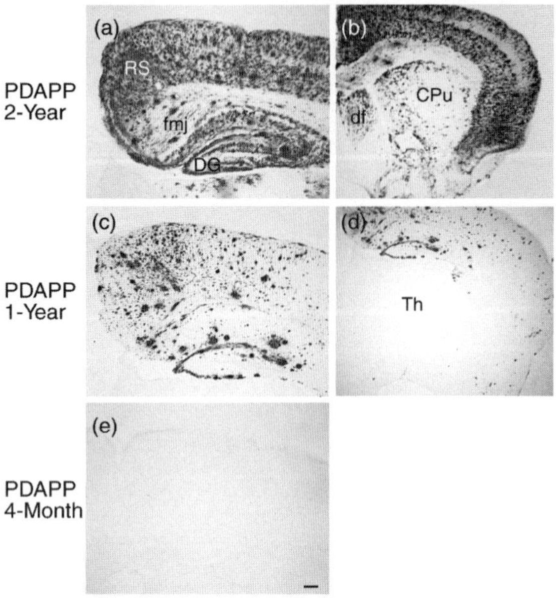

FIGURE 16.1. Aβ-containing plaques accumulate with age in the PDAPP mouse brain. Brain sections were stained with an antibody against human Aβ. At 2 years of age, there are many mature and diffuse plaques in the cerebral cortex and hippocampus (A). A lower number of compacted plaques are also found in subcortical regions (B), such as the caudate-putamen, and in white matter regions. Compared with the 2-year-old mouse, the number of compacted plaques is less in the 1-year-old PDAPP mouse cortex and hippocampus (C), and there are no plaques in subcortical regions like the thalamus (D). There are very few Aβ-containing plaques in the 4-month-old PDAPP brain (E) and none in the 2-year-old non-transgenic control brain (not illustrated). Abbreviations: CPu, caudate-putamen; df, dorsal fornix; DG, dentate gyrus; fmj; forceps major corpus callosum; RS, retrosplenial cortex; Th, thalamus. Marker, 150 μm in (A), (C), (E), and 300 μm in (B) and (D). Reproduced from German et al. [69].

pocampus, there is an activation of microglia in regions containing neuritic plaques (Fig. 16.2) [35, 56, 57]. Also, there is activation of astrocytes in the region of Aβ-containing plaques. Even in the models that lack mutant APP, astrocytes are still activated [41]. These data suggest that glial activation and inflammation are not solely related to the presence of neuritic plaques.

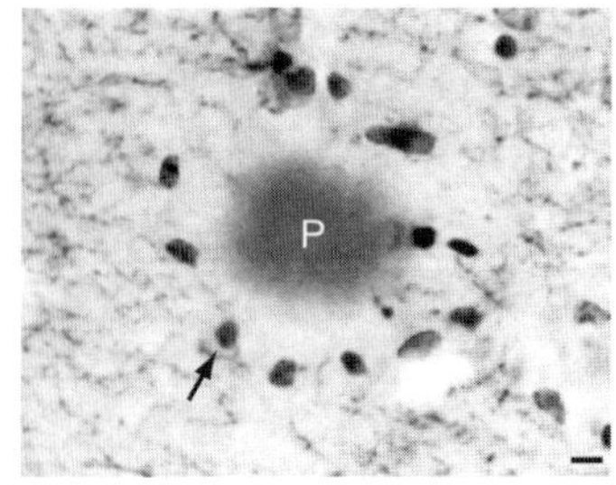

FIGURE 16.2. Microglial cells surround neuritic plaques in the PDAPP mouse cerebral cortex. Notice the numerous microglial cells (arrow points to one of several) surrounding the neuritic plaque (P). This section is stained with an antibody against ChAT (black fibers), and the section is counterstained with cresyl violet. Marker, 6 μm.

16.3.4 Hippocampal and Cortical Cell Loss

Modest neuron loss in hippocampus and cortex has been reported in some AD mouse models. Hemizygous 18-month-old PDAPP mice have been examined for cortical cell loss, but there was none even in regions that contained a high density of plaques [58]. However, Calhoun et al. [59] reported a moderate loss of cortical neurons in old APP23 mice. The cell loss was correlated with amyloid plaque density in this study. In the PS cDKO mouse, there is an age-related cortical atrophy and thinning of the cortical mantle, although no detailed quantitative cell counts have yet been reported [41]. In addition, a hallmark of AD, a marked shrinkage of the hippocampus, has been observed in the PDAPP mouse [60, 61]. There is also a loss of CA1 neurons in the hippocampus in APP+PS1 mutant animals [38].

Neurodegeneration becomes prominent in APP mouse models with impaired PS function. Using a mouse model that expresses mutations in both APP (KM670/671NL and V717I) and PS1 (M146L), significant neurodegeneration has been reported in the hippocampal CA1 region. The neurodegeneration appears to be age-related [38], and the neurons that are destined to degenerate accumulate Aβ protein within the somata [62]. These data suggest that neurodegeneration can occur from intracellular accumulation of Aβ protein. In mice with mutant tau and APP, there are NFTs in entorhinal cortex and hippocampus CA1 that increase in number with age, especially in female transgenic animals [39]. Detailed cell counting was not performed in this study, however. It will be interesting to make quantitative measurements of neurodegeneration in cortical, hippocampal, and subcortical regions in animal models that exhibit NFTs to determine whether the NFTs play a role in the degeneration of these neurons.

16.3.5 Cholinergic Cell Loss

Cholinergic nerve terminal abnormalities are common in the hippocampus and cortex of APP mouse models [63–69]. Cholinergic degenerative changes occur specifically in regions that eventually exhibit neuritic plaque deposition (Fig. 16.3). In 2-year-old homozygous PDAPP mice, for example, there is a very high density of Aβ-containing neuritic plaques in the cingulate cortex but only a low density in the striatum. At this same time point, there is a significant reduction in cholinergic enzyme activity in the cingulate cortex, but no significant reduction in enzyme activity or cholinergic cell density in the striatum [69].

Neocortical cholinergic nerve terminals degenerate prior to Aβ plaque deposition. There is a significant reduction in the number of cholinergic nerve terminal varicosities in young homozygous PDAPP mice versus age-matched controls, at a time when only a very few Aβ plaques are present [69]. Other types of studies support this conclusion. For example, behavioral impairments [70, 71], synaptic transmission deficits [72], and loss of cortical nerve terminal markers in the PDAPP mouse [73] precede the formation of neuritic plaques in APP mouse models of AD. These findings are consistent with the hypothesis that *nerve terminal* toxicity comes from extracellular soluble forms of Aβ.

There are markedly swollen ChAT-containing cholinergic nerve terminal varicosities in proximity to mature Aβ-containing plaques. The morphological similarity to the APP-positive neuritic plaques found in the PDAPP mouse [35] and human AD tissue [74] indicates that neuritic dystrophy associated with Aβ deposition affects cortical cholinergic nerve terminals. The swollen cholinergic nerve terminals are more than twice the normal size, and their density is extensive within the cortex and hippocampus of 2-year-old homozygous PDAPP mice. Similar morphological abnormalities have been observed in cholinergic synapses in mice carrying a mutation in APP [64, 68] and double mutations in APP and PS1 [63, 66, 67]. Likewise, the swollen cholinergic nerve terminals have been identified using antibodies against ChAT [66, 68, 69], the p75 nerve growth factor (NGF) receptor [67], the vesicular acetylcholine transporter [63], and immunostaining for acetylcholinesterase [64]. The swelling may be related to the induction of brain-derived neurotrophic factor in plaque-associated glial cells in the APP mouse models [75].

Because cholinergic synaptic transmission is important for learning and memory [76, 77] reductions in cholinergic nerve terminals may play a part in the learning deficits observed in APP-transgenic mice [78, 79] and in the PDAPP mouse [80]. The severe cholinergic pathology in the PDAPP mouse is similar to that in end-stage AD postmortem brain where there are marked decreases in the density of cholinergic nerve terminals and ChAT enzyme activity [81, 82].

Neurodegeneration of the basal forebrain cholinergic neurons is one of the cardinal features of AD; however, in AD mouse models these neurons do not degenerate. In the PDAPP mouse, there is no reduction in the number of basal forebrain cholinergic somata in the aged homozygous PDAPP mouse (Fig. 16.4) [69]. At 2 years of age, there are a similar number of basal forebrain cholinergic somata in homozygous PDAPP mice versus 2-month-old homozygous PDAPP mice. The basal forebrain cholinergic somata collectively within the medial septal nucleus and in the vertical and horizontal limbs of the diagonal band of Broca project to the cingulate cortex and hippocampus in the rodent [83, 84], both of which are regions that contain dense accumulations of Aβ-containing neuritic plaques in the 2-year-old animals. In hemizygous APP transgenic mice, there is also no loss of basal forebrain cholinergic neurons [64, 68], nor in $APP_{SWE}/PS1_{M146L}$ transgenic mice [67]. The lack of reduction in the number of basal forebrain cholinergic somata in the APP mouse models differ from that observed in AD patients, perhaps because the pathologic process in the animals lasts for a much shorter time period than is typical in man. It is also possible that expression of genes or activation of proteins that play a role in neuroprotection occur in the APP mouse models

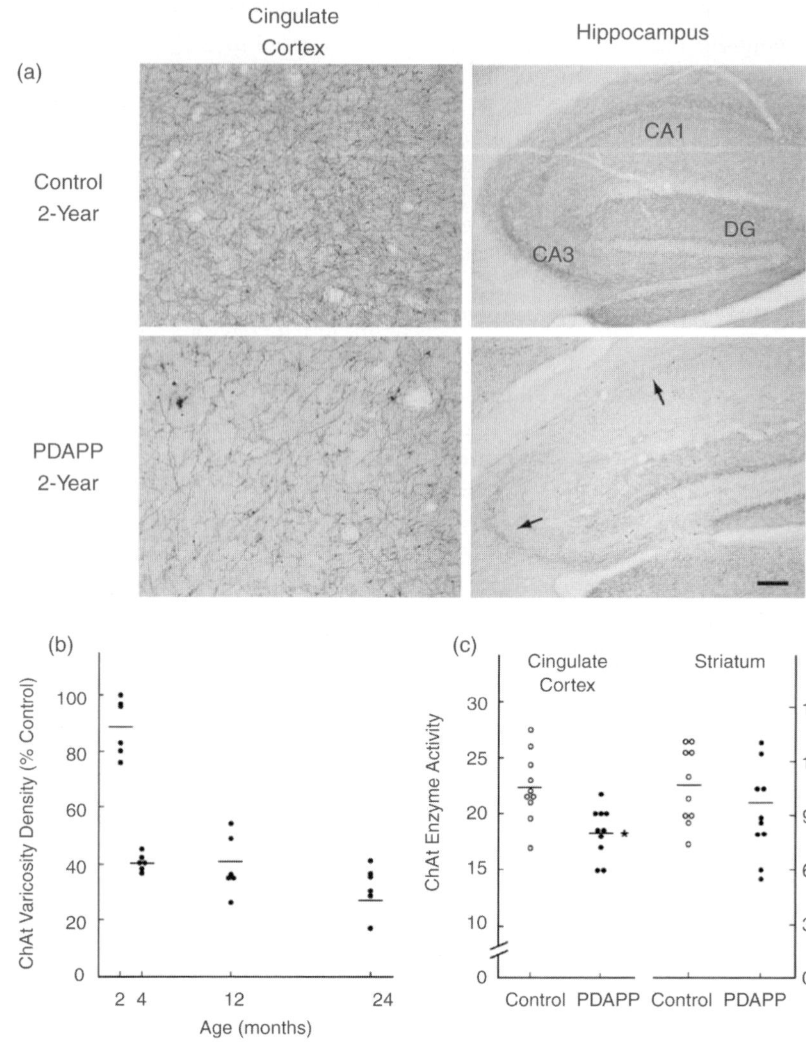

FIGURE 16.3. There is a marked decrease in cortical cholinergic markers in the PDAPP mouse. (A) The density of nerve fibers, immunostained for ChAT, is decreased in the cingulate cortex and hippocampus of the 2-year-old PDAPP mouse. ChAT fiber density is illustrated in the control 2-year-old mouse and in a 2-year-old PDAPP animal in both the cingulate cortex and hippocampus. Arrows in the hippocampus of the PDAPP mouse illustrate CA1 and CA3 regions, which contain clear losses of ChAT immunostained fibers. Abbreviations: CA1, CA1 field of the hippocampus; CA3, CA3 field of the hippocampus; DG, dentate gyrus. Marker, 70 μm. (B) There is an age-related decrease in the density of cholinergic varicosities in the PDAPP mouse. Homozygous PDAPP mice and age-matched control mice were examined at 2 months, 4 months, 1 year, and 2 years of age. Data represent ChAT varicosity density (varicosities $\times 10^6/mm^3$) for individual animals in the cingulate cortex as a percent of the age-matched control mice. (C) ChAT enzyme activity is significantly decreased in the cingulate cortex, but not in the striatum, of 2-year-old PDAPP compared with age-matched control mice. Data represent values for individual mice (nmol mg $protein^{-1}$ h^{-1}). There was a significant 18% average reduction (asterisk) in enzyme activity in the cingulate cortex (Student's t =3.27, $p < 0.04$), but no change in enzyme activity in the striatum (Student's t = 1.42). From German et al. [69].

that counter the neurotoxic effects of Aβ, as reported for the APP_{sw} mouse model of AD [85, 86]. It is also possible that NFTs are important for neurodegeneration to occur, and thus it will be interesting to determine whether the cholinergic neurons degenerate in mouse models that have NFTs [39, 40].

The loss of cholinergic nerve terminals in AD mouse models, without a loss of basal forebrain

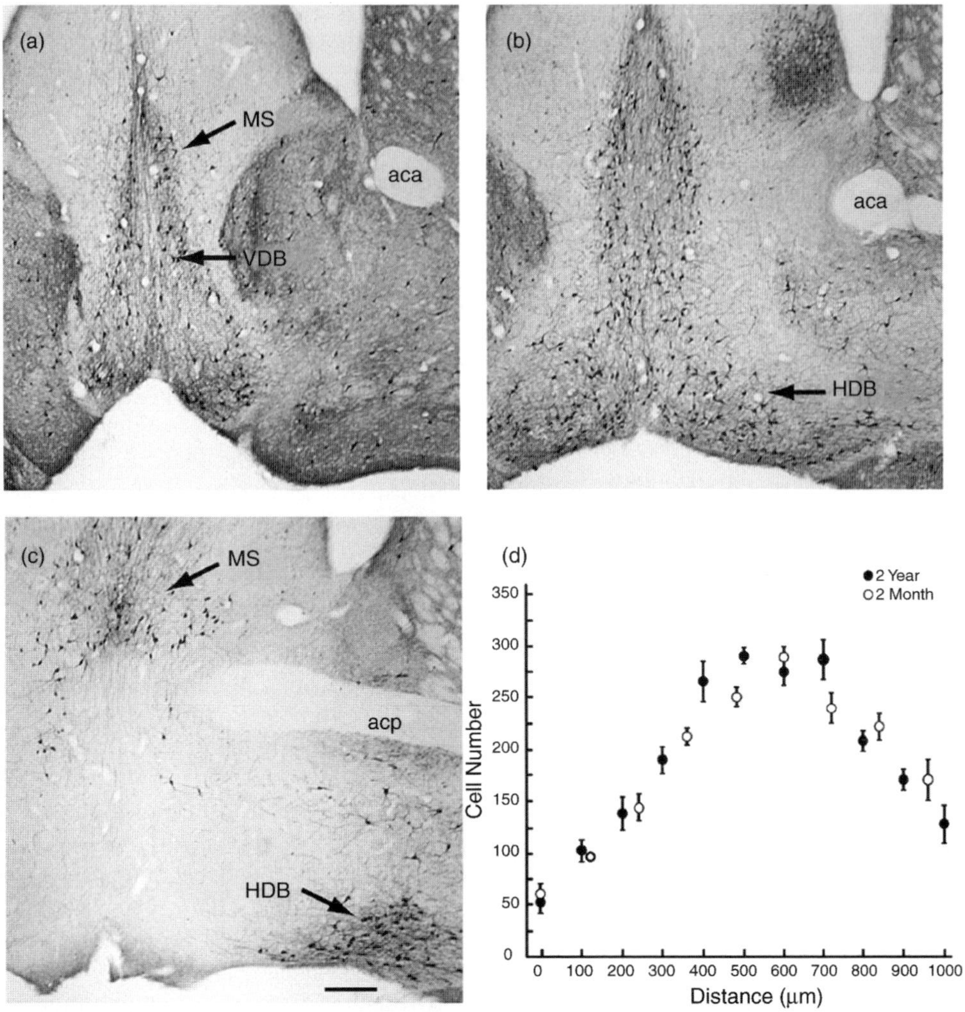

FIGURE 16.4. There is no age-related change in the number of basal forebrain cholinergic neurons in the PDAPP mouse. Basal forebrain cholinergic neurons were examined in the medial septal (MS) nucleus, and in the vertical (VDB) and horizontal limb (HDB) of the diagonal band of Broca. Representative sections, immunostained with an antibody against ChAT, are illustrated at rostral (A), middle (B), and caudal (C) locations where the basal forebrain cells were counted. Abbreviations: aca, anterior commissure, anterior; acp, anterior commissure, posterior. Marker, 300 μm. (D) There is no difference in the number of basal forebrain somata per tissue section throughout the rostral-caudal 1.0 mm of the basal forebrain in 2-month-old versus 2-year-old PDAPP mice. Illustrated are the mean total number of somata per tissue section ± SEM (n = 6/group) for sections from rostral (0 μm distance) to caudal (1000 μm) within the basal forebrain complex of the two mouse groups. From German et al. [69].

cholinergic somata, is consistent with the hypothesis that the neuropathology begins in the cerebral cortex and hippocampus prior to spreading in a retrograde fashion to subcortical regions [87]. The density of cholinergic nerve terminals in the cortex is reduced by approximately 65% in the 2-year-old PDAPP mouse versus age-matched non-transgenic controls, yet there is no reduction in the number of basal forebrain cholinergic somata that innervate this cortical region [69]. Likewise, in 2-year-old APP23 mutant mice, which carry a lower Aβ burden than in the homozygous PDAPP mice, there is a 29% reduction in total cholinergic fiber length in the cerebral cortex and no loss of basal forebrain cholinergic somata [68].

16.3.6 Noradrenergic Cell Loss

There is significant loss of LC neurons in AD [12–14], however, it is not found in the one AD mouse model reported to date, the PDAPP mouse [88]. Comparing 2-year-old homozygous animals with 2-month-old homozygous animals, the rostral-caudal distribution of LC neurons is similar. It is interesting that there is a cell shrinkage selectively within the region of the LC where cells reside that project to the cortex and hippocampus [88], suggesting that these neurons are in the early stage of degeneration. It will be interesting to determine whether AD mouse models that exhibit NFTs will exhibit loss of LC neurons that project selectively to the forebrain regions where Aβ-pathology exists, as in AD [13]. The NFTs, however, do not appear to be responsible for all of the neurodegeneration that occurs in mouse models as some loss of hippocampal neurons occurs in APP mouse models that do not express NFTs [38, 59]. In addition, in Neimann-Pick type C (NPC) disease, there is neurodegeneration and NFT formation in man [89]; however, in the NPC mouse there is marked neurodegeneration without tangle formation [90, 91].

16.3.7 Neurogenesis

With AD mouse models, changes in adult hippocampal neurogenesis can actually be quantified, in contrast with the qualitative approach required in human postmortem studies. Using quantitative analysis, adult neurogenesis has been observed to be decreased in several AD mouse models. Neurogenesis is decreased in an APP mouse model of AD (Tg2576 mouse) in the subependymal zone, a region of the brain that gives rise to olfactory neurons [92]. Notably, adult neurogenesis is also decreased in the hippocampal subgranular zone (SGZ), which gives rise to dentate gyrus neurons, in three different AD mouse models [93–95]. The Tg2576 mouse [93] and the PDAPP mouse [95] show an age-related decrease in SGZ neurogenesis. In the homozygous PDAPP mouse, neurogenesis is markedly decreased in the hippocampus of 1-year-old animals, and there is a 38% decrease in the number of granule cells in the dentate gyrus [95]. Given that the PDAPP mouse model of AD shows decreased hippocampal volume, an age-related loss of cholinergic input to the cortex and hippocampus (e.g., Ref. 69), and deficits in hippocampal function [78, 96], it will be interesting to determine whether treatments that restore learning and memory and reduce Aβ-plaque neuropathology can ameliorate the deficit in hippocampal neurogenesis.

16.4 Future Treatment Possibilities

At least six strategies have been proposed for the treatment of AD, which have been tested in AD mouse models. The first potential therapeutic treatment for AD used the PDAPP mouse model and demonstrated that *immunization* with the human $A\beta_{42}$peptide caused a marked reduction in plaque pathology when given to older animals. In addition, when immunization was given to young animals, it blocked the development of plaque pathology as the animals aged [97]. Aβ-immunization also reduces amyloid deposition in the Tg2576 mouse model of AD [98]. Similar findings were reported after immunization with antibodies against $A\beta_{42}$. For example, Janus et al. [99] found that Aβ antibody immunization reduced memory impairment and plaque pathology in an AD mouse model, and Dodart et al. [49] found that immunization with Aβ-antibody m266 reversed the memory impairment in the PDAPP mouse even before there were reductions in Aβ-plaque neuropathology. Kotilinek et al. [100] demonstrated that immunization with Aβ-antibody BAM10 reversed the memory impairment in the Tg2576 mouse model of AD. Because the cognitive impairments are improved after such a short antibody treatment, it is unlikely that the improvement was due to structural changes in the brain and perhaps reflects removal of extracellular $A\beta_{42}$ oligomers from the synaptic environment [47].

When the Aβ peptide immunization approach was used on AD patients, aseptic meningoencephalitis occurred in 6% of the patients, and the trial was stopped [101, 102]. However, recent data from a group of the immunized patients indicate that after 1 year, the patients still had high levels of $A\beta_{42}$ antibody in blood, and the "dementia score" was no different from a year previously versus a decline in dementia score in control patients that were not immunized [103]. These data suggest that some form of immunization therapy may be of benefit to AD patients; however, the success may depend upon the degree of cerebral amyloid angiopathy (CAA) in specific patients. Recent data suggest that the antibody target (N-terminal vs. central domain directed) has an effect on the induction of CAA in the PDAPP mouse [104], which may provide insight into the optimal design for future Aβ-antibodies for immunization therapy.

Epidemiological data indicate that long-term nonsteroidal anti-inflammatory drug (NSAID) treatment has dramatic effects on the incidence of AD [105] resulting in a reduction of risk by as much as 60–80% [106, 107]. The NSAID ibuprofen has been used in the Tg2576 mouse model of AD and found to significantly decrease Aβ-neuritic plaques, and decrease brain levels of $A\beta_{42}$ peptide, by a mechanism independent of its anti-inflammatory effects [56, 108]. Similar beneficial effects of reducing AD neuropathology have been found with different NSAID drugs in an APP mouse model (e.g., Ref. 109). However, additional work is needed to identify which NSAIDs will provide anti-AD effects because some compounds (e.g., celecoxib) *increase* brain $A\beta_{42}$ in the brains of Tg2576 mice via effects of γ-secretase [110].

Treatments have been proposed that would slow the production of the $A\beta_{42}$ peptide. Inhibitors of the two proteases, β- and γ-secretase, which cleave Aβ from APP have been developed. However, the current β-secretase inhibitors do not easily cross the blood-brain barrier, and γ-secretase inhibition can potentially inhibit Notch signaling [111] and produce adverse effects. In mice that have significantly reduced levels of PS function, there is seborrheic keratosis and autoimmune disease [112]. This treatment strategy will require careful testing in AD mouse models.

Another approach for the treatment of AD involves modulation of cholesterol homeostasis. Chronic use of cholesterol-lowering drugs, the statins, is associated with a lowered incidence of AD [113, 114]. High-cholesterol diets have been found to increase Aβ neuropathology in APP mouse models [115, 116], and cholesterol-lowering drugs reduce neuropathology in APP mice [117]. However, a recent study questions the use of statins in females because although lovastatin lowered cholesterol in both male and female Tg2576 mice, it *increased* the number of plaques in the hippocampus and cortex of females but not males [118]. In addition, the beneficial effects of statins for AD may also derive from their ability to reduce the microglial inflammatory response [119].

Another strategy for lowering Aβ concentrations in brain is based on the observation that Aβ aggregation is partly dependent upon the metal ions Cu^{2+} and Zn^{2+}. Aβ deposition was reduced in APP transgenic mice treated with the antibiotic clioquinol, which is a chelator of Cu^{2+} and Zn^{2+} [120]. Human clinical trials with clioquinol are in progress.

Recent studies have also examined the effects of environmental enrichment and dietary supplements on AD neuropathology in mouse models of the disease. Two studies have examined whether voluntary exercise has an effect on Aβ plaque load and brain peptide levels and also cognitive function [121, 122]. One of the studies used the TgCRND8 mouse, which expresses two mutations in APP, and found that 5 months of voluntary exercise decreased amyloid plaque load and improved cognitive function, and the effect was related to altered APP processing [121]. The other study used the Tg2576 mouse model of AD and found that 6 months of voluntary exercise improved cognitive function, but amyloid plaque pathology was *enhanced* [122]. The latter study demonstrates that cognitive function is not positively correlated with plaque pathology, and both studies support clinical data showing that people leading a physically active life have a lower incidence of AD. Finally, using the aged Tg2576 mouse model of AD, it has been demonstrated that increased intake of the omega-3 polyunsaturated fatty acid docosahexaenoic acid reduces brain levels of Aβ [123].

The current AD mouse models are being used for testing putative AD therapies and their effects on specific aspects of AD neuropathology. Several AD mouse models exhibit an age-related reduction in the density of cholinergic nerve terminal varicosities without a reduction in the numbers of basal forebrain cholinergic somata (e.g., Ref. 69). Will early administration of therapies that reduce plaque pathology and restore learning/memory in AD mouse models, like NSAIDs and immunization with $A\beta_{42}$ peptides, block cholinergic nerve terminal degeneration? In the bigenic AD mouse model of Schmitz et al. [38], which exhibits degeneration of CA1 hippocampal neurons, will some of the above AD therapies block and/or reduce the magnitude of NFTs and neurodegeneration? Because adult hippocampal neurogenesis is abnormal in AD [32] and abnormal in APP mouse models [93, 95], will therapies that reduce brain concentrations of $A\beta_{42}$ normalize neurogenesis? Once a mouse model is developed that mimics all of the major neuropathologic features of the human disease (Aβ-plaques, NFTs, and neurodegeneration), these and numerous other questions can be more fully addressed in the process of finding novel therapies for the treatment of the human condition.

Acknowledgments This work was supported by The Carl J. and Hortense M. Thomsen Chair in Alzheimer's Disease Research and the NIH/NIA Center.

References

1. Wong PC, Cai H, Borchelt DR, et al. Genetically engineered mouse models of neurodegenerative diseases. Nat Neurosci 2002; 5:633-39.
2. Higgins GA, Jacobsen H. Transgenic mouse models of Alzheimer's disease; phenotype and application. Behav Pharmacol 2003; 14:419-38.
3. German DC, Eisch AJ. Mouse models of Alzheimer's disease: insight into treatment. Rev Neurosci 2004; 15:353-69.
4. Braak H, Braak E. Neuropathological stageing of Alzheimer-related changes. Acta Neuropathol 1991; 82:239-59.
5. Bowen DM, Smith CB, White P, et al. Neurotransmitter-related enzymes and indices of hypoxia in senile dementia and other abiotrophies. Brain 1976; 99:459-96.
6. Davies P, Maloney AJ. Selective loss of central cholinergic neurons in Alzheimer's disease. Lancet 1976; 2:1403.
7. Perry EK, Perry RH, Blessed G, et al. Necropsy evidence of central cholinergic deficits in senile dementia. Lancet 1977; 1:189.
8. Whitehouse PJ, Price DL, Struble RG, et al. Alzheimer's disease and senile dementia: loss of neurons in the basal forebrain. Science 1982; 215:1237-39.
9. Mesulam M-M, Geula C. Nucleus basalis (Ch4) and cortical cholinergic innervation in the human brain: observations based on the distribution of acetylcholinesterase and choline acetyltransferase. J Comp Neurol 1988; 275:216-40.
10. Yamamoto T, Hirano A. Nucleus raphe dorsalis in Alzheimer's disease: neurofibrillary tangles and loss of large neurons. Ann Neurol 1985; 17:573-77.
11. Mann DMA, Yates PO, Marcyniuk B. Dopaminergic neurotransmitter systems in Alzheimer's disease and in Down's syndrome at middle age. J Neurol Neurosurg Psychiatry 1987; 50:341-44.
12. Iversen LL, Rossor MN, Reynolds GP, et al. Loss of pigmented dopamine-β-hydroxylase positive cells from locus coeruleus in senile dementia of Alzheimer's type. Neurosci Lett 1983; 39:95-100.
13. German DC, Manaye KF, Smith WK, et al. Disease-specific patterns of locus coeruleus cell loss. Ann Neurol 1992; 32:667-76.
14. Chan-Palay V, Asan E. Alterations in catecholamine neurons of the locus coeruleus in senile dementia of the Alzheimer's type and in Parkinson's disease with and without dementia and depression. J Comp Neurol 1989; 287:373-92.
15. Kalaria RN. Microglia and Alzheimer's disease. Curr Opin Hematol 1999; 6:15-24.
16. Altman J, Das GD. Autoradiographic and histological evidence of postnatal hippocampal neurogenesis in rats. J Comp Neurol 1965; 124:319-35.
17. Ericksson PS, Perfilieva E, Bjork-Eriksson T, et al. Neurogenesis in the adult human hippocampus. Nat Med 1998; 4:1313-317.
18. Cameron HA, Woolley CS, McEwen BS, et al. Differentiation of newly born neurons and glia in the dentate gyrus of the adult rat. Neuroscience 1993; 56:337-44.
19. Hastings NB, Gould E. Rapid extension of axons into the CA3 region by adult-generated granule cells. J Comp Neurol 1999; 413:146-54.
20. Markakis EA, Gage FH. Adult-generated neurons in the dentate gyrus send axonal projections to field CA3 and are surrounded by synaptic vesicles. J Comp Neurol 1999; 406:449-60.
21. Cameron H A, McKay RD. Adult neurogenesis produces a large pool of new granule cells in the dentate gyrus. J Comp Neurol 2001; 435:406-17.
22. Gould E, Reeves AJ, Fallah M, et al. Hippocampal neurogenesis in adult old world primates. Proc Natl Acad Sci U S A 1999; 96:5263-67.
23. Gould E, Vail N, Wagers M, et al. Adult-generated hippocampal and neocortical neurons in macaques have a transient existence. Proc Natl Acad Sci USA 2001; 98:10910-17.
24. Kornack DR, and Rakic P. Continuation of neurogenesis in the hippocampus of the adult macaque monkey. Proc Natl Acad Sci U S A 1999; 96:5768-73.
25. Boonstra R, Galea L, Matthews S, et al. Adult neurogenesis in natural populations. Can J Physiol Pharmacol 2001; 79:297-302.
26. Eisch AJ, Nestler EJ. To be or not to be: adult neurogenesis and psychiatry. Clin Neurosci Res 2001; 2: 93-108.

27. Eisch AJ. Adult neurogenesis: implications for psychiatry. Prog Brain Res 2002; 138:317-44.
28. Abrous DN, Adriani W, Montaron MF, et al. Nicotine self-administration impairs hippocampal plasticity. J Neurosci 2002; 22:3656-62.
29. Kempermann G, Kuhn HG, Gage FH. Experience-induced neurogenesis in the senescent dentate gyrus. J Neurosci 1998; 18:3206-12.
30. Kuhn HG, Dickinson-Anson H, Gage FH. Neurogenesis in the dentate gyrus of the adult rat: age-related decrease of neuronal progenitor proliferation. J Neurosci 1996; 16:2027-33.
31. Vallieres L, Campbell IL, Gage FH et al. Reduced hippocampal neurogenesis in adult transgenic mice with chronic astrocytic production of interleukin-6. J Neurosci 2002; 22:486-92.
32. Jin K, Peel AL, Mao XO, et al. Increased hippocampal neurogenesis in Alzheimer's disease. Proc Natl Acad Sci U S A 2004; 101:343-47.
33. Jordan-Sciutto KL, Dorsey R, Chalovich EM, et al. Expression patterns of retinoblastoma protein in Parkinson's disease. J Neuropathol Exp Neurol 2002; 62:68-74.
34. Curtis MA, Penney EB, Pearson AG, et al. Increased cell proliferation and neurogenesis in the adult human Huntington's disease brain. Proc Natl Acad Sci U S A 2003; 100:9023-27.
35. Games D, Adams D, Alessandrini R, et al. Alzheimer-type neuropathology in transgenic mice overexpressing V717F β-amyloid precursor protein. Nature 1995; 373:523-27.
36. Hsiao K, Chapman P, Nilsen S, et al. Correlative memory deficits, Abeta elevation, and amyloid plaques in transgenic mice. Science 1996; 274:99-102.
37. Borchelt DR, Ratovitski T, van Lare J, et al. Accelerated amyloid deposition in the brains of transgenic mice coexpressing mutant presenilin 1 and amyloid precursor proteins. Neuron 1997; 19: 939-45.
38. Schmitz C, Rutten BPF, Pielen A, et al. Hippocampal neuron loss exceeds amyloid plaque load in a transgenic mouse model of Alzheimer's disease. Am J Pathol 2004; 164:1495-502.
39. Lewis J, Dickson DW, Lin W-L, et al. Enhanced neurofibrillary degeneration in transgenic mice expression mutant tau and APP. Science 2001; 293:1487-91.
40. Oddo S, Caccamo A, Shepherd JD, et al. Triple-transgenic model of Alzheimer's disease with plaques and tangles: intracellular Aβ and synaptic dysfunction. Neuron 2003; 39:409-21.
41. Saura CA, Choi S-Y, Beglopoulos V, et al. Loss of presenilin function causes impairments of memory and synaptic plasticity followed by age-dependent neurodegeneration. Neuron 2004; 42:23-36.
42. Murrell J, Farlow M, Ghetti B, et al. A mutation in the amyloid precursor protein associated with hereditary Alzheimer's disease. Science 1991; 254:97-99.
43. Lazarov O, Lee M, Peterson DA, et al. Evidence that synaptically released [beta]amyloid accumulates as extracellular deposits in the hippocampus of transgenic mice. J Neurosci 2002; 22:9785-93.
44. Sheng JG, Price DL, Kotiatsos VE. Disruption of cortico-cortical connections ameliorates amyloid burden in terminal fields in a transgenic model of A[beta] amyloidosis. J Neurosci 2002; 22:9794-99.
45. Walsh DM, Hartley DM, Kusumoto Y, et al. Amyloid β-protein fibrillogenesis: structure and biological activity of protofibrillar intermediates. J Biol Chem 1999; 274:25945-52.
46. Hartley DM, Walsh DM, Ye CP, et al. Protofibrillar intermediates of amyloid β-protein induce acute electrophysiological changes and progressive neurotoxicity in cortical neurons. J Neurosci 1999; 19:8876-84.
47. Walsh DM, Klyubin I, Fadeeva JV, et al. Naturally secreted oligomers of amyloid beta protein potently inhibit hippocampal long-term potentiation in vivo. Nature 2002; 416:535-39.
48. Klyubin I, Walsh DM, Lemere CA et al. Amyloid β protein immunotherapy neutralizes Aβ oligomers that disrupt synaptic plasticity in vivo. Nat Med 2005; 11:556-61.
49. Dodart J-C, Bales KR, Gannon KS, et al. Immunization reverses memory deficits without reducing brain Aβ burden in Alzheimer's disease model. Nat Neurosci 2002; 5:452-57.
50. Zohar O, Cavallaro S, Agata VD, et al. Quantification and distribution of β-secretase alternative splice variants in the rat and human brain. Mol Brain Res 2003; 115:63-68.
51. Masliah E, Sisk A, Mallory M, et al. Neurofibrillary pathology in transgenic mice overexpressing V717F beta-amyloid precursor protein. J Neuropathol Exp Neurol 2001; 60:357-68.
52. Tomidokoro Y, Ishiguro K, Harigaya Y, et al. Abeta amyloidosis induces the initial stage of tau accumulation in APP(Sw) mice. Neurosci Lett 2001; 299: 169-72.
53. Sturchler-Pierrat C, Abramowski D, Duke M, et al. Two amyloid precursor protein transgenic mouse models with Alzheimer's disease-like pathology. Proc Natl Acad Sci U S A, 1997; 94:13287-92.
54. Kurt MA, Davies DC, Kidd M, et al. Hyperphosphorylated tau and paired helical filament-like structures in the brains of mice carrying mutant amyloid precursor protein and mutant presenilin-1 transgenes. Neurobiol Dis 2003; 14:89-97.
55. Schwab C, Hosokawa M, McGeer PL. Transgenic mice overexpressing amyloid beta protein are an

incomplete model of Alzheimer's disease. Exp Neurol 2004; 188:52-64.
56. Yan Q, Zhang J, Liu H, et al. Anti-inflammatory drug therapy alters β-amyloid processing and deposition in an animal model of Alzheimer's disease. J Neurosci 2003; 23:7504-09.
57. Richards JG, Higgins GA, Ouagazzal A-M, et al. PS2APP transgenic mice, coexpressing hPS2mut and hAPPswe, show age-related cognitive deficits associated with discrete brain amyloid deposition and inflammation. J Neurosci 2003; 23:8989-03.
58. Irizarry MC, Soriano F, McNamara M, et al. Aβ deposition is associated with neuropil changes, but not with overt neuronal loss in the human amyloid precursor protein V717F (PDAPP) transgenic mouse. J Neurosci 1997; 17:7053-59.
59. Calhoun ME, Wiederhold K-H, Abramowski D, et al. Neuron loss in APP transgenic mice. Nature 1998; 395:755-56.
60. Gonzales-Lima F, Berndt JD, Valla JE, et al. Reduced corpus callosum, fornix and hippocampus in PDAPP transgenic mouse model of Alzheimer's disease. NeuroReport 2001; 12:2375-79.
61. Redwine JM, Kosofsky B, Jacobs RE, et al. Dentate gyrus volume is reduced before the onset of plaque formation in PDAPP mice: a magnetic resonance microscopy and stereologic analysis. Proc Natl Acad Sci U S A 2003; 100:1381-86.
62. Casas C, Sergeant N, Itier J-M, et al. Massive CA1/2 neuronal loss with intraneuronal and N-terminal truncated Abeta42 accumulation in a novel Alzheimer transgenic model. Am J Pathol 2004; 165: 1289-300.
63. Wong TP, Debeir T, Duff K, et al. Reorganization of cholinergic terminals in the cerebral cortex and hippocampus in transgenic mice carrying mutated presenilin-1 and amyloid precursor protein transgenes. J Neurosci 1999; 19:2706-16.
64. Bronfman FC, Moechars D, Van Leuven F. Acetylcholinesterase-positive fiber deafferentation and cell shrinkage in the septohippocampal pathway of aged amyloid precursor protein London mutant transgenic mice. Neurobiol Dis 2000; 7:152-68.
65. Sturchler-Pierrat C, Staufenbiel M. Pathogenic mechanisms of Alzheimer's disease analyzed in the APP23 transgenic mouse model. Ann N Y Acad Sci 2000; 920:134-39.
66. Hernandez D, Sugaya K, Qu T, et al. Survival and plasticity of basal forebrain cholinergic systems in mice transgenic for presenilin-1 and amyloid precursor protein mutant genes. NeuroReport 2001; 12: 1377-84.
67. Jaffar S, Counts SE, Ma SY, et al. Neuropathology of mice carrying mutant APP_{swe} and/or $PS1_{M146L}$ transgenes: alterations in the $p75^{NTR}$ cholinergic basal forebrain septohippocampal pathway. Exp Neurol 2001; 170:227-43.
68. Boncristiano S, Calhoun ME, Kelly PH, et al. Cholinergic changes in the APP23 transgenic mouse model of cerebral amyloidosis. J Neurosci 2002; 22: 3234-43.
69. German DC, Yazdani U, Speciale SG, et al. Cholinergic neuropathology in a mouse model of Alzheimer's disease. J Comp Neurol 2003; 462: 371-81.
70. Holcomb L, Gordon MN, McGowan E, et al. Accelerated Alzheimer-type phenotype in transgenic mice carrying both mutant amyloid precursor protein and presenilin 1 transgenes. Nat Med 1998; 4:97-100.
71. Moechars D, Dewachter I, Lorent K, et al. Early phenotypic changes in transgenic mice that overexpress different mutants of amyloid precursor protein in brain. J Biol Chem 1999; 274:6483-92.
72. Hsia A, Masliah E, McConlogue L, et al. Plaque-independent disruption of neural circuits in Alzheimer's disease mouse models. Proc Natl Acad Sci U S A 1999; 96:3228-33.
73. Mucke L, Masliah E, Yu GQ, et al. High-level neuronal expression of $A\beta_{1-42}$ in wild-type human amyloid protein precursor transgenic mice: synaptotoxicity without plaque formation. J Neurosci 2000; 20:4050-58.
74. Joachim C, Games D, Morris J, et al. Antibodies to non-beta regions of the beta-amyloid precursor protein detect a subset of senile plaques. Am J Pathol 1991; 138:373-84.
75. Burbach GJ, Hellweg R, Haas CA, et al. Induction of brain-derived neurothophic factor in plaque-associated glial cells in aged APP23 transgenic mice. J Neurosci 2004; 24:2421-30.
76. Coyle JT, Price DL, DeLong MR. Alzheimer's disease: a disorder of cortical cholinergic innervation. Science 1983; 219:1184-90.
77. Fibiger HC. Cholinergic mechanisms in learning, memory and dementia: a review of recent evidence. Trends Neurosci 1991; 14:220-23.
78. Chapman PF, White GL, Jones MW, et al. Impaired synaptic plasticity and learning in aged amyloid precursor protein transgenic mice. Nat Neurosci 1999; 2:271-76.
79. Morgan D, Diamond DM, Gottschall PE, et al. Aβ peptide vaccination prevents memory loss in an animal model of Alzheimer's disease. Nature 2000; 408: 982-85.
80. Chen G, Chen KS, Knox J, et al. A learning deficit related to age and β-amyloid plaques in a mouse model of Alzheimer's disease. Nature 2000; 408: 975-79.

81. DeKosky ST, Ikonomovic MD, Styren SD, et al. Upregulation of choline acetyltransferase activity in hippocampus and frontal cortex of elderly subjects with mild cognitive impairment. Ann Neurol 2002; 51:145-55.
82. Geula C, Mesulam M-M. Systematic regional variations in the loss of cortical cholinergic fibers in Alzheimer's disease. Cerebral Cortex 1996; 6: 165-77.
83. McKinney M, Coyle JT, Hedreen JC. Topographic analysis of the innervation of the rat neocortex and hippocampus by the basal forebrain cholinergic system. J Comp Neurol 1983; 217:103-21.
84. Rye DB, Wainer BH, Mesulam MM, et al. Cortical projections arising from the basal forebrain: a study of cholinergic and noncholinergic components employing combined retrograde tracing and immunohistochemical localization of choline acetyltransferase. Neuroscience 1984; 13:627-43.
85. Stein TD, Johnson JA. Lack of neurodegeneration in transgenic mice overexpressing mutant amyloid precursor protein is associated with increased levels of transthyretin and the activation of cell survival pathways. J Neurosci 2002; 22:7380-88.
86. Stein TD, Anders NJ, DeCarli C et al. Neutralization of transthyretin reverses the neuroprotective effects of secreted amyloid precursor protein (APP) in APP_{SW} mice resulting in tau phosphorylation and loss of hippocampal neurons: support for the amyloid hypothesis. J Neurosci 2004; 24:7707-717.
87. Saper CB, Wainer B, German DC. Axonal and transneuronal transport in the transmission of neurological disease: potential role in system degenerations, including Alzheimer's disease. Neuroscience 1987; 23:389-98.
88. German DC, Nelson O, Liang F, et al. The PDAPP mouse model of Alzheimer's disease: locus coeruleus neuronal shrinkage. J Comp Neurol 2005; 492:469-76.
89. Love S, Bridges LR, Case CP. Neurofibrillary tangles in Niemann-Pick disease type C. Brain 1995; 118: 119-29.
90. Tanaka J, Nakamura H, Miyawaki S. Cerebellar involvement in murine sphingomyelinosis: a new model of Niemann-Pick disease. J Neuropathol Exp Neurol 1988; 47:291-00.
91. German DC, Quintero EM, Liang C-L, et al Selective neurodegeneration, without neurofibrillary tangles, in a mouse model of Niemann-Pick C disease. J Comp Neurol 2001; 433:415-25.
92. Haughey NJ, Nath A, Chan SL, et al. Disruption of neurogenesis by amyloid β-peptide, and perturbed neural progenitor cell homeostasis, in models of Alzheimer's disease. J Neurochem 2002; 83: 1509-24.
93. Haughey NJ, Liu D, Nath A, et al. Disruption of neurogenesis in the subventricular zone of adult mice, and in human cortical neuronal precursor cells in culture, by amyloid beta-peptide: implications for the pathogenesis of Alzheimer's disease. Neuromol Med 2002; 1:125-35.
94. Feng R, Rampon C, Tang YP, et al. Deficient neurogenesis in forebrain-specific presenilin-1 knockout mice is associated with reduced clearance of hippocampal memory traces. Neuron 2001; 32:911-26.
95. Donovan MH, Yazdani U, Norris RD, et al. Decreased adult hippocampal neurogenesis in the PDAPP mouse model of Alzheimer's disease. J Comp Neurol 2006; 495:70-83.
96. Lanz TA, Carter DB, Merchant KM. Dendritic spine loss in the hippocampus of young PDAPP and Tg2576 mice and its prevention by the ApoE2 genotype. Neurobiol Dis 2003; 13:246-53.
97. Schenk D, Barbour R, Dunn W, et al. Immunization with amyloid-beta attenuates Alzheimer's disease-like pathology in the PDAPP mouse. Nature 1999; 400:173-77.
98. Das P, Howard V, Loosbrock N, et al. Amyloid-β immunization effectively reduces amyloid deposition in $FcR\gamma^{-/-}$ knock out mice. J Neurosci 2003; 23:8532-38.
99. Janus C, Pearson J, McLaurin J, et al. Aβ peptide immunization reduces behavioural impairment and plaques in a model of Alzheimer's disease. Nature 2000; 408:979-82.
100. Kotilinek LA, Bacskai B, Westerman M, et al. Reversible memory loss in a mouse transgenic model of Alzheimer's disease. J Neurosci 2002; 22: 6331-335.
101. Schenk D. Amyloid-β immunotherapy for Alzheimer's disease: the end of the beginning. Nat Rev Neurosci 2002; 3:824-28.
102. Orgogozo J-M, Gilman S, Dartigues J-F, et al. Subacute meningoencephalitis in a subset of patients with AD after $A\beta_{42}$ immunization. Neurology 2003; 61:46-54.
103. Hock C, Konietzko U, Streffer JR, et al. Antibodies against β-amyloid slow cognitive decline in Alzheimer's disease. Neuron 2003; 38:547-54.
104. Racke MM, Boone LI, Hepburn DL, et al. Exacerbation of cerebral amyloid antiopathy-associated microhemorrhage in amyloid precursor protein transgenic mice by immunotherapy is dependent on antibody recognition of deposited forms of amyloid β. J Neurosci 2005; 25:629-36.
105. McGeer PL, Schulzer M and McGeer EG. Arthritis and anti-inflammatory agents as possible protective factors for Alzheimer's disease: a review of 17 epidemiologic studies. Neurology 1996; 47:425-32.

106. Stewart WF, Kawas C, Corrada M, et al. Risk of Alzheimer's disease and duration of NSAID use. Neurology 1997; 48:626-32.
107. in t' Veld BA, Ruitenberg A, Hofman A, et al. Nonsteroidal anti-inflammatory drugs and the risk of Alzheimer's disease. N Engl J Med 2001; 345: 1515-21.
108. Lim GP, Yang F, Chu T, et al. Ibuprofen suppresses plaque pathology and inflammation in a mouse model of Alzheimer's disease. J Neurosci 2000; 20: 5709-14.
109. Jantzen PT, Connor KE, DiCarlo G, et al. Microglial activation and β-amyloid deposit reduction caused by a nitric oxide-releasing nonsteroidal anti-inflammatory drug in amyloid precursor protein plus presenilin-transgenic mice. J Neurosci 2002; 22:2246-54.
110. Kukar T, Murphy MP, Eriksen JL, et al. Diverse compounds mimic Alzheimer's disease-causing mutations by augmenting Aβ42 production. Nat Med 2005; 11:545-50.
111. Haass C, DeStrooper B. The presenilins in Alzheimer's disease -proteolysis holds the key. Science 1999; 286:916-19.
112. Tournoy J, Bossuyt X, Snellinx A, et al. Partial loss of presenilins causes seborrheic keratosis and autoimmune disease in mice. Hum Mol Genet 2004; 13:1321-31.
113. Wolozin B, Kellman W, Ruosseau P, et al. Decreased prevalence of Alzheimer's disease associated with 3-hydroxy-3-methyglutaryl coenzyme A reductase inhibitors. Arch Neurol. 2000; 57: 1439-443.
114. Jick H, Zornberg GL, Jick SS, et al. Statins and the risk of dementia. Lancet 2000; 356:1627-31.
115. Sparks DL, Kuo YM, Roher A, et al. Alterations of Alzheimer's disease in the cholesterol-fed rabbit, including vascular inflammation. Preliminary observations. Ann N Y Acad Sci 2000; 903:335-44.
116. Refolo LM, Malester B, LaFrancois J, et al. Hypercholesterolemia accelerates the Alzheimer's amyloid pathology in a transgenic mouse model. Neurobiol Dis 2000; 7:321-31.
117. Refolo LM, Pappolla MA, LaFrancois J, et al. A cholesterol-lowering drug reduces beta-amyloid pathology in a transgenic mouse model of Alzheimer's disease. Neurobiol Dis 2001; 8:890-99.
118. Park IH, Hwang EM, Hong HS, et al. Lovasatin enhances Abeta production and senile plaque deposition in female Tg2576 mice. Neurobiol Aging 2003; 24:637-43.
119. Cordle A, Landreth G. 3-Hydroxy-3-methylglutaryl-coenzyme A reductase inhibitors attenuate β-amyloid-induced microglial inflammatory responses. J Neurosci 2005; 25:299-07.
120. Cherny RA, Atwood CS, Xilinas ME, et al. Treatment with a copper-zinc chelator markedly and rapidly inhibits beta-amyloid accumulation in Alzheimer's disease transgenic mice. Neuron 2001; 30:665-76.
121. Adlard PA, Perreau VM, Pop V, et al. Voluntary exercise decreases amyloid load in a transgenic model of Alzheimer's disease. J Neurosci 2005; 25: 4217-4221.
122. Jankowshy JL, Melnikova T, Fadale DJ, et al. Environmental enrichment mitigates cognitive deficits in a mouse model of Alzheimer's disease. J Neurosci 2005; 25:5217-224.
123. Lim GP, Calon F, Morihara T, et al. A diet enriched with the Omega-3 fatty acid docosahexaenoic acid reduces amyloid burden in an aged Alzheimer's mouse model. J Neurosci 2005; 25:3032-40.
124. Quon D, Wang Y, Catalano R, et al. Formation of beta-amyloid protein deposits in brains of transgenic mice. Nature 1991; 352:239-41.
125. Higgins LS, Holtzman DM, Rabin J, et al. Transgenic mouse brain histopathology resembles early Alzheimer's disease. Ann Neurol 1994; 35: 598-607.
126. Higgins LS, Rodems JM, Catalano R, et al. Early Alzheimer's disease-like histopathology increases in frequency with age in mice transgenic for beta-APP751. Proc Natl Acad Sci U S A 1995; 92: 4402-406.
127. Dodart JC, Mathis C, Saura J, et al. Neuroanatomical abnormalities in behaviorally characterized APP(V717F) transgenic mice. Neurobiol Dis 2000; 7:71-85.
128. Luth HJ, Apelt J, Ihunwo AO, et al. Degeneration of beta-amyloid-associated cholinergic structures in transgenic APP SW mice. Brain Res 2003; 977:16-22.
129. Bondolfi L, Calhoun M, Ermini F, et al. Amyloid-associated neuron loss and gliogenesis in the neocortex of amyloid precursor protein transgenic mice. J Neurosci 2002; 22:515-22.
130. Chishti MA, Yang DS, Janus C, et al. Early-onset amyloid deposition and cognitive deficits in transgenic mice expressing a double mutant form of amyloid precursor protein 695. J Biol Chem 2001; 276:21562-70.
131. Takeuchi A, Irizarry MC, Duff D, et al. Age-related amyloid beta deposition in transgenic mice overexpressing both Alzheimer mutant presenilin 1 and amyloid beta precursor protein Swedish mutant is not associated with global neuronal loss. Am J Pathol 2000; 157:331-39.

132. Urbanc B, Cruz L, Le R, et al. Neurotoxic effects of thioflavin S-positive amyloid deposits in transgenic mice and Alzheimer's disease. Proc Natl Acad Sci U S A 2002; 99:13990-95.
133. Duff K, Eckman C, Zehr C, et al. Increased amyloid-beta42(43) in brains of mice expressing mutant presenilin-1. Nature 1996; 383:710-13.
134. Cheng IH, Palop JJ, Esposito LA, et al. Aggressive amyloidosis in mice expressing human amyloid peptides with the Arctic mutation. Nat Med 2004; 10:1190-192.

Subject Index

Page numbers followed by f and t indicate figures and tables, respectively.

Author Index

D

E

M

Printed in the United States of America